Springer Series in **Nonlinear Dynamics**

Springer Series in **Nonlinear Dynamics**

Series Editors: F. Calogero, B. Fuchssteiner, G. Rowlands, M. Wadati, and V. E. Zakharov

Solitons – Introduction and Applications
Editor: M. Lakshmanan

What Is Integrability?
Editor: V. E. Zakharov

Rossby Vortices and Spiral Structures
By M. V. Nezlin and E. N. Snezhkin

Algebro-Geometric Approach to Nonlinear Integrable Equations
By E. D. Belokolos, A. I. Bobenko, V. Z. Enol'skii, A. R. Its and V. B. Matveev

Darboux Transformations and Solitons
By V. B. Matveev and M. A. Salle

Optical Solitons
By F. Abdullaev, S. Darmanyan and P. Khabibullaev

Wave Turbulence Under Parametric Excitation
Applications to Magnets
By V. S. L'vov

Kolmogorov Spectra of Turbulence
Wave Turbulence
By V. E. Zakharov, V. S. L'vov, G. Falkovich

Victor S. L'vov

Wave Turbulence Under Parametric Excitation

Applications to Magnets

With 69 Figures

Springer-Verlag
Berlin Heidelberg New York
London Paris Tokyo
Hong Kong Barcelona
Budapest

Professor Dr. Victor S. L'vov
Dept. of Physics, Weizmann Institute of Science
76100 Rehovot, Israel

Series Editor:
Professor Dr. V. E. Zakharov
Landau Institute for Theoretical Physics
Academy of Sciences
ul. Kosygina 2
117940 Moscow V-334
Russia

ISBN 3-540-51991-2 Springer-Verlag Berlin Heidelberg New York
ISBN 0-387-51991-2 Springer-Verlag New York Berlin Heidelberg

Cip-data applied for

This work is subject to copyright. All rights are reserved, whether the whole or part of the material is concerned, specifically the rights of translation, reprinting, reuse of illustrations, recitation, broadcasting, reproduction on microfilms or in other ways, and storage in data banks. Duplication of this publication or parts thereof is only permitted under the provisions of the German Copyright Law of September 9, 1965, in its current version, and a copyright fee must always be paid. Violations fall under the prosecution act of the German Copyright Law.

© Springer-Verlag Berlin Heidelberg 1994
Printed in Germany

The use of registered names, trademarks etc. in this publication does not imply, even in the absence of a specific statement, that such names are exempt from the relevant protective laws and regulations and therefore free for general use.

Camera ready by author
SPIN: 10016207   57/3140-5 4 3 2 1 0 – Printed on acid-free paper

Preface

WAVE TURBULENCE is a state of a system of many simultaneously excited and interacting waves characterized by an energy distribution which is *not* in any sense close to thermodynamic equilibrium. Such situations arise, for example, in a choppy sea, in a hot plasma, in dielectrics under a powerful laser beam, in magnets placed in a strong microwave field, etc. Among the great variety of physical situations in which wave turbulence arises, it is possible to select two large limiting groups which allow a detailed analysis. The first is *fully developed wave turbulence* arising when energy pumping and dissipation have essentially different space scales. In this case there is a wide power spectrum of turbulence. This type of turbulence is described in detail e.g. in *Zakharov* et al.[1] In the second limiting case the scales in which energy pumping and dissipation occur are the same. As a rule, in this case a narrow, almost singular spectrum of turbulence appears which is concentrated near surfaces, curves or even points in k-space. One of the most important, widely investigated and instructive examples of this kind of turbulence is parametric wave turbulence appearing as a result of the evolution of a parametric instability of waves in media under strong external periodic modulation (laser beam, microwave electromagnetic field, etc.).

The present book deals with parametric wave turbulence. The first investigations of this subject appeared in the West in the late fifties – early sixties. Considerable theoretical and experimental progress has been achieved during the last two decades by researchers in Russia and the Ukraine including the author. Their results may be less known in the West. This book summarizes all these developments from a common point of view. Since a large part of the book is fairly detailed, it could be used as an introductory textbook to the physics of weakly interacting waves. It should be useful for:

• students, young researchers or anybody who seeks an introduction to the ideas and methods of the physics of nonlinear waves and wave turbulence;

• experts in solid state physics, hydrodynamics, physics of magnetism and plasma physics studying various nonlinear phenomena which might have similar features.

---

[1] V. Zakharov, V. L'vov and G. Falkovich: *Kolmogorov Spectra of Turbulence, Wave Turbulence,* (Springer, Berlin Heidelberg 1992)

In writing this book I have pursued three aims. The first was to present parametric wave turbulence as an essential and instructive part of nonlinear wave physics. For this reason Chap. 1 contains a detailed description of the classical Hamiltonian formalism which has become now a universal language of nonlinear wave physics. The whole book is written in this language. Chapters 1 and 3 constitute a general introduction to the physics of nonlinear waves. They contain a description of dynamical processes playing an important role in nonlinear physics in general and, in particular, in wave turbulence. These include three- and four-wave elementary processes of interaction, decay and modulation instabilities, self-focusing, collapses of waves, etc.

The second and the main aim of the book is to present a consistent theory of parametric wave turbulence from the very beginning, from the basic physical ideas and simple approximations up to the advanced theory describing fine features of parametric wave turbulence in various physical situations (Chaps. 5 – 8, 10). The basic physical phenomenon is a correlation of the phases $\varphi(\boldsymbol{k})$ and $\varphi(-\boldsymbol{k})$ in pairs of $\pm\boldsymbol{k}$-waves (having wave vectors $\boldsymbol{k}$ and $-\boldsymbol{k}$) caused by a coherent pumping field. As a result the behavior of parametric waves is in some respects similar to that of paired superconducting $\pm\boldsymbol{k}$-electrons near the Fermi surface. The formulation of the so-called *S-theory* developed by *Zakharov, Starobinets* and the author[2],[3] is given in Chap. 5. This chapter is central to this part of the book. The S-theory describes the turbulence of parametrically excited waves in the framework of the *mean-field approximation*, similar to the approximations in the Curie-Weiss theory of a molecular field, the Landau theory of second-order phase transition as well as the BCS-theory of superconductivity. In the S-theory the role of the parametric wave interaction is reduced to renormalizing the coefficients of the linear equations (wave frequency and pumping amplitudes) describing the parametric instability. For simple physical situations the ground state of the interacting parametric waves at various pumping powers is described in terms of the S-theory.

The Advanced S-theory developed in Chap. 6 takes into account additional physical factors which might be important: nonlinear damping of waves, non-zero temperature of the thermal bath, linewidth of the pumping, etc. Non-stationary phenomena, such as transient processes and resonance of collective oscillations in the system of interacting parametric waves are discussed in Chap. 7. Chapter 8 deals with *secondary parametric wave turbulence*, i.e. chaotic auto-oscillations of the double correlation functions of the system of parametrically excited waves. The notion of *secondary turbulence* emphasizes the chaotic behavior of correlation functions. These are the result of averaging over the *primary* turbulent ensemble of parametrically

---

[2] V. Zakharov, V. L'vov and S. Starobinets: *Spin-Wave Turbulence Beyond the Parametric Excitation Threshold*, Sov. Phys.-Usp., **17**, No. 6 (1975) pp. 896-919
[3] V. L'vov: *Nonlinear Spin Waves* (Nauka, Moscow 1987) [In Russian]

excited waves. In particular, we discuss the route to "secondary" chaos and the geometry of its attractors.

The final chapter is devoted to phenomena beyond the mean-field approximation. An advanced theory of parametric wave turbulence is developed which takes into account not only the self-consistent interaction of wave pairs but also the scattering of individual waves and their interactions with the thermal bath and with the inhomogeneities of the media. The formalism of this theory is by far more complicated than that of the S-theory. Hence, Chap. 10 presumes a higher level of knowledge that the rest of the book. In particular the reader should be familiar with the ideas of the Feynman diagram technique.

The third and last aim of the book is to present numerous and instructive experimental results on parametric wave turbulence in a systematic way and in comparison with the consistent theory developed. The main experimental data were obtained under parametric excitation of spin waves (magnons) in magnetically ordered dielectrics. This is due to the fact that such experimental investigations are particularly easy to perform. It is usually possible to experiment with magnons at room temperature, in the conventional range of microwave frequencies (5 to 50 GHz) using perfect monocrystals. It is easy to excite magnons by a microwave magnetic field up to distinctly nonlinear levels when their behavior is determined completely by their interactions[4]. The theory of parametric wave turbulence was developed first as an attempt to explain the experimental data on the parametric excitation of spin waves (in particular, the stationary level of excitation, nature of auto-oscillations). Later a set of experiments on nonlinear spin waves (e.g. double parametric resonance) was performed in order to check the predictions of the S-theory. Some of these experiments revealed new phenomena of major importance; for instance, the finite frequency width of parametric wave distribution. Attempts to understand these phenomena led to essential progress in the development of the theory. Thus, nonlinear spin waves play a crucial role in the physics of nonlinear waves and especially in parametric wave turbulence. Therefore, I felt it necessary to include in the book Chap. 2 presenting a short review of the necessary background on the physics of magnets and Chaps. 3 and 4 describing the linear and nonlinear properties of spin waves in the Hamiltonian language.

Rehovot
Summer 1993

Victor S. L'vov

---

[4] V. L'vov and L. Prozorova: "Spin Waves Above the Threshold of Parametric Excitations", in *Spin Waves and Magnetic Excitations*, Vol. 1, ed. by A.S. Borovik-Romanov and S.K. Sinha (Elsevier Sci. Publ., Amsterdam 1988) pp. 233–285

Acknowledgements

I am greatly indebted to many people. My thanks go first of all to the many authors whose contributions were considered or cited in the book. Among them I especially acknowledge my friends V. Zakharov and S. Starobinets who introduced me to the problems of nonlinear waves and parametric turbulence. I am also extremely grateful to V. Zautkin, G. Melkov, L. Prozorova, A. Smirnov, V. Cherepanov, G. Falkovich and E. Podivilov for permanent discussions and cooperation in trying to understand the parametric spin-wave turbulence. The contribution of all of these colleagues to the subject made the realization of the book possible.

I am greatly indebted to Inna Pertsovskaya who translated this book into English. I am uniquely responsible for the remaining mistakes due to last minute modifications.

The financial assistance from the Meyerhoff Foundation of the Weizmann Institute of Science is also gratefully acknowledged.

My son Sergei was of great help in preparing the manuscript.

Finally I would like to express my gratitude to my wife Ludmila whose help, moral support and sympathy played an important role in the production of this book.

Contents

1 **Introduction to Nonlinear Wave Dynamics** 1
  1.1 Hamiltonian Method for Description of Waves
     in a Continuous Medium 2
    1.1.1 Hamiltonian Equations of Motion 3
    1.1.2 Transfer to Complex Variables 4
    1.1.3 Hamiltonian Structure Under Small Nonlinearity 5
    1.1.4 Dynamic Perturbation Theory. Elimination
          of "Non-Resonant" Terms from the Hamiltonian 10
  1.2 Dimensional Estimation of Hamiltonian Coefficients 11
  1.3 Dynamic Equations of Motion
     for Weakly Non-Conservative Wave Systems 16
    1.3.1 Taking into Account Linear Wave Damping 17
    1.3.2 Allowing for Thermal Noise 18
    1.3.3 Nonlinearity of Wave Damping 19
  1.4 Three-Wave Processes 20
    1.4.1 Confluence of Two Waves
          and Other Induced Processes 20
    1.4.2 Decay Instability 22
    1.4.3 Interaction of Three Waves with Finite Amplitude 24
    1.4.4 Explosive Three-Wave Instability 26
  1.5 Four-Wave Processes 27
    1.5.1 Modulation Instability of the Plane Wave 28
    1.5.2 Equation for the Envelopes 30
    1.5.3 Package Evolution in Unbounded Media 31

2 **The General Properties of Magnetodielectrics** 35
  2.1 Classification of Substances by Their Magnetic Properties 35
    2.1.1 Diamagnets 35
    2.1.2 Superconductors 35
    2.1.3 Paramagnets 36
    2.1.4 Magnetically Ordered Substances (Magnets) 36
  2.2 Nature of Interaction of Magnetic Moments 38
    2.2.1 Exchange Interaction in the Hydrogen Molecule . 38
    2.2.2 Interatomic Exchange 40
    2.2.3 Interatomic Exchange of Large Spins 41
    2.2.4 Indirect Exchange Interactions 42

		2.2.5	Relativistic Interactions	43
	2.3	Energy of Ferromagnets in the Continuum Approximation		45
	2.4	Magnetic and Crystallographic Structure of Some Magnets		49
		2.4.1	Crystals with Spinel Structure	50
		2.4.2	Crystals with Garnet Structure	51
		2.4.3	Crystals with Hexagonal Structure	52
		2.4.4	Crystals with Rhombohedral Structures	53
3	**Spin Waves (Magnons) in Magnetically Ordered Dielectrics**			55
	3.1	Hamiltonian of Magnons in Ferromagnets (FM)		56
		3.1.1	Spectrum of Magnons in Cubic Ferromagnets	57
		3.1.2	Amplitudes of Three- and Four-Magnon Interaction	60
		3.1.3	Three-Magnon Hamiltonian	62
		3.1.4	Four-Magnon Interaction Hamiltonian	63
	3.2	Hamiltonian Function of Magnons in Antiferromagnets		65
		3.2.1	Magnon Spectrum in Antiferromagnets (AFM)	65
		3.2.2	Interaction Hamiltonian in "Easy-Plane" Antiferromagnets	68
		3.2.3	Nuclear Magnons in "Easy-Plane" Antiferromagnets	69
	3.3	Comments at the Road Fork		70
	3.4	Calculation of Magnon Hamiltonian		70
		3.4.1	Equation of Motion of Magnetic Moment	70
		3.4.2	Canonical Variables for Spin Waves in Ferromagnets (FM)	72
		3.4.3	Calculation of Frequencies and Interaction Amplitudes of Waves	73
4	**Nonlinear Dynamics of Narrow Packets of Spin Waves**			77
	4.1	Elementary Processes of Spin Wave Interaction		77
		4.1.1	Three-Magnon Processes	77
		4.1.2	Modulation Instability of Spin Waves	80
	4.2	Self-Focusing of Magnetoelastic Waves in Antiferromagnets (AFM)		82
		4.2.1	Structure of Basic Equations	82
		4.2.2	Properties of Unidimensional Equations	84
		4.2.3	Stability of Solitons and Self-Focusing Theorem	84
		4.2.4	Evolution of Magnetoelastic Waves in the Absence of a Linear Bond Between Magnons and Phonons	86
	4.3	Methods of Parametric Excitation of Spin Waves		87
		4.3.1	Transverse Pumping of Spin Waves in FM	87

	4.3.2	Parallel Pumping of Spin Waves in FM	90
	4.3.3	"Oblique" Pumping of Spin Waves in FM	91
	4.3.4	Suhl Instability of the Second Order in FM	91
	4.3.5	Parallel Pumping in "Easy-Plane" Antiferromagnets	92
	4.3.6	Parametric Pumping of Nuclear Magnons	94

5 Stationary Nonlinear Behavior of Parametrically Excited Waves. Basic S-Theory ... 93

5.1	History of the Problem		95
5.2	Statement of a Problem of Nonlinear Wave Behavior		98
5.3	Phase Relations and Mechanisms for Amplitude Limitation		100
	5.3.1	Analysis of Phase Relations	100
	5.3.2	Nonlinear Mechanisms for Limiting Parametric Instability	101
5.4	Basic Equations of Motion in the S-Theory		102
	5.4.1	Statistical Properties of a Non-Interacting Field	102
	5.4.2	Mean-Field Approximation	103
	5.4.3	General Analysis of Basic Equations of S-Theory	106
5.5	Ground State of System of Interacting Parametric Waves		108
	5.5.1	Stationary States and Analysis of Instability	108
	5.5.2	Ground State Under Low Supercriticality	111
	5.5.3	Threshold of Generation of Second Group of Pairs	114
	5.5.4	Ground State Under High Supercriticality	116
	5.5.5	Nonlinear Susceptibilities of Parametric Waves	118

6 Advanced S-Theory: Supplementary Sections ... 121

6.1	Ground State Evolution of System with Increasing Pumping Amplitude		123
	6.1.1	Ground State of Parametric Waves for Complex Pair Interaction Amplitudes	124
	6.1.2	The Second and Intermediate Thresholds	125
	6.1.3	Nonlinear Behavior of Non-Analytic Pair Interaction Amplitudes	128
6.2	Influence of Nonlinear Damping on Parametric Excitation		132
	6.2.1	Simple Theory	132
	6.2.2	Influence of Non-Analyticity on Nonlinear Damping	135
6.3	Parametric Excitation Under the Feedback Effect on Pumping		139
	6.3.1	Hamiltonian of the Problem	139

	6.3.2	General Analysis of the Equations of Motion ...	141
	6.3.3	First-Order Processes	143
	6.3.4	Second-Order Processes	146
6.4	Nonlinear Theory of Parametric Wave Excitation at Finite Temperatures		147
	6.4.1	Different Time Correlators and Frequency Spectrum	147
	6.4.2	Basic Equations of Temperature S-Theory	148
	6.4.3	Separation of Waves into Parametric and Thermal	150
	6.4.4	Two-Dimensional Reduction of Basic Equations .	151
	6.4.5	Distribution of Parametric Waves in k	152
	6.4.6	Spectrum of Parametric Waves	153
	6.4.7	Heating Below Threshold	153
	6.4.8	Influence of Thermal Bath on Total Characteristics	153
6.5	Introduction to Spatially Inhomogeneous S-Theory		155
	6.5.1	Basic Equations	155
	6.5.2	Parametric Threshold in Inhomogeneous Media .	157
	6.5.3	Stationary State in Non-Homogeneous Media ...	160
6.6	Nonlinear Behavior of Parametric Waves from Various Branches. Asymmetrical S-Theory		165
	6.6.1	Derivation of Basic Equations	165
	6.6.2	Stationary States in Isotropic Case	167
6.7	Parametric Excitation of Waves by Noise Pumping		172
	6.7.1	Equations of S-Theory Under Noise Pumping ..	173
	6.7.2	Distribution of Parametric Waves Above Threshold	175

7 Non-Stationary Behavior of Parametrically Excited Waves .. 179

7.1	Spectrum of Collective Oscillations (CO)		179
	7.1.1	Spectrum of Spatially Homogeneous CO in the Non-Dissipation Limit	179
	7.1.2	Influence of Wave Damping on the CO Spectrum	181
	7.1.3	Spectrum of Spatially Non-Homogeneous CO ...	182
7.2	Linear Theory of CO Resonance Excitation		184
	7.2.1	Basic Equations and Their Solution	184
	7.2.2	CO Excitation by a Microwave Field	185
	7.2.3	Direct CO Excitation by a Radio Frequency Field	187
	7.2.4	Coupled Motions of Collective Excitations of Parametric Waves and Sound	188
7.3	Threshold Under Periodic Modulation of Dispersion Law		189
7.4	Large-Amplitude Collective Oscillations and Double Parametric Resonance		193
	7.4.1	Stationary State Under Periodic Modulation ...	193

		7.4.2	Parametric Excitation of CO of Parametric Wave System	194

- 7.5 Transient Processes when Pumping is Turned on 194
  - 7.5.1 Small Supercriticality Range 194
  - 7.5.2 High Supercriticality Range 198
- 7.6 Parametric Excitation Under Sweeping of Wave Frequency 200
  - 7.6.1 Qualitative Analysis of the Problem 200
  - 7.6.2 Basic Equations of S-Theory Under Frequency Sweeping 204
  - 7.6.3 Solution of S-Theory Equations 205
  - 7.6.4 Dependence of the Number of Waves on the Pumping Amplitude 206
- 7.7 Problems 209

8 Secondary Parametric Wave Turbulence 213
- 8.1 Instability of Ground State and Auto-Oscillations 214
  - 8.1.1 Properties and Nature of Spin Wave Oscillations 214
  - 8.1.2 Numerical Simulation of Auto-Oscillation in the S-Theory 215
  - 8.1.3 Conditions for Excitation of Auto-Oscillations .. 217
- 8.2 Route to Chaos in Dynamic Systems 216
  - 8.2.1 Introduction 219
  - 8.2.2 Elementary Concepts of Theory of Dynamic Chaos 221
  - 8.2.3 Chaos of Parametric Magnons in $CsMnF_3$ 225
- 8.3 Geometry of Attractors of Secondary Parametric Turbulence of Magnons 229
  - 8.3.1 Effective Phase Space and Dimensionality of Inclusion 229
  - 8.3.2 Experimental Study of Attractor Structure in $CsMnF_3$ 230
- 8.4 Secondary Turbulence and Collapses in Narrow Parametric Wave Packets 233
  - 8.4.1 Equations for Envelopes 233
  - 8.4.2 Stationary Solitons 235
  - 8.4.3 Average Characteristics of Secondary Turbulence 236
  - 8.4.4 Destruction of Parametric Solitons with Large Amplitude 237
  - 8.4.5 Soliton Mechanism of Amplitude Limitation 239

9 Experimental Investigations of Parametrically Excited Magnons 243
- 9.1 Experimental Investigations of Parametric Instability of Magnons 243
  - 9.1.1 Methods and Materials Investigated 243
  - 9.1.2 Measurements of Constants in Spin Wave Spectra 244

	9.1.3	Spin Wave Damping	245
9.2	Nonlinear Behavior of Parametric Magnons – General Information		250
	9.2.1	Measuring Technique for Susceptibilities χ' and χ''	250
	9.2.2	Comparison of S-Theory and Experiment for Susceptibilities	252
	9.2.3	Measurements of Interaction (Frequency Shift) Amplitude	255
	9.2.4	Nonlinear Ferromagnetic Resonance	257
9.3	Investigations of Stationary State With One Group of Pairs		258
	9.3.1	Nonlinear Susceptibility in the One-Group State	259
	9.3.2	Direct Measurement of Pair Phase	260
9.4	Electromagnetic Radiation of Parametric Magnons		262
	9.4.1	Frequency of Parametric Magnons	262
	9.4.2	Frequency Width of Parametrically Excited Magnons	263
9.5	Collective Resonance of Parametric Magnons		266
	9.5.1	Experimental Technique	267
	9.5.2	Frequency of Collective Resonance	269
	9.5.3	Susceptibility to Field of Weak Microwave Signal	272
	9.5.4	Linewidth of Collective Resonance	273
	9.5.5	Oscillations of Longitudinal Magnetization	274
	9.5.6	Other Methods for Excitation of Collective Oscillations	275
9.6	Stepwise Excitation in YIG		276
	9.6.1	Re-Radiation into the Transverse Channel	277
	9.6.2	Interaction of Second-Group Magnons and Transverse Signal	278
9.7	Conditions of Excitation of Auto-Oscillations of Magnons		281
	9.7.1	Experimental Setup	283
	9.7.2	Intensive Auto-Oscillations of Mode $m = 0$	284
	9.7.3	Crossing the Instability Boundary and Spatially Inhomogeneous Auto-Oscillations	286
	9.7.4	Instability of Higher Collective Modes	288
9.8	Effect of Radio-Frequency Field Modulation on Parametric Resonance		289
	9.8.1	Suppression of Parametric Instability by Modulation	289
	9.8.2	Stationary State of Parametric Magnons Under Modulation of Their Frequency	291

9.9	Double Parametric Resonance and Inhomogeneous Collective Oscillations of Magnons .	293
9.10	Parametric Excitation of Magnons Under Noise Modulation of their Frequencies	294
	9.10.1 Threshold Amplitude of Noise Pumping	294
	9.10.2 Efficiency of Phase Mechanism Under Noise Pumping	296

10 Nonlinear Kinetics of Parametrically Excited Waves 301

- 10.1 General Equations 301
- 10.2 Limit of the S-Theory 305
  - 10.2.1 Form of the Green's Function 305
  - 10.2.2 Separation of the Waves into Parametric and Thermal 306
- 10.3 Nonlinear Theory of Parametric Excitation of Waves in Random Media 307
  - 10.3.1 General Equations in the S, g^2-Approximation .. 308
  - 10.3.2 Distribution Function of Parametric Waves 309
  - 10.3.3 Behavior of Parametrically Excited Waves Beyond the Threshold 310
- 10.4 Consistent Nonlinear Theory for Parametric Excitation of Waves 311
  - 10.4.1 Spectral Density of Parametrically Excited Waves 311
  - 10.4.2 Structure of the Distribution Function in k-Space 313

References 317

Subject Index 327

List of Main Symbols with Short Comments

All symbols are fully defined where they are first introduced in the text. For the convenience of the reader some of the most frequently used symbols are collected here.

$a(\boldsymbol{k},t) = a_k$, $b(\boldsymbol{k},t) = b_k$,
$c(\boldsymbol{k},t) = c_k$, $d(\boldsymbol{k},t) = d_k$ — canonical variables – complex amplitudes of $\boldsymbol{k}$-waves (having wave vector $\boldsymbol{k}$). These are classical analogues of the Bose operators, so-called c-numbers

$\mathcal{H} = \mathcal{H}_2 + \mathcal{H}_{\text{int}}$ — Hamiltonian function or Hamiltonian which is the classical analogue the Hamiltonian operator in quantum mechanics. Here:

$\mathcal{H}_2 = \sum_k \omega_k b_k^* b_k$ — Hamiltonian of non-interacting $\boldsymbol{k}$-waves having frequency ω_k, quadratic in canonical variables, and

$\mathcal{H}_{\text{int}} = \mathcal{H}_3 + \mathcal{H}_4 + \ldots$ — interaction Hamiltonian. Here:

$\mathcal{H}_3$, $\mathcal{H}_4$ — the parts of $\mathcal{H}_{\text{int}}$, of the third- and fourth-order in canonical amplitudes which describe the three- and four-wave processes of interaction

$\mathcal{H}_{\text{p}}$ — pumping Hamiltonian describing the interaction of waves with an external alternating pumping field. As a rule a microwave magnetic field plays the role of pumping of spin waves (magnons) in magnetics.

$\boldsymbol{H}$, H — static magnetic field, as a rule, spatially homogeneous

H_{a}, H_{D}, H_{ex} — anisotropy field, Dzyaloshinsky field and exchange field in magnetics

h and ω_{p} — amplitude and frequency of pumping field for parametric excitation of waves (in magnetics usually a microwave magnetic field).

h_{th} and $p = (h/h_{\text{th}})^2$ — threshold amplitude and "supercriticality" for parametric excitation of waves. For typical experiments in magnetics $h_{\text{th}} < 1$ Oe

List of Main Symbols with Short Comments

$\omega(\boldsymbol{k}) = \omega_k$	frequency of non-interacting $\boldsymbol{k}$-wave or dispersion law
$\omega_{\mathrm{NL}}(\boldsymbol{k})$	frequency of interacting $\boldsymbol{k}$-wave – nonlinear dispersion law
$\omega_{\mathrm{ex}} = gH_{\mathrm{ex}}$	exchange frequency in magnetics. In the order of magnitude $\hbar\omega_{\mathrm{ex}} \simeq T_{\mathrm{C}}$, were T_{C} is the Curie temperature
$g \approx \mu_{\mathrm{B}}/\hbar$	magnetic-to-mechanical ratio ($\approx 2\pi \cdot 2.8$ MHz/Oe)
μ_{B}	Bohr magneton
$\boldsymbol{M}, M$	magnetization – magnetic moment per unit volume
$\omega_{\mathrm{M}} = 4\pi g M$	characteristic frequency of magnetic dipole–dipole interaction
$V(\boldsymbol{k}) = V_k$	amplitudes of $\pm\boldsymbol{k}$-pair interaction with pumping field
$V(\boldsymbol{k}_1; \boldsymbol{k}_2, \boldsymbol{k}_3) = V_{1,23}$	amplitudes of three-wave interaction of type $\boldsymbol{k}_1 \Leftrightarrow \boldsymbol{k}_2 + \boldsymbol{k}_3$
$T(\boldsymbol{k}_1, \boldsymbol{k}_2; \boldsymbol{k}_3, \boldsymbol{k}_4) = T_{12,34}$	amplitudes of four-wave interaction of type $\boldsymbol{k}_1 + \boldsymbol{k}_2 \Leftrightarrow \boldsymbol{k}_3 + \boldsymbol{k}_4$
$T(\boldsymbol{k}, \boldsymbol{k}') = T_{kk'}$	amplitudes of four-wave interaction of type $\boldsymbol{k} + \boldsymbol{k}' \Leftrightarrow \boldsymbol{k} + \boldsymbol{k}'$, describing the nonlinear frequency shift
$S(\boldsymbol{k}, \boldsymbol{k}') = S_{kk'}$	amplitudes of four-wave interaction of type $\boldsymbol{k} + (-\boldsymbol{k}) \Leftrightarrow \boldsymbol{k}' + (-\boldsymbol{k}')$, describing the self-consistent interaction of pairs of wave $\pm\boldsymbol{k}$ and $\pm\boldsymbol{k}'$
$\gamma(\boldsymbol{k}) = \gamma_k$	damping of $\boldsymbol{k}$-wave
$n(\boldsymbol{k}, t) = n_k$	simultaneous double correlation function of $\boldsymbol{k}$-waves, "number" of $\boldsymbol{k}$-waves
$N = \sum_k n_k$	total number of parametric (parametrically excited) waves
$\sigma(\boldsymbol{k}, t) = \sigma_k$	simultaneous double *anomalous* correlation function of $\pm\boldsymbol{k}$-pairs of waves, "anomalous correlator" of parametric pairs
$\varphi(\boldsymbol{k}) = \varphi_k$	phase of $\boldsymbol{k}$-waves
$\Psi(\boldsymbol{k}) = \Psi_k = \varphi_k + \varphi_{-k}$	phase of $\pm\boldsymbol{k}$-pairs of parametrically excited waves
Ω, Ω	direction of $\boldsymbol{k}$, solid angle
Θ, φ	polar and azimuthal angles
$\propto, \approx$	proportional, approximately equal
$\simeq$	of the same order

1 Introduction to Nonlinear Wave Dynamics

Our aim is to describe the nonlinear properties of a system of interacting waves in various media from a common point of view. We would like to abstain from using the features of these media until it is absolutely necessary for studying a particular problem. On the other hand, the method of description must be sufficiently well-known and convenient. The classical Hamiltonian description method of nonlinear waves (including spin waves in magnetodielectrics) described in Sect. 1.1 seems to meet these requirements. By this method, the Hamiltonian equations for the canonical variables, i.e. complex wave amplitudes in continuous media, will be obtained, the general structure of the Hamiltonian function (*Hamiltonian*) for small-amplitude waves, irrespective of their physical nature, will be studied, and the dynamic perturbation theory will be formulated, so that nonresonant terms from the Hamiltonian of interaction should be excluded.

We should leave aside unwieldy calculations of the Hamiltonian function of the spin waves in magnetodielectrics until later. On the other hand, the general theory of nonlinear waves is better illustrated by simple specific examples. Therefore in Sect. 1.2, the dimensional analysis yields expressions for the frequency and interaction amplitudes of sound waves, of waves on a liquid surface, and of the simplest type of spin waves in ferromagnets. The following sections of the present chapter discuss self-action and interaction of almost monochromatic wave packets. They may be described by dynamic equations of motion for complex amplitudes. Phase relations in the equations are essential. Therefore these processes can naturally be called *dynamic*. On the other hand, interactions of wave packets broad in k-space with almost random phases may be described by the kinetic equations for wave population number. Such processes can be called *kinetic*. These will be treated later, in Chap. 10.

In Section 1.3 of this chapter we shall add a phenomenological term to the Hamiltonian equations. This term describes a weak interaction of the dynamic wave system with the environment which serves as a thermal bath. This makes it possible to obtain from the very beginning much more realistic approximations in describing the wave dynamics than it is possible under the purely Hamiltonian approach (which is applicable to systems conserving their energy). This procedure will be rigorously proven by means of the diagrammatic technique. Its applicability domain was given, for example, in Chap. 7 of my Russian book [1.1].

Sections 1.4 and 1.5 of this chapter describe some dynamic processes (confluence of two waves into one; generation of the second harmonic; decay of one wave into two various four-wave processes, including self-focusing and collapse) irrespective of the wave type and the medium in which the waves propagate. Only in the third chapter we shall study the specific characteristics of these processes using the explicit form of dispersion laws and the interaction Hamiltonian of spin waves in ferromagnets and antiferromagnets.

1.1 Hamiltonian Method for Description of Waves in a Continuous Medium

The Hamiltonian method is applicable to a wide group of weakly interacting and weakly dissipative wave systems and is used to reveal the common properties of such system. Indeed, equations of motion for waves may vary considerably if these equations are written in terms of natural medium variables. For instance, the Bloch equations describing magnetic moment motion do not resemble the Maxwellian equations for a nonlinear dielectric. The latter differ drastically from the Eulerian equations for a compressible liquid. At the same time spin electromagnetic and sound waves are primarily waves, i.e. oscillations of the medium transferred from one of its points to another. If we are interested only in the propagation peculiarities of waves with small amplitudes, e.g. in the diffraction phenomenon, it is not important whether it is the it magnetic moment, electric field or medium density that oscillates. To study how non-interacting waves propagate in the medium it is quite sufficient to use information given by the dispersion law $\omega(\boldsymbol{k})$. Similarly, the other expansion coefficients of the Hamiltonian provide all necessary (and almost superfluous) information for investigating the nonlinear properties of the wave system. It is therefore clear that two wave systems with similar dispersion laws and $\boldsymbol{k}$-dependences of "essential" Hamiltonian coefficients will show a similar nonlinear behavior. Their equations of motion in natural variables may at the same time look completely different.

The method of second quantization can also reveal the common properties of various wave systems. In this method, operators characterizing the natural variables of the medium are represented in terms of creation and annihilation Bose operators. This quantum-mechanical method is commonly used in the physics of nonlinear waves, in spite of the fact that the values of population numbers under which nonlinear effects are significant are usually great and the wave system is classical. Therefore when the powerful method of second quantization is used at the stage of problem statement, the Planck constant has to appear. In the resulting expressions it is afterwards cancelled. Such inconsistency must be due to the fact that modern physicists are taught quantum mechanics in much more detail than classical mechanics.

In our opinion, the problems of nonlinear wave dynamics are much easier to solve using the classical Hamiltonian method [1.1-5]. One can avoid unnecessary difficulties arising from the non-commutativity of operators a and a^+. The Hamiltonian method provides as simple an interpretation as the method of second quantization since the complex canonical variables a and a^* are classical analogues (c-numbers) of the Bose operators. That is why we shall use the classical term of *spin waves* and the quantum-mechanical term of *magnons* as near synonyms.

1.1.1 Hamiltonian Equations of Motion

The Hamiltonian equations for systems with a single degree of freedom, as is well known, have the form [1.2]:

$$\frac{dq}{dt} = \frac{\partial \mathcal{H}}{\partial p}, \qquad \frac{dp}{dt} = -\frac{\partial \mathcal{H}}{\partial q}. \tag{1.1.1}$$

Here the Hamiltonian function $\mathcal{H}$ (henceforth called *Hamiltonian*) depends on the canonical variables, i.e. the generalized coordinate q and generalized momentum p. Usually, $\mathcal{H}$ is the system energy expressed in terms of the canonical variables. Systems with n degrees of freedom are characterized by pairs of canonically conjugated variables $q_1, q_2, ... q_n$ and $p_1, p_2, ... p_n$. They satisfy the equations

$$\frac{dq_i}{dt} = \frac{\partial \mathcal{H}}{\partial p_i}, \qquad \frac{dp_i}{dt} = -\frac{\partial \mathcal{H}}{\partial q_i}, \tag{1.1.2}$$

where $\mathcal{H}$ is a function of all q_i and p_i. In the simplest case the continuous medium can be characterized by a pair of canonical variables $q(\mathbf{r}, t)$, $p(\mathbf{r}, t)$ at each point $\mathbf{r}$. The equations of motion for $q(\mathbf{r}, t)$, $p(\mathbf{r}, t)$ are obtained by generalizing (1.1.2):

$$\frac{\partial q(\mathbf{r}, t)}{\partial t} = \frac{\delta \mathcal{H}}{\delta p(\mathbf{r}, t)}, \qquad \frac{\partial p(\mathbf{r}, t)}{\partial t} = -\frac{\delta \mathcal{H}}{\delta q(\mathbf{r}, t)}. \tag{1.1.3}$$

The Hamiltonian $\mathcal{H}$ is dependent on $p(t)$ and $q(t)$ taken at all the points $\mathbf{r}$, i.e. it is a functional of $q(\mathbf{r}, t)$, $p(\mathbf{r}, t)$. The symbols $\delta/\delta p$ and $\delta/\delta q$ designate variational derivatives which generalize notions of partial derivatives for the case of continuous degrees of freedom.

It must be noted that (1.1.3) specify the dynamics of only one wave type, e.g. the sound. To allow for several wave types or polarizations the medium must be characterized at each point $\mathbf{r}$ by several pairs of variables $q_j(\mathbf{r}, t)$, $p_j(\mathbf{r}, t)$, $j = 1...$. The respective generalization of the equations of motion (1.1.3) yields

$$\frac{\partial q_j(\boldsymbol{r},t)}{\partial t} = \frac{\delta \mathcal{H}}{\delta p(\boldsymbol{r},t)}, \qquad \frac{\partial p_j(\boldsymbol{r},t)}{\partial t} = -\frac{\delta \mathcal{H}}{\delta q_j(\boldsymbol{r},t)}. \tag{1.1.4}$$

1.1.2 Transfer to Complex Variables

A formal advantage of the Hamiltonian method is the symmetrical representation of the equation for the coordinate q and the momentum p. To this end, we first change over to new canonical variables $Q(\boldsymbol{r}) = \lambda q(\boldsymbol{r}), P(\boldsymbol{r}) = p(\boldsymbol{r})/\lambda$, the dimension factor λ being chosen in such a way that P and Q have the same dimension. Then we introduce complex variables:

$$a_j = (Q_j + iP_j)/\sqrt{2}, \qquad a_j^* = (Q_j - iP_j)/\sqrt{2} \tag{1.1.5}$$

with equation of motion

$$\sqrt{2}\frac{\partial a_j}{\partial t} = \frac{\delta \mathcal{H}}{\delta P_j} - i\frac{\delta \mathcal{H}}{\delta Q_j}, \qquad \sqrt{2}\frac{\partial a_j^*}{\partial t} = \frac{\delta \mathcal{H}}{\delta P_j} + i\frac{\delta \mathcal{H}}{\delta Q_j}.$$

Substituting in the above expression $\mathcal{H}(a,a^*)$ we obtain

$$i\frac{\partial a_j}{\partial t} = \frac{\delta \mathcal{H}}{\delta a_j^*}, \qquad -i\frac{\partial a_j^*}{\partial t} = \frac{\delta \mathcal{H}}{\delta a_j}. \tag{1.1.6}$$

The second equation may be obtained from the first one by complex conjugation and, consequently, instead of two real equations (1.1.3) we obtain one complex equation. In quantum mechanics there is a corresponding change from the coordinate–momentum representation to the representation of creation and annihilation Bose operators. Their classical analogues are the complex canonical variables.

Canonical variables (1.1.5) are by no means unique. There is a wide range of possible changes from the canonical variables (a, a^*) to other variables (b, b^*) where the equations of motion retain their canonical form:

$$i\frac{\partial b(\boldsymbol{r},t)}{\partial t} = \frac{\delta \mathcal{H}}{\delta b^*(\boldsymbol{r},t)}. \tag{1.1.7}$$

Such transformations are called *canonical* [1.2].

The possibility of choosing various canonical variables is an important advantage of the Hamiltonian method. Thus, the variables suitable for a given problem can be chosen.

1.1.3 Hamiltonian Structure Under Small Nonlinearity

For a wide variety of problems of nonlinear wave dynamics the wave amplitude may be characterized by some natural dimensionless parameter x. For sound waves the parameter is the relation between the density variation in a sound wave and the mean medium density; for surface waves in liquids it is the relation between the vertical declination and the wavelength. For spin waves the angle of the precession of the magnetic moment serves as a dimensionless parameter x. If the parameter x is of the order unity it leads to phenomena specific to a given problem, e.g. sound is transformed into shock waves, "white horses" appear on the liquid surface, magnetization in ferromagnets is reversed, producing a domain wall which, generally speaking, is able to move. Obviously, it is not worth considering all these phenomena from a single viewpoint. However, if the parameter of wave nonlinearity is small, the specific features of the medium are no longer essential and the wave dynamics can be described in general terms (irrespective of these features) i.e. in terms of the dispersion law $\omega(\boldsymbol{k})$, group and phase velocity, probabilities of elementary processes of interaction involving three, four waves, etc. In this section it will be shown how these notions are introduced in the scope of the Hamiltonian formalism under small nonlinearity x.

The canonical variables a_j, a_j^* will be chosen so as to characterize the wave amplitude and become zero in the absence of a wave. The index $j = 1,...n$ determines the wave polarization or type, since, generally speaking, various waves can be simultaneously excited in media: sound, light, etc. Assuming a_j, a_j^* to be small, let us expand the functional $\mathcal{H}\{a_j(\boldsymbol{r},t), a_j^*(\boldsymbol{r},t)\}$ in a power series in a_j, a_j^*. We are not interested in the zero term in $\mathcal{H}\{0,0\}$, since it does not enter into the equations of motion: $\delta\mathcal{H}\{0,0\}/\delta a = 0$. The terms of the first order $\mathcal{H}$ are zero, since we take the medium to be in an equilibrium state in absence of waves, consequently, with a minimum at $a_j = a_j^* = 0$. Therefore, the expansion of $\mathcal{H}$ begins with the terms of the second order

$$\mathcal{H} = \mathcal{H}_2 + \mathcal{H}_{\text{int}} . \tag{1.1.8}$$

The most general form of $\mathcal{H}_2$ is

$$\begin{aligned}\mathcal{H}_2 = \sum_{i,j=1}^{n} \int \Big\{ &A_{ij}(\boldsymbol{r},\boldsymbol{r}')a_i(\boldsymbol{r},t)a_j^*(\boldsymbol{r}',t) \\ &+ \frac{1}{2}[B_{ij}^*(\boldsymbol{r},\boldsymbol{r}')a_i(\boldsymbol{r},t)a_j(\boldsymbol{r}',t) + \text{c.c.}]\Big\} d\boldsymbol{r}d\boldsymbol{r}' .\end{aligned} \tag{1.1.9}$$

Here "c.c" denotes complex conjugate. As $\mathcal{H}$ must be Hermitian

$$A_{ij}(\boldsymbol{r},\boldsymbol{r}') = A_{ji}^*(\boldsymbol{r}',\boldsymbol{r}), B_{ij}(\boldsymbol{r},\boldsymbol{r}') = B_{ji}(\boldsymbol{r}',\boldsymbol{r}) . \tag{1.1.10}$$

In a spatially homogeneous medium A and B depend only on the coordinate difference $\boldsymbol{R} = \boldsymbol{r} - \boldsymbol{r}'$ and if there is inversion symmetry, they are even functions of the difference $\boldsymbol{R}$. Then it follows from (1.1.10) that

$$A_{ij}(\boldsymbol{R}) = A_{ji}^*(-\boldsymbol{R}), \qquad B_{ij}(\boldsymbol{R}) = B_{ji}(-\boldsymbol{R}) . \tag{1.1.11}$$

The Hamiltonian can be significantly simplified by the Fourier transform

$$\begin{aligned} a(\boldsymbol{k},t) = a_k &= \frac{1}{V_s} \int a(\boldsymbol{r},t) \exp(-i\boldsymbol{k}\boldsymbol{r}) \, d\boldsymbol{r} \ , \\ a(\boldsymbol{r},t) &= \sum_k a(\boldsymbol{k},t) \exp(i\boldsymbol{k}\boldsymbol{r}) \ , \end{aligned} \tag{1.1.12}$$

where V_s is the sample volume (of the medium in which the waves propagate). We consider the wave vector $\boldsymbol{k}$ to be a discrete variable. If necessary, one may pass from summation over $\boldsymbol{k}$ to integration

$$(2\pi)^3 \sum_k = V_s \int d\boldsymbol{k} \tag{1.1.13}$$

in the conventional way. The Fourier transform (1.1.12) is canonical but not unimodal. This implies that the Hamiltonian equation retains its canonical form but the new Hamiltonian differs from the previous one by a factor: it is divided by the sample volume

$$i\partial a_k/\partial t = \delta \mathcal{H}(a_k, a_k^*)/\delta a_k^*, \quad \mathcal{H}(a_k, a_k^*) = \mathcal{H}(a(\boldsymbol{r}), a^*(\boldsymbol{r}))/V_s . \tag{1.1.14}$$

It is significant that in the new variables $a(\boldsymbol{k},t)$ the quadratic part of the Hamiltonian is a sum over $\boldsymbol{k}$ and contains no summation over $\boldsymbol{k}'$:

$$\begin{aligned} \mathcal{H}_2 = \sum_{i,j} \sum_k &\Big\{ A_{ij}(\boldsymbol{k}) a_i(\boldsymbol{k},t) a_j^*(\boldsymbol{k},t) \\ &+ \frac{1}{2} [B_{ij}^*(\boldsymbol{k}) a_i(\boldsymbol{k},t) a_j(-\boldsymbol{k},t) + \text{c.c.}] \Big\} , \end{aligned} \tag{1.1.15}$$

$$\begin{aligned} A_k &= A(\boldsymbol{k}) = \int A(\boldsymbol{R}) \exp(i\boldsymbol{k}\boldsymbol{R}) \, d\boldsymbol{R} , \\ B_k &= B(\boldsymbol{k}) = \int A(\boldsymbol{R}) \exp(i\boldsymbol{k}\boldsymbol{R}) \, d\boldsymbol{R} . \end{aligned} \tag{1.1.16}$$

Obviously, the spatial homogeneity of the medium is responsible for this fact.

In some cases the Hamiltonian (1.1.15) may be diagonalized in "wave types" by means of a linear transformation

$$b_i(\boldsymbol{k}) = \sum_j [u_{ij}(\boldsymbol{k}) a_j(\boldsymbol{k}) + v_{ij}(\boldsymbol{k}) a_j^*(-\boldsymbol{k})] . \tag{1.1.17}$$

The matrices $[u]$ and $[v]$ may be chosen so that the Hamiltonian $\mathcal{H}$ takes the form

$$\mathcal{H}_2 = \sum_j \sum_{\boldsymbol{k}} \omega_j(\boldsymbol{k}) b_j(\boldsymbol{k}) b_j^*(\boldsymbol{k}) \,. \tag{1.1.18}$$

Diagonal elements of the matrix $[u]$ will henceforth be considered real. This can always be arranged through suitable choice of phases for complex variables $b(\boldsymbol{k}, t)$.

Let us consider in more detail the case when waves of a single type propagate in the medium. Taking in (1.1.15, 17, 18) $i = j = 1$, substitute (1.1.17) into (1.1.15). By the condition of coincidence of (1.15) and (1.18) we obtain:

$$\omega(\boldsymbol{k})[u^2(\boldsymbol{k}) + |v(\boldsymbol{k})|^2] = A(\boldsymbol{k})\,, \qquad 2\omega(\boldsymbol{k})u(\boldsymbol{k})v(\boldsymbol{k}) = B^*(\boldsymbol{k})\,, \tag{1.1.19a, b}$$

$$2\omega(\boldsymbol{k})u(\boldsymbol{k})v^*(\boldsymbol{k}) = B(\boldsymbol{k})\,, \qquad u^2(\boldsymbol{k}) - |v(\boldsymbol{k})|^2 = 1 \,. \tag{1.1.19c, d}$$

If the second relation is satisfied, the transformation is canonical. Comparison of (1.1.19b) and (1.1.19c) shows that obtaining a diagonalizing transformation is possible only if $\omega(\boldsymbol{k})$ is real. Note that in the variables $b_j(\boldsymbol{k})$ the equations of motion (1.1.7) with the Hamiltonian (1.1.18) take a trivial form

$$\partial b_j(\boldsymbol{k}, t)/\partial t + i\omega_j(\boldsymbol{k}, t) b(\boldsymbol{k}, t) = 0\,, \tag{1.1.20a}$$

and have a solution

$$b_j(\boldsymbol{k}, t) = b(\boldsymbol{k}, 0) \exp[i\omega_j(\boldsymbol{k}) t] \,. \tag{1.1.20b}$$

Therefore the condition of $\omega(\boldsymbol{k})$ being real simply implies the medium's stability with respect to an exponential increase of the wave amplitudes. $A(\boldsymbol{k})$, as follows from (1.1.19a), must in this case also be real.

From (1.1.18) transformation coefficients can be easily obtained

$$u(\boldsymbol{k}) = \sqrt{\frac{A(\boldsymbol{k}) + \omega(\boldsymbol{k})}{2\omega(\boldsymbol{k})}} \,, \qquad v(\boldsymbol{k}) = -\frac{B(\boldsymbol{k})}{|B(\boldsymbol{k})|} \sqrt{\frac{A(\boldsymbol{k}) + \omega(\boldsymbol{k})}{2\omega(\boldsymbol{k})}} \tag{1.1.21}$$

as well as two expressions of the opposite sign for the frequency

$$\omega(\boldsymbol{k}) = \pm\sqrt{A^2(\boldsymbol{k}) - |B(\boldsymbol{k})|^2} \,. \tag{1.1.22}$$

According to the relation (1.1.19a) we choose from these the expression of the same sign as $A(\boldsymbol{k})$ (*Zakharov* [1.3]).

If there are several oscillation branches, finding a diagonalizing representation is not easy, though the frequencies $\omega(\boldsymbol{k})$ can be determined without it. The equations of motion (1.1.14) with the Hamiltonian (1.1.15) must be written for the variables $a_j(\boldsymbol{k}, t)$:

$$\frac{\partial a_j(\boldsymbol{k},t)}{\partial t} + i\sum_i [A_{ij}(\boldsymbol{k})a_i(\boldsymbol{k},t) + B_{ij}(\boldsymbol{k})a_i^*(-\boldsymbol{k},t)] = 0,$$

$$\frac{\partial a_j^*(-\boldsymbol{k},t)}{\partial t} - i\sum_i [A_{ij}^*(\boldsymbol{k})a_i^*(-\boldsymbol{k},t) + B_{ij}^*(-\boldsymbol{k})a_i(\boldsymbol{k},t)] = 0.$$

Substituting a, $a^* \propto \exp(-i\omega t)$ into this expression yields an algebraic "secular" equation for obtaining the frequencies:

$$\begin{bmatrix} A_{ij} - \omega\delta_{ij}, & B_{ij} \\ B_{ij}^*, & A_{ij}^* - \omega\delta_{ij} \end{bmatrix} = 0. \tag{1.1.23}$$

Note that the diagonalization of the Hamiltonian (1.1.17) is possible if all the roots of (1.1.23), i.e. the wave frequencies $\omega_j(\boldsymbol{k})$, are real.

The variables $b_j(\boldsymbol{k},t)$ are normal variables of the linear theory and therefore are especially suitable for solving nonlinear problems. All the "linear" difficulties of the medium model are tackled in the single step of going over to the b_j variables. In these variables the linearized equations of motion become trivial (see (1.1.20a)). They describe the propagation of free waves obeying the dispersion laws $\omega_j(\boldsymbol{k})$. All the "linear" information required for the investigation of nonlinear problems is contained in the functions $\omega_j(\boldsymbol{k})$. All necessary data on the interaction of waves can be found in the other coefficients of the expansion of the Hamiltonian $\mathcal{H}$ in powers of b_j:

$$\mathcal{H} = \mathcal{H}_2 + \mathcal{H}_{\text{int}}, \qquad \mathcal{H}_{\text{int}} = \mathcal{H}_3 + \mathcal{H}_4 + ... \tag{1.1.24}$$

The physical meaning of $\mathcal{H}_3$, $\mathcal{H}_4$ can be easily explained by analogy with quantum mechanics. The Hamiltonian $\mathcal{H}_3$ describes three-wave processes. In a simple case when all the waves are of a single type

$$\mathcal{H}_3 = \frac{1}{2}\sum_{123}(V_q b_1 b_2 b_3^* + \text{c.c.})\delta(1+2-3)$$
$$+ \frac{1}{3}\sum_{123}(U_q^* b_1^* b_2^* b_3^* + \text{c.c.})\delta(1+2+3). \tag{1.1.25}$$

Hereafter we shall use the following short notation $b_1, b_2, ...$ to denote $b(\boldsymbol{k}_1,t), b(\boldsymbol{k}_2,t)$; $\delta(...)$ is the Kronecker symbol, $\delta(1+2+3)$ designates $\delta(\boldsymbol{k}_1+\boldsymbol{k}_2+\boldsymbol{k}_3)$ which represents the law of conservation of momentum. The multi index q denotes $(\boldsymbol{k}_1, \boldsymbol{k}_2, \boldsymbol{k}_3)$, therefore $V_q = V(q) = V(\boldsymbol{k}_1, \boldsymbol{k}_2, \boldsymbol{k}_3)$, and finally

$$\sum_{123} = \sum_{\boldsymbol{k}_1 \boldsymbol{k}_2 \boldsymbol{k}_3}, \qquad \sum_{1+2=3} = \sum_{\boldsymbol{k}_1 \boldsymbol{k}_2 \boldsymbol{k}_3} \delta(\boldsymbol{k}_1+\boldsymbol{k}_2-\boldsymbol{k}_3).$$

The Hamiltonian $\mathcal{H}_4$ describes the four-wave processes

$$\mathcal{H}_4 = \frac{1}{4} \sum_{1+2=3+4} W_p b_1^* b_2^* b_3 b_4 + \frac{1}{6} \sum_{1=2+3+4} (G_p b_1 b_2^* b_3^* b_4^* + \text{c.c.})$$
$$+ \frac{1}{6} \sum_{1+2+3+4=0} (R_p b_1 b_2 b_3 b_4 + \text{c.c.}) , \quad p = (\boldsymbol{k}_1, \boldsymbol{k}_2, \boldsymbol{k}_3, \boldsymbol{k}_4) . \quad (1.1.26)$$

The following question may arise: to which order in b, b^* is the expansion of the Hamiltonian $\mathcal{H}_{\text{int}}$ to be taken? This question is rather general and can be answered with similar generality: $\mathcal{H}_{\text{int}}$ and the terms of higher order, generally speaking, need not be included. This can be supported as follows: Since expansion into a series is performed in terms of a small parameter, each subsequent term is less that its antecedent and the dynamics of the wave system will be determined by the very first expansion term in $\mathcal{H}_{\text{int}}$, i.e. $\mathcal{H}_3$. However, three-wave processes may be *nonresonant* or, equivalently, *forbidden*. This means that the condition of spatio-temporal synchronism (or, in terms of quasi-particles, the law of energy-momentum conservation):

$$\omega(\boldsymbol{k} + \boldsymbol{k}_1) = \omega(\boldsymbol{k}) + \omega(\boldsymbol{k}_1) \quad (1.1.27)$$

may be impossible to satisfy. Let d denote the dimension of the medium under study, $\boldsymbol{k}$ be a vector of the d-dimensional space ($d > 1$). Equation (1.1.27) determines a $2d - 1$ hypersurface in the $2d$-dimensional space of the vectors $\boldsymbol{k}$, $\boldsymbol{k}_1$. If this surface actually exists (i.e. $\omega(\boldsymbol{k})$ is real), then the law of dispersion $\omega(\boldsymbol{k})$ is called a *decay* law and three-wave processes are allowed. If (1.1.27) has no real solutions, three-wave processes are forbidden and the law of dispersion is *non-decaying*.

In isotropic media $\omega(\boldsymbol{k})$ is a function only of the amplitude k. In this case for Goldstone modes (for which $\omega(0) = 0$) a simple criterion can be formulated: the dispersion law is decaying if $\omega''(k) > 0$ and is non-decaying if $\omega''(k) < 0$ [1.3]. In particular, if $\omega(k) \propto k^z$, dispersion laws where $z > 1$ are decaying. If the dispersion law is quadratic with a gap

$$\omega(k) = \omega_0 + (sk)^2 , \quad (1.1.28)$$

then for small k three-wave decaying processes (1.1.27) are forbidden since the wave of minimum energy can no longer decaying. More detailed analysis of (1.1.27,28) shows that three-wave processes of decay are forbidden only for the waves where $(sk)^2 < 2\omega_0$.

It is important that 4-wave processes described by the Hamiltonian are always allowed. It is obvious from the conservation law for scattering processes

$$\omega(\boldsymbol{k}) + \omega(\boldsymbol{k}') = \omega(\boldsymbol{k} + \boldsymbol{\kappa}) + \omega(\boldsymbol{k}' - \boldsymbol{\kappa}), \quad (1.1.29)$$

which are allowed at $\kappa \to 0$ under any law of dispersion. Consequently, $\mathcal{H}_4$ will specify the dynamics of the wave system with non-decay spectrum and the expansion terms $\mathcal{H}_5$, $\mathcal{H}_6$, ... will describe small and, as a rule, insignificant corrections.

Obviously, this formal scheme is oversimplified. Reality again proves to be much more complex. For example, for spin waves in ferromagnets even in the decay part of the spectrum not only $\mathcal{H}_3$ but also $\mathcal{H}_4$ must be allowed for. The hamiltonian $\mathcal{H}_3$ can only be due to the magnetic dipole–dipole interaction and is relatively small as compared to the Hamiltonian $\mathcal{H}_4$ describing the exchange interaction. In some problems of wave dynamics on a liquid surface $\mathcal{H}_5$ has to be taken into consideration. Nevertheless, it is generally sufficient to allow only for three-wave processes $\mathcal{H}_3$ in the decay part of the spectrum and for four-wave processes $\mathcal{H}_4$ in the non-decay spectrum.

1.1.4 Dynamic Perturbation Theory. Elimination of "Non-Resonant" Terms from the Hamiltonian

Consider a non-decay dispersion law when three-wave processes are forbidden by conservation laws. In this case the Hamiltonian of interaction $\mathcal{H}_3$ describing these processes has to be a certain extent non-essential. Let us show that in this case we can pass to new canonical variables $c(\boldsymbol{k},t)$, $c^*(\boldsymbol{k},t)$ such that $\mathcal{H}_3\{c(\boldsymbol{k}),c^*(\boldsymbol{k})\} = 0$. The quadratic part of the Hamiltonian therewith retains its previous form

$$\mathcal{H}_2 = \sum_{\boldsymbol{k}} \omega(\boldsymbol{k},t) c^*(\boldsymbol{k}) c(\boldsymbol{k},t) \,, \tag{1.1.30}$$

and in the four-wave Hamiltonian of interaction there appear some additional terms quadratic in the amplitudes $\mathcal{H}\{b(\boldsymbol{k}), b^*(\boldsymbol{k})\}$ in the old variables. To this end, a quasi-linear transformation is performed

$$\begin{aligned} b_{\boldsymbol{k}} = c_{\boldsymbol{k}} - \sum_{1,2} &\left[\frac{U_{\boldsymbol{k},1,2}\, c_1^* c_2^* \delta(\boldsymbol{k}+1+2)}{\omega_{\boldsymbol{k}} + \omega_1 + \omega_2} \right. \\ &\left. - \frac{(V_{\boldsymbol{k};1,2}\, c_1 c_2 - 2V_{2;\boldsymbol{k},-1} c_{-1} c_2^*)\delta(\boldsymbol{k}-1-2)}{\omega_{\boldsymbol{k}} - \omega_1 - \omega_2} \right] + O(c^3) \,. \end{aligned} \tag{1.1.31}$$

The necessity of cubic terms $O(c^3)$ in the transformation to obtain the correct value of the four-wave interaction amplitudes was first pointed out by *Krasitskii* [1.6]. The transformation (1.1.31) (with cubic terms!) is approximately canonical, with sufficient accuracy [1.6].

Substitution of (1.1.31) into the Hamiltonian (1.1.24) shows that $\mathcal{H}\{c^*,c\}$ retains the form (1.1.30), $\mathcal{H}_3 = 0$ and additional terms appear in matrix elements T_p. Such terms will be written down only for the scattering processes of the type $2 \Rightarrow 2$ which will later be most important for us

$$\mathcal{H}_4 = \frac{1}{4} \sum_{1+2=3+4} T_p c_1^* c_2^* c_3 c_4 \,, \quad p = (\boldsymbol{k}_1, \boldsymbol{k}_2; \boldsymbol{k}_3, \boldsymbol{k}_4), \tag{1.1.32}$$

$$T_p = W_p + \tilde{W}_p$$

$$\tilde{W}_p - \frac{U_{(\bar{1}+\bar{2}),1,2}U^*_{(\bar{3}+\bar{4}),3,4}}{\omega_{(3+4)} + \omega_3 + \omega_4} - \frac{V^*_{(1+2),1,2}V_{3+4;3,4}}{\omega_{(1+2)} - \omega_1 - \omega_2} - \frac{V^*_{1;3,(1-3)}V_{4;2,(4-2)}}{\omega_{(4-2)} + \omega_2 - \omega_4}$$

$$- \frac{V^*_{2;4,(2-4)}V_{3;1,(3-1)}}{\omega_{(3-1)} + \omega_1 - \omega_3} - \frac{V^*_{2;3,(3-2)}V_{4;1,(4-1)}}{\omega_{(4-1)} + \omega_1 - \omega_4} - \frac{V^*_{1;4,(1-4)}V_{3;2,(2-4)}}{\omega_{(3-2)} + \omega_2 - \omega_3}.$$

Here $(\boldsymbol{j} \pm \boldsymbol{i}) = \boldsymbol{k}_j \pm \boldsymbol{k}_i$. Additional terms in T_p describe the processes of scattering that occur in the second order of perturbation theory for three-wave processes. In this case a "virtual" induced wave appears at the intermediate stage; for this wave the resonance condition is not satisfied.

There is an important property of scattering processes: they do not alter the total number of waves. Therefore the equations of motion corresponding to the Hamiltonian (1.1.32) retain not only the energy integral but also the following integral

$$N = \int c^*(\boldsymbol{k},t) c(\boldsymbol{k},t)\, d\boldsymbol{k} , \qquad (1.1.33)$$

which has a meaning of the total number of quasi-particles or the integral of the wave action. Within the scope of the total system allowing also for small effects caused by higher order processes the value N is an adiabatic invariant.

In conclusion it must be recalled that the described procedure is fundamentally based on the fact that the transformation (1.1.31) is almost linear, which is possible if the frequency denominator does not become zero. In this case the wave spectrum must be non-decay. In other words, canonical transformations enable us to eliminate only such nonresonant terms of the Hamiltonian of interactions for which the conservation laws of energy-momentum are not valid. Thus the best quasi-linear canonical variables are those for which the Hamiltonian includes no terms corresponding to the forbidden processes. Note that the above-described transformation is analogous to transformation of the Hamiltonians to their normal forms, i.e., in the vicinity of fixed points in classical analytical mechanics [1.2].

1.2 Dimensional Estimation of Hamiltonian Coefficients

It was shown above that when the coefficient of nonlinearity is small, the Hamiltonian of a system of interacting waves has a standard form (1.1.18, 25 and 26). The question is whether the functions ω, V_q, T_p can be estimated without going carefully into each specific problem and whether we can understand their dependence on wave vectors. The answer: It can be done using dimensional considerations unless the characteristic parameters of these waves can be combined in such a way that the resulting product has the dimension of length. It is said in such a case that the problem is *completely self-similar* (has *self-similarity of the first type*).

First of all, let us find the dimensions of the canonical variables $b(\boldsymbol{k},t)$ and matrix elements of the Hamiltonian of interaction V_q, T_p. The dimension of $b(\boldsymbol{k},t)$ is obtained from (1.1.18), taking into account that $\mathcal{H}$ has the dimension of energy density, and ω that of frequency:

$$[\mathcal{H}] = \mathrm{g\,cm}^{2-d}\,\mathrm{s}^{-2},\ [\omega(\boldsymbol{k})] = \mathrm{s}^{-1}\ ,\ [b(\boldsymbol{k})] = \mathrm{g}^{1/2}\mathrm{cm}^{1-d/2}\mathrm{s}^{-1/2}, \qquad (1.2.1)$$

where d is the dimension of the medium. Taking into consideration that $[\omega(\boldsymbol{k})] = [V_q b] = [T_p b^2]$ we readily obtain

$$[b(\boldsymbol{k})] = \mathrm{g}^{-1/2}\mathrm{cm}^{d/2-1}\mathrm{s}^{-1/2}\ ,\qquad [T_{1,2;3,4}] = \mathrm{g}^{-1}\mathrm{cm}^{d-2}\ . \qquad (1.2.2)$$

As should be expected, the dimension $[V_s b^2]$ (here V_s is system volume) is equal to that of Planck's constant $\hbar$. Evidently, our classical approach is valid when quantum mechanical population numbers $N(k) = V b^2/\hbar$ are much greater than unity. On the other hand, the amplitudes of the waves $b(\boldsymbol{k},t)$ must not be too large for the Hamiltonian of interaction $\mathcal{H}_{\text{int}}$ to remain small compared to $\mathcal{H}_2$. This yields an upper bound on $b(\boldsymbol{k},t)$, wich schematically can be written as

$$\omega(\boldsymbol{k}) < V_{k,k,k} \sum_{k'} b(\boldsymbol{k}',t)\ . \qquad (1.2.3)$$

Introducing the dimensionless wave amplitude

$$x(\boldsymbol{k},t) = b(\boldsymbol{k},t)/B(\boldsymbol{k})\ ,\qquad B(\boldsymbol{k}) = |\omega(\boldsymbol{k})/V(\boldsymbol{k},\boldsymbol{k},\boldsymbol{k})|\ , \qquad (1.2.4)$$

the condition of small nonlinearity can be written as

$$x(\boldsymbol{k}) \ll 1\ . \qquad (1.2.5)$$

Now we can consider some examples.

Sound in continuous medium. Only the medium density ρ and elasticity coefficient κ (with dimensions $[\rho] = \mathrm{g\,cm}^{-3}$, $[\kappa] = \mathrm{g\,cm}^{-1}\mathrm{s}^{-2}$) can enter as parameters into the equations of motion for this problem. These values and the wave vector $\boldsymbol{k}$ can unambiguously enter into the combination with the dimension of frequency $[\omega(\boldsymbol{k})] = \mathrm{s}^{-1} = [\rho^x \kappa^y k^z] = \mathrm{g}^{x+y}\,\mathrm{cm}^{-(3x+y+z)}\,\mathrm{s}^{-2y}$. Equating exponents of g, cm, s, we obtain three equations $x+y=0$, $3x+y+z=0$, $2y=1$. Hence $x=-1/2$, $y=1/2$, $z=1$. Therefore, dimensional considerations result in the linear dispersion law

$$\omega(\boldsymbol{k}) = c_s k\ ,\qquad c_s = a\sqrt{\kappa/\rho}\ . \qquad (1.2.6)$$

Here c_s is the sound velocity, and a is a dimensionless parameter of order unity. Parameters of our problem similarly can produce the combination $B(\mathbf{k})$ with dimension equal to that of the canonical variable $b(\mathbf{k},t)$

$$B(\mathbf{k}) = \sqrt{\rho c_s/k} \tag{1.2.7}$$

and three-wave interaction amplitudes

$$V_{1,2,3} = \sqrt{k_1 k_2 k_3 c_s/\rho}\, f(\mathbf{k}_1/k_1, \mathbf{k}_2/k_1, \mathbf{k}_3/k_1) \,. \tag{1.2.8}$$

The dimensionless function f depends in this case on eight dimensionless arguments, i.e. 2 ratios k_2/k, k_3/k and six angular variables giving the directions of the three vectors. Actually, there are only three essential angular variables, i.e. $\cos\theta_{12}$, $\cos\theta_{23}$ and $\cos\theta_{31}$ ($\cos\theta_{ij} = (\mathbf{k}_i \cdot \mathbf{k}_j)/(k_i k_j)$) since in our problem there is no preferential direction.

In the Hamiltonian description the wave amplitude is proportional to $b(\mathbf{k})$. In the case of the sound wave, density and velocity of the medium oscillate. Writing the density as a sum of the constant level ρ_0 and the oscillating component $\rho_1(\mathbf{r},t)$, one has

$$\begin{aligned}\rho_1(\mathbf{r},t) &= \mathrm{Re}\{\rho(\mathbf{k})\exp[i\mathbf{k}\mathbf{r} - i\omega(\mathbf{k})t]\} \,,\\ \mathbf{v}(\mathbf{r},t) &= \mathrm{Re}\{\mathbf{v}(\mathbf{k})\exp[i\mathbf{k}\mathbf{r} - i\omega(\mathbf{k})t]\} \,,\end{aligned} \tag{1.2.9}$$

where $\rho(\mathbf{k})$, $\mathbf{v}(\mathbf{k})$ denotes the wave amplitude in the natural variables. The relation between natural and normal canonical variables can be easily obtained in linear approximation using dimensional considerations

$$\rho(\mathbf{k},t) \simeq \sqrt{k\rho_0/c_s}\, b(\mathbf{k},t) \,, \qquad v(\mathbf{k},t) \simeq \sqrt{kc_s/\rho_0}\, b(\mathbf{k},t) \,. \tag{1.2.10}$$

The condition of small nonlinearity in terms of canonical variables can be rewritten as follows:

$$x(\mathbf{k}) \simeq \rho(\mathbf{k})/\rho_0 \simeq v(\mathbf{k})/c_s \ll 1 \,. \tag{1.2.11}$$

Gravitational waves on liquid surface. Gravitational waves are sufficiently long waves for which the surface tension is not essential and the restoring force tending to make the surface plane is due to gravitation. Clearly, essential parameters must include besides the liquid density ρ also the free fall acceleration g, $[g] = \mathrm{cm\,s^{-2}}$. Using a procedure similar to that employed in the previous example and taking into account that the problem is plane ($d = 2$) we have

$$\omega(\mathbf{k}) \simeq \sqrt{gk} \,, \qquad B(\mathbf{k}) \simeq (\rho^2 g k^{-5})^{1/4} \,. \tag{1.2.12}$$

Clearly in this case the dispersion law is non-decaying: $\omega(\mathbf{k}) \simeq k^z$, $z = 1/2 < 1$. Therefore, the principal interaction is the four-wave interaction with the following amplitude element:

$$T(\boldsymbol{k},2;3,4) = \frac{k^3}{\rho} f(\frac{\boldsymbol{k}_1}{k}, \frac{\boldsymbol{k}_2}{k}, \frac{\boldsymbol{k}_3}{k}, \cos\theta_{k1}, \cos\theta_{k2}, \cos\theta_{k3}) \ . \tag{1.2.13}$$

The natural variable describing waves on the surface of water is the deviation of the liquid surface from the plane $\mu(\boldsymbol{r},t)$. The dimensionless wave amplitude is $x(\boldsymbol{k},t) = \mu(\boldsymbol{k},t)k = b(\boldsymbol{k},t)/B(\boldsymbol{k})$, hence the relation of $\mu(\boldsymbol{k},t)$ and $b(\boldsymbol{k},t)$ is obtained

$$\mu(\boldsymbol{k},t) \simeq (k/\rho^2 g)^{1/4} b(\boldsymbol{k},t) \ . \tag{1.2.14}$$

Capillary waves. For sufficiently short waves the restoring force must be fully determined by the surface tension. The essential parameters in this case will include instead of g the coefficient of the surface tension σ with dimension equal to that surface energy density $[\sigma] = \mathrm{g\,s^{-2}}$. Therefore,

$$\begin{aligned}&\omega(\boldsymbol{k}) \simeq \sqrt{\sigma k^3/\rho}\ , \quad B(k) \simeq (\rho\sigma/k^3)^{1/4}\ , \\ &\mu(\boldsymbol{k},t) \simeq (\rho\sigma k)^{-1/4} b(\boldsymbol{k},t) \ .\end{aligned} \tag{1.2.15}$$

The dispersion law of capillary waves is decay, i.e. $z = 3/2 > 1$. Therefore the principal interaction is the 3-wave interaction

$$V(\boldsymbol{k},\boldsymbol{k}_1,\boldsymbol{k}_2) = (\sigma k^9/\rho^3)^{1/4} f(\frac{\boldsymbol{k}_1}{k}, \frac{\boldsymbol{k}_2}{k}, \cos\theta_{k1}, \cos\theta_{k2}) \ . \tag{1.2.16}$$

Comparing dispersion laws of capillary waves (1.2.15) and gravitational waves (1.2.12) the boundary value of the wave vector under which those frequencies are equal can be easily found

$$k_* \simeq \sqrt{\rho g/\sigma} \ . \tag{1.2.17}$$

At $k \ll k_*$ the gravitational energy of the wave is greater than the surface tension energy, and the latter can be neglected. Thus, waves of long wavelength on the liquid surface will be gravitational. Correspondingly, with $k \gg k_*$ surface waves will be capillary ones with the law of dispersion (1.2.15). It can be shown that at arbitrary k the dispersion law on the surface of a deep liquid has the form [1.4]:

$$\omega(\boldsymbol{k}) = \sqrt{gk + \sigma k^3/\rho} \ . \tag{1.2.18}$$

In spite of the fact that dimensional estimates yield results accurate up to a dimensionless factor of order unity, the dispersion laws (1.2.12, 15) are correct for long or short waves respectively.

For the waves on the water surface at room temperature we have $k_* = 4$ cm ($\rho = 1$ g/cm, $\sigma = 70$ g/s). The respective wavelength $\lambda = 2\pi/k_* \simeq 1.6$ cm and the frequency $f_* = \omega/2\pi \simeq 0.2$ Hz.

Spin waves in the Heisenberg ferromagnet. We shall proceed from the following expression for the energy of the Heisenberg ferromagnet (see, e.g., [1.1] or (2.3.6))

$$\mathcal{H}_{\text{ex}} = \alpha \int (\partial M_i / \partial x_j)^2 \, d\boldsymbol{r} \, . \tag{1.2.19}$$

Here α is the parameter of nonhomogeneous exchange, $\boldsymbol{M}$ denotes magnetization. The physical relevance of these values will be discussed in the next chapter. Now we shall only give their dimensions $[\alpha] = \text{cm}^2$, $[M] = \text{g}^{1/2}\,\text{cm}^{-1/2}\,\text{s}^{-1}$. There can be no combination of those values and the wave vector with dimension equal to that of the frequency. Therefore one more dimensional parameter must enter into the problem of the spin waves. This parameter determines the dynamics of the electron spin system in the magnetic field and is the ratio of the magnetic moment of the electron $\mu_B/2$ to its mechanical moment $\hbar/2$:

$$g = \mu_B/\hbar \simeq 2\pi\, 2.8 \text{ Hz/Oe} \, . \tag{1.2.20}$$

It should not be confused with the dimensionless g-factor of the electron which approximates two.

The values α, M, g and k can be combined to build up frequency. This construction is ambiguous since the combination αk^2 is dimensionless and dimensional considerations do not determine with what exponent the combination αk^2 must enter into various expressions. If we take into account that $\mathcal{H}_{\text{ex}} \propto \alpha$ and impose a condition that $\omega(\boldsymbol{k}) \simeq \alpha$ we obtain

$$\omega(k) \simeq gM\alpha k^2 \, . \tag{1.2.21}$$

Later the dynamics of spin waves in the Heisenberg (exchange) approximation will be shown to be determined by four-wave scattering processes $2 \Rightarrow 2$. The coefficients of the Hamiltonian of interaction $T(\boldsymbol{1},\boldsymbol{2};\boldsymbol{3},\boldsymbol{4})$ as well as the frequencies $\omega(\boldsymbol{k})$ must be proportional to α. The other factors are found from the analysis of dimensions

$$T(\boldsymbol{1},\boldsymbol{2};\boldsymbol{3},\boldsymbol{4}) \simeq g^2 \alpha k^2 \, . \tag{1.2.22}$$

Spin waves are oscillation of magnetization. Consequently, the canonical amplitude of the spin wave $b(\boldsymbol{k},t)$ in the linear approximation must be proportional to $\boldsymbol{m}(\boldsymbol{k},t)$, i.e. to the space Fourier transform of the variable part of the magnetization. On the other hand, α does not have to enter into the expression for the $b(\boldsymbol{k},t)$ and $\boldsymbol{m}(\boldsymbol{k},t)$ relation, since this relation does not depend on the specific form of the expression for energy. Assuming $\boldsymbol{m}(\boldsymbol{k},t) = b(\boldsymbol{k},t)M^x g^y k^z$ we obtain from dimensional analysis

$$m(\boldsymbol{k},t) \simeq \sqrt{gM}\, b(\boldsymbol{k},t) \, . \tag{1.2.23}$$

The condition of small nonlinearity in terms of canonical variables for the spin waves is rewritten as follows

$$x(\boldsymbol{k}) \simeq m(\boldsymbol{k})/M \simeq \sqrt{g/M}b(\boldsymbol{k}) \ll 1 \ . \tag{1.2.24}$$

The dimensionless parameter $x(\boldsymbol{k})$ (1.2.24) describes the precession angle of the magnetic moment.

1.3 Dynamic Equations of Motion for Weakly Non-Conservative Wave Systems

In Sect. 1.2. we considered the simplest case of a Hamiltonian equation of motion for the wave amplitudes when the medium of the wave propagation is spatially homogeneous, and at the same time conservative; then the interaction of the waves with the medium may be neglected. These equations may be represented as:

$$\partial b(\boldsymbol{k},t)/\partial t + i\omega(\boldsymbol{k})b(\boldsymbol{k},t) = -i\delta\mathcal{H}_{\mathrm{int}}/\delta b^*(\boldsymbol{k},t) \ . \tag{1.3.1}$$

There are, of course, no completely isolated systems and even their weak interaction with the medium may sometimes prove significant. To understand the role of such an interaction and describe it properly the problem must be formalized and reduced.

Let us subdivide the physical system under consideration (ferromagnet, liquid, plasma, etc.) into two parts: the dynamic subsystem of nonlinear waves and the remaining medium. The state of the nonlinear wave system may be strongly excited. These waves will be described in detail by means of dynamic equations of motion for the wave amplitudes $b(\boldsymbol{k},t)$. Let us assume that the "medium" weakly interacts with nonlinear waves and is in thermodynamic equilibrium. Then it can act as a *thermal bath* for the nonlinear wave systems.

The physical characteristics of the remaining part of the medium that we assume to be a thermal bath may be absolutely different for specific cases. If our aim is a detailed description of long-wave length motions of the continuous medium (liquid or plasma) by means of dynamical equations, small-scale non-cooperative motions of the individual particles (liquid molecules, electrons and ions in plasma) will act as a thermal bath. In the dynamic description of long spin waves in ferromagnets the thermal bath can be formed by short spin waves in thermodynamic equilibrium. If the excitation level of the spin waves is so great that the thermodynamic equilibrium of the short spin waves is significantly disturbed then the subsystem of phonons which are weakly connected with the subsystem of spin waves starts to act as a thermal bath. The spin system can evidently be affected so that the equilibrium of phonons will also be drastically disturbed. Then the medium in which the sample is immersed (air, liquid nitrogen, helium, etc.) can be treated as a thermal bath. The above examples show that the subdivision of the considered physical system into two parts, the nonlinear dynamic subsystem and the thermal bath, is fairly arbitrary. How a particular physical system will be subdivided will depend on its specific features, the way and level of its excitation and the intended accuracy of the description.

1.3.1 Taking into Account Linear Wave Damping

The interaction of the waves with the thermal bath leads to damping. For sound in fluid this is viscous damping, which converts wave energy into heat. For long spin waves in ferromagnets damping is mostly caused by their interaction with the "magnon thermal bath" – the thermally excited reservoir of the other magnons. An important role, especially in antiferromagnets, can be played by the interaction of spin waves with phonons (i.e. with sound). Spin waves may also be damped as a result of their interactions with different impurities, crystal defects, pores and other inhomogeneities.

Damping of small amplitude waves is known in most cases to be exponential

$$|b(\bm{k},t)| = |b(\bm{k},0)| \exp(-\gamma(\bm{k})t) . \qquad (1.3.2)$$

The decrement $\gamma(\bm{k})$ depends on the wave vector $\bm{k}$, medium temperature T, magnetic field $\bm{H}$ and other experimental conditions. In good samples of magnetodielectrics (YIG, $MnCO_3$, $CsMnF_3$, etc.) the spin wave damping rate is small compared to their frequency: the relation $\omega(\bm{k})/\gamma(\bm{k}) \simeq 10^3 - 10^4$. Therefore the Hamiltonian equations (1.3.1) can be used in the zeroth approximation not allowing for the damping. In the first approximation the small damping can be taken into account by adding to each equation the term $\gamma(\bm{k})b(\bm{k},t)$ which results in an exponentially decreasing amplitude (1.3.2)

$$\partial b(\bm{k},t)/\partial t + [i\omega(\bm{k}) + \gamma(\bm{k})]b(\bm{k},t) = -i\delta\mathcal{H}_{\text{int}}/\delta b^*(\bm{k},t) . \qquad (1.3.3)$$

This phenomenological procedure allowing for the small wave damping as an imaginary addition to the frequency is commonly used in theoretical physics and normally causes no objection. For (1.3.3) to be valid not only the damping and interaction must be small (these conditions may be written as $\gamma(\bm{k}) \ll \omega(\bm{k})$, $\mathcal{H}_{\text{int}} \ll \mathcal{H}_2$) but the wave amplitudes must be bounded below and above. We actually think the whole world consists of two parts: the system of waves (with the amplitudes $b(\bm{k},t)$ and Hamiltonian $\mathcal{H}$, which we are going to describe in detail dynamically using (1.3.3)) and thermal bath, which we shall assume to be in thermodynamic equilibrium with the temperature T. The term $\gamma(\bm{k})b(\bm{k},t)$ describes "the wave friction on the thermal bath", i.e. energy flux to the thermal bath, and results in the wave amplitude b gradually vanishing. These amplitudes, however, must relax not to zero, but to the thermodynamical equilibrium value which is specified by the Rayleigh-Jeans distribution

$$|b_0(\boldsymbol{k})|^2 = n_0(\boldsymbol{k}) = T/\omega(\boldsymbol{k}), \tag{1.3.4}$$

which holds true in the classical limit $T \ll \hbar\omega(\boldsymbol{k})$. Therefore (1.3.3) is true when $|b(\boldsymbol{k})| \ll |b_0(\boldsymbol{k})| = \sqrt{T/\omega(\boldsymbol{k})}$.

1.3.2 Allowing for Thermal Noise

If the inequality $|b| \gg |b_0|$ is not satisfied the random Langevin force $f(\boldsymbol{k},t)$ must be added to (1.3.3)

$$\frac{\partial b(\boldsymbol{k},t)}{\partial t} + [i\omega(\boldsymbol{k}) + \gamma(\boldsymbol{k})]b(\boldsymbol{k},t) = -i\frac{\delta \mathcal{H}_{\text{int}}}{\delta b^*(\boldsymbol{k},t)} + f(\boldsymbol{k},t) \ . \tag{1.3.5}$$

This force imitates thermal noise - chaotic shocks on the part of the thermal bath which usually cause an increase in wave energy $\propto b^*(\boldsymbol{k},t)b(\boldsymbol{k},t)$. As is known from statistical physics, random forcing is Gaussian forcing with the correlator

$$\langle f(\boldsymbol{k},t)f^*(\boldsymbol{k}',t')\rangle = 2\delta(\boldsymbol{k}-\boldsymbol{k}')\delta(t-t')\gamma(\boldsymbol{k})n_0(\boldsymbol{k}) \ . \tag{1.3.6}$$

The properties of the thermal noise (1.3.6) are sufficiently clear: it is not correlated at different moments of time and for waves with different $\boldsymbol{k}$. The factor $2\gamma(\boldsymbol{k})n_0(\boldsymbol{k})$ has been selected so that the numbers of waves $n(\boldsymbol{k})$ should relax to the equilibrium value $n_0(\boldsymbol{k})$. Equations (1.3.5, 6) at $\mathcal{H}_{\text{int}} = 0$ quite readily yield

$$\frac{1}{2}\frac{\partial n(\boldsymbol{k},t)}{\partial t} = -\gamma(\boldsymbol{k})[n(\boldsymbol{k},t) - n_0(\boldsymbol{k})] \ , \tag{1.3.7}$$

$$\langle b(\boldsymbol{k})b^*(\boldsymbol{k}')\rangle = n(\boldsymbol{k})\delta(\boldsymbol{k}-\boldsymbol{k}') \tag{1.3.8}$$

describing this phenomenon. The term "number of waves" is used here by analogy with quantum mechanics. It must, however, be borne in mind that in our classical approach the value $n(\boldsymbol{k})$ has dimensions of the action (erg·s), while quantum-mechanical occupation numbers $n_{\text{qm}}(\boldsymbol{k})$ are dimensionless. At $n_{\text{qm}}(\boldsymbol{k}) \gg 1$ they are correlated:

$$n(\boldsymbol{k},t) = \hbar n_{\text{qm}}(\boldsymbol{k},t) \ . \tag{1.3.9}$$

1.3.3 Nonlinearity of Wave Damping

One more restriction on the validity regime of (1.3.3) is due to the assumption that $\gamma(\boldsymbol{k})$ is completely independent of all the other numbers of waves $n(\boldsymbol{k}')$. Generally speaking, this is not the case. But under small $n(\boldsymbol{k}')$, $\gamma(\boldsymbol{k})$ may be expanded in terms of $n(\boldsymbol{k}')$ and we may restrict ourselves to the first terms

$$\gamma(\boldsymbol{k}, n(\boldsymbol{k}')) = \gamma_0(\boldsymbol{k}) + \sum_{\boldsymbol{k}'} \mu(\boldsymbol{k}, \boldsymbol{k}') n(\boldsymbol{k}') \; . \tag{1.3.10}$$

If $\mu(\boldsymbol{k}, \boldsymbol{k}') > 0$, the nonlinear damping is commonly called positive; for under $\mu(\boldsymbol{k}, \boldsymbol{k}') < 0$ it is called negative. The nature of the damping nonlinearity will be discussed in Chap. 10. At the moment note only that the function $\mu(\boldsymbol{k}, \boldsymbol{k}')$ may be of any sign and be either symmetrical under transposition $\boldsymbol{k} \Leftrightarrow \boldsymbol{k}'$ or antisymmetric. It may be also characterized by no symmetry at all. At the first stage (and throughout Chap. 1) we shall assume that the numbers $n(\boldsymbol{k}')$ are sufficiently small for the nonlinearity of wave damping to be neglected.

In conclusion it must be emphasized that (1.3.3, 5) can be employed to describe various nonlinear wave phenomena. To this end, we must only put the original equations of medium motion into the Hamiltonian form. Unfortunately, there is as yet no sufficiently effective general method of obtaining canonical variables, but for many general cases the canonical variables have already been found. In [1.7], for example, well-known canonical Clebsch variables describing eddy flows and potential barotropic flows of the ideal liquid are given. *Zakharov* and *Filonenko* [1.8] have found canonical variables for waves on the liquid surface, and *Pokrovskii* and *Khalatnikov* introduced four pairs of canonical variables describing the flow of normal and superfluid components of liquid helium with the framework of two-liquid hydrodynamics. In his survey [1.3] *Zakharov* describes canonical variables for relativistic hydrodynamics, for the hydrodynamics of charged liquid (plasma) interacting with the electromagnetic field, and for magnetic hydrodynamics. This is by no means a complete list of canonical variables for different nonlinear media.

One more remark: Equations (1.3.3, 5) are approximate. In some specific cases they may be invalid even if the conditions which we discussed above are satisfied. This is true, for instance, for the problem of the parametric excitation of waves which are scattered by static defects. In such cases fine nonlinear effects must be investigated more consistently. All the interactions, including interactions resulting in the damping of the nonlinear waves, must be treated in the greatest detail, which enables one to abandon the somewhat unjustified subdivision of the system into nonlinear waves and thermal bath. This microscopic approach based on the Wyld diagrammatic technique [1.9] is described in my book [1.1]. At present, at this first stage of investigating nonlinear phenomena we shall confine ourselves to the simple phenomenological Eqs. (1.3.3, 5).

1.4 Three-Wave Processes

Substitution of the Hamiltonian $\mathcal{H}$ given by (1.1.25) into (1.3.3) yields

$$\partial b(\boldsymbol{k},t)/\partial t + [\omega(\boldsymbol{k}) + i\gamma(\boldsymbol{k})]b(\boldsymbol{k},t)$$
$$= -\frac{i}{2}\sum_{1+2=k} V^*_{k,1,2}b_1 b_2 - i\sum_{k+2=1} V_{1,k,2}b_1 b_2^* - i\sum_{1+2+k=0} U_{k,1,2}b_1^* b_2^* . \quad (1.4.1)$$

This is the basic equation for describing different three-wave processes. Some of these will be discussed below.

1.4.1 Confluence of Two Waves and Other Induced Processes

Let two monochromatic waves propagate in some medium:

$$b(\boldsymbol{k},t) = b_1 \delta(\boldsymbol{k}-\boldsymbol{k}_1)\exp[-i\omega(\boldsymbol{k}_1)t] + b_2\delta(\boldsymbol{k}-\boldsymbol{k}_2)\exp[-i\omega(\boldsymbol{k}_2)t]. \quad (1.4.2)$$

By virtue of (1.4.1) this will lead to the emergence of three additional waves

$$\begin{aligned}b(\boldsymbol{k},t) =& b_3\delta(\boldsymbol{k}-\boldsymbol{k}_1-\boldsymbol{k}_2)\exp[-i\omega(\boldsymbol{k}_1)t-i\omega(\boldsymbol{k}_2)t]\\ &+ b_4\delta(\boldsymbol{k}+\boldsymbol{k}_2-\boldsymbol{k}_1)\exp[-i\omega(\boldsymbol{k}_1)t+i\omega(\boldsymbol{k}_2)t]\\ &+ b_5\delta(\boldsymbol{k}+\boldsymbol{k}_2+\boldsymbol{k}_1)\exp[i\omega(\boldsymbol{k}_1)t+i\omega(\boldsymbol{k}_2)t]\end{aligned} \quad (1.4.3)$$

with the amplitudes

$$\begin{aligned}b_3 &= \frac{V^*(1+2,1,2)b(1)b(2)}{2[\omega(\boldsymbol{k}_1)+\omega(\boldsymbol{k}_2)-\omega(\boldsymbol{k}_1+\boldsymbol{k}_2)-i\gamma(\boldsymbol{k}_1+\boldsymbol{k}_2)]},\\ b_4 &= -\frac{V^*(1,(1-2),2)b(1)b(2)}{[\omega(\boldsymbol{k}_1)-\omega(\boldsymbol{k}_2)-\omega(\boldsymbol{k}_1-\boldsymbol{k}_2)-i\gamma(\boldsymbol{k}_1-\boldsymbol{k}_2)]},\\ b_5 &= -\frac{U(-1-2,1,2)b^*(1)b^*(2)}{[\omega(\boldsymbol{k}_1)+\omega(\boldsymbol{k}_2)-\omega(\boldsymbol{k}_1+\boldsymbol{k}_2)+i\gamma(\boldsymbol{k}_1+\boldsymbol{k}_2)]}.\end{aligned} \quad (1.4.4)$$

If the frequencies $\omega(\boldsymbol{k}_1)$ and $\omega(\boldsymbol{k}_2)$ satisfy the condition

$$\omega(\boldsymbol{k}_1) + \omega(\boldsymbol{k}_2) = \omega(\boldsymbol{k}_1 + \boldsymbol{k}_2), \quad (1.4.5)$$

the amplitude b_3 is large in comparison with b_4 and b_5. In this case we deal with the resonance process of confluence of two waves. If instead of (1.4.5) the condition

$$\omega(\boldsymbol{k}_1) - \omega(\boldsymbol{k}_2) = \omega(\boldsymbol{k}_1 - \boldsymbol{k}_2) \quad (1.4.6)$$

is satisfied then, on the contrary, $|b_4| \gg |b_{3,5}|$. From the viewpoint of quantum mechanics and quasi-particles, this process cannot be understood easily, but classically, in terms of amplitudes, it can be accounted for by resonance excitation by a wave with the wave vector $\bm{k}_4 = \bm{k}_1 - \bm{k}_2$ at one of the combination frequencies $\omega_4 = \omega(\bm{k}_1) - \omega(\bm{k}_2)$. If the medium in which the waves are propagating is in a state of thermodynamic equilibrium these waves have positive frequencies. Consequently, the condition

$$\omega(\bm{k}_1) + \omega(\bm{k}_2) + \omega(-\bm{k}_1 - \bm{k}_2) = 0 \qquad (1.4.7)$$

required for the amplitude b_5 in (1.4.4) to be high is not satisfied under equilibrium. These processes can be neglected if the Q-factor of the waves is large

$$Q(\bm{k}) = \omega(\bm{k})/\gamma(\bm{k}) \gg 1 \,. \qquad (1.4.8)$$

From the above examples we can draw the very important conclusion that the parameter $1/Q(\bm{k})$ is small and nonresonant processes can be neglected in most cases, if we truncate the corresponding terms at the very first stage of the problem solution when formulating the interaction Hamiltonian. Only in studying relatively weak four-wave processes (see Sect. 1.5) must the contribution of non-resonance three-wave processes be allowed for in the second order of perturbation theory.

Let us consider the process (1.4.5) of two-wave confluence in more detail. It is clear from the first expression of (1.4.4) that the resonance curve is Lorentzian. If the function $V(\bm{3},\bm{1},\bm{2})$ is real, the phase shift in the resonance is $\Delta\varphi = \varphi_1 + \varphi_2 - \varphi_3 = \pi/2$. Far from the resonance $\Delta\varphi = 0$. In order to evaluate the efficiency of the two-wave confluence process, we must know the value of the interaction amplitude $V(\bm{3},\bm{2},\bm{1})$, the law of wave dispersion $\omega(\bm{k})$ and the decrement $\gamma(\bm{k})$. By way of example consider first the confluence of two sound waves for the case when the condition of 3-wave resonance (1.4.5) is satisfied. Let two sound waves $\bm{k}_1$ and $\bm{k}_2$ be excited in a continuous medium. They cause density oscillations of the continuous medium

$$\rho(\bm{r},t) = \rho_1 \cos(\omega_1 t - \bm{k}_1 \bm{r}) + \rho_2 \cos(\omega_2 t - \bm{k}_2 \bm{r}) \,.$$

Employing first the connection (1.2.10) between natural $\rho_{1,2}$ and canonical $b_{1,2}$ variables and also the estimate (1.2.8) for the 3-phonon interaction amplitude $V(\bm{1};\bm{2};\bm{3})$ we obtain from (1.4.4) at resonance:

$$x_3 \simeq x_1 x_2 (\omega/\gamma_3) \,, \qquad x_j = \rho_j/\rho_0 \,. \qquad (1.4.9a,b)$$

The ratio $Q = \omega/\gamma$ (for sound waves in acoustic media the Q-factor may reach $10^3 \div 10^4$ and more) describes resonance amplification of processes of wave conversion. Making use of (1.2.15, 16) and (1.4.4) similar estimations may be carried out for resonance three-wave conversion of the capillary waves on the liquid surface. Two capillary waves with the amplitudes μ_1,

μ_2 (μ denotes the deviation of the liquid surface from the unperturbed state) in resonance produce a third wave with the amplitude μ_3:

$$\mu_3 \simeq \mu_1 \mu_2 (\omega/\gamma_3) , \qquad \omega = \sqrt{ck^3/\rho} . \qquad (1.4.10a,b)$$

If the wave amplitude is specified by a dimensionless parameter $x = k\mu$ (x is the angle of the deflection of the normal to the liquid surface from the vertical line), then the formula (1.4.10a) for capillary waves may be rewritten so that its form coincides with (1.4.9a) for sound:

$$x \simeq x_1 x_2 (\omega/\gamma_3) , \qquad x_j = k_j \mu_j . \qquad (1.4.11a,b)$$

It must be emphasized that expressions similar to (1.4.9a, 11a) are, generally speaking, common to any problem of resonance conversion of waves, provided it contains no dimensionless parameters characterizing the relations of different interactions. In particular, the exact formula (4.1.4) for the resonance conversion of magnons which will be obtained in Chap. 4 coincides with (1.4.9a, 11a) to an accuracy of numeric coefficients and angular dependence, if x denotes the precession angle of the magnetic moment.

1.4.2 Decay Instability

Here we consider the instability of a plane monochromatic wave $b(\mathbf{k},t)$ with respect to decay into two other waves. Examples of such a process are the decay of homogeneous precession of magnetization (or electromagnetic radiation of microwave frequencies) into two magnons, decay of a phonon into two magnons, etc. Among these processes are also included induced light scattering: photon decaying into a photon and phonon, into a photon and magnon, and a number of other processes. Thus, let a monochromatic wave with the frequency $\omega_0 = \omega(\mathbf{k}_0)$ and with the amplitude $b(\mathbf{k},t)$ propagate in the medium:

$$b(\mathbf{k},t) = b\, \delta(\mathbf{k}-\mathbf{k}_0) \exp(-i\omega_0 t) . \qquad (1.4.12)$$

Then (1.4.1) for small amplitude waves has the following form:

$$\partial b_1/\partial t + [\gamma_1 + i\omega(\mathbf{k}_1)]b_1 + iV b b_2^* \exp(-i\omega_0 t) = 0 ,$$

$$\partial b_2^*/\partial t + [\gamma_2 + i\omega(\mathbf{k}_2)]b_2^* - iV^* b^* b_1 \exp(-i\omega_0 t) = 0 . \qquad (1.4.13)$$

The second nonlinear term in (1.4.1) is retained here. The first and third terms for the decay process of interest for us are not resonant, and with small $1/Q$-factor (1.4.8) they can be neglected. A solution of (1.4.13) is sought in the following form:

$$b_1(t) = b_1 \exp[(\nu - i\omega_1)t] , \qquad b_2^*(t) = b_2^* \exp[(\nu - i\omega_2)t] , \qquad (1.4.14)$$

$$\omega_1 + \omega_2 = \omega_0 . \qquad (1.4.15)$$

The instability exponent ν in (1.4.14b) is not complex conjugate, even if it is a complex value. Only in this case and if the condition (1.4.15) is satisfied the substitution (1.4.14) converts the differential equation (1.4.11) to an algebraic one with constant coefficients. It has non-zero solutions if

$$\begin{bmatrix} \gamma_1 + \nu + i[\omega(\boldsymbol{k}_1) - \omega_1] & iVb \\ -iVb^* & \gamma_2 + \nu - i[\omega(\boldsymbol{k}_2) - \omega_2] \end{bmatrix} = 0 . \quad (1.4.16)$$

It must be noted that the condition (1.4.15) determines only the sum of $\omega_1 + \omega_2$, whereas their difference remains arbitrary. This arbitrariness implies that if a certain frequency δ is added to ω_1 and subtracted from ω_2, then in accordance with (1.4.14) this will lead to the substitution $\nu \to \nu - i\delta$. We may use this arbitrariness in order to simplify the analysis of the (1.4.16). Thus, ω_1 and ω_2 will be selected so that

$$(\gamma_1 + \nu)[\omega(\boldsymbol{k}_2) - \omega_2] = (\gamma_2 + \nu)[\omega(\boldsymbol{k}_1) - \omega_1] . \quad (1.4.17)$$

The imaginary part in (1.4.16) then becomes zero and for ν from (1.4.16, 17) we obtain:

$$\begin{aligned} 2\nu &= -\gamma_1 - \gamma_2 + \sqrt{B + \sqrt{B^2 + 2(\Delta\gamma\Delta\omega)^2}} , \\ 2B &= 4|Vb|^2 + (\Delta\gamma)^2 - 4(\Delta\omega)^2 , \\ \Delta\gamma &= \gamma_1 + \gamma_2 , \quad 2\Delta\omega = \omega(\boldsymbol{k}_1) + \omega(\boldsymbol{k}_2) - \omega_0 . \end{aligned} \quad (1.4.18)$$

Then, substituting ν from (1.4.18) into (1.4.17) ω_1 and ω_2 can be found from the solution of the system (1.4.15, 17). It is clear that ω_1 will approximate $\omega(\boldsymbol{k}_1)$, and ω_2 will be close to $\omega(\boldsymbol{k}_2)$. For the simple case when $\gamma_1 = \gamma_2 = \gamma$,

$$\begin{aligned} \nu &= -\gamma + \sqrt{|Vb|^2 - (\Delta\omega)^2} , \\ \omega_1 &= \omega(\boldsymbol{k}_1) - \Delta\omega/2 , \quad \omega_2 = \omega(\boldsymbol{k}_2) + \Delta\omega/2 . \end{aligned} \quad (1.4.19)$$

At $\operatorname{Re}\nu > 0$ the wave amplitudes exponentially increase in time (according to (1.4.16)). The source wave in this case is said to be unstable with respect to decay into two waves b_1 and b_2. ν is called an increment of decay instability. This instability is often (especially if $\boldsymbol{k}_0 = 0$) also called parametric. The instability increment is maximum at resonance $\Delta\omega = 0$:

$$2\nu_{\max} = -(\gamma_1 + \gamma_2) + \sqrt{4|Vb|^2 + (\Delta\gamma)^2} . \quad (1.4.20)$$

There is a threshold of the decay instability due to the damping of the waves

$$|Vb_{\mathrm{cr}}|^2 = \gamma_1\gamma_2 . \quad (1.4.21)$$

If the amplitudes V are known, by measuring b we may find the wave damping. In this way spin wave damping in ferromagnets is usually measured. The interface of the instability region can be found from (1.4.18) at $\nu = 0$. If $\gamma_1 = \gamma_2$ the instability region is

$$(\Delta\omega)^2 < 4|Vb|^2 - \gamma^2 \ . \tag{1.4.22}$$

Remember that the resonance conditions for the decay

$$\omega(\boldsymbol{k}) = \omega(\boldsymbol{k}_1) + \omega(\boldsymbol{k}_1 - \boldsymbol{k}_2) \ . \tag{1.4.23}$$

are not always satisfied and accordingly the laws of dispersion $\omega(\boldsymbol{k})$ are subdivided into decay and nondecay ones.

As an example, consider the decay instability in two systems: (i) sound in the continuum; (ii) capillary waves on the liquid surface. Using (1.2.6, 8, 9) for the sound and (1.2.15, 16) for the capillary waves we estimate the increment of the decay instability:

$$\begin{aligned}\nu(\boldsymbol{k},\boldsymbol{\kappa}) &\simeq |V(\boldsymbol{k},\boldsymbol{k}_+,\boldsymbol{k}_-)b(\boldsymbol{k})| \simeq \gamma(\boldsymbol{k})\omega(\boldsymbol{k}) \ , \\ \boldsymbol{k}_\pm &= \boldsymbol{k}/2 \pm \boldsymbol{\kappa} \ , \quad \kappa \leq k \ .\end{aligned} \tag{1.4.24}$$

For sound the wave frequency $\omega(\boldsymbol{k})$ and the dimensionless amplitude of the wave $x(\boldsymbol{k})$ are given by (1.2.6) and (1.4.9b), for the capillary waves they are specified by (1.4.10a, 11b).

In Chap. 3 we shall deal with the decay instability of magnons in magnetodielectrics in detail. The exact formulae obtained there agree with the results of (1.4.24), if $\omega(\boldsymbol{k})$ denotes the frequency of magnons and $x(\boldsymbol{k})$ designates the angle of magnetic moment precession.

1.4.3 Interaction of Three Waves with Finite Amplitude

Consider the interaction of three wave packages with narrow spectrum whose characteristic vectors $\boldsymbol{k}_1, \boldsymbol{k}_2, \boldsymbol{k}_3$ and frequencies $\omega(\boldsymbol{k}_1), \omega(\boldsymbol{k}_2), \omega(\boldsymbol{k}_3)$ satisfy the resonance condition $\boldsymbol{k}_1 = \boldsymbol{k}_2 + \boldsymbol{k}_3$ (*Bloembergen* problem [1.10]),

$$\omega(\boldsymbol{k}_1) = \omega(\boldsymbol{k}_2) + \omega(\boldsymbol{k}_3) \ . \tag{1.4.25a}$$

Assume

$$b(\boldsymbol{k}) = \sum_{j=1}^{3} b_j(\boldsymbol{k}) \tag{1.4.25b}$$

and take b_j to be non-zero for small $\boldsymbol{\kappa} = \boldsymbol{k} - \boldsymbol{k}_j$. Within each packet the strong $\boldsymbol{r}, t$-dependence is excluded by the substitution

$$a_j(\boldsymbol{\kappa}) = b_j(\boldsymbol{k}_j + \boldsymbol{\kappa}) \exp i[\omega(\boldsymbol{k}_j)t - \boldsymbol{k}_j \boldsymbol{r}] \ . \tag{1.4.26}$$

Equations (1.4.1) are reduced by taking

$$\omega(\boldsymbol{k}_j + \boldsymbol{\kappa}) = \omega(\boldsymbol{k}_j) + (\boldsymbol{\kappa} \boldsymbol{v}_j) \ . \tag{1.4.27}$$

Further change over to $\boldsymbol{r}$-representation according to the formula

$$a_j(r) = (2\pi)^{-3/2} \int a_j(k) \exp(ikr)\,dk \ . \tag{1.4.28}$$

Then

$$\left(\frac{\partial}{\partial t} + \gamma_1 + \boldsymbol{v}_1 \cdot \nabla\right) a_1(r,t) = -iV a_2(r,t) a_3(r,t) \ ,$$

$$\left(\frac{\partial}{\partial t} + \gamma_2 + \boldsymbol{v}_2 \cdot \nabla\right) a_2(r,t) = -iV a_1(r,t) a_3^*(r,t) \ ,$$

$$\left(\frac{\partial}{\partial t} + \gamma_3 + \boldsymbol{v}_3 \cdot \nabla\right) a_3(r,t) = -iV a_2^*(r,t) a_1(r,t) \ , \tag{1.4.29}$$

where $V = (2\pi)^{3/2} V(\mathbf{1};\mathbf{2},\mathbf{3})$ and the variables $a(r,t)$ denote complex envelopes of the waves $\exp[i(\boldsymbol{k}_i r - \omega_i t)]$. Equation (1.4.29) under $\gamma_i = 0$ is a Hamiltonian equation with the Hamiltonian

$$\mathcal{H} = \sum_{j=1}^{3} \frac{\boldsymbol{v}_j}{2i} \int [a_j^*(r)\nabla a_j(r)]\,dr + V \int [a_1^* a_2 a_3 + \text{c.c.}]\,dr \ . \tag{1.4.30}$$

In addition to $\mathcal{H}$, Eq. (1.4.29) under $\gamma_i = 0$ has two other independent integrals (the *Manly-Row* integrals)

$$N_1 + N_2 = \text{const.}\,, \quad N_2 + N_3 = \text{const.}\,, \quad N_j = \int a_j^*(r) a_j(r)\,dr \ . \tag{1.4.31}$$

The values N_i designate the total number of particles of j type. Employing three motion integrals enables one to study these equations effectively (see, e.g., [1.10]). Furthermore, the method of the *inverse scattering problem* (see, for instance, the survey by *Zakharov* [1.11], *Novikov* et al. [1.12]) makes it possible to study the evolution of three waves within (1.4.29) for arbitrary, localized initial conditions. Here again we consider a simple but interesting example, the generation of the second harmonic ($a_2 = a_3$) in the stationary case. Let us assume for the sake of simplicity $\gamma_j = 0, \boldsymbol{v}_j = \boldsymbol{v}$ and at $z = 0$, $a_1 = 0$, $a_2 = a_3 = A_2$ and is real. Substituting into (1.4.29) $a_1 = iA_1$, $a_2 = a_3 = A_2$ yields

$$v\partial A_1/\partial z = V A_2^2\,, \quad v\partial A_2/\partial z = -V A_1 A_2 \ . \tag{1.4.32}$$

Solution of these equations with conditions $A_1(0) = 0$, $A_2(0) = B$ on the interface ($z = 0$) has the following form:

$$A_1(z) = B \tanh(VBz/v)\,, \quad A_2(z) = B/\cosh(VBz/v) \ . \tag{1.4.33}$$

The characteristic length $L = v/(VB)$ is called the length of interaction. At $z \ll L$ all the energy of the initial wave is transferred into the second harmonic A_1. Experimental generation of the second harmonic is the most impressive experiment of nonlinear optics: a red beam enters a crystal and leaves it as a green beam. Complete conversion is impossible because the conditions of resonance are not completely satisfied, etc. In the most successful experiment $K_{tz} \simeq 90\%$.

1.4.4 Explosive Three-Wave Instability

If waves with negative energy can exist in the medium (active medium), the resonance conditions may be satisfied for "three-wave generation from the vacuum"

$$\omega(\boldsymbol{k}_1) + \omega(\boldsymbol{k}_2) + \omega(\boldsymbol{k}_3) = 0, \qquad \boldsymbol{k}_1 + \boldsymbol{k}_2 + \boldsymbol{k}_3 = 0. \qquad (1.4.34)$$

This process is described by the last term in (1.4.1). Retaining only this term in the vicinity of the resonance (1.4.34) we seek the solution in the form of

$$b(\boldsymbol{k},t) = \sum_{j=1}^{3} a_j(t)\delta(\boldsymbol{k} - \boldsymbol{k}_j)\exp[-i\omega(k_j)t].$$

The fastest increase of the wave amplitudes will take place if $\omega(\boldsymbol{k}_j)$ and $\boldsymbol{k}_j$ exactly satisfy the condition (1.4.34). In this case (1.4.1) yields

$$\begin{aligned}
\partial a_1/\partial t + \gamma_1 a_1 &= -2iU a_2^* a_3^*, \\
\partial a_2/\partial t + \gamma_2 a_2 &= -2iU a_1^* a_3^*, \\
\partial a_3/\partial t + \gamma_3 a_3 &= -2iU a_1^* a_2^*.
\end{aligned} \qquad (1.4.35)$$

At $\gamma_j = \gamma$ these equations have a solution $|A_j| = A$, where

$$A(t) = A_0\gamma\exp(-\gamma t)\{\gamma + 2UA[\exp(-\gamma t) - 1]\}^{-1}. \qquad (1.4.36)$$

If the initial amplitude A_0 is sufficiently large $2UA_0 > \gamma$, the amplitude $A(t)$ becomes infinite over a finite time t, under small $t_0 = 1/(2UA_0)$. Such an instability is called *explosive*. In this simplest solution "explosion" occurs simultaneously over the whole space. Actually, any spatial inhomogeneity of initial conditions will lead to "explosions" at isolated points. This phenomenon has been consistently explained theoretically by *Zakharov* and *Manakov* [1.13] within the frame of equations in partial derivatives similar to the (1.4.9) and also integrated by the inverse scattering problem.

Note that the complete realization of conditions (1.3.4) for explosive instability is impossible in magnetodielectrics. However, for small magnetic fields in ferromagnets the conditions under which homogeneous precession of magnetization decays into three magnons

$$\omega_0 = \omega(\boldsymbol{k}_1) + \omega(\boldsymbol{k}_2) + \omega(\boldsymbol{k}_3), \qquad \boldsymbol{k}_1 + \boldsymbol{k}_2 + \boldsymbol{k}_3 = 0 \qquad (1.4.37)$$

can be fulfilled. This process is described by a four-magnon Hamiltonian

$$\mathcal{H}_4 = \frac{1}{3}\sum_{1+2+3=0}(T_{0;1,2,3}b_0 b_1^* b_2^* b_3^* + \text{c.c.}). \qquad (1.4.38)$$

Evidently the amplitudes of the magnons b_1, b_2, b_3 when b_0 is large will evolve according to (1.4.36) (with the substitution $U \Rightarrow Tb_0$) until it is possible to neglect the inverse influence of magnons on the homogeneous precession. The curious reader could obtain a fourth equation to (1.4.35) for the amplitude of the homogeneous precession in the external field h and take this inverse influence into account.

1.5 Four-Wave Processes

When three-wave processes are forbidden by conservation laws the equation of motion (1.3.1) must allow for additional terms due to the four-wave interaction Hamiltonian $\mathcal{H}_{\text{int}}$ (1.1.26). One cannot completely avoid taking into account the three-wave terms, since according to (1.1.32) in the second order of the perturbation theory they result in allowed four-wave processes. It would be interesting to find out which contribution into the four-wave interaction amplitudes is larger: the original W or that arising because of the interaction of the three-wave processes $\tilde{T} \simeq V^2/\Delta\omega$. If the initial Hamiltonian is expanded in terms of a single small parameter, then $\Delta\omega W \simeq V^2$. Then the contributions under consideration are of the same order of magnitude. This is the case (for instance) for the dipole–dipole contribution to $\mathcal{H}_{\text{int}}$ for magnons in ferromagnets. When the situation is more complicated, any of these contributions may prevail for special reasons.

Recall one of the results of Sect. 1.1.4: via an appropriate nonlinear canonical transformation (similar to (1.1.31)) all the terms of $\mathcal{H}_{\text{int}}$ which are responsible for the processes forbidden by conservation laws can be eliminated. Therefore the four-wave scattering of the $2 \Rightarrow 2$ type waves is of particular interest

$$\omega(\boldsymbol{k}_1) + \omega(\boldsymbol{k}_2) = \omega(\boldsymbol{k}_3) + \omega(\boldsymbol{k}_4), \qquad \boldsymbol{k}_1 + \boldsymbol{k}_2 = \boldsymbol{k}_3 + \boldsymbol{k}_4, \qquad (1.5.1)$$

since it is allowed by all dispersion laws if all $\boldsymbol{k}_j$ are close. Henceforth, only such processes will be considered. Equations (1.3.3) in this case are:

$$\begin{aligned}\frac{\partial b(\boldsymbol{k},t)}{\partial t} &+ [i\omega(\boldsymbol{k}) + \gamma(\boldsymbol{k})]b(\boldsymbol{k},t) \\ &= -\frac{1}{2}\sum_{123} T_{\boldsymbol{k},1;2,3} b_1^* b_2 b_3 \delta(\boldsymbol{k}+\boldsymbol{1}-\boldsymbol{2}-\boldsymbol{3}),\end{aligned} \qquad (1.5.2)$$

where $T = W + \tilde{T}$ is the total amplitude of the processes (1.1.32b). Let us begin the consideration of these equations with the simplest problem.

1.5.1 Modulation Instability of the Plane Wave

Let $b(\mathbf{k}) = b_0 \delta(\mathbf{k} - \mathbf{k}_0)$ be a wave of finite amplitude propagating in a medium. Equations (1.5.2) have the form (at $\gamma(\mathbf{k}) = 0$):

$$\partial b_0/\partial t + i\omega_{\rm NL}(\mathbf{k}_0)b_0 = 0 , \qquad (1.5.3)$$

$$\omega_{\rm NL}(\mathbf{k}_0) = \omega(\mathbf{k}_0) + T_{00}|b_0|^2 , \quad T_{00} = \frac{1}{2}T(\mathbf{k}_0, \mathbf{k}_0; \mathbf{k}_0, \mathbf{k}_0), \qquad (1.5.4)$$

where $\omega_{\rm NL}(\mathbf{k}_0)$ is the frequency of the wave b_0 dependent on its amplitude. In the presence of the wave b_0 write (1.5.2) for the waves with small amplitudes whose wave vectors and frequencies satisfy the resonance conditions:

$$2\omega(\mathbf{k}_0) = \omega(\mathbf{k}_1) + \omega(\mathbf{k}_2) , \qquad 2\mathbf{k}_0 = \mathbf{k}_1 + \mathbf{k}_2 , \qquad (1.5.5)$$

$$\begin{aligned}\partial b_1/\partial t + i\omega_{\rm NL}(\mathbf{k}_1)b_1 + iT_{1,2}\,b_0^2 b_2^* = 0 , \\ \partial b_2/\partial t + i\omega_{\rm NL}(\mathbf{k}_2)b_2 + iT_{1,2}\,b_0^* b_1^2 = 0 ,\end{aligned} \qquad (1.5.6)$$

Here $b_j = b(\mathbf{k}_j)$, $T_{10} = T_{10,10}/2$, $T_{12} = T_{00,12}/2$, $T_{20} = T_{20,20}/2$ and

$$\omega_{\rm NL}(\mathbf{k}_1) = \omega(\mathbf{k}_1) + 2T_{10}|b_0|^2 , \quad \omega_{\rm NL}(\mathbf{k}_2) = \omega(\mathbf{k}_2) + 2T_{20}|b_0|^2 , \qquad (1.5.7)$$

are dependences of the frequencies of b_1, b_2 -waves (i.e. waves with amplitudes b_1 and b_2) on the amplitude of another wave b_0. Note that (1.5.7) differs from (1.5.4) in factor 2. The instability increment is calculated similar to the three-wave case. Comparing (1.5.6) and (1.4.11) we may immediately write (by analogy with (1.4.17)):

$$\nu^2 = |T_{12}\,b^2|^2 - [\omega_{\rm NL}(\mathbf{k}_2) - 2\omega_{\rm NL}(\mathbf{k}_0)]^2/4 . \qquad (1.5.8)$$

If $\mathbf{k}_1$ and $\mathbf{k}_2$ differ significantly from $\mathbf{k}_0$ and at the same time $\Delta\omega = 0$, then the order of magnitude of the instability increment $\nu = |T_{12}||b_0|^2$ also coincides with the nonlinear addition to the wave frequency. In a dissipative medium when the damping decrement of waves $\mathbf{k}_{1,2}$ is $\gamma_{1,2}$ an instability threshold arises. The threshold wave amplitude (similar to (1.4.19)) is

$$|T_{12}\,b^2| = \sqrt{\gamma_1\gamma_2} . \qquad (1.5.9)$$

When $\mathbf{k}_1$ and $\mathbf{k}_2$ approximate $\mathbf{k}_0$ the conservation law (1.5.5) must allow for nonlinear additions to the frequency

$$\omega_{\rm NL}(\mathbf{k}_{1,2}) = \omega(\mathbf{k}_0) \mp \mathbf{v}\cdot\boldsymbol{\kappa} + \frac{\partial^2\omega}{2\partial k_i \partial k_j}\kappa_i\kappa_j + 2T_{00}|b_0|^2 , \qquad (1.5.10)$$

where $\mathbf{k}_{1,2} = \mathbf{k}_0 \mp \boldsymbol{\kappa}$, $\mathbf{v} = (\partial\omega/\partial\mathbf{k})_{\mathbf{k}=\mathbf{k}_0}$. The function $\omega_{\rm NL}(\mathbf{k}_{1,2})$ was expanded in terms of $\boldsymbol{\kappa}$; and $\mathbf{k}$-dependence in the matrix elements of T has been neglected. Now

$$\nu^2(\boldsymbol{\kappa}) = -\hat{L}\kappa^2(\hat{L}\kappa^2 + 2T_{00}|b_0|^2) , \quad \hat{L}\kappa^2 = \frac{1}{2}\sum_{i,j}\frac{\partial^2\omega(\mathbf{k})}{\partial k_i \partial k_j}\kappa_i\kappa_j . \qquad (1.5.11)$$

If $T_{00}\hat{L}\kappa^2 > 0$, then $\nu < 0$, and there is no instability. This is the difference between the considered situation and the case of the three waves when instability emerged already when the conservation laws were satisfied and the sign of the matrix element was of no significance. When the quadratic form of (1.5.11) is not of fixed sign, the instability criterion

$$T_{00}\hat{L}\kappa^2 < 0 \tag{1.5.12}$$

can be obtained for any sign of T_{00} by changing the direction of $\boldsymbol{\kappa}$. If the form $\hat{L}\kappa^2$ is of fixed sign, the instability region is restricted with respect to κ from 0 to $\hat{L}\kappa^2 = -4T_{00}|b_0|^2$. The instability is maximum for $\hat{L}\kappa^2 = -2T_{00}|b_0|^2$. As the instability (1.5.11) of the $\boldsymbol{k}$-wave increases, new waves arise with $\boldsymbol{k}_{1,2}$ approximating $\boldsymbol{k}_0$. This may be interpreted as the appearance of modulation of the amplitude and the phase of the initial $\boldsymbol{k}_0$-wave. Therefore, such an instability is called by *Zakharov* in [1.14] a *modulation instability*.

Such an instability is a common phenomenon. For instance, for Langmuir waves in plasma $\hat{L}\kappa^2 > 0$, $T_{00} < 0$, which brings about modulation instability leading to Langmuir collapse. In nonlinear dielectrics where $\partial n/\partial |E|^2$ (refractive index n increases with the increase of the electric field strength E), $T_{00} < 0$, since $\omega = kc/\sqrt{n}$ and $\partial \omega/\partial |E|^2 \simeq -\partial n/\partial |E|^2$. Considering that for the linear law of dispersion $\omega'' > 0$ and $T_{00}\omega'' < 0$, this leads to modulation instability whose increase results in self-focusing of light.

By way of example consider the modulation instability of gravitational waves on the liquid surface. The law of dispersion of these waves (see (1.4.12)) has the form $\omega(\boldsymbol{k}) = \sqrt{gk}$. Find the square form of (1.5.12):

$$\hat{L}\kappa^2 = \left(\kappa_\parallel^2 - \frac{\kappa_\perp^2}{2}\right)\frac{\omega(k)}{2k^2}, \quad \boldsymbol{\kappa}_\parallel = \frac{\boldsymbol{k}(\boldsymbol{\kappa}\cdot\boldsymbol{k})}{k^2}, \quad \boldsymbol{\kappa}_\perp = \boldsymbol{\kappa} - \boldsymbol{\kappa}_\parallel. \tag{1.5.13}$$

It can be seen that for $\boldsymbol{\kappa} = \boldsymbol{\kappa}_\perp$, $\hat{L}\kappa^2 < 0$ and at $\boldsymbol{\kappa} = \boldsymbol{\kappa}''$, $\hat{L}\kappa^2 > 0$. Thus, the square form is not of fixed sign and modulation instability will increase whatever the sign of T_{00}.

We now estimate the threshold amplitude of the modulation instability μ_{cr} and its increment ν. It follows from (1.2.13, 14) and (1.5.9) that

$$\omega(k)(k\mu_{\text{cr}})^2 \simeq \gamma(k). \tag{1.5.14}$$

Here $\gamma(k)$ is the damping decrement of the gravitational waves. For estimation take $\gamma(k)$ as the damping decrement $\gamma = 2\nu_c k^2$ due to viscosity (see Sect. 25 in [1.15]). The kinematic viscosity ν_c is equal to 10^{-2} cm^2 s^{-1} for water. Hence and from (1.5.14) for waves of the wavelength 100 cm we obtain $\mu_{\text{cr}} \simeq (10^{-2} - 10^{-3})$ cm. Thus, the threshold amplitude of the wave is vanishingly small and modulation instability of long sea waves must always be in progress. From (1.2.13, 14) and (1.5.8) the increment is easily estimated

$$\nu \simeq \omega(k)(k\mu)^2. \tag{1.5.15}$$

For $\lambda = 100$ cm and $\mu \simeq (10-20)$ cm the increment of the modulation instability (1.5.15) approximates the wave frequency, i.e. the instability is fast increasing. What is the characteristic modulation period L? It may be estimated as $2\pi/\kappa_m$, where κ_m corresponds to the maximum $\nu(\kappa)$. It follows from (1.5.11) and (1.5.13) that in the plane $\kappa_\parallel$, $\kappa_\perp$ the instability region is in a narrow band in the vicinity of the lines $\sqrt{2}\kappa_\perp = \pm\kappa_\parallel$ which bend at κ comparable to the value of k. To find the region of maximum increment on this band (see (1.5.7, 8)) the explicit $\boldsymbol{k}$-dependence of T_{12} must be known. It can be shown that $\kappa_m \simeq 0.1k$. This implies that modulation instability results in long-period longitudinal and lateral modulation of the sea wave amplitude which is known as the "tenth wave" phenomenon.

Later in Chap. 3 we shall treat the modulation instability of spin waves in ferromagnets and antiferromagnets.

1.5.2 Equation for the Envelopes

In studying the nonlinear stage of modulation instability one must take advantage of the fact that all the secondary waves are in a narrow region near $\boldsymbol{k}$, i.e. the propagation of a narrow wave package must be studied or, equivalently, one must study the slow changes of the complex amplitude (*envelope*) of a monochromatic wave.

Here we shall obtain an equation for the envelope. In (1.5.6) $\omega(\boldsymbol{k})$ will be expanded in terms of κ up to the terms $\propto \kappa^2$ and we shall take

$$T(1,2;3,4) = T(\boldsymbol{k}_0, \boldsymbol{k}_0; \boldsymbol{k}_0, \boldsymbol{k}_0) = T/4\pi^3 \;. \tag{1.5.16}$$

This approach is correct if there are no long-range forcing in the medium and if $T(1,2;3,4)$ is a continuous function of its arguments. Next, as in (1.4.3), we pass to the slow variables

$$a(\boldsymbol{k}) = b(\boldsymbol{k}+\boldsymbol{\kappa})\exp[i(\boldsymbol{k}\boldsymbol{r}-\omega_0 t)] \tag{1.5.17}$$

and then to $\boldsymbol{r}$-representation according to (1.4.28). We have as a result:

$$\left[i\left(\frac{\partial}{\partial t}+\boldsymbol{v}\cdot\nabla\right)+\hat{L}-T|a|^2\right]a(\boldsymbol{r},t) = 0 \;, \tag{1.5.18}$$

$$\hat{L} = \frac{1}{2}\frac{\partial^2\omega(\boldsymbol{k})}{\partial k_i \partial k_j}\frac{\partial^2}{\partial x_i \partial x_j} \;. \tag{1.5.19}$$

In an isotropic medium $\omega(\boldsymbol{k}) = \omega(k)$ and

$$\hat{L} = \frac{\omega''}{2}\frac{\partial^2}{\partial z^2}+\frac{v}{2k}\Delta_\perp \;, \qquad \Delta_\perp = \frac{\partial^2}{\partial x^2}+\frac{\partial^2}{\partial y^2} \;. \tag{1.5.20}$$

Here $v = \partial\omega/\partial k$, $\omega'' = \partial^2\omega/\partial k^2$. Equation (1.5.18) describes the evolution of a narrow wave package: the term $\boldsymbol{v}\nabla$ describes its motion as a whole with a group velocity, the terms $\partial^2/\partial z^2$ and $\Delta_\perp$ describe the dispersion and the diffraction, and the last term describes the nonlinear self-action of the waves in the package. Optically, this term describes the dependence of the refractive index of the medium on the square of amplitude of the electric field. When (1.5.18) is treated as a Schrödinger equation the nonlinear term may be interpreted as a potential of a self-consistent attraction (at $T < 0$) or repulsive potential (at $T > 0$), which is proportional to the density of particles. Equations (1.5.18, 19) are Hamiltonian equations with the Hamiltonian

$$\mathcal{H} = \frac{1}{2} \int [i\boldsymbol{v}(a^*\nabla a - a\nabla a^*) + \omega'' |\frac{\partial a}{\partial z}|^2 + \frac{v}{k}|\Delta_\perp a|^2 + 2T|a|^4]\, d\boldsymbol{r} \ . \quad (1.5.21)$$

This equation has one more integral of motion $N = \int |a|^2 d\boldsymbol{r}$, meaning "the total number of particles".

There are two problems to which (1.5.19) applies. Firstly the wave amplitude may be given at a boundary of the medium, e.g. light is incident on a nonlinear dielectric. In this case the term $\omega'' \partial^2 a/\partial z^2$ may be neglected compared to $v\partial a/\partial z$ (z is the $\boldsymbol{v}$-direction). Secondly, if the problem is stationary, then $\partial a/\partial t = 0$. In this case:

$$\left[i v \frac{\partial}{\partial z} + \frac{v}{2k}\Delta_\perp - T|a|^2\right] a(\boldsymbol{r},t) = 0 \ . \quad (1.5.22)$$

Considering $\tau = z/v$ as a new time (1.5.22) can be treated as a two-dimensional Schrödinger equation ($\boldsymbol{r} = x, y$).

1.5.3 Package Evolution in Unbounded Media

Let us change over to a reference system moving with a group velocity. Then the term $\boldsymbol{v} \cdot \nabla a$ vanishes. The dispersion term ω'' must be retained. Thus (1.5.18) takes the form:

$$\left[i \frac{\partial}{\partial t} + \frac{\omega''}{2}\frac{\partial^2}{\partial z^2} + \frac{v}{2k_0}\Delta_\perp - T|a|^2\right] a(\boldsymbol{r},t) = 0 \ . \quad (1.5.23)$$

Consider qualitatively the solution of this problem in the cases of three dimensions ($d = 3$), two dimensions ($d = 2$) and one dimension. The case $d = 3$ within (1.5.23) corresponds to the evolution of the 3-dimensional bunch moving with a group velocity. If $d = 2$ we deal with the problem of the plane package $a(x,y,t)$ moving with a group velocity or the problem of stationary self-focusing of the package $a(x,y,t)$ within the scope of (1.5.22) after the substitution $z/v = t$. The case $d = 1$ corresponds, for example, to the stationary plane self-focusing $a(x,z)$. Further on, for definiteness we shall always assume $\omega'' > 0$ and will consider the case of particles "attraction" $T < 0$. Let the package (at a certain time) have a characteristic size

ℓ and amplitude in the center a. The number of particles in it is given by $N = \int |a|^2 d\boldsymbol{r} \simeq |a|^2 \ell^d$, where d is the dimensionality of the package. Since N is an integral of motion,

$$a(t) = \sqrt{N} \ell^{-d}(t) \; . \tag{1.5.24}$$

Estimate with the help of (1.5.21) and (1.5.24) the energy of the package:

$$\mathcal{H} \simeq \omega'' N \ell^{-2} - |T| N^2 \ell^{-d} \; . \tag{1.5.25}$$

It is clear that in $d = 1$ there exists a stationary solution with

$$\ell = \ell_0 \simeq \omega''/(|T|N) \simeq \omega''/(|Ta^2|\ell_0) \; , \tag{1.5.26}$$

minimizing the energy (1.5.25). In this case the pressure of the particles due to their motion in the potential well balances the attractive force. In the case of 3 dimensions as $\ell \to 0$ the pressure increases more slowly than the attractive force which leads to collapse, and the particles fall into the center over some finite time. In this case the amplitude $a(t)$ and the size $\ell(t)$ are connected by the relation (1.5.24) where N is the number of the particles involved in collapse. The energy (1.5.25) of the collapsing particle decreases. Evidently the total energy (including the kinetic energy of the particles moving away, not drawn into the collapse) is retained. In the case of two dimensions the fate of the package of particles is determined by the initial conditions: at $\omega'' > TN \simeq T|a|^2/\ell$ the minimum (1.5.25) is achieved as $\ell \to \infty$, i.e. the particles are moved away. Under $\omega'' < TN$ as in the case of three dimensions, part of the package is involved in the collapse process.

Let us treat the stationary solutions of (1.5.18) in the case of one dimension in more detail. Substituting for $a(z,t)$ as $a(z,t) = \varphi(z) \exp(i\lambda^2 t)$, we obtain

$$\begin{aligned} \omega'' \varphi_{zz} &= 2(\lambda^2 \varphi - |T|\varphi^3) = -\partial U/\partial \varphi \; , \\ U &= |T|\varphi^4/2 - \lambda^2 \varphi^2 \; , \quad \varphi_{zz} = \partial^2 \varphi/\partial z^2 \end{aligned} \tag{1.5.27}$$

After transformation $z \to t$ this equation may be treated as a Newtonian equation for particles with mass ω'' and coordinate γ moving in the field $U(\gamma)$. It has an integral of motion corresponding to the "energy":

$$E = \frac{1}{2}\omega'' \varphi_{zz}^2 + \frac{1}{2}|T|\varphi^4 - \lambda^2 \varphi^2 \; . \tag{1.5.28}$$

The simplest solution is a particle at rest at the bottom of the well $\gamma = $ constant, $\lambda = |T|\gamma^2$ corresponds in the initial equation (1.5.23) to a plane wave with a nonlinear frequency shift $-|T|\gamma^2$. When $E > E_{\min}$ it results in periodical oscillations of a particle in the well, i.e. to periodical modulation of the wave $a(z)$. The solution with $E = 0$ is of great interest. It can be obtained from (1.5.28) at $E = 0$

$$\gamma(z,t) = \sqrt{\frac{2}{|T|}} \frac{\lambda \exp(i\lambda^2 t)}{\cosh(\sqrt{2}\lambda z/\sqrt{\omega''})} . \tag{1.5.29}$$

The characteristic size ℓ_0 of this localized solution, i.e. *soliton*, is given, as can be readily seen, by the expression (1.5.25). In addition, while the soliton (1.5.29) is at rest (in the reference system moving with group velocity) there exist also moving solitons as well as solutions with two, three, etc. solitons. N-soliton solutions are of great importance. Employing the method of the *inverse scattering problem* for the nonlinear Schrödinger equation (1.5.22, 23) *Zakharov* and *Shabat* [1.16] showed that an arbitrary (fairly smooth) initial distribution of $a(z,0)$ as $t \to \infty$ generally breaks into N solitons whose number, amplitudes and velocities may be obtained from $a(z,0)$. Solitons are often stable formations: on collision, both solitons retain their velocities and amplitudes.

Finally, the qualitative conclusion that in two- and three-dimensional cases the initial package is destroyed over the finite time will be verified by a direct calculation. For simplicity (1.5.23) will be written in a dimensionless form. At $\omega'' = 0$, $T < 0$

$$\begin{aligned} i\frac{\partial \Psi}{\partial t} &= \frac{\delta \mathcal{H}}{\delta \Psi}, \qquad \mathcal{H} = \frac{1}{2}\int \left[|\nabla \Psi|^2 - \frac{1}{2}|\Psi|^4\right] d\boldsymbol{r} , \\ i\frac{\partial \Psi}{\partial t} &+ \Delta \Psi + |\Psi|^2 \Psi = 0 . \end{aligned} \tag{1.5.31}$$

Following the example of *Vlasov* et al. [1.17] we calculate

$$\begin{aligned} \frac{\partial^2}{\partial t^2}\langle R^2\rangle &= \frac{\partial^2}{\partial t^2}\int r^2 |\Psi|^2 d\boldsymbol{r} = i\frac{\partial}{\partial t}\int \sum_j x_j^2 \nabla(\Psi^*\nabla\Psi - \Psi\nabla\Psi^*)d\boldsymbol{r} \\ &= -2i\frac{\partial}{\partial t}\sum_j x_j \left(\Psi\frac{\partial \Psi^*}{\partial x_j} - \Psi^*\frac{\partial \Psi}{\partial x_j}\right) d\boldsymbol{r} . \end{aligned}$$

Making use once more of the equation of motion and integrating by parts so that no terms proportional to x should be retained we obtain

$$\frac{\partial^2}{\partial t^2}\langle R^2\rangle = 4\int |\nabla\Psi|^2 d\boldsymbol{r} = 8\mathcal{H} + 2(2-d)\int |\Psi|^4 d\boldsymbol{r} , \tag{1.5.32}$$

where d is the dimensionality of the space, $\mathcal{H}$ is the Hamiltonian (1.5.30). Thus, at $d > 2$ and $8\mathcal{H} > \partial^2\langle R^2\rangle/\partial t^2$, we have

$$\langle R^2\rangle < 4\mathcal{H}t^2 + C_1 t + C_2 . \tag{1.5.33}$$

If at the initial time $\mathcal{H} < 0$, i.e. $2\int |\Delta\Psi|^2 d\boldsymbol{r} < \int |\Psi|^4 d\boldsymbol{r}$, then over a finite time the solution becomes singular and collapse takes place because $\langle R^2\rangle$ goes to zero. The stationary two-dimensional solution of (1.5.23) (round wave guide) would be of great practical interest since it would enable one to transmit the energy of laser radiation over large distances without the losses caused by diffraction divergence. Such a solution, as it clear from (1.5.25), corresponds to $\mathcal{H} = 0$. Unfortunately, this solution is not stable. If at the initial time $\mathcal{H}$ proves positive because of the fluctuations, then by virtue of (1.5.33) (it should be recalled that at $d = 2$ there is an equality sign in (1.5.33)) $\langle R^2 \rangle$ will increase and the waveguide will "blur". In the opposite case $\mathcal{H} < 0$, field intensity in the center will be very high. As a rule, this leads to irreversible destruction of the nonlinear dielectric accompanied by losses of radiation energy. For details see the review by *Akhmatov* et al. [1.18], *Zakharov* et al. [1.19].

2 The General Properties of Magnetodielectrics

University and college departments of physics traditionally do not include advanced courses in magnetism in their curriculum. Therefore we offer the reader this short review chapter presenting necessary data on magnetically ordered dielectrics. This saves the reader the necessity of referring to many books on magnetism [2.1–13].

2.1 Classification of Substances by Their Magnetic Properties

All substances have more or less pronounced magnetic properties in the sense that their properties change to some degree under the influence of a magnetic field. One main parameter quantifying this influence is the *magnetic susceptibility* χ, the derivative of the magnetization M with respect to the strength of the magnetic field H. All substances fall into one of four groups according to the magnitude and sign of χ:

2.1.1 Diamagnets

Diamagnets have a magnetic susceptibility which is negative and small compared to unity $|\chi| \simeq 10^{-6}$. Atoms with no magnetic moment of their own are usually diamagnetic. Diamagnetism can be explained by the Lenz rule according to which a current arising in the system of charges (electron shells of atoms) placed into the magnetic field tends to reduce this field. Therefore their resulting magnetic moment is directed opposite to the applied field.

2.1.2 Superconductors

From the viewpoint of their magnetic properties, superconductors constitute a special group. They eject the magnetic field into which they are placed when its strength is below some critical value. This phenomenon is called the *Meissner effect*. It results in *strong diamagnetism* when $\chi = -1/4\pi$.

2.1.3 Paramagnets

These are substances with a low positive susceptibility. They comprise ions with uncompensated magnetic moments of electrons μ that have a spin nature: $\mu = \mu_B S$ where μ_B is the Bohr magneton, S is the spin of an atom (ion). Such ions are termed *paramagnetic*. Examples of such paramagnetic ions are elements with incomplete d-shells (elements of the Ferrum group: Fe, Ni, Co, Mn), f-shells (rare earth elements Sm, Dy, Yb, Lu, Tm), etc. The magnetic moment of the paramagnets is a result of two competing factors, the external magnetic field and temperature. The first factor tends to produce parallel orientation of the magnetic moment of ions: under the influence of the second one the magnetic moment of ions becomes chaotic. At high temperatures the resulting magnetic moment is small $\chi \simeq n\mu/T$ (here n is the concentration of paramagnetic atoms with moment μ, T is the temperature in energy units). At room temperature $\chi \simeq 10^{-3}$. This picture is observed if the interaction between the magnetic moments of the ions is sufficiently small compared with the temperature. If the temperature is decreased the substance passes to a *magnetically ordered state*.

2.1.4 Magnetically Ordered Substances (Magnets)

These have many typically magnetic properties and are characterized in particular by an additional type of collective excitations, *spin waves*, whereby the magnetic moments oscillate with respect to the ordered orientation. Nonlinear properties of spin waves will be the main object of study in this book. Spin wave properties are, with other magnetic properties, largely determined by the type of *magnetic ordering*, or, in other words, by the magnetic structure of magnets. Let us consider the types of magnetic structure.

1 Ferromagnets – FM. According to modern classification these are substances in which the magnetic moments of all the atoms have parallel orientation. This results in a macroscopic magnetic moment equal to the sum of the magnetic momentum of all the atoms. Without the external magnetic field such a fully ordered state often proves to be thermodynamically unstable and the FM breaks into *domains*, i.e. macroscopic regions in which the moments of the electrons are parallel. The resultant moment of the sample then approximates zero. Then the ferromagnet (e.g. iron) is said to be not magnetized, or in the multidomain state. To attain a one-domain state, an external magnetic field should be applied. Lest this state be broken without the external magnetic field, the motion of the domain walls and the rotation of magnetic moments inside the domains must be specially prevented. For more details on the domain structure see [2.4, 9]. In the general case magnetic ordering is accompanied by the appearance of several *magnetic sublattices*, each of which is a group of ions with similar magnetic moments.

To be more exact, magnetic ions of the same sublattice are *translationally invariant*, i.e. can be replaced by each other via an integer number of elementary translations of the crystal lattice. The number of magnetic sublattices is in this case determined by the number of magnetic atoms (ions) in an elementary cell allowing for the magnetic order. Such a magnetic cell may comprise an additional elementary crystal cell. From this viewpoint FMs are substances with a single magnetic sublattice.

2 Ferrimagnets. These are substances with several magnetic sublattices (with magnetization $\boldsymbol{M}_j$). Interaction between the sublattices results in an orientation such that the total magnetic moment $\boldsymbol{M} = \sum_j \boldsymbol{M}_j$ is nonzero. In the simplest case there are two collinear sublattices with different magnetic moments $M_1 \neq M_2$ with antiparallel orientation so that $\boldsymbol{M} = \boldsymbol{M}_1 - \boldsymbol{M}_2 \neq 0$. For three or more sublattices the orientation is not necessarily collinear. A classical example of a ferrimagnet is the Yttrium-Iron Garnet (YIG) $Y_3Fe_5O_{12}$ with twenty magnetic sublattices.

3 Antiferromagnets – AFM. Unlike the ferrimagnet, the sum of magnetic moments of the AFM lattices is equal to zero. The simplest case of an AFM has two equivalent antiparallel sublattices $\boldsymbol{M}_1 = -\boldsymbol{M}_2$. Antiferromagnetic ordering is typical, for instance in oxides MnO, FeO, CoO, and fluorides MnF_2, CoF_2, FeF_2.

4 Antiferromagnets with weak ferromagnetism. Unlike pure AFMs the sublattices of these substances are weakly uncollinear owing to a specific relativistic Dzyaloshinsky–Moria interaction [2.9, 16]. Since such interaction is significantly weaker than the exchange interaction, the resultant magnetic moment $\boldsymbol{M}$ is much less than the magnetization of each sublattice. Examples of such a structure are $\alpha-Fe_2O_3$, NiF_2, $MnCO_2$ and $CoCO_2$.

5 Helicoidal structures or helicomagnets. These substances are characterized by much more complicated ordering and cannot be described in terms of magnetic sublattices. If there is one magnetic atom per elementary crystal cell (in the paramagnetic phase) such structure may be represented, say, as a "stationary wave of the magnetic moment"

$$\mu_{jz} = \mu_z, \quad \mu_{jx} = \mu\cos(\boldsymbol{q}\boldsymbol{R}_j), \quad \mu_{jy} = \mu\sin(\boldsymbol{q}\boldsymbol{R}). \qquad (2.1.1)$$

Here $\boldsymbol{\mu}_j$ is the ion magnetization in the cell $\boldsymbol{R}_j$, $\boldsymbol{q}$ is the wave vector of the spiral incommensurable with the period of the reciprocal lattice. In the so-called simple S–S spiral $\mu_z = 0$, and the resultant magnetic moment of the crystal is equal to zero. Such an ordering is antiferromagnetic. If $\boldsymbol{\mu}_z \neq 0$ the spiral is called *conical* or *ferromagnetic*. There are more complicated spirals in which μ_z depends periodically on the ion number. Helicoid magnetic structure at low temperature is typical for many rare earth metals. Such structures have also been observed in alloys of MnAu$_2$, the compound MnI$_2$, etc.

6 Magnetic glasses. This is a wide class of substances, which are of great interest for theoreticians and promising for practical applications. Magnetic glasses can be called congealed paramagnets with some magnetic structure but without long-range magnetic ordering. Their sample average magnetic moment is zero, but thermodynamic average of the magnetic moment of each paramagnetic atom differs from zero. As in magnetically ordered substances, magnetic glasses may support the propagation of long spin waves. There is an extensive literature on magnetic glasses and *amorphous magnets* (see, for example, [2.14]).

In conclusion of this section it must be noticed that the above classification is fairly schematic since it is always difficult to fit reality into formulae. For instance, there is no strict borderline between AFMs with weak ferromagnetism and ferrimagnets, between helimagnets with the spiral period in commensurable with the crystallographic period and multi-sublattice AFMs. For more details on these problems – types of magnetic ordering, its relation to crystal symmetry, etc. – see [2.9, 12], and the works by *Landau* [2.15] and *Dzyalochinsky* [2.16].

2.2 Nature of Interaction of Magnetic Moments

It is common knowledge that the strongest interaction determining the type of magnetic structure is usually the *exchange interaction*. It is electrostatic in nature and is due to the Pauli principle which states that no two electrons can exist in identical quantum states. The origin of the exchange interaction can best be illustrated by the simple example of the hydrogen molecule.

2.2.1 Exchange Interaction in the Hydrogen Molecule

The Hamiltonian $\mathcal{H}$ of this system consists of the unperturbed Hamiltonians of the two atoms a and b, $\mathcal{H}_a$ and $\mathcal{H}_b$, and the Hamiltonian of their interaction

$$\mathcal{H}_{ab} = e^2 \left[\frac{1}{R_{ab}} + \frac{1}{R_{12}} - \frac{1}{R_{a2}} - \frac{1}{R_{b1}} \right], \qquad (2.2.1)$$

where R_{ab} is the distance between the protons, R_{12} denotes the distance between the electrons, R_{a2} and R_{b1} is the distance between electron of the atom a and the proton of the atom b and vice versa. Since $\mathcal{H}_{ab}$ does not depend on the spin variables S_1, S_2 the multiplication form of the wave function of the molecule must be sought. In accordance with the Pauli principle the wave function Ψ must be antisymmetric under exchange of electrons. As in the state with total spin of the system $S = 1$, the electron spins are parallel, the spin term of Ψ will be symmetrical. Its coordinate term φ will therefore be antisymmetric. Accordingly, in the state with antiparallel spins, $S = 0$, the function φ will be symmetric. To first order in the perturbation theory in the interaction Hamiltonian $\mathcal{H}_{ab}$ we construct the functions of the zeroth approximation $\varphi_\pm(\boldsymbol{R}_1, \boldsymbol{R}_2)$ using the wave functions of the hydrogen atoms $\varphi(\boldsymbol{R}_{ci})$ composed of the i-th electron ($i=1, 2$) and c-th nucleus ($c = a, b$):

$$\varphi_\pm(\boldsymbol{R}_1, \boldsymbol{R}_2) = \frac{[\varphi(\boldsymbol{R}_{a1})\varphi(\boldsymbol{R}_{b2}) \pm \varphi(\boldsymbol{R}_{a2})\varphi(\boldsymbol{R}_{b1})]}{2\sqrt{1-\delta^2}} , \qquad (2.2.2)$$

where δ is the overlap integral. In this approximation the energy E_+ of hydrogen molecule in the state with $S = 1$ and the energy E_- of the molecule with $S = 0$ have the following forms:

$$E_\pm = \int \varphi_\pm^2(\boldsymbol{R}_1, \boldsymbol{R}_2) \mathcal{H}_{ab} \, d\boldsymbol{R}_1 d\boldsymbol{R}_2 = [A(R_{ab}) \pm B(R_{ab})]/(1+\delta^2) , (2.2.3)$$

$$\begin{aligned} A(R_{ab}) &= \int \mathcal{H}_{ab} \, \varphi^2(\boldsymbol{R}_{a1}) \varphi^2(\boldsymbol{R}_{b2}) \, d\boldsymbol{R}_1 d\boldsymbol{R}_2 , \\ B(R_{ab}) &= \int \mathcal{H}_{ab} \, \varphi(\boldsymbol{R}_{a1})\varphi(\boldsymbol{R}_{b1})\varphi(\boldsymbol{R}_{a2})\varphi(\boldsymbol{R}_{b2}) \, d\boldsymbol{R}_1 d\boldsymbol{R}_2 . \end{aligned} \qquad (2.2.4)$$

Computation using these formulas shows that the function $E_+(R_{ab}) > 0$ and decreases monotonically as the distance between the nuclei R_{ab} increases and the function $E_-(R_{ab})$ has a sharp minimum under some $R_{ab} = R_0$. It is important that $E_-(R_{ab}) < 0$. This means that two hydrogen atoms can form a molecule only in a state with antiparallel spins. The quantum-mechanical explanation of the homopolar chemical bond presented in our book was given by *Heitler* and *London* in 1927 [2.17]. As can readily be seen from (2.2.4), the function $A(R_{ab})$ determines the Coulomb energy of two atoms on the assumption that the first electron "belongs" only to the nucleus a, and the second one belongs to the nucleus b. As the distance between the atoms increases this energy diminishes as $1/R_{ab}$. The function $B(R_{ab})$ is a nondiagonal matrix element of the electrostatic energy $\mathcal{H}_{ab}$ between the state $\langle a1, b2|$ in which the first electron belongs to the atom a and the second electron belongs the atom b, and the state $|a2, b1\rangle$ in which the electrons exchange their positions. Since the function $B(R_{ab})$ is non-zero only due to the overlap of electron shells it decreases exponentially as the distance increases.

This simple example shows that the pure quantum effect of electron exchange leads to an interaction between the atoms determined by the total spin of the atom S in spite of the fact that the Hamiltonian of the interaction is independent of the spin variables. The two expressions (2.2.3) for the binding energy $E_\pm$ in the state with total spin $S = 0$ and $S = 1$ may be written uniformly using the spin variables $\boldsymbol{S}_1, \boldsymbol{S}_2$.

$$E_\pm = E - J\boldsymbol{S}_1\boldsymbol{S}_2 \,, \quad E = 3E_-/4 + E_+/4 \,, \quad J = E_- - E_+ \,. \qquad (2.2.5)$$

Here it has been considered that the values of the operator $\boldsymbol{S}_1\boldsymbol{S}_2$ are given by 3/4 in the singlet state (when $S_1 = 1$) and 1/4 in the triplet state (when $S = 0$). The spin-dependent term in (2.2.5) is called the *Hamiltonian of exchange interaction* and is denoted as

$$\mathcal{H}_{ex} = -J\boldsymbol{S}_1\boldsymbol{S}_2 \,. \qquad (2.2.6)$$

The function $J(R_{12})$ is called the *exchange integral*. If $J < 0$ (as in our example) the antiparallel spin orientation is desirable, then the exchange interaction is said to be *antiferromagnetic*. Under $J > 0$ parallel orientation of spins is established and the exchange is called *ferromagnetic*.

The operator (2.2.6) is commonly known as the *Heisenberg Hamiltonian*, though in this form it was first obtained by *Dirac* and first used in the theory of magnetism by *Van Flek*. But it was *Heisenberg* who showed that interactions resulting in magnetic ordering had quantum-mechanical nature due to the Pauli principle. The original idea that spontaneous magnetization is due to the specific interaction between the magnetic atoms was first suggested by *Weiss* in 1907. His estimation of the temperature of transition based on the magnetic dipole interaction, however, was 0.25 K, which is less than its real value by a factor of 10^4. But in spite of such a failure he prove to be an incorrigible optimist, saing that "this difficulty must be treated not as an argument against this theory of molecular field, but as an incentive for looking for new ideas in the theory of the atomic structure". Such ideas were indeed formulated 20 years later!

2.2.2 Interatomic Exchange

As a rule, magnetic atoms have several valence electrons, therefore the question of exchange interaction arises. This case differs from the above-mentioned in the first place in the fact that binding energy includes only interaction between the electrons

$$\mathcal{H}_{ab} = e^2/R_{12} \qquad (2.2.7)$$

and in the second place due to the fact that the coordinate terms of various quantum-mechanical states of electrons of a single atom m and $n - \varphi_m(\boldsymbol{R})$ and $\varphi_n(\boldsymbol{R})$ are orthogonal. Therefore their overlap integral equals zero and the expression (2.2.3) is reduced:

$$E_\pm = A \pm B \,. \qquad (2.2.8)$$

Then from (2.2.4, 5) follows a simple expression for J:

$$J = 2B = \int \rho^*(\boldsymbol{R}_1)(e^2/R_{12})\rho(\boldsymbol{R}_2)\,d\boldsymbol{R}_1 d\boldsymbol{R}_2 \; , \qquad (2.2.10)$$
$$\rho(\boldsymbol{R}) = \varphi_m(\boldsymbol{R})\varphi_n^*(\boldsymbol{R}) \; .$$

Employing the Fourier transform we can obtain

$$J = \frac{e^2}{2\pi^2} \int \frac{|\rho(\boldsymbol{k})|^2}{k^2}\,d\boldsymbol{k} \; . \qquad (2.2.11)$$

Thus, the interatomic exchange is a ferromagnetic one, and it brings about parallel spin orientation of the valence electrons of one atom. Hence follows the well-known *Hund rule*: among all the states of an atom with the same configuration (the number of valence electrons) the state with the maximum possible spin will have the lowest energy value. The exchange integral (2.2.10) has no small factor (due to the overlap of the wave functions). Its order of magnitude approximates the Coulomb energy of the two electrons on the same center, i.e. it has an atomic scale of $1\;eV$. This contribution to the energy is usually much greater than all the others and the magnetic moments of the valence electrons may be considered rigidly connected to one another. They are manifested in interatomic interactions as a single big magnetic moment with respective atomic spin $S > 1/2$.

2.2.3 Interatomic Exchange of Large Spins

Consider the exchange interaction of two atoms in the state with a total spin $S_a > 1/2$ and $S_b > 1/2$. The wave functions are antisymmetric under rearrangement of N electrons on the atom a and N electrons on the atom b. Using these, a completely antisymmetric wave function must be constructed. Here each of the allowed types of rearrangement symmetry of the space wave function is associated with a certain value of the total spin S (for details see, for instance, [2.12]). Consequently, each S is associated with its own energy value of the two-atom system $E(S)$. The total spin of the system may take $2S_b + 1$ values (for definiteness we take $S_a > S_b$) from $S_a - S_b$ to $S_a + S_b$ and accordingly the energy of the two atoms can assume $2S + 1$ values. They can be described by an effective spin Hamiltonian having $2S_b + 1$ constant:

$$E(\boldsymbol{S}) = E - J_{ab}^{(1)}\boldsymbol{S}_a\boldsymbol{S}_b - J_{ab}^{(2)}(\boldsymbol{S}_a\boldsymbol{S}_b)^2 - \ldots - J_{ab}^{(2S)}(\boldsymbol{S}_a\boldsymbol{S}_b)^{2S_b} \; . \qquad (2.2.12)$$

This implies that the Hamiltonian of the exchange interaction of two multi-electron atoms, generally speaking, can have no simple Heisenberg form (2.2.6). It must be emphasized, though, that to the first approximation in the overlap of the wave functions of atoms a and b (when only the interatomic exchange of a simple electron pair is allowed for) all the $J_{ab}^{(n)}$ except $J_{ab}^{(1)}$ are equal to zero (*Nedlin* [2.18]). For this reason, the Heisenberg approximation (2.2.6) can generally be used also for the exchange interaction between multielectron atoms. Only in some rare cases the term proportional to $J_{ab}^{(2)}$ (called the *biquadratic exchange*) has to be taken into account.

2.2.4 Indirect Exchange Interactions

It must be noted that in magnetic dielectrics, paramagnetic ions usually are not nearest neighbor, but are separated by a diamagnetic ion (oxygen, fluoride). Therefore their wave functions do not overlap and the above discussed direct exchange interaction is absent.

Nevertheless, exchange interaction arises in higher orders of the perturbation theory due to the overlap integrals between the wave functions of the paramagnetic and diamagnetic ions. One of the mechanisms of such an exchange suggested by *Kramers* in 1934 and *Anderson* in 1963 (see, e.g. [2.9, 11, 12]) is due to the mixing of states of the magnetic and nonmagnetic ions. Indeed, electrons are not fully localized on the ions even in ion crystals. Therefore there is a non-zero probability of the transition of one of the electrons belonging to the diamagnetic ion c (it will be denoted $1c$) to the neighboring paramagnetic ion a. Since the initial Hamiltonian of the system is independent of spins, this transition will not be accompanied by an overturn of the spin. As a result, the spin of the electron $1c$ remains antiparallel to the spin of the second, already unpaired electron of the ion c (electron $2c$). Thus, two exchange interactions (namely the exchange of the unpaired electron of the ion a with the incoming $1c$ and the exchange of the electron of the other paramagnetic ion b with the unpaired electron $2c$ in the ion c) result in the dependence of the total energy of the three ions $a-c-b$ on the relative orientation of the spins S_a and S_b. This suggests that the exchange interaction of the paramagnetic ions a and b separated by the diamagnetic ion c does exist. Such interaction was named the *super-exchange* or *indirect exchange*. There are other mechanisms of indirect exchange corresponding to different terms of the perturbation theory for the energy of the three ions $a-c-b$. (See, for instance, [2.12, 17, 20], etc.).

Obviously, the calculation of the exchange integrals for particular crystals will be too cumbersome a task requiring computer processing, data bases on ion spectroscopy for a reasonable choice of the initial wave functions, etc. All this, of course, is beyond the scope of our work. Henceforth the exchange integral J_{ab} between the ions of the ions a and b of the crystal will be taken to be a phenomenological constant whose value could be experimentally obtained. Studying nonlinear properties of the spin waves in the present book we shall proceed from the Heisenberg approximation of the exchange Hamiltonian of the magnetic:

$$\mathcal{H}_{\text{ex}} = \sum J_{ab} S_1 S_2 \ . \tag{2.2.13}$$

All of the above should not be understood as the derivation of the formula for $\mathcal{H}_{\text{ex}}$. We intended to illustrate the only following two facts: firstly, that the exchange interaction has a simple nature: it is simply the part of the Coulomb interaction between the electrons due to the Pauli principle;

secondly, that the actual calculation of the exchange integrals in real crystals is not only difficult but unnecessary for the investigation of nonlinear properties of the spin wave.

2.2.5 Relativistic Interactions

The exchange interaction is the strongest, but not the only interaction between the magnetic ions of the magnets. Of fundamental importance are also relativistic interactions: *dipole–dipole* interactions between the magnetic moments of the electrons and the *spin–orbit* interaction between the spin and orbital moments of the electron. In the first place, these result in an effective *anisotropy energy* which determines the energetically advantageous directions of magnetization about the crystallographic axes. In the second place, they lead to processes where the number of the spin waves is not conserved. This results in the complete thermodynamic equilibrium in the system. Later, these cases will be discussed in more detail. Now we shall only write down and consider the expression for the energy of those interactions. We start from the magnetic dipole–dipole interaction:

$$\mathcal{H}_{\mathrm{dd}} = 2\mu_{\mathrm{B}}^2 \sum_{a,b} \frac{1}{R_{ab}^5}\left[(\boldsymbol{S}_a\boldsymbol{S}_b)R_{ab}^2 - 3(\boldsymbol{S}_a\boldsymbol{R}_{ab})(\boldsymbol{S}_b\boldsymbol{R}_{ab})\right] . \qquad (2.2.14)$$

Here, as above, $\boldsymbol{R}_{ab} = \boldsymbol{R}_a - \boldsymbol{R}_b$ is the distance between the atoms a and b. The energy of the dipole–dipole interaction calculated per atom is $4S^2\mu^2/a$, where a denotes the lattice constant. The corresponding value of the magnetic ordering temperature is of the order of 0.25 K. It is much less than the experimentally observed Curie temperature due to exchange interaction, which is normally between 10 and 1000 K.

A further type of relativistic interaction is the *spin–orbit interaction* whose operator has the form:

$$\mathcal{H}_{\mathrm{so}} = \lambda(\boldsymbol{LS}) . \qquad (2.2.15)$$

Here $\boldsymbol{L}$ and $\boldsymbol{S}$ designate the operators of the mechanical and spin atomic moments, and λ is the coupling parameter. The order of magnitude of $\lambda \simeq (v/c)^2 E_a$, where v is the characteristic velocity of an electron in an atom, c is the velocity of light, and E_a is the atomic energy, of the order of one Rydberg, i.e. 10 eV. Ordinarily, $\lambda \simeq 10^{-2}$ eV. The energy (2.2.15) is determined by orientation of $\boldsymbol{S}$ relative to $\boldsymbol{L}$. On the other hand, the energy of an ion in the "crystal" electric field depends upon the orientation of $\boldsymbol{L}$ relative to the lattice, since different $\boldsymbol{L}$ are associated with different wave functions. Therefore, in the second order of the perturbation theory with respect to the Hamiltonian (2.2.15) the energy of the ion becomes dependent on its spin orientation about the crystallographic axes. It can be described by the effective spin Hamiltonian

$$\mathcal{H}_{\text{eff}}^{(2)} = \sum_{i,j} \lambda_{i,j}^{(2)} S_i S_j , \qquad i,j = x,y,z . \tag{2.2.16}$$

Here $\lambda_{ij}^{(2)} \simeq \lambda^2/\triangle$, where $\triangle$ is the splitting energy of the levels with different L by the crystal field. Since $\triangle \simeq 1$ eV, $\lambda^{(2)} \simeq 10^{-4}$ eV $\simeq 1$ K. If the ground state of the ion is the singlet ($L = 0$), then the value of $\lambda^{(2)}$ is lower still, because anisotropy arises due to the admixture of states with $L = 0$ in the ground state. As a result, in various crystals $\lambda^{(2)} \simeq 0.01 - 1$ K, which corresponds to the effective field of the crystallographic anisotropy $\mathcal{H}_a \simeq 0.1 \div 10$ kOe. In uniaxial crystals:

$$\lambda_{ij}^{(2)} = \lambda_1 \delta_{ij} + \lambda_2 \delta_{zz} .$$

For $\lambda_2 < 0$ the anisotropy field tends to orient spins along the z-axis. Then the crystal is said to have the anisotropy of the *"easy-axis type"*. At $\lambda_2 > 0$ the anisotropy orients the spins perpendicular to the z-axis. This is referred to as anisotropy of the *easy-plane type*.

If the nearest neighbors of the ion have cubic symmetry then $\lambda_2 = 0$ and $\mathcal{H}_{\text{eff}}^{(2)} = $ const. Then there is no crystallographic anisotropy in the second order of the perturbation theory in the spin–orbit interaction (2.2.15). It arises only in the fourth order. In this case

$$\mathcal{H}_{\text{eff}}^{(4)} = \sum_{ijkl} \lambda_{ijkl}^{(4)} S_i S_j S_k S_l , \tag{2.2.17}$$

where $\lambda^{(4)} \simeq \lambda^4/\triangle^3 \ll \lambda^{(2)} \ll \lambda$. The corresponding field of the cubic anisotropy will be less still than the field of a uniaxial anisotropy: H_a is of the order of single to tens of Oersteds. Depending on the sign of some combination of constants in (2.2.17) the easy axis will be either the direction [111] or [100].

In conclusion it must be noted that allowing for the overlap of the wave functions of the ions as well as for the spin–orbit interaction results in the effective Hamiltonian of the form of (2.2.16) where spins belong to the neighboring atoms a and b, i.e. $\mathcal{H}_D \simeq S_{ai} S_{bj}$. As has already been noted, such an interaction can lead to weakly noncollinear sublattices [2.20].

2.3 Energy of Ferromagnets in the Continuum Approximation

Recall that in the spin system of the magnetodielectrics the strongest interaction is the exchange interaction which provides for the very existence of magnetic order. The main part of this interaction is described by the Heisenberg Hamiltonian $\mathcal{H}_{\text{ex}}$ (2.2.13). In the dynamics of long spin waves the dipole–dipole interaction $\mathcal{H}_{\text{dd}}$ (2.2.14) is also of great importance. The interaction of the magnetic moments of electrons with a homogeneous external magnetic field $\boldsymbol{H}$ must also be taken into account:

$$\mathcal{H}_m = -2\mu_{\text{B}} \sum_a (\boldsymbol{H}\boldsymbol{S}_a) \,. \tag{2.3.1}$$

In the previous section it has been shown that the spin–orbit interaction results in the emergence of the effective energy of the magnetic anisotropy. In uniaxial crystals (see (2.2.16)):

$$\mathcal{H}_a^{(2)} = \lambda^{(2)} \sum_a (S_{a,z})^2 \,. \tag{2.3.2}$$

In ferromagnets with cubic symmetry (see (2.2.17)):

$$\mathcal{H}_a^{(4)} = \lambda^{(4)} \sum_a \left[(S_{a,x})^4 + (S_{a,y})^4 + (S_{a,z})^4 \right] \,. \tag{2.3.3}$$

These expressions contain only the terms of the second and fourth orders of the perturbation theory in the spin–orbit interaction. Sometimes still weaker terms of sixth order are taken into account. In uniaxial hexagonal crystals such terms bring about the anisotropy in the *basal plane* perpendicular to the 6-fold rotational axis of symmetry.

The total Hamiltonian of the ferromagnet spin system contains, generally speaking, all the above-mentioned terms:

$$\mathcal{H} = \mathcal{H}_{\text{ex}} + \mathcal{H}_{\text{dd}} + \mathcal{H}_m + \mathcal{H}_a \,. \tag{2.3.4}$$

This expression does not contain the interaction of the spin subsystem of a crystal with its other subsystems – phonons, excitons, etc.

We now turn to a discussion of the classical approximation for the ferromagnetic energy assuming the magnetic moments of electrons on the neighboring sites to be almost parallel. Formally, the procedure must be as follows. First, the classical expression for the ferromagnetic energy must be written, which corresponds to the Hamiltonian (2.3.4). Second, the notion of the averaged density of the magnetic moment $\boldsymbol{M}(\boldsymbol{r})$ must be introduced. To this end the magnetic moment of the electron must be "spread" over the unit cell. Third, we must change from a summation over the sites to an integration over space.

This procedure can be most easily performed for the part of the Hamiltonian containing a single summation over sites. It is from (2.3.1-3) that the well-known expressions for W_m and W_a are most readily obtained:

$$W_m = -\int \boldsymbol{H} \boldsymbol{M}(\boldsymbol{r}) d\boldsymbol{r} \; , \qquad W_a^{(2)} = \frac{\lambda^{(2)} v_0}{4\delta \mu_B^2} \int M_z^2(\boldsymbol{r}) d\boldsymbol{r} \; , \qquad (2.3.5a,b)$$

$$W_a^{(4)} = K_4 \int [M_x^4(\boldsymbol{r}) + M_y^4(\boldsymbol{r}) + M_z^4(\boldsymbol{r})] d\boldsymbol{r} \; , \qquad K_4 = (\lambda^4 v_0^3 / 16 \mu_B^4) \; . (2.3.6)$$

$\boldsymbol{M}(\boldsymbol{r})$ here denotes the density of the magnetic moment, K_4 is the cubic constant of the 4-th order anisotropy, and v_0 designates the volume of the unit cell. The transformation of (2.2.13) for the exchange energy proves somewhat more complicated. Here only the final expression for W_{ex} will be shown. The reader can do the necessary calculations independently, or can see it, for instance, in [2.8]. The result is

$$W_{\text{ex}} = \frac{\alpha_{ik}}{2} \int \frac{\partial M_j}{\partial x_i} \frac{\partial M_j}{\partial x_k} d\boldsymbol{r} \; , \qquad \alpha_{ik} = \frac{1}{8\mu_B^2} \sum_n J(r_n) r_i(\boldsymbol{n}) r_k(\boldsymbol{n}) \; . \quad (2.3.7)$$

Here $r_j(\boldsymbol{n})$ is the i-projection ($i = x, y, z$) of the $\boldsymbol{r}$-coordinate of the magnetic atoms $\boldsymbol{n}$ in the lattice. Expressions (2.3.5-7) were derived under the assumption that the distance l over which the exchange interval is essentially changed is small in comparison with the distance at which an essential change $\boldsymbol{M}(\boldsymbol{r})$ takes place.

Note that the expression (2.3.7) for the exchange energy has a great generality. To a great extent it is independent of model assumptions: spin magnitude, dependence of the exchange integral on the distance between atoms, etc. It may be treated as a phenomenological expression for the non-homogeneous exchange energy. Indeed if it is assumed, first, that this energy is independent of magnetization orientation about the crystallographic axis, second, that inversion is an element of crystal symmetry and, third, that there exists a quadratic dependence of the energy on magnetization, then from symmetry considerations the expression (2.3.6) can be obtained. The proof of this simple statement will not be given here. The reader can find it in [2.8]. We shall only note that the first of the above assumptions is general and follows from the invariance of the exchange interaction with respect to the rotation of all spins. The second assumption is based on the fact that such inversion is characteristic of most magnetically ordered dielectrics. As for the last assumption, strictly speaking it is correct only for the case $S = 1/2$ when the exchange interaction operator has the Heisenberg form (2.2.13). Under $S > 1/2$ the approximation (2.2.13) sometimes proves inadequate and the additional term must be allowed for, i.e., the so-called *"biquadratic exchange"* whose operator is given by (2.2.12). There is a corresponding additional term in the energy of the inhomogeneous exchange:

2.3 Energy of Ferromagnets in the Continuum Approximation

$$W_{\text{ex}}^{(4)} = \frac{\bar{\alpha}_{ik}}{2} \int M_l \frac{\partial M_l}{\partial x_i} M_m \frac{\partial M_m}{\partial x_k}\, d\mathbf{r}\,, \tag{2.3.8}$$

$$\bar{\alpha}_{ik} = \frac{1}{8\mu_{\text{B}}^4} \sum_{\mathbf{a}} J^{(2)}(\mathbf{a}) a_i a_k\,. \tag{2.3.9}$$

The expression (2.3.9) is obtained under the assumption that only the "nearest neighbor" interactions are essential. Than $J^{(2)}(\mathbf{a})$ is biquadratic excange integral $J^{(2)}(\mathbf{r})$ (2.2.12). For the general case the expression (2.3.8) can be obtained from symmetry considerations, thus taking the exchange interaction to be invariant with respect to the rotations of $\mathbf{M}$ and the presence of inversion.

Now we consider the energy of dipole–dipole interactions. The following macroscopic energy corresponds to the Hamiltonian $\mathcal{H}_{\text{dd}}$ (2.2.14):

$$W_{\text{dd}} = \frac{v_0^2}{2} \sum_{a \neq b} \frac{1}{R_{ab}^5}\left[(\mathbf{M}_a \cdot \mathbf{M}_b)R_{ab}^2 - 3(\mathbf{M}_a \cdot \mathbf{R}_{ab})(\mathbf{M}_b \cdot \mathbf{R}_{ab})\right]. \tag{2.3.10}$$

Passing to integration in (2.3.10) is not so trivial, though the resulting expression for W_{dd} is simple:

$$W_{\text{dd}} = -\frac{1}{2}\int\left[\mathbf{M}\cdot\mathbf{H}_m + \frac{4\pi M^2}{3} + a_{ik} M_i M_k\right] d\mathbf{r}\,, \tag{2.3.11}$$

$$a_{ik} = v_0 \sum_b \frac{\partial^2 r}{\partial R_{ab}^i \partial R_{ab}^k}\left(\frac{1}{R_{ab}}\right)\,. \tag{2.3.12}$$

Here $\mathbf{H}_m$ is the static magnetic field due to the magnetic moment $\mathbf{M}(\mathbf{r})$. Explicit and detailed derivation of these expressions is presented in [2.8]. Therefore here we shall discuss only the physical meaning of these expressions. The first term in (2.3.11) is a well-known expression of the magnetic energy under continuous space distribution of the magnetic dipoles. The others term account for the discrete structure of the magnets. They are due to the difference between the true value of the magnetic field in dipole locations and the averaged value $\mathbf{H}_m$. The second term in (2.3.11) approximately allows for this difference, as it is done in the calculation of Lorentz-Lorenz correction to the permittivity of the crystal(see [2.18]). Specifically, in calculating the magnetic field $\mathbf{H}$ acting upon the dipole, the dipole is taken to be inside the spherical cavity (the Lorentz sphere) in the continuum with a continuous distribution of the magnetization $\mathbf{M}$. It can be shown (see, e.g., [2.3]) that

$$\mathbf{H} = \mathbf{H}_m + 4\pi\mathbf{M}/3\,. \tag{2.3.13}$$

Taking this into account, it becomes clear that the first two terms in (2.3.11) are simply the energy of the dipole interaction $-\mathbf{M}\cdot\mathbf{H}/2$. Finally, the last term in (2.3.11) allows for the difference between the real crystal discreteness and the approximate model using the Lorentz sphere. From symmetry considerations in cubic crystals $a_{ij} = a\delta_{ij}$. On the other hand, it follows from (2.3.12) that $\text{Tr}\{a_{ij}\} = 0$, and, as a result, $a = 0$. This means that for cubic crystals the results of the approximate model with the Lorentz sphere are exact. In uniaxial crystals at the same time $a_{ik} = a_i \mathcal{E}_{ik}$, $a_z \neq a_x = a_y$, and the last term in (2.3.11) does not become zero. It can be represented as:

$$-\frac{a_z - a_x}{2} \int M_z^2(\boldsymbol{r}) \, d\boldsymbol{r} - \frac{a_x}{2} \int M^2(\boldsymbol{r}) \, d\boldsymbol{r} \; . \tag{2.3.14}$$

The form of the first term coincides here with the expression (2.3.5) for $W_a^{(2)}$ and therefore represents the dipole–dipole contribution to the energy of crystallographic anisotropy. Henceforth this term will be considered to have been accounted for in (2.3.5) by corresponding renormalization of the coefficient $\lambda^{(2)}$. The remaining terms in (2.3.11) and (2.3.14) proportional to $M^2(\boldsymbol{r})$ represent the constant contribution to the energy W which does not depend on the orientation of $\boldsymbol{M}$ and is of no interest for us. Therefore in (2.3.11) we can retain only the first term.

For this term to be expressed only in terms of magnetization the following equations of magnetostatics must be solved

$$\text{curl}\,\boldsymbol{H} = 0 \; , \qquad \text{div}\,\boldsymbol{B} = 0 \; . \tag{2.3.15}$$

We will be interested in the following two cases:

1. The Plane Wave: $\boldsymbol{M}(\boldsymbol{r}) = \langle \boldsymbol{M} \rangle + [\boldsymbol{m}_k \exp(i\boldsymbol{k}\boldsymbol{r}) + \text{c.c.}]$, the wavelength of the spin wave $2\pi/k$ being much less than the sample size.

2. The sample is shaped as an ellipsoid and $\boldsymbol{M}(\boldsymbol{r}) = \boldsymbol{M} = \text{const}$. As is well known, (2.3.15) has exact solutions in both cases. In the first case

$$\boldsymbol{H}(\boldsymbol{r}) = [\boldsymbol{H}_m(\boldsymbol{k}) \exp(i\boldsymbol{k}\boldsymbol{r}) + \text{c.c.}] \; , \qquad \boldsymbol{H}(\boldsymbol{k}) = -4\pi\boldsymbol{k}(\boldsymbol{k} \cdot \boldsymbol{m}_k)/k^2 \; .$$

In the second case $H_{m,i} = -4\pi N_{ik} M_k$, $\text{Tr}\{N_{ij}\} = 1$, where N_{ij} is the *tensor of demagnetizing factors*, depending on the shape of the ellipsoid. Using these results the energy of the dipole–dipole interaction (2.3.11) can be expressed in terms of $\boldsymbol{M}$. To this end, represent $\boldsymbol{M}(\boldsymbol{r})$ as an expansion in terms of plane waves

$$\boldsymbol{M}(\boldsymbol{r}) = \langle \boldsymbol{M} \rangle + \sum_{\boldsymbol{k}} \boldsymbol{m}(\boldsymbol{k}) \exp(i\boldsymbol{k} \cdot \boldsymbol{r}) \; . \tag{2.3.16}$$

Here we take $\boldsymbol{M}(\boldsymbol{k}) = 0$ at $\boldsymbol{k} = 0$, and the mean magnetization value in the sample is given by the formula

$$\langle M \rangle = \frac{1}{V_s} \int M(r)\, dr\;, \tag{2.3.17}$$

where V_s is the volume of the sample. Substituting the expansion (2.3.16) into the formulae to (2.3.11) and using the expression for $H(r)$ and for $H_{m,i}$ we obtain

$$W_{dd} = 2\pi V N_{ik}\langle M_i\rangle\langle M_k\rangle + 2\pi V \sum_k [k\cdot m(k)][k\cdot m(-k)]/k^2\;. \tag{2.3.18}$$

Recall that this expression was derived under the assumption that the minimum linear dimension of the sample is much greater than $2\pi/k$. This means that we do not allow for the *Walker modes* which are magnetostatic eigenmodes of magnetization oscillations with characteristic dimension on the order of the sample size (see, for instance, Sect. 11 in [2.8]). The only Walker mode taken into account by the first term in (2.3.18) is the *uniform precession of magnetization* (UP). This limitation of (2.3.18) is, however, not very important since in most experiments treated any modes below the Walker modes (except for UP) are not excited.

Summing up the present section we may say that the energy of the ferromagnet W is the sum

$$W = W_m + W_a + W_{ex} + W_{dd}\;, \tag{2.3.20}$$

where W_m is the energy of interaction with the external field, W_a is the energy of magnetic anisotropy, W_{ex} is the exchange energy and W_{dd} is the energy of the dipole–dipole interaction, with expressions given by respectively by (2.3.5–9, 18). These expressions will be used for the calculation of the spin waves' dispersion law and the functions characterizing their interaction.

2.4 Magnetic and Crystallographic Structure of Some Magnets

We shall not give a detailed description or classification of magnetic crystals. The reader is referred to the specialized treatises [2.22], reviews [2.11], chapters in books [2.9]. In the present book we give only the structures of those magnets that are most often employed in experiments.

2.4.1 Crystals with Spinel Structure

Such crystals are of special interest, since they were used for the first verification of the Néel theory of ferromagnetism, which generalized the Weiss theory of the molecular field to the case of several magnetic sublattices. The spinel structure is typical of dioxides of transition metals having the chemical formula $A^{2+}B_2^{3+}O_4$. Here A^{2+} and B^{3+} stand for two- and three-charge ions of the metals. Spinels are also formed by almost densely packed oxygen.

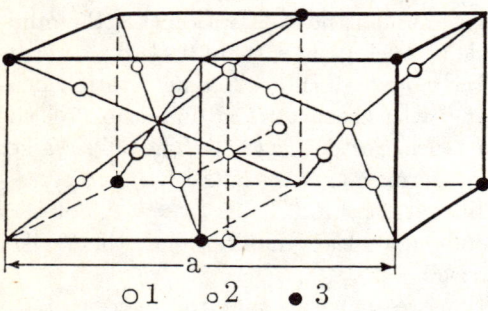

Fig. 2.1. Two of eight octants of the elementary cell of spinels: (1) oxygen, (2) metal in the octahedral interstices, (3) metal in the tetrahedral interstices

The ions of a metal are located in octahedral and tetrahedral interstices surrounded respectively by six and four oxygen ions (Fig. 2.1). The unit cell of the spinel consists of eight basic formula units (AB_2O_4) and comprises 32 oxygen atoms forming 16 octahedric and 8 tetrahedric interstices. They are designated respectively by the letters d and a. The cell a is a cube of edge length $a_0 = 8.5$ Å) consisting of eight octants (small cubes $a_0/2$ on edge) of two types, differing in the arrangements of their ions. The octants of different types have a common edge. In the so-called *normal spinels* the cations A are located in the tetrahedral sites and the cations B are in the octahedral sites. This is characteristic, for instance, of the ferrites $ZnFe_2O_4$, $CdFe_2O_4$.

In the *inverted spinels* the cations A and half of the cations B are located in octahedrons and the other half of the cations B are in tetrahedrons. This refers, for example, to manganese, lithium, and nickel ferrites. The ferrite $MgFe_2O_4$ has an intermediate (between the normal and the inverted spinels) type of cation localization.

If both metals A and B are paramagnetic the spinel is as a rule ferrimagnetic with predominating intersublattice exchange $|J_{AB}| \gg |J_{AA}|, |J_{BB}|$, with the antiferromagnetic sign. In this case the magnetizations of the sublattices A and B are antiparallel.

Normal spinels $ZnFe_2O_4$ and $CdFe_2O_4$ where the magnetic ions occupy only equivalent (octahedric) positions are antiferromagnetic. This is due to the antiferromagnetic sign of the B–B exchange subdividing the ensemble of the sites B into two magnetic sublattices.

2.4.2 Crystals with Garnet Structure

The Yittrium–Iron Garnet (YIG) is a wonderful phenomenon. Its role in the physics of magnets is comparable to the part germanium played in the physics of semiconductors and quartz in acoustics of crystals. This can be explained by the following. Firstly, it has a high Curie temperature $T_c = 560$ K which enables one to experiment at room temperature. On the other hand, although the unit cell consists of 80 atoms, each of which must assume its unique "correct" position, the technology of YIG crystal growing has been developed so well that the sound damping decrement is less than in quartz crystals. Finally it has the narrowest known line of ferromagnetic resonance and the smallest damping decrement of the spin waves. All this makes it indispensable not only in SHF engineering but also in the experimental physics of magnets studying new effects and phenomena.

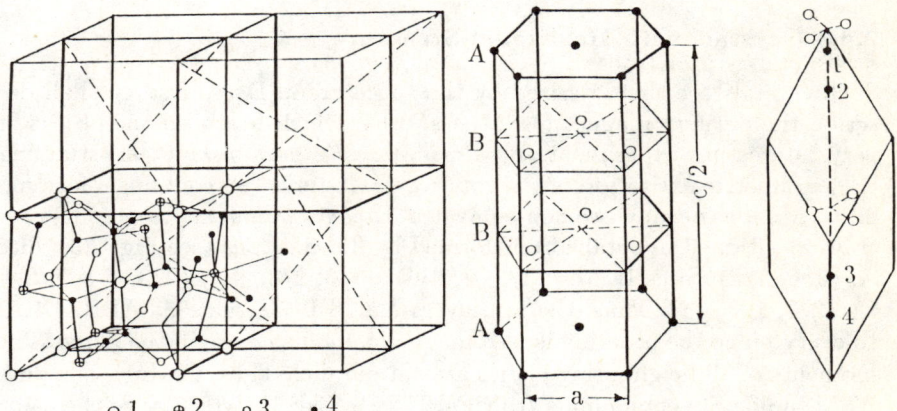

Fig. 2.2. (left) The elementary cell of YIG: (1), (2) and (3) are the positions a, c and d respectively, (4) are the oxygen ions

Fig. 2.3. (middle) One half of the CsMnF$_3$ elementary cell (A and B are the Mn ions); (right) an elementary cell of Fe$_2$O$_3$ and Cr$_2$O$_3$: (1), (2), (3) and (4) are the ions of Fe or Cr

The YIG unit cell is a half-cube $a_0 = 12.38$ Å on edge consisting of 4 identical octants, i.e. small cubes $a_0/2$ on edge comprising one formula unit (Fig. 2.2). The arrangement of atoms in the octant has no cubic symmetry. In the unit cell, however, the orientation of octants makes it practically cubic of the O_h symmetry group. The apices and centers of every octant are occupied by Fe^{3+} cations surrounded by octahedrons (the so-called a-site). Their coordinates in the first octant can be written as $(0,0,0)$; $(0,1/2,1/2)$; $(1/2,0,1/2)$; $(1/2,1/2,0)$; $(1/2,1/2,1/2)$; $(1/4,1/4,1/4)$. A unit cell comprises 8 a-sites. There are also 12 tetrahedral sites ($12d$) occupied by

Fe^{2+} anions and 12 dodecahedral sites (12c), occupied by Y anions. Their coordinates in the first octant of the site (12d) are (3/8, 0, 1/4); (1/4, 3/8, 0); (0, 1/4, 3/8); and in the site (12c) (1/8, 0, 1/4); (1/4, 1/8, 0); (0, 1/4, 1/8).

The oxygen anions occupy the common sites. The sites of all the 40 atoms in a unit cell (of $Y_3Fe_2^{3+}Fe_3^{2+}O_{12}$) are considered, e.g., in [2.12]. Thus, there are two non-equivalent sites of paramagnetic ferrum anions in YIG, i.e. a and d. Therefore, in a rough approximation YIG is sometimes considered a ferrimagnet with 2 sublattices. This approximation, however, has no field of applicability (*Kolokolov* et al. [2.21, 22]). At $T < 150$ K only one branch of the spin wave spectrum is excited and YIG can be considered a ferromagnetic. At large T the excitations of at least four or more spectrum branches must be taken into account.

In addition to YIG there are rare earth ferrogarnets $R_3Fe_5O_{12}$, where R = Sm, Dy, Yb, Lu, Tm. Their rare earth sublattices are also magnets and the number of magnetic sublattices is larger than in YIG.

2.4.3 Crystals with Hexagonal Structure

Such crystals are characterized by the space group D_{6h}^4. First we shall describe the antiferromagnet (AFM) $CsMnF_3$, which has been the object of very interesting experiments by *Prosorova, Kotyuzhansky* et al. studying the parametric excitation of the spin waves. Its unit cell contains 6 formula units. Mn ions occupy two non-equivalent sites A (2 ions) and B (4 ions) centrally positioned in octahedrons formed by fluorine ions (see Fig. 2.3). Site A: (0, 0, 1/2). Sites B are : (1/3, 2/3, i); (1/3, 2/3, $-i$); (2/3, 1/3, 1/2 + i); (2/3, 1/3, $-i$). Thus, the Mn layers A and B alternate as ABBABBA. In every plane the ordering is ferromagnetic; moments are in the plane. The moments of all neighboring planes are antiparallel. Thus, $CsMnF_3$ is a pure AFM, unlike its compounds $RbNiF_3$, $TlNiF_3$, $CsFeF_3$ which have the same structure, but are ferrimagnets. In these crystals intensive antiferromagnetic A–B exchange and a weak ferromagnetic B–B exchange are observed. Therefore these crystals may be treated as quasi-two-dimensional ones formed by triples, i.e. by groups of three strongly coupled BAB-planes: (BAB)(BAB)... All the moments are in the basal plane; magnetic ordering of the planes is of the type (+–+)(+–+)(+–+).... In the structure under consideration, the crystallographic surrounding of each magnetic ion has a six-fold symmetry axis. Anisotropy in the basal plane arises only in the sixth order of perturbation theory in the spin–orbit interaction and thus appears rather small: $H = 1$ Oe. Also, uniaxial anisotropy (of the easy-plane type) arising in the second order of perturbation theory is rather big – the field of anisotropy is of the order of 10 kOe.

2.4.4 Crystals with Rhombohedral Structures

The two most remarkable representatives are $\alpha-\text{Fe}_2\text{O}_3$ and Cr_2O_3. Dzyaloshinsky's theory, stating the connection of weak ferromagnetism arising in AFMs with magnetic ordering symmetry was verified employing these crystals by authors of [2.12, 17]. The space group of these crystals is D_{3d}^6. The unit cell comprises two formula units, i.e. four magnetic ions located on the axis of the 3-fold rotation symmetry (Fig. 2.3, right). They are antiferromagnetically ordered. Under a certain sequence of alternating spin signs in a unit cell weak ferromagnetism is possible, only when the moments are in the basal plane [2.12]. These conditions exist in MnCO_3 at temperatures $950 > T > 250$ K. Indeed, this substance has been experimentally proven to have a weak ferromagnetic moment within the above mentioned temperature range. As for Cr_2O_3, the spin alternation in it is of another kind and in full accordance with the theory that it contains no magnetic moment $\alpha-\text{Fe}_2\text{O}_3$ (AFM with anisotropy of the easy–plane type) is also a rhombohedric crystal. Its Néel temperature is 32 K. This magnetic is employed in experiments on the parametric excitation of spin waves. Unlike CsMnF_3 it is characterized by weak ferromagnetism.

No detailed description of the other magnetic structures will be given. We shall only name some crystals of perovskite structure (CaTiO_3) whose composition can be represent as ABX where X is oxygen or fluorine, and A and B designate metals [2.12]. These are YFeO_3, KMnF_3, RbCoF_3. Magnetic crystals with a structure of rock salt are worth mentioning, i.e. AFM MnO, FeO, CoO, NiO, and ferromagnets EuO, EuS, EuSe (at $T < 2.8$ K). It must be emphasized that in the long list of the named crystals only the three europium compounds are simple ferromagnets with one magnetic lattice.

3 Spin Waves (Magnons) in Magnetically Ordered Dielectrics

The previous two chapters were introductory. Chapter 1 treated the dynamic processes of nonlinear wave interactions within the scope of the classical Hamiltonian formalism. At that stage we refrained from analyzing the nature of particular waves (electromagnetic waves, sound waves, spin waves etc.), their polarization and the type of medium in which they propagate. All the information required for the study of nonlinear dynamics and kinetics of specific wave types under certain physical conditions and nonlinear media is contained in the Hamiltonian function of the waves, i.e. in the dispersion laws $\omega_j(\boldsymbol{k})$ and functions $V(q)$ and $T(p)$ describing the amplitudes of interaction. Therefore, in order to study the nonlinear behavior of spin waves (magnons) in magnets within the frame of the classical Hamiltonian method one should first of all obtain (or know) their Hamiltonian function, namely the law of dispersion of spin waves, and the amplitudes of their interaction. This chapter mainly presents the summarized results of calculation of the Hamiltonian functions of spin waves accompanied by short comments.

Sect. 3.1 deals with the magnons in ferromagnets, Sect. 3.2 treats the magnons in antiferromagnets. At the first stage of the study of the nonlinear magnons it is sufficient to read Sects. 3.1, 2 of this chapter. One can skip the details of the "internal structure" of the magnets which are given in Sects. 3.3, 4 and required for obtaining the Hamiltonian coefficients. For general information on magnets required for the further understanding of the physics of the considered nonlinear phenomena the reader is referred to the previous chapter. It must be emphasized that we can (as has been already mentioned) employ the Hamiltonian of a magnetic in the physics of nonlinear magnons irrespective of its origin because in the Hamiltonian approach the following two problems are quite independent. The first one is the investigation of the nonlinear behavior of a particular wave type (e.g. spin waves) through its known Hamiltonian function and the second one is the investigation of different classes of magnetodielectrics and possible types of interaction in order to obtain the Hamiltonian function of the spin subsystem.

Readers interested primarily in the problems of nonlinear physics of different waves of which spin waves are only one of the interesting examples can simply skip the Sects. 3.3-4 devoted to the calculation of the spin Hamiltonian. Likewise expert cooks can ignore the details of butchering or grape picking if they are satisfied with the quality of the foodstuffs. But the great chefs are not content with a mere standard, they seek perfection and know how to use the finest flavors depending on the way, time and place of the food product primary procession. It is for such "chefs" of physics that the last part of this chapter is intended.

It will be of interest also for the researchers using nonlinear processes as a method of studying the magnetodielectrics. Sections 3.3.1 gives and comments upon the "verbal" derivation of the equations of motion for magnetization known as "Bloch equations". Section 3.3.2 presents the relation between the observable medium variables (magnetization) and canonical variables. In these variables the Bloch equations are re-expressed as canonical Hamiltonian equations, and the expression for the energy of the spin subsystem of a magnetic (obtained in Sect. 2.3) becomes a Hamiltonian function. Afterwards it is formally simple to calculate the coefficients of the expansion of this function into a power series of normal canonical variables (i.e. to obtain the dispersion laws of the spin waves and the coefficients of their interaction Hamiltonian), although actually it is a rather awkward procedure. The general outline of these calculations is given in Sect. 3.4.

The readers who will set off through the obstacle race of this chapter with its heap of facts and formulae and get to the end must have their reward. First of all they will be proud to have won the battle with the chapter. Secondly, they may acquire a new viewpoint on the spin waves in magnetodielectrics as an interesting and relatively simple subject of inquiry in the physics of nonlinear wave phenomena. And they will attain an ultimate understanding of the fact that things are actually much more complicated than they seem. For example, the applicability range of the described approach to the nonlinear magnons under finite temperatures is yet to be found. We do not know the exact degree of error in obtaining the amplitudes of the interaction $V(q)$ and $T(p)$ at $T \simeq T_c/2$. There are other facts that are not clear. The detailed treatment of those fine and complicated problems is yet to be carried out. We can now only express our certainty that the simple Hamiltonian method of describing the dynamics of magnetodielectrics will reveal much necessary information about nonlinear magnons.

3.1 Hamiltonian of Magnons in Ferromagnets (FM)

As it has already been mentioned in Chap. 2 Yttrium Iron Garnet (YIG), which is a cubic ferromagnet, serves as a classic object of experimental studies in nonlinear dynamics and kinetics of magnons. At low temperatures the relative motions of magnetic sublattices (optical magnons) are not excited and YIG can be considered a ferromagnet. Therefore we shall begin our study with the spin Hamiltonian in ferromagnets with cubic symmetry. In typical experiments the external magnetic field $\boldsymbol{H}$ is spatially homogeneous. Therefore we will study this particular case. Secondly, we can assume the sample shape to be ellipsoid. In this situation the magnetic field inside of sample $\boldsymbol{H}_1$ is also spatially homogeneous [3.1]:

$$H_{1.i} = H_1 - 4\pi \sum_j N_{ij} M_j . \tag{3.1.1}$$

Here $\boldsymbol{M}$ is the magnetization and N_{ij} is the tensor of demagnetizing coefficients; its trace is equal to unity. Most practicable are samples shaped as a sphere, long cylinder or thin discs. For the sphere, long (along the z-axis) cylinder and for the thin (along the z-axis) disc we have respectively

$$N_{ij} = \frac{1}{3}\delta_{ij} , \tag{3.1.2}$$

$$N_z = N_{zz} = 0 , \quad N_{ij} = \frac{1}{2}\delta_{ij} , \quad i,j = x,y , \tag{3.1.3}$$

$$N_z = N_{zz} = 1 , \quad N_{ij} = 0 , \quad i,j = x,y . \tag{3.1.4}$$

And, finally, let us confine ourselves to the case when the field $\boldsymbol{H}$ is oriented along one of the principal axes of the ellipsoid which, in turn, coincides with one of the axes of symmetry of the crystal sample: 4-fold - [100], 3-fold [111] or 2-fold [110] axes. It is sufficient for the magnetization $\boldsymbol{M}$ in unexcited ferromagnets to be parallel to $\boldsymbol{H}$.

3.1.1 Spectrum of Magnons in Cubic Ferromagnets

The frequency of the magnons (obtained in Sect. 3.4 with the help of information from Sects. 1.1, 2) are:

$$\omega^2(\boldsymbol{k}) = A^2(\boldsymbol{k}) - |B(\boldsymbol{k})|^2 , \tag{3.1.5}$$

$$A(\boldsymbol{k}) = \omega_\mathrm{H} - \omega_\mathrm{M} N_z + \alpha\omega_\mathrm{a} + \omega_\mathrm{ex}(ak)^2 + \frac{1}{2}\omega_\mathrm{M}\sin^2\Theta ,$$
$$B(\boldsymbol{k}) = \frac{1}{2}\omega_\mathrm{M}\sin^2\Theta \exp(2i\varphi) + \beta\omega_\mathrm{a} . \tag{3.1.6}$$

Here Θ and φ are polar and azimuthal angles of the vector $\boldsymbol{k}$ in the spherical coordinates with z-axis oriented along $\boldsymbol{M}$, and the angle taken φ, which is respect to the direction [100] in the plane perpendicular to $\boldsymbol{M}$, ω_M is the circular frequency of the precession of the magnetic moment in the external field $\boldsymbol{H}$

$$\omega_\mathrm{H} = gH , \tag{3.1.7}$$

where $g = \mu_\mathrm{b}/\hbar$ is the magneto-mechanical ratio for the electron; μ_b is Bohr magneton; $g \simeq 2\pi \cdot 2.8$ MHz/Oe is a dimensional value. It must not be confused with the dimensionless g-factor approximately equal to 2 for an electron. The frequencies ω_M and ω_H characterize the magnetic dipole–dipole interaction and crystallographic anisotropy

$$\omega_M = 4\pi g M , \quad \omega_a = g H_a , \quad H_a = K_4 M^3 . \tag{3.1.8}$$

The field of anisotropy H_a is proportional to the conventional constant of the cubic anisotropy of the fourth order K_4, specified by (2.3.5). Coefficients α and β depend on the orientation of the magnetic field $\boldsymbol{H}$ (and the co-inciding magnetization direction) with respect to the crystallographic axes (Table 3.1). The frequency ω_{ex} characterizes the exchange interaction, the order of magnitude $\hbar\omega_{ex}$ approximates T_c, where T_c is the temperature of magnetic ordering, usually called the Curie temperature, and a is the lattice constant. The exchange frequency ω_{ex} is proportional to the constant of nonhomogeneous exchange α_{ij} given by the formula (2.3.6). In cubic ferromagnets:

$$\alpha_{ik} = \alpha \delta_{ik} , \quad \omega_{ex} a^2 = 2 g M \alpha . \tag{3.1.9}$$

Table 3.1. Coefficients characterizing the anisotropy energy contribution to the frequency of the magnons (3.1.5) and to the interaction amplitudes (3.1.25)

Magnetization orientation	α	β	δ
[100]	1	0	-9
[111]	-2/3	0	6
[110]	-1/2	-3/2	9/2

The specially selected notation of (3.1.5) makes it convenient to compare the contributions of different interactions to the spin wave frequency $\omega(\boldsymbol{k})$ (and later also to the amplitudes of H_{int}). The frequency $\omega_M = g H_M$ and magnetic field $H_M = 4\pi M$ can be used as convenient characteristic scales. For YIG at room temperature $4\pi M = 1.75$ kOe, $H_{ex} = \omega_{ex}$ kOe, $H_a = 0.084$ kOe. Analyzing experimental results it is useful to remember that the linear frequency associated with the field $H_a = 1$ kOe is $f = \omega/2\pi$, equal to $2.8 \cdot 10^9$ Hz, the circular frequency associated with the field is $1.76 \cdot 10^{10}$ s^{-1} and the associated temperature is 0.13 K. For the selected scale the cubic anisotropy is obviously small: $\omega_a/\omega_M = 0.05$. It is small also in some other cubic ferromagnets. In the first approximation such crystals can be assumed to be isotropic. This, however, does not imply that the crystallographic anisotropy can be completely neglected. Later some striking manifestations of nonlinear dynamics of magnons will be shown to be direct results of the presence of anisotropy.

The exchange frequency ω_{ex} significantly exceeds ω_M. For YIG $\omega_{ex}/\omega_M \simeq 220$. Formula (3.1.5) shows that the characteristic value of the wave vector

k_M can be determined at which the contributions of the dipole–dipole and exchange interactions become equal

$$\omega_\mathrm{ex}(ak_\mathrm{M})^2 = \omega_\mathrm{M} \ .$$

For YIG $ak_\mathrm{M} = 0.07$ and correspondingly $k_\mathrm{M} = 7 \cdot 10^5$ cm^{-1}. In usual experiments on parametric magnon excitation the pumping frequency approximates 10 or 36 GHz (1 GHz = 10^9 Hz), in this case the wave vector of the magnons may vary depending on the value of H, from $5 \cdot 10^3$ to $5 \cdot 10^3$ cm^{-1}. Consequently, both dipole–dipole and the exchange interactions are significant for us. The $\boldsymbol{k}$-dependence (3.1.5) of the magnon frequency $\omega(\boldsymbol{k})$ is rather complicated. It may be graphically shown (Fig. 3.1) as a family of surfaces with a constant frequency. Under $\omega(\boldsymbol{k}) \gg \omega_\mathrm{M}$ their shape is approximately spherical:

$$\omega(\boldsymbol{k}) = \omega_\mathrm{H} - \omega_\mathrm{M} N_z + \omega_\mathrm{ex}(ak)^2 + \frac{1}{2}\omega_\mathrm{M} \sin^2 \Theta \ . \qquad (3.1.10)$$

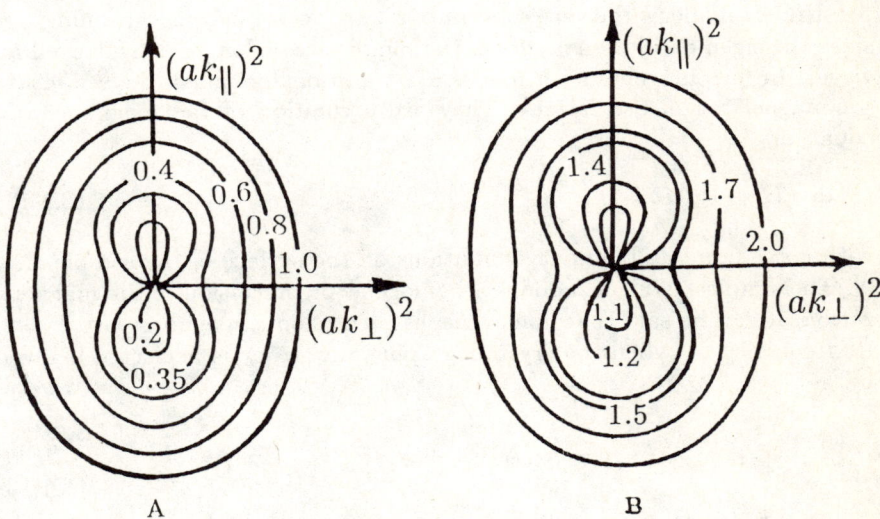

Fig. 3.1. Family of surfaces of constant frequency for an isotropic ferromagnet: the internal field $H_1 = H - 4\pi N_z M$ was chosen equal to one tenth of $H_\mathrm{M} = 4\pi M$ (a); $H_1 = H_\mathrm{M}$ (b). On the y-axis the value $(ak_\parallel)$, on the x-axis the value $(ak_\perp)$. The numbers on the curves designate the ratio $\omega(\boldsymbol{k})/\omega_\mathrm{M}$

Under $\omega(\boldsymbol{k}) \simeq \omega_\mathrm{M}$ a necking arises at the equator of the surface $\omega(\boldsymbol{k}) =$ const and it resembles a dumb-bell. Under further decrease of k this surface resembles two rounded cones with contacting points. Under $k \leq k_\mathrm{M}$, as can be seen, the dependence of $\omega(\boldsymbol{k})$ on the direction $\boldsymbol{k}$ at $k \to 0$ plays an important role. This nonanalyticity is due to the long-range influence of the dipole–dipole interaction.

Concluding the analysis of the dispersion law of magnons (3.1.5) we should like to make two more remarks. First, there is a gap in the $\omega(\boldsymbol{k})$-dependence. The minimum frequency characterizes magnons with $k \to 0$, $\Theta = 0$. Neglecting the anisotropy

$$\omega_{\min} = \omega_H - \omega_M N_z = g(H - 4\pi N_z M) . \tag{3.1.11}$$

As can be seen from (3.1.1), this frequency corresponds to the precession of the magnetic moment in the internal field $\boldsymbol{H}_1$: $\omega_M = gH_1$. Under $H < 4\pi N_z M$ formula (3.1.11) is no longer valid. It follows from (3.1.5) in this case that $\omega^2(\boldsymbol{k})$ becomes negative under $k \to 0$, $\Theta = 0$ and, consequently, the frequency of magnons becomes purely imaginary. This in turn implies an exponential increase of amplitudes, since $a(\boldsymbol{k},t) \sim \exp[-i\omega(\boldsymbol{k})t]$. Therefore, under $H < 4\pi N_z M$ the homogeneously magnetized state of the ferromagnet proves to be unstable and the sample is broken into domains.

The second remark concerns the applicability of (3.1.5). The formula holds true only when $kL \gg 1$, where L is the characteristic size of the sample. Under smaller k the very concept of a wave vector becomes meaningless, since the eigen-oscillation modes of the sample are no longer plane travelling waves. In ferromagnets such modes are traditionally called *Walker modes* (of magnetization oscillation). They are a solution of the magnetostatic equation

$$\operatorname{div}(\boldsymbol{H} + 4\pi \boldsymbol{M}) = 0 , \tag{3.1.12}$$

with corresponding boundary conditions on the surface of the sample [3.1, 2]. The simplest Walker mode is a homogeneous precession of the magnetization. It can be said that homogeneous precession is magnons with $\boldsymbol{k} = 0$. Its frequency ω_0 when the crystallographic anisotropy is neglected is given by

$$\omega_0^2 = \left[\omega_H - \omega_M N_z + \frac{1}{2}\omega_M(N + x + N_y)\right]^2 - \left[\frac{1}{2}\omega_M(N_x - N_y)\right] . \tag{3.1.13}$$

An especially simple expression can be obtained for a sphere: $\omega_0 = gH$. Expressions for the frequencies of other Walker modes can be found in [3.1].

3.1.2 Amplitudes of Three- and Four-Magnon Interaction

Earlier in Sect. 1.2.4 we estimated the amplitudes of four-wave interactions $T(12, 34)$ for the Heisenberg ferromagnet whose energy is given by the exchange Hamiltonian

$$T_{\text{ex}}(\mathbf{12}, \mathbf{34}) \simeq \omega_{\text{ex}}(ak)^2 g/2M \tag{3.1.14}$$

at $k_1 \simeq k_2 \simeq k_3 \simeq k_4 \simeq k$. This formula has been derived from (1.2.22) by substituting for α via ω_{ex} in accordance with (3.1.9).

Now let us find the contribution to the interaction Hamiltonian resulting from the magnetic dipole–dipole interaction. The energy of this interaction is determined only by the magnetization $\boldsymbol{M}$. In addition to $\boldsymbol{M}$, the dipole–dipole Hamiltonian can include the dimensional parameter g, specifying the dynamics of the electron spin system in the magnetic field; this parameter is the ratio of the magnetic moment of the electron to the mechanical moment (1.2.19). Using the two dimensional values M and g a value with dimensionality of frequency can be constructed uniquely: $[gM] = \text{s}^{-1}$ (here, as in Sect. 1.2, square brackets denote the dimensionality of the enclosed value). As can be seen from (3.1.7) the frequency ω_M characterizing the value of the magnetic dipole–dipole interaction differs from this combination by a numeric factor 4π. This factor may be due to the fact that the oscillation of the magnetic moment with the amplitude $\boldsymbol{m}$ generates a magnetic field $\boldsymbol{h}$ with the amplitude $4\pi\boldsymbol{m}$.

Likewise by dimensional analysis we can estemate the contribution of the dipole–dipole interactions to the amplitudes of the 3-magnon interaction $V_M(1, 23)$ and to the 4-magnon interaction $T_M(12, 34)$. Out of the values M and g the combinations with dimensionality V and T can be constructed: $[V_M] = [g^{1/2} M^{1/2}]$, $[T_M] = [g^2]$. Hence the dimensional estimates for these amplitudes are

$$V_M \simeq \sqrt{gM} \to \omega_M \sqrt{g/2M} ,\tag{3.1.15}$$

$$T_M \simeq g^2 \to \omega_M(g/2M) = 2\pi g^2 .\tag{3.1.16}$$

Behind the arrows more exact estimates of the interaction amplitudes are presented with the numerical factors obtained by calculations (Sect. 3.1.3).

Expressions (3.1.14) and (3.1.16) for T_{ex} and T_M on comparison manifest their similar structure: $T \simeq \omega_{\text{eff}}(g/2M)$ where ω_{eff} is the effective frequency of interactions equal to $\omega_{\text{ex}}(ak)^2$ for the exchange interaction and to ω_M for the magnetic dipole–dipole interaction. Evidently, the contribution of the energy of the crystallographic anisotropy (2.3.5) to the interaction Hamiltonian must be proportional to the constant of the anisotropy K_4 (2.3.5) or to the frequency of anisotropy ω_a (3.1.8). Therefore the expressions for the contribution of the crystallographic anisotropy energy V_a and T_a to the amplitudes of 3- and 4-magnon interactions V and T can be obtained from (3.1.15) and (3.1.16) by the substitution $\omega_M \to \omega_a$:

$$V_a \simeq \omega_a gM \to 0 ,\tag{3.1.17a}$$

$$T_a \simeq \omega_a(g/M) \to \delta\omega_a(g/2M) .\tag{3.1.17b}$$

On the right of the arrow are the results of calculating these amplitudes. The interactions of the anisotropy do not generally result in 3-magnon processes so $V_a = 0$. The numerical factor δ depends on the orientation of the equilibrium magnetization with respect to the crystallographic axes. The coefficients δ for the three most symmetric orientations are presented in Table 3.1.

The obtained estimates of the interaction amplitudes are sufficient to calculate the magnon characteristics, showing the integral dependence on the amplitudes of the $\mathcal{H}_{\text{int}}$, e.g. temperature dependence of the magnetization, frequency and attenuation decrement of magnons. However, other nonlinear characteristics of the spin dynamics, which are discussed below, may depend significantly on the fine properties of the amplitudes of the $\mathcal{H}_{\text{int}}$, i.e., the details of their angular dependences, etc. Therefore we have no way out but to calculate $\mathcal{H}_{\text{int}}$ (not the most pleasant of procedures) taking into account all more or less significant interactions - exchange, dipole interactions, anisotropy, etc. The scheme of these calculations is given in Sects. 1.1, 2.3 and 3.4. Here only the results are presented.

3.1.3 Three-Magnon Hamiltonian

As the calculations show, the amplitudes of the Hamiltonian $\mathcal{H}_{\text{int}}$ given by the formula (1.1.25) have the following form:

$$V_{1,23} = \frac{1}{2}[(V_3 + V_2)u_1u_2u_3 + (V_2^* + V_3^*)v_1^*v_2v_3 \\ + v_1^*(V_2u_2v_3 + V_3v_2u_3) + u_1(V_2^*v_2u_3 + V_3^*u_2v_3)] \\ + (V_1v_1^* + V_1^*u_1)(u_2v_3 + v_1u_3) , \tag{3.1.18}$$

$$U_{123} = \frac{1}{2}[V_1u_1(v_2u_3 + u_2v_3) + V_2u_2(u_1v_3 + v_1u_3) \\ + V_2u_3(v_1u_2 + u_1v_2) + V_1^*v_1(v_2u_3 + u_1v_2) \\ + V_2^*v_2(u_1v_3 + v_1u_3) + V_3^*v_3(u_1v_2 + v_1u_2)] , \tag{3.1.19}$$

$$V(\boldsymbol{k}) = -Vk_z(k_x + ik_y)/k^2 = -V\sin\Theta_{\boldsymbol{k}}\cos\Theta_{\boldsymbol{k}}\exp(i\varphi_{\boldsymbol{k}}) ,$$
$$V = \omega_{\text{M}}(g/2M) , \quad V_j = V(\boldsymbol{k}_j) . \tag{3.1.20}$$

Here u_k and v_k are the coefficients of the linear canonical diagonalizing transforms (1.1.19) where the frequency of the magnons $\omega(\boldsymbol{k})$ and coefficients $A(\boldsymbol{k})$ and $B(\boldsymbol{k})$ are given by (3.1.5, 6). The order of magnitude of $\tilde{V}(\boldsymbol{1}, \boldsymbol{23})$ has been estimated above (3.1.15). Nevertheless the exact expressions (3.1.18–20) are essential for our further study. These give the angular dependences $\tilde{V}(\boldsymbol{1}, \boldsymbol{23})$, and, in particular, reflect the fact that $\tilde{V}(\boldsymbol{1}, \boldsymbol{23})$ vanishes if all the wave vectors $\boldsymbol{k}_1, \boldsymbol{k}_2, \boldsymbol{k}_3$ are orientated along or across $\boldsymbol{M}$, i.e. $\Theta(\boldsymbol{k}) = 0$ or $\Theta(\boldsymbol{k}) = \pi/2$.

3.1.4 Four-Magnon Interaction Hamiltonian

The general expression for the amplitudes of the $\mathcal{H}_4$ (1.1.26) is too complicated to be presented here. For this reason we shall study several limiting cases which will be of further use. Firstly we shall consider the situation when ω_M is small as compared to ω_H or $\omega_{ex}(ak)^2$. Here $A(\boldsymbol{k}) \gg |B(\boldsymbol{k})|$, thus the u,v-transformation in $\mathcal{H}_4$ is not necessary and one can take $a(\boldsymbol{k}) = b(\boldsymbol{k})$. Then

$$W_{12,34} = 2(E_{12} + E_{34}) + (C_{13} + C_{14} + C_{23} + C_{24}) \\ - \frac{1}{2}(D_1 + D_2 + D_3 + D_4) + W_a, \tag{3.1.21}$$

$$E_{kk'} = -\omega_{ex}a^2 \boldsymbol{k} \cdot \boldsymbol{k}' g/2M, \quad C_{kk'} = C(\boldsymbol{k} - \boldsymbol{k}'),$$
$$C(\boldsymbol{k}) = (g/M)\omega_M(k_z/k)^2, \quad C(0) = \omega_M N_z g/M,$$
$$D_k = |B_k|g/M = (g/M)\omega_M(k_+/k)^2, \quad k_+ = k_x + ik_y,$$
$$W_a = \delta\omega_a g/2M. \tag{3.1.22}$$

Values for δ are given in Table 3.1.

$$G_{1,234} = -g(B_2 + B_3 + B_4)/3M. \tag{3.1.23}$$

Abandoning the assumption $A(\boldsymbol{k}) \gg |B(\boldsymbol{k})|$ we shall write the expressions for $W(12,34)$ in the case when the wave vectors $\boldsymbol{k}_1$, $\boldsymbol{k}_2$, $\boldsymbol{k}_3$ and $\boldsymbol{k}_4$ satisfy the particular relations:

$$\begin{aligned} S_W(\boldsymbol{k}, \boldsymbol{k}') &= W(\boldsymbol{k}, -\boldsymbol{k}; \boldsymbol{k}', -\boldsymbol{k}')/2 \\ &= W_1\left[u^2(\boldsymbol{k})u^2(\boldsymbol{k}') + v^2(\boldsymbol{k})v^2(\boldsymbol{k}')\right] + 4W_2 u(\boldsymbol{k})u(\boldsymbol{k}')v(\boldsymbol{k})v(\boldsymbol{k}') \\ &\quad - \frac{g}{2M}\Big\{v(\boldsymbol{k})u(\boldsymbol{k}')[u(\boldsymbol{k})u(\boldsymbol{k}')[B^*(\boldsymbol{k}) + 2B^*(\boldsymbol{k}')] \\ &\quad + v(\boldsymbol{k})v^*(\boldsymbol{k}')[B(\boldsymbol{k}') + 2B(\boldsymbol{k})]] \\ &\quad + v(\boldsymbol{k}')u(\boldsymbol{k})[u(\boldsymbol{k})u(\boldsymbol{k}')(2B(\boldsymbol{k}) + B(\boldsymbol{k}')) \\ &\quad + v(\boldsymbol{k})v(\boldsymbol{k}')(2B(\boldsymbol{k}') + B(\boldsymbol{k}))]\Big\}, \end{aligned} \tag{3.1.24a}$$

$$\begin{aligned} T_W(\boldsymbol{k}, \boldsymbol{k}') &= [W(\boldsymbol{k}, \boldsymbol{k}'; \boldsymbol{k}, \boldsymbol{k}') + W(\boldsymbol{k}, -\boldsymbol{k}'; \boldsymbol{k}, -\boldsymbol{k}')]/4 \\ &= 2W_1 \operatorname{Re}\{u(\boldsymbol{k})u(\boldsymbol{k}')v(\boldsymbol{k})v(\boldsymbol{k}')\} \\ &\quad + W_2\left[u(\boldsymbol{k})^2 + |v(\boldsymbol{k})|^2\right]\left[(u(\boldsymbol{k}'))^2 + |v(\boldsymbol{k}')|^2\right] \\ &\quad - \frac{g}{2M}\left[(u(\boldsymbol{k})^2 + |v(\boldsymbol{k})|^2\right]\operatorname{Re}\left\{[B(\boldsymbol{k}) + 2B(\boldsymbol{k}')]v^*(\boldsymbol{k}')u(\boldsymbol{k})\right\} \\ &\quad + (u(\boldsymbol{k}')^2 + |v(\boldsymbol{k}')|^2)\operatorname{Re}\{[2B(\boldsymbol{k}) + B(\boldsymbol{k}')]v^*(\boldsymbol{k})u(\boldsymbol{k}')\}], \end{aligned} \tag{3.1.24b}$$

$$\begin{aligned} W_1 &= [\omega_{ex}(ak)^2 - \omega_M]g/2M + W_a, \\ W_2 &= \omega_M(N_z - 1)g/2M + W_a. \end{aligned} \tag{3.1.25}$$

Besides, an essential contribution to the amplitudes of the four-magnon interaction is made by the amplitudes of the 3-wave interaction in the second order of perturbation theory. The general expression for this contribution is given in Sect. 1.1.4. Now we are interested only in its particular cases at $k_2 = -k_1$, $k_4 = -k_3$ and $k_2 = k_3$, $k_1 = k_4$:

$$S_{12} = S_W(k_1, k_2) - \frac{2U^*(k_0, k_1, -k_1)U(k_0, k_2, -k_2)}{\omega(k_0) + 2\omega(k)}$$
$$- \frac{2\mathrm{Re}\{V^*(k_0, k_1, -k_1)V(k_0, k_2, -k_2)\}}{\omega(k_0) - 2\omega(k)}$$
$$- \frac{4V(k_1; k_2, k_5)V^*(k_2; k_1, -k_5)}{\omega(k_5)}$$
$$- \frac{4V(k_1; -k_2, k_6)V^*(k_2; k_1, -k_6)}{\omega(k_6)}, \qquad (3.1.26)$$

$$T_{12} = T_W(k_1, k_2) - \frac{|U(k_2, k_1, -k_6)|^2}{\omega(k_6) + 2\omega(k_1)}$$
$$- \frac{|U(-k_1, k_2, k_5)|^2}{\omega(k_5) + 2\omega(k)} - \frac{|V(k_6; k_1, k_2)|^2}{\omega(k_6) - 2\omega(k)}$$
$$- \frac{|V(k_5; -k_1, k_2)|^2}{\omega(k_5) - 2\omega(k)} - \frac{4\mathrm{Re}\{V(k_1; k_1, 0)V(k_2; k_2, 0)\}}{\omega(k_0)}$$
$$- \frac{2[|V(k_1; k_2, k_5)|^2 + |V(k_2; k_1, -k_5)|^2]}{\omega(k_5)}$$
$$- \frac{2[|V(k_1; -k_2, k_6)|^2 + |V(k_2; -k_1, k_6)|^2]}{\omega(k_6)}. \qquad (3.1.27)$$

Here $k_5 = k_1 - k_2$, $k_6 = k_1 + k_2$, ω_0 is the frequency of the homogeneous precession and it is assumed that $\omega(k_1) = \omega(k_2) = \omega(k)$. The second and third terms in (3.1.27) are due to the interaction of magnon pairs with $\pm k_1$ and $\pm k_2$ via "virtual" homogeneous precession; the last two terms are caused by the interaction via "virtual" magnons with $k = k_1 \pm k_2$. The order of magnitude of the terms (3.1.26-27) is the same as that of the direct contribution of the dipole–dipole interaction to W, i.e. similar to the order of magnitude of πg^2 in full accordance with the estimation of (3.1.16). However, in an important specific case when $k_1 \perp M$ and $k_2 \perp M$ (i.e. $\Theta(k_1) = \Theta(k_2) = \pi/2$) terms of this order of magnitude make no contribution to $S(k, k')$ and $T(k, k')$ at all.

It must be noted that in (3.1.24-28) the frequency of the anisotropy ω_a is multiplied by the numerical factor 9 – 11 and at the same time ω_M is divided by two. Thus, the relative contribution of ω_a compared to ω_M becomes 20 times as great which leads to a strong crystallographic anisotropy of the four-wave interaction in "almost isotropic ferromagnets" with a small ratio ω_a/ω_M. Note that in the classical ferrimagnet YIG the ratio ω_a/ω_M is about 1/20.

The expression for W includes the demagnetization factor N_z (z is the magnetization direction). As a result W is dependent on the shape of the sample and magnetization direction. Thus, for a normally magnetized disc $N_z = 1$; for a tangentially magnetized disc $N_z = 0$. All of the above-mentioned offers experimenters a rare opportunity to change at will the amplitudes of the interaction Hamiltonian of the magnons. For the benefit of the theoreticians cooperating with the experimenters one may even provide the knobs of the installation with special inscriptions notifying what particular coefficients of the Hamiltonian are being changed.

3.2 Hamiltonian Function of Magnons in Antiferromagnets

3.2.1 Magnon Spectrum in Antiferromagnets (AFM)

The simplest AFMs have two magnetic sublattices and accordingly two branches of magnon spectra. The quadratic part of the Hamiltonian is of a standard form (1.1.18):

$$\mathcal{H}_2 = \sum_{\bm{k}}[\omega(\bm{k})a^*(\bm{k})a(\bm{k}) + \Omega(\bm{k})b^*(\bm{k})b(\bm{k})] \,. \tag{3.2.1}$$

It is given in order to define the designations of magnon frequencies in the two branches of spectra $\omega(\bm{k})$ and $\Omega(\bm{k})$ and of the normal canonical variables $a(\bm{k})$, $b(\bm{k})$ where the quadratic part of the Hamiltonian is diagonal. In uniaxial antiferromagnets with anisotropy of the "easy axis" type (AFM EA) the field of crystallographic anisotropy $\bm{H}_a$ tends to maintain magnetization parallel to this axis (it is usually called the z-axis).

By analogy with ferromagnets the frequencies of magnons with $k \to 0$ may be conjectured to correspond to the frequencies of the precession of sublattices magnetization in the field $\bm{H}_a$, i.e. $\omega_0 = \Omega_0 = \omega_a$, where

$$\omega_a = gH_a \,. \tag{3.2.2}$$

This, however, is wrong. Upward oriented magnetization of one sublattice $\bm{M}_1$ is influenced by the field of anisotropy $\bm{H}_{a1}$, also directed upwards: $\bm{H}_{a1} = \bm{H}_1$. Another field ($\bm{H}_{a2} = -\bm{H}_a$) acts upon the other sublattice with $\bm{M}_2 = -\bm{M}_1$ directed downward. As a result the sublattices "desire" to be oppositely precessed. In this case antiparallelism of $\bm{M}_1$ and $\bm{M}_2$ is inevitably violated, impeded by a strong exchange interaction between the sublattices. As a result it turns out that [3.3]

$$\omega_0^2 = \Omega_0^2 = (\omega_a + \omega_{ex})^2 - \omega_{ex}^2 \simeq 2\omega_a\omega_{ex} \,, \tag{3.2.3}$$

where the frequency $\omega_{ex} = gH_{ex}$ characterizes the value of the antiferromagnetic exchange between sublattices. Sometimes the dimensionless "exchange constant" B acts as such a characteristic: $H_{ex} = BM_{ex}$.

If the external magnetic field is applied to the AFM EA along the main axis, the equivalence of the sublattices bringing about the coincidence of the frequencies ω_0 and Ω_0 is violated. Here

$$\begin{aligned} \Omega_0 &= \sqrt{2\omega_a \omega_{ex}} + \omega_H\,, \\ \omega_0 &= \sqrt{2\omega_a \omega_{ex}} - \omega_H\,, \quad \omega_H = gH\,. \end{aligned} \quad (3.2.4)$$

In the critical field $H_{cr} = \sqrt{2H_a H_{ex}}$ the frequency ω_0 becomes zero and the instability of magnons increases, which leads to a changed ground state, i.e. to the overturning of the sublattices [3.1].

In the uniaxial antiferromagnets with the anisotropy of the "easy plane" type (AFM EP) the anisotropy develops the moments $\boldsymbol{M}_1$ and $\boldsymbol{M}_2$ into the plane perpendicular to the axis $\boldsymbol{c}$. Since the moments in this plane can oscillate almost freely one of the magnon frequencies under $k \to 0$ proves to be small – of the order of gH. Calculations following the below described procedure show

$$\begin{aligned} \omega^2(\boldsymbol{k}) =& \omega_H(\omega_H + \omega_D) + 2\omega_{ex}(\omega_n + \omega_{ph}) + v_1^2 k_z^2 + v_2^2 k^2\,, \\ \Omega^2(\boldsymbol{k}) =& 2\omega_a \omega_{ex} + \omega_D(\omega_H + \omega_D) \\ &+ 2\omega_{ex}(\omega_n + \omega_{ph}) + v_1^2 k_z^2 + v_2^2 k^2\,. \end{aligned} \quad (3.2.5)$$

Here $\boldsymbol{H}$ is perpendicular to the direction of the main axis; v_1 and v_2 are the longitudinal and lateral velocities of magnons with order of magnitude $v_{1,2} \simeq \omega_{ex} a$, where a is the lattice constant; $\omega_D = gH_D$, H_D is the Dzyaloshinsky field due to the specific relativistic interaction between the sublattices. The value of H_D is usually of the order of several kOe. In some AFM because of symmetry $H_D = 0$.

Table 3.2. Phenomenological constants determining magnon frequencies in antiferromagnets with the anisotropy of the "easy plane" type

Crystal	H_{ex}, kOe	H_a, kOe	H_D, kOe	H_Δ^2, (kOe)^2
MnCO$_3$	320	3.04	4.4	5.8/T(K)
CsMnF$_3$	350	2.48	0	6.4/T(K)+0.3
FeBO$_3$	3000	5.3	100	4.9

The two terms in (3.2.5) $\omega_{ex}\omega_n$ and $\omega_{ex}\omega_{ph}$ are due to the smallness of respective electron exchange with nuclear spins and crystal lattice. However, in (3.2.5) these frequencies are multiplied by the greatest frequencies ω_{ex}, and the contribution of these small interactions to frequency $\omega(\boldsymbol{k})$ may appear quite significant. This phenomenon in AFM is traditionally named the *exchange amplification of small interactions* AFM. The frequency ω_n corresponds to the precession of the magnetic moment of an electron in an effective magnetic field $\boldsymbol{H}$ due to hyperfine interaction of electrons with the nucleus

$$\omega_n = gH_n = A<I>/\hbar . \qquad (3.2.6)$$

A designates the constant of the superfine interaction, $<I_z>$ is the mean value of the nuclear spin inversely proportional to the temperature. For manganese nuclei, for instance, $H_\Delta \simeq (10/T)$ Oe and "the nuclear gap" $H_\Delta = \sqrt{2H_{ex}H_n}$ is of the order of several kOe under $T = 4$ K (Table 3.2). The question may arise as to the cause of the gap in the magnon spectrum, which is equal to gH_Δ, if the problem in this case has an axial symmetry with respect to the rotation of the movements of sublattices in the easy plane. The answer is as follows: The frequency of the nuclear spin precession ω_n is much smaller than the frequencies of the electron subsystem precession gH_Δ, gH_D... Therefore, nuclear spins do not partake in the movement of the electron spins. They are arranged along the mean values of magnetizations of the sublattices. Therefore, the axial symmetry of the crystal is broken which gives rise to effective magnetic fields $\boldsymbol{H}_{n1} = \boldsymbol{H}_n$ and $\boldsymbol{H}_{n2} = -\boldsymbol{H}_n$ acting respectively on the sublattices $\boldsymbol{M}_1$ and $\boldsymbol{M}_2$. These fields by (3.2.3) for AFM EA bring about electron spin precession with frequency $\sqrt{2\omega_{ex}\omega_n}$. The same is true about magnetostriction leading to the uniaxial deformation of the crystal which fails to follow the precession of the sublattice moments. This deformation in its turn results in the emergence of effective fields $\boldsymbol{H}_{ph1} = -\boldsymbol{H}_{ph2} = \boldsymbol{H}_{ph}$ in the easy plane. The field $\boldsymbol{H}_{ph}$ is small; it is proportional to the square of the magnetostriction constant. The exchange amplification may result in a considerable contribution of the magnetostriction to the magnon frequency.

In the cubic AFM (e.g. in RbMnF$_3$) the expressions for magnon frequencies have the form

$$\begin{aligned}\omega^2(\boldsymbol{k}) &= 3\omega_a\omega_{ex} + \omega_\Delta^2 + v^2k^2 , \quad \omega_\Delta = gH_\Delta , \\ \Omega^2(\boldsymbol{k}) &= \omega^2(\boldsymbol{k}) - (3/2)\omega_a\omega_{ex} + \omega_\Delta^2 + v^2k^2 .\end{aligned} \qquad (3.2.7)$$

It is assumed here that $\boldsymbol{H}\|[100]$ and its value is greater than the overturn field $\sqrt{(3/2)H_aH_{ex}}$.

In all the expressions for magnon frequencies in AFM the dipole–dipole interaction has not been allowed for. The external magnetic field and relativistic interactions result in the small magnetic moment in AFM $\boldsymbol{M} = \boldsymbol{M}_1 + \boldsymbol{M}_2$. Accordingly, the relative contribution of the dipole–dipole interaction to the magnon frequencies as a rule proves to be H_{ex}/H times smaller; this factor amounts to $10^3 \div 10^4$ for most AFM. However this interaction can prove essential when studying fine nonlinear properties of antiferromagnetic magnons.

3.2.2 Interaction Hamiltonian in "Easy Plane" Antiferromagnets

Experiments on nonlinear properties of magnons in the "easy plane" antiferromagnets (AFM with the anisotropy of "the easy–plane" type) and in cubic AFM are of particular interest due to low-lying magnon branches whose frequencies are in the convenient range below 50 gHz. The calculation of the Hamiltonian is the simplest for AFM EP where the second branch of magnons lies much higher than the other one: $\Omega(\boldsymbol{k}) \gg \omega(\boldsymbol{k})$. For AFM EP we shall write the results of such calculations for the part of $\mathcal{H}_{\text{int}}$ of great interest for us:

$$\mathcal{H}_3 = \sum_{1=2+3} [\frac{1}{2} V_q^{(1)} b_1 a_2^* a_3^* + V_q^{(2)} a_1 b_2^* a_3^* + \text{c.c.}]$$
$$+ \frac{1}{2} \sum_{1+2+3=0} [U_q b_1^* a_2^* a_3^* + \text{c.c.}] + \dots , \quad q = (\boldsymbol{k}_1, \boldsymbol{k}_2, \boldsymbol{k}_3) , \tag{3.2.8}$$

$$\mathcal{H}_4 = \frac{1}{2} \sum_{1+2=3+4} W_p a_1^* a_2^* a_3 a_4 , \quad p = (\boldsymbol{k}_1, \boldsymbol{k}_2, \boldsymbol{k}_3, \boldsymbol{k}_4) , \tag{3.2.9}$$

$$V_q^{(1)} = -\frac{\sqrt{g\omega_{\text{ex}}}}{4\sqrt{M\omega_2\omega_3\Omega_1}} \omega_H (\Omega_1 + \omega_2 + \omega_3) ,$$

$$V_q^{(2)} = -\frac{\sqrt{g\omega_{\text{ex}}}}{4\sqrt{M\omega_1\omega_3\Omega_2}} \omega_H (\omega_1 - \Omega_2 + \omega_3) , \tag{3.2.10}$$

$$U_q = -\frac{\sqrt{g\omega_{\text{ex}}}}{4\sqrt{M\omega_2\omega_3\Omega_1}} \omega_H (\Omega_1 - \omega_2 - \omega_3) ,$$

$$W_p = \frac{9\omega_{\text{ex}}}{4M\sqrt{\omega_1\omega_2\omega_3\omega_4}} [\omega_1^2 + \omega_2^2 + \omega_3^2 + \omega_4^2$$
$$- \omega_{1-3}^2 - \omega_{2-3}^2 - \omega_{1+2}^2 - \omega_{3+4}^2] . \tag{3.2.11}$$

As before, we used the shorthand notation

$$\Omega_j = \Omega(\boldsymbol{k}_j) , \quad \omega_j = \omega(\boldsymbol{k}_j) , \quad \omega_{i-j} = \omega(\boldsymbol{k}_i - \boldsymbol{k}_j) .$$

It should be recalled that for describing scattering processes of the $2 \to 2$ type not only the Hamiltonian $\mathcal{H}_4$ is to be taken into consideration, but also the Hamiltonian $\mathcal{H}_3$ in the second order of perturbation theory. It can be shown that the contribution of the processes with the participation of the virtual waves from the upper branch is as great as others in $\omega_{\text{ex}}/\omega(\boldsymbol{k})$ times. The Hamiltonian of the processes is explicitly written in (3.2.8). Using (1.1.32) we obtain the expression for the effective amplitudes of 4-magnon processes

$$T_{\mathrm{p}} = W_{\mathrm{p}} - \frac{V^{(1)}(5;1,2)V^{(1)*}(5;3,4)}{\omega(\boldsymbol{k}_1+\boldsymbol{k}_2) - 2\omega(\boldsymbol{k})}$$
$$+ \frac{U(-5,2,1)U^*(-5,3,4)}{\omega(\boldsymbol{k}_5) + 2\omega(\boldsymbol{k})}$$
$$+ \frac{V^{(2)}(1;6,3)V^{(2)*}(4;6,-2) + V^{(2)*}(3;-6,1)V^{(2)}(2;6,4)}{\omega(\boldsymbol{k}_1-\boldsymbol{k}_3)} \quad (3.2.12)$$
$$+ \frac{V^{(2)}(2;7,1)V^{(2)*}(4;6,2) + V^{(2)*}(3;6,-1)V^{(2)}(1;-7,2)}{\omega(2\boldsymbol{k})} .$$

In this expression $\boldsymbol{k}_5 = \boldsymbol{k}_1+\boldsymbol{k}_2$, $\boldsymbol{k}_6 = \boldsymbol{k}_1-\boldsymbol{k}_3$, $\boldsymbol{k}_7 = \boldsymbol{k}_2-\boldsymbol{k}_1$. From (3.2.10–12) the diagonal contribution to T_p due to $\mathcal{H}_3$ is of the same order of magnitude as W_p. The general expression for T is rather cumbersome. We shall write it for the most interesting case when $\omega(1) = \omega(2) = \omega(\boldsymbol{k})$ and $\boldsymbol{k}_1 = \boldsymbol{k}_3$, $\boldsymbol{k}_2 = \boldsymbol{k}_4$ or $\boldsymbol{k}_1 + \boldsymbol{k}_2 = 0$:

$$S(\boldsymbol{k},\boldsymbol{k}') = T(\boldsymbol{k},-\boldsymbol{k};\boldsymbol{k}',-\boldsymbol{k}')/2 \,, \quad T(\boldsymbol{k},\boldsymbol{k}') = T(\boldsymbol{k}\boldsymbol{k}',\boldsymbol{k}\boldsymbol{k}')/2 \,,$$
$$S(\boldsymbol{k},\boldsymbol{k}') = T(\boldsymbol{k},\boldsymbol{k}') =$$
$$- \frac{g^2\omega_{\mathrm{ex}}}{8M\omega^2(\boldsymbol{k})}\left[\omega_0^2 + \omega_{\mathrm{H}}^2 \frac{3\Omega_0^2 - 4\omega^2(\boldsymbol{k})}{\Omega_0^2 - 4\omega^2(\boldsymbol{k})}\right] . \quad (3.2.13)$$

3.2.3 Nuclear Magnons in "Easy Plane" Antiferromagnets

The concept of *Nuclear Magnons* (NM) was first put forward by *de Gennes* et al. [3.3]. NM are the collective oscillations of spins of nuclei, e.g. Mn nuclei in AFM $MnCO_3$. There is, of course, no direct exchange interaction between the nuclear spins and their magnetic dipole interaction is negligibly small because of the smallness of the magnetic moment of the nuclei. Nuclear spin motion becomes collective as a result of the indirect exchange interaction described by *Suhl* and *Nakamura* in 1958. It arises in the second order of perturbation theory in the hyperfine interaction (SFI) of nuclear spins with electron spins. The spin deflection of the i-th nucleus deflects under SFI the electron spin of the i-th atom; this perturbation propagates in the electron spin system and, in its turn, acts on other nuclear spins through SFI. The resulting correlation of the nuclear spin motion is the nuclear magnon. A detailed presentation of theoretical concept and experimental data on NM is given, for example, by *Tulin* [3.4].

In two-sublattice AFM EP there are two branches of electron magnons $\omega_e(\boldsymbol{k})$ and $\Omega_e(\boldsymbol{k})$, and, respectively, two NM branches $\omega_n(\boldsymbol{k})$ and $\Omega_n(\boldsymbol{k})$:

$$\omega_n^2(\boldsymbol{k}) = \Omega_n^2(0)[1 - \Delta^2/\omega_e^2(\boldsymbol{k})] \,, \quad (3.2.14)$$
$$\Omega_n^2(\boldsymbol{k}) = \Omega_n^2(0)[1 - \Delta^2/\Omega_e^2(\boldsymbol{k})] \,. \quad (3.2.15)$$

$\omega_n^2(0)$ here designates the frequency of the magnetic resonance in the hyperfine field of the electron ($\Omega_n(0)/2\pi$ for AFM $CsMnF_3$, $MnCO_3$ and $RbMnF_3$

equals, respectively, 666 MHz, 640 MHz and 686 MHz), $\Delta^2 = 2\omega_{ex}\omega_n$, where ω_{ex} is the frequency of the inter-sublattice exchange and ω_n is the precession frequency of the electron magnetic moment in the effective magnetic field due to SFI and to the ordering of nuclear spins

$$\omega_n = gH_n = A<I_z>/\hbar \simeq AH/T \ .$$

The value Δ is the nuclear gap (which is amplified by exchange) in the spectrum of electron magnons (see (3.2.6) and Table 3.2).

The Hamiltonian of NM interaction is of no use for us and will therefore not be described here.

3.3 Comments at the Road Fork

Dear reader, dear colleague! Before we go further, let us take a short rest and look back: in Chap. 1 we got to know the classical Hamiltonian formalism for nonlinear waves of arbitrary nature and wrote the Hamiltonian equations of motion for the weakly nonlinear waves in almost conservative media. We understood that all the required information (and almost nothing superfluous) is contained in the Hamiltonian of the nonlinear wave system. The second chapter briefly outlined the general information on the physics of magnetodielectrics useful when discussing the nonlinear behavior of their spin system. As for the Hamiltonian function of the magnons in ferro-, antiferro- and ferrimagnets it was presented in Sects. 3.1–3 in a form sufficient for further consideration and to start the study of the nonlinear dynamics of magnons in Chap. 4. Thus, the end of the present Chapter (Sect. 3.4) can be skipped by those readers who are not interested in the recipes of the magnetic "cuisine". We invite others to consider in the following section the equations of motion for the spin waves and to study the simplest way of calculating the magnon Hamiltonian.

3.4 Calculation of Magnon Hamiltonian

3.4.1 Equation of Motion of Magnetic Moment

To derive consistently equations of motion of the magnetic moment $M(r,t)$ at finite temperatures is an extremely complicated task even for the long-wave variations. To solve this problem we would have to determine the temperature dependences of the amplitudes of interaction, magnon frequencies, to find the magnon damping and to obtain a host of other useful data. If we had made this attempt strictly and consistently we would have got stuck in the very beginning and given up our ultimate aim, i.e. to obtain a clear and simple method for describing nonlinear dynamics of the magnons under low and intermediate temperatures. Consequently, we shall make several assumptions difficult to check but simple and highly "natural" and therefore traditional. Let us cite from [3.1], p.44: "Owing to strong exchange interaction between the spins of individual atoms of the ferromagnet its magnetic moment is in good approximation "rigid" if only the ferromagnetic temperature is sufficiently low". In other words, the modulus of the magnetic moment density vector can, although very weakly, depend on time. Therefore, for the first approximation the time variation in the density of the magnetic moment must to be of precession character, i.e. must obey the law

$$\partial \boldsymbol{M}(\boldsymbol{r},t)/\partial t = g[\boldsymbol{H}_{\text{eff}}(\boldsymbol{r},t) \times \boldsymbol{M}(\boldsymbol{r},t)] \,, \tag{3.4.1}$$

where $\boldsymbol{H}_{\text{eff}}(\boldsymbol{r},t)$ is a certain vector which will be henceforth referred to as the *effective magnetic field*. Likewise, at the accepted level of rigor, we can assert that the equations of motion (3.4.1) must be conservative, i.e. conserve the total energy of the ferromagnet W. Indeed, if in our model there is nothing but the magnons described by (3.4.1), their total energy must be conserved. Since W is a functional of magnetization, we have that

$$\frac{dW}{dt} = \int \frac{\delta W}{\delta \boldsymbol{M}(\boldsymbol{r},t)} \cdot \frac{\partial \boldsymbol{M}(\boldsymbol{r},t)}{\partial t} d\boldsymbol{r} \,.$$

Substituting here $\partial \boldsymbol{M}/\partial t$ from the equations of motion (3.4.1), we have

$$\frac{dW}{dt} = g \int \frac{\delta W}{\delta \boldsymbol{M}} \cdot [\boldsymbol{M} \times \boldsymbol{H}_{\text{eff}}]) d\boldsymbol{r} \,.$$

For this integral to be zero under any dependence $\boldsymbol{M}(\boldsymbol{r})$, the vector $\boldsymbol{H}_{\text{eff}}$ must be parallel to the vector $\delta W/\delta \boldsymbol{M}$. The proportionality coefficient can be found by considering some simple situation with the present equation of motion, for example, the uniform precession of the magnetic moment in a homogeneous field $\boldsymbol{H}$. Then W is the energy of the interaction with the external field

$$W = -\int \boldsymbol{H} \cdot \boldsymbol{M}(\boldsymbol{r}) d\boldsymbol{r} \,. \tag{3.4.2}$$

Calculating the functional derivative by the rule (A 1.2) we obtain

$$\delta W/\delta \boldsymbol{M} = -\boldsymbol{H} \,. \tag{3.4.3}$$

It implies that the sought-for factor is minus one

$$\boldsymbol{H}_{\text{eff}} = -\delta W/\delta \boldsymbol{M} \,. \tag{3.4.4}$$

Therefore we obtain the equations of motions

$$\frac{\partial \boldsymbol{M}(\boldsymbol{r},t)}{\partial t} = g[\frac{\delta W}{\delta \boldsymbol{M}} \times \boldsymbol{M}] \,. \tag{3.4.5}$$

Certainly, the easy attitude to manipulations leading to (3.4.5), an attitude I share with the authors of [3.1], by no means guarantees their correctness. All of us took great pains to sweep under the carpet many delicate and complicated problems. Some of them will be discussed later. Now we shall proceed from (3.4.5) assuming that they give a good first approximation for the description of the magnons (spin waves).

3.4.2 Canonical Variables for Spin Waves in Ferromagnets (FM)

In (3.4.5) we shall pass to the circular variables

$$M_\pm = M_x \pm i M_y \,, \quad M_z^2 = M_0^2 - M_+ M_- \,,$$

$$\partial M_+(\boldsymbol{r},t)/\partial t = 2ig M_z(\boldsymbol{r},t)(\delta W/\delta M_-) \,. \tag{3.4.6}$$

The z-axis direction should be selected along the equilibrium value of the magnetization. Then at small variations in amplitudes of the magnetic moment the M_+-values will be small and M_z will approach the length of the vector M, i.e. M_0. Comparison of (1.1.7) and (3.4.6) shows that these equations, in the approximation linear in M, have the Hamiltonian form if the following values a and a^* are taken as canonical variables:

$$M_+/\sqrt{2gM_0} \,, \quad M_-/\sqrt{2gM_0} \,.$$

Therefore, the canonical variables should better be sought in the following form:

$$M_+ = a f(a^* a)\sqrt{2gM_0} \,, \quad M_- = a^* f(a^* a)\sqrt{2gM_0} \,, \tag{3.4.7}$$

where $f(0) = 1$. Substituting (3.4.7) into (3.4.6) we obtain an equation for $a(\boldsymbol{r},t)$ which must coincide with the canonical one (1.1.7). The resulting differential equation for $f(|a|^2)$

$$f^2(x) + 2x f(x)[df(x)/dx] = 2g\sqrt{M_0^2 - x f^2(x)}$$

has a unique solution satisfying the condition $f(0) = 1$,

$$f(x) = \sqrt{1 - gx/2M_0} \,.$$

Thus we have expressed the "natural variables" of ferromagnets $\boldsymbol{M}$ in terms of the canonical variables

$$M_+ = a\sqrt{2gM_0(1 - ga^*a/2M_0)} \,,$$

$$M_- = M_+^* \,, \quad M_z = M_0 - g a^* a \,. \tag{3.4.8}$$

The energy of the ferromagnet W expressed in terms of the canonical variables becomes the Hamiltonian function (Hamiltonian) $\mathcal{H}(a^*, a)$. It can be seen easily that the transformation (3.4.8) is a classical analog of the Holdstein-Primakoff transformation [3.1]. They were first employed in 1960 for analyzing the nonlinear dynamics of magnons by *Schlömann* et al. [3.5].

This choice of the canonical variables is certainly not the only possible one. Thus, it can be shown (see Problem 3.1) that for (3.4.5) there exist other canonical variables and in terms of these variables the vector $\boldsymbol{M}$ could be expressed as follows

$$M_z + iM_x = M_0 \sqrt{1 + g|b^* - b|^2/2M_0} \exp\{i(b^* + b)\sqrt{g/2M_0}\} ,$$

$$M_y = i(b^* - b)\sqrt{gM_0/2} . \tag{3.4.9}$$

These formulations are a classical analogue of the representation of the spin operator in terms of Bose-operators suggested by *Bar'yakhtar* and *Yablonsky* [3.6].

Comparison of (3.4.8) and (3.4.9) shows that the variables of Holstein-Primakoff a, a^* and Bar'yakhtar and Yablonsky b, b^* coincide in the linear approximation. The advantages of the various representations the magnetization in terms of canonical variables for the solution of specific problems will be discussed later. Now we only want to emphasize once more that there is a wide choice of variables in which the equations of motion have a canonical form.

3.4.3 Calculation of Frequencies and Interaction Amplitudes of Waves

Now we are ready to obtain the frequencies of the magnons and the amplitudes of the three- and four-magnon interactions in ferro-, antiferro- and ferrimagnets. Summing up the above, we may say that it is a problem which can be solved in 7 steps.

The first step is to find the canonical variables. The solution of this problem is not formalized and to solve it is to a certain extent an art. However, for a wide scope of interesting physical situations the canonical variables have already been obtained. For references see Sect. 1.3.3. For spin waves in a magnetic the canonical variables have been presented in (3.4.8–9).

The second step is to obtain the Hamiltonian function (Hamiltonian) of the system. As a rule this Hamiltonian is the energy of the system expressed in terms of the canonical variables. This refers also to the spin system of the magnetodielectrics. Phenomenological expression for the energy of the antiferromagnets includes additional terms of relativistic origin. It will be given later (3.4.10).

The third step is to find an equilibrium state and to expand the Hamiltonian in a power series in the canonical variables $a_j(\boldsymbol{r})$, $a_j(\boldsymbol{r})$, describing deviations from the state.

The fourth step is almost as simple. We must pass to the $\boldsymbol{k}$-representation (to the variables $a_j(\boldsymbol{k})$) by means of the Fourier-transform (1.1.12). After that, however, do not fail to divide the Hamiltonian by the volume of the sample.

The fifth step is to find the normal canonical variables $b_j(\boldsymbol{k})$, where the quadratic part of the Hamiltonian is diagonal. Generally, this problem can be solved via a linear canonical u,v-transformation (1.1.17). For the ferromagnets which have only one magnon branch the expression for the coefficients $u(\boldsymbol{k})$ and $v(\boldsymbol{k})$ is given by (1.1.21). For two-sublattice antiferromagnets (where there are two branches of magnons) the problem is reduced to obtaining the diagonalizing u,v-transformation for the Hamiltonian $[A,B]$-matrix with the dimensions 4×4. This transformation is also a 4×4 matrix and to obtain it is not trivial. A little trick to achieve this will be described later. In a twenty-sublattice ferrimagnet YIG there are twenty magnon branches. Therefore the Hamiltonian $[A,B]$-matrix, generally speaking, has the dimensionality 40×40 and the u,v-matrix diagonalizing it must have the same dimensionality. Although the dimensionality of both matrices can be reduced to 20×20 (due to the collinearity of the sublattices), the "direct" obtaining of the $[u,v]$-matrix is technically impossible. However, in [3.7] we show that in this problem there are latent numerically small parameters that make it possible to obtain analytically with acceptable accuracy both the diagonalizing $[u,v]$-matrix and the dispersion laws of all the twenty branches of the spin waves in YIG. The exact procedure of that is given in [3.7].

The sixth step is to change over in the interaction Hamiltonian from the initial canonical variables $a_j(\boldsymbol{k})$ to the normal canonical variables $b_j(\boldsymbol{k})$. For multisublattice magnets this procedure, though generally simple, becomes very cumbersome. Additional problems of computation due to the cancellation of big terms in the final expression arise for antiferromagnets.

The seventh and the last step is to change over the interaction Hamiltonian via nonlinear canonical transformation (1.1.31) from the variables $b_j(\boldsymbol{k})$ to the variables $c_j(\boldsymbol{k})$ where there are no non-resonance terms in the three-magnon Hamiltonian. This step is required for calculation of the contribution of the three-magnon processes to the four-magnon Hamiltonian in the second order of the perturbation theory.

The successive performance of all the 7 steps for ferromagnets presents no difficulties. The results of these calculations are presented in Sect. 3.1.

3.4 Calculation of Magnon Hamiltonian

On calculating the frequencies of magnons and the amplitudes of their interaction in AFM EP one can proceed from the following phenomenological expression for the energy in a two-sublattice model

$$W = \frac{1}{2} \int \{[BM^2 + \alpha_1 |\frac{\partial \boldsymbol{L}}{\partial \boldsymbol{r}}|^2 + \alpha_2 |\frac{\partial \boldsymbol{M}}{\partial \boldsymbol{r}}|^2] + bM_z^2 + aL_z^2 - 2\beta(\boldsymbol{n}\cdot[\boldsymbol{M}\times\boldsymbol{L}]) - 2\boldsymbol{M}\cdot\boldsymbol{H}\} d\boldsymbol{r} \ . \quad (3.4.10)$$

Here $\boldsymbol{n} = \boldsymbol{r}/r$, $\boldsymbol{M}_1$ and $\boldsymbol{M}_2$ designate the magnetization of sublattices, $M_1 = M_2 = M_0$, $\boldsymbol{M} = \boldsymbol{M}_1 + \boldsymbol{M}_2$, $\boldsymbol{L} = \boldsymbol{M}_1 - \boldsymbol{M}_2$, $\alpha_{1,2} > 0$ are exchange constants, $\beta > 0$ denotes the Dzyaloshinsky constant [3.1], $a > 0$ and b are the anisotropy constants.

It is convenient to proceed as follows: first find the configuration of the moments $\boldsymbol{M}_1$ and $\boldsymbol{M}_2$ in the ground state from the minimum energy condition; second, in two axillary coordinate systems oriented along the respective magnetizations $\boldsymbol{M}_j$ ($j = 1, 2$) pass to the canonical variables (3.4.8) $c_j(\boldsymbol{r})$ and $c_j^*(\boldsymbol{r})$ ($j = 1, 2$) in each sublattice as in FM. Third, to change over to Fourier-representation: $c(\boldsymbol{r}) \to c(\boldsymbol{k})$, and go to the symmetric variables $d_{1,2} = (c_1 \pm c_2)/2$. In this case the quadratic part of the Hamiltonian will be quasi-diagonal, the variables d_1 and d_1^* will describe one magnon branch and d_1 and d_1^* will describe the other one; then through the u, v-transformation in each branch the quadratic parts of the Hamiltonian must be diagonalized; and, finally, the expressions for $\omega(\boldsymbol{k})$, $\Omega(\boldsymbol{k})$ and $\mathcal{H}_{\text{int}}$ will be obtained in the variables described in Sect. 3.2.

It is essential in the u, v-transformation that $u > 1$, $|v| > 1$ and at the same time $|u|^2 - |v|^2 = 1$. It will lead in $\mathcal{H}_{\text{int}}$ to the cancellation of the greatest terms proportional to Bu^4, Bu^2v^2, Bv^4. Therefore it is necessary to do the cumbersome calculations very accurately, retaining in the initial expression for the energy small terms whose contribution after multiplication by u^4, v^4 and u^2v^2 may prove essential. There exists the opinion that the magnetic dipole interaction can thus result in an anomalously big contribution to $\mathcal{H}_4$. However, the calculations by *Lutovinov* and *Safonov* [146] have shown that this does not take place.

Problem

Show that for the Bloch equation (3.4.5) the canonical variables coordinate and momentum are

$$q = M_x/g, \quad p = \varphi = \arctan(M_x/M_z).$$

Hint: Directly differentiating allowing for the bond $M^2(p,q) = M_0^2$ make sure that (3.4.5) in the variables (3.5.1) are reduced to the form (1.1.6). Further, following the procedure described in Sect. 1.1 one may pass from p, q to the complex variables b, b^*. The vector $\boldsymbol{M}$ via these variables will be represented by (3.4.9).

4 Nonlinear Dynamics of Narrow Packets of Spin Waves

Chapter 1 contained the introduction to the general theory of nonlinear wave dynamics within the classical Hamiltonian formalism. "General" here implies that this theory describes nonlinear processes irrespective of the nature of waves and the type of the nonlinear media where their propagation takes place. Clearly, this approach, alongside with the evident advantages has certain limitations. To avoid them one must use the concrete laws of waves dispersion as well as the k-dependences of the interaction Hamiltonian amplitudes providing specific information about a practical problem. In this chapter the attempt is made to overcome such limitations on the description of the nonlinear dynamics of spin waves in magnetically ordered dielectrics. In this attempt we shall proceed from the general theory developed in Chap. 1 and use the magnon Hamiltonians calculated in Chap. 3.

Section 4.1 describes elementary interaction processes involving three and four magnons in ferromagnets. This section illustrates the general theory of the three- and four-wave processes developed in Sects. 1.4.1, 2 and 1.5.1. Section 4.2 is much more independent. Here, the theory of wave self-focusing (an elementary introduction to this theory is given in Sects. 1.5.3, 4) is presented in connection with magnetoelastic waves in antiferromagnets. This problem is very interesting from the experimental point of view and rather peculiar from the theoretical angle. In Sect. 4.3 various methods of parametric magnon excitation in ferro- and antiferromagnets are described. They have all been given thorough experimental study.

4.1 Elementary Processes of Spin Wave Interaction

4.1.1 Three-Magnon Processes

Two simple examples of the behavior of magnetodielectrics will be considered here to illustrate the general theory of three-wave processes.

The first example is the confluence of two magnons in an isotropic ferromagnetic. Let two spin waves with wave vectors k_1, k_2 and canonical amplitudes b_1 and b_2 be excited in the ferromagnet. Then according to the first formula of (1.4.4) they bring about the excitation of the third magnon with the canonical amplitude b_3:

$$b_3 = \frac{V_{3,12}^* b_1 b_2}{2[\omega_1 + \omega_2 - \omega_3 - i\gamma_3]} . \tag{4.1.3}$$

By substituting the explicit expression (3.1.18, 20) for the amplitude $V(\mathbf{3, 12})$ into (4.1.3) and expressing the canonical amplitudes $b(\mathbf{k})$ in terms of the amplitudes of magnetization oscillations $m(\mathbf{k})$ using (3.8.8) we obtain

$$\frac{m_3}{M} = \frac{m_1 m_2}{4M^2} \left[\frac{k_{1z} k_{1+}}{k_1^2} + \frac{k_{2z} k_{2+}}{k_2^2} \right] \frac{M}{\omega_1 + \omega_2 - \omega_3 - i\gamma_3} . \tag{4.1.4}$$

For simplicity we take in this formula $\omega_{\mathrm{ex}}(ak_j)^2 \gg \omega_{\mathrm{M}}$. This enables us to simplify the expressions for $V(\mathbf{3, 12})$ assuming in (3.1.19, 20) $v_j = 0$ and $u_j = 1$. It should be recalled that (4.1.4) is in agreement with the general (whatever the nature of the waves) estimation (1.4.9, 11) of the wave conversion under the resonance condition: $x_3 \simeq x_1 x_2 (\omega/\gamma)$, if x_j denotes the dimensionless amplitude of a spin wave $x_j = m_j/M$, and if ω is replaced by ω_{M}, the frequency of the dipole–dipole interaction is responsible for the conversion process. The exact formula (4.1.4) as compared with the estimation for x_3 gives the value of the numerical factor and determines how the efficiency of the transformation depends on the orientation of the wave vectors $\mathbf{k}_1$ and $\mathbf{k}_2$ of the initial waves. In particular, if $\mathbf{k}_1$ and $\mathbf{k}_2$ are directed along the magnetization ($\mathbf{k}_j \| \mathbf{M}$, $k_{j+} = k_{jx} + k_{jy} = 0$) or transverse to the magnetization direction ($\mathbf{k} \perp \mathbf{m}$, $k_{jz} = 0$) the conversion efficiency becomes zero.

It must be noted that to observe the nonlinear process described here experimentally is not easy, first because it is rather difficult to generate and record monochromatic spin waves and second because even at low temperatures their path length is fairly short, of the order of a few millimeters. It is much easier to observe the confluence of two Walker modes into one. Their amplitudes in resonance will be connected by a relation similar to (4.1.4) to an accuracy of the numerical factor of the order unity.

The second example is the decay instability of the monochromatic spin wave which splits into two waves. The conservation laws in this process have the form:

$$\omega(\mathbf{k}_1) = \omega(\mathbf{k}_2) + \omega(\mathbf{k}_3) , \quad \mathbf{k}_1 = \mathbf{k}_2 + \mathbf{k}_3 . \tag{4.1.5a, b}$$

The resonance conditions (4.1.5), as it will be recalled, are not always fulfilled in the decay processes. Accordingly, the laws of wave dispersion are subdivided into decay and non-decay. If the dispersion law $\omega(\mathbf{k})$ is not of the form k^x, then one part of the spectrum may turn out to be a decay one and the other a non-decay one. This is the case, for instance, of Langmuir waves in non-isothermal plasma, of waves on a liquid surface with the dispersion law (1.2.18) and for spin waves in cubic ferromagnets with the dispersion law (3.1.5, 6). For the interpretation of the experimental results it will later be important to know the values of the wave vector k_S determining the boundary of the decay processes in cubic ferromagnets.

Let us calculate k_S, confining ourselves to the cases when the magnetization $\boldsymbol{M}\|[100]$ or $\boldsymbol{M}\|[111]$. Then in the law of dispersion (3.1.5) the factor $\beta = 0$ (see Table 3.1). First consider processes where (4.1.5) $\boldsymbol{k}_2 = \boldsymbol{k}_3 = 2\boldsymbol{k}_1$. Then from (4.1.5a) and (3.1.5) we obtain the expression for $\boldsymbol{k}_1$

$$\omega_{\text{ex}}(ak_1^2) = 2\sqrt{A^2 - B^2} \, , \tag{4.1.6}$$

where $A = \omega_H - \omega_M N_z - \alpha\omega_a + B$, and $B = [\omega_M \sin^2 \Theta]/2$. A more detailed analysis of the dispersion law (3.1.5) reveals that the value of the wave vector $\boldsymbol{k}_1$ (4.1.6) is boundary in the fact that under $k < k_1 = k_S$ decay processes for spin waves with $\omega(\boldsymbol{k}_3)$ are forbidden, while they are allowed under $k > k_S$. Sect. 4.3, which is devoted to the methods of parametric excitation of magnons in magnets, will give a detailed treatment of the decay instability of a magnon with $k = 0$, i.e. homogeneous precession of the magnetization into two other magnons, into the nuclear magnons in antiferromagnets, etc. By way of example, the decay instability of a monochromatic spin wave into two spin waves will be briefly outlined. Under resonance (4.1.4) the threshold amplitude $b_{1\text{th}}$ according to (1.4.19) is $|V_{1,23}b_{1,th}|^2 = \gamma_2\gamma_3$; γ_2 and γ_3 here are the damping decrement in the final state. Substituting here the interaction amplitudes $V(\boldsymbol{1}, \boldsymbol{2}, \boldsymbol{3})$ (3.1.18, 20) and taking for simplicity $\boldsymbol{k}_1$ (the wave vector of the initial wave with the finite amplitude b_1) to be perpendicular to $\boldsymbol{M}(\sin\Theta = 1)$, we obtain the critical value of b_1:

$$|b_{1,\text{th}}| = \frac{2\gamma}{\omega_M}\sqrt{\frac{2M}{g}} \, , \quad \gamma = \sqrt{\gamma_2\gamma_3} \tag{4.1.7}$$

corresponding to the threshold of decay into two magnons propagating through the optimum angles $\Theta_2 = \pi - \Theta_3 = \pi/4$, $\varphi_2 = \varphi_3$. To the amplitude (4.1.7) corresponds the critical value of the precession angle

$$\Psi_{1\text{th}} = m_{1\text{th}}/M = b_{1\text{th}}\sqrt{2g/M} = 4\gamma/\omega_M \, . \tag{4.1.8}$$

Thus a spin wave with amplitude b and the precession angle Ψ exceeding the critical values of (4.1.7, 8) is unstable with respect to its splitting into two others.

4.1.2 Modulation Instability of Spin Waves

The linear theory of the modulation instability of a plane wave of arbitrary nature has been considered in Sect. 1.5.1. This is a decay instability of the second order of the wave (with the wave vector $\boldsymbol{k}_0$) splitting into two other waves (with the wave vectors $\boldsymbol{k}_1$ and $\boldsymbol{k}_2$). This instability is due to the four-wave processes

$$\omega(\boldsymbol{k}_0) + \omega(\boldsymbol{k}_0) = \omega(\boldsymbol{k}_1) + \omega(\boldsymbol{k}_2) ,$$
$$\boldsymbol{k}_0 + \boldsymbol{k}_0 = \boldsymbol{k}_1 + \boldsymbol{k}_2 , \quad \boldsymbol{k}_{1,2} = \boldsymbol{k}_0 \pm \boldsymbol{\kappa} . \tag{4.1.9}$$

The threshold amplitude of the initial wave $b_{0\text{th}}$ according to (1.5.9) is determined by the damping of the waves in the final state γ_1 and γ_2:

$$|T_{00,12} b_{0,\text{th}}^2| = \sqrt{\gamma_1 \gamma_2} . \tag{4.1.10}$$

The modulation instability is not inevitable. In addition to the usual threshold condition $|b_0| > |b_{0\text{th}}|$ the criterion (1.5.13) must be fulfilled:

$$T_{00,00} \hat{L} \kappa^2 < 0 , \tag{4.1.11}$$

where $\hat{L}\kappa^2$ is the quadratic form (1.5.11) determined by the dispersion law $\omega(\boldsymbol{k})$. On the basis of the specific dispersion laws of magnons and the four-magnon amplitudes (presented in Chap. 3) the modulation instability of magnons in ferro- and antiferromagnets can be studied [4.1].

It can be done most easily for the case of easy-plane antiferromagnets. The dispersion law of their magnons is nearly isotropic (and, accordingly, $\hat{L}\kappa^2 > 0$) and the four-magnon amplitudes T are negative (see (3.2.12)) and are continuous functions of the wave vectors. The above developed theory can therefore be directly applied, and according to the criterion (4.1.11) spin waves of finite amplitude in the easy-plane antiferromagnets are unstable.

Also unstable are spin waves in ferromagnets whose Hamiltonian includes only the exchange interaction. It can be seen from (3.1.10, 21) that L in this case is given by

$$\hat{L}\kappa^2 = \omega_{\text{ex}}(a\kappa)^2 > 0 , \quad T_{00} = -(g/M)\omega_{\text{ex}}(ak_0)^2 \tag{4.1.12}$$

and the instability criterion (4.1.11) is fulfilled. It follows from (1.5.11) and (4.1.12) that the instability is maximum when

$$(\kappa/k_0)^2 = -2|b|^2 T_{00}/L \simeq 2g|b|^2/M \simeq 2\Delta M_z/M , \tag{4.1.13}$$

where ΔM_z is the variation of the static magnetization caused by the existence of the initial wave b_0. The threshold amplitude of the spin waves is given by (4.1.10) and, consequently,

$$(\Delta M_z/M)_{\text{th}} = \gamma g/(2\sqrt{2} M T_{00}) = \gamma/[2\sqrt{2}\omega_{\text{ex}}(ak_0)^2] . \tag{4.1.14}$$

Thus, if its amplitude is big enough the "exchange" spin wave in ferromagnets is always unstable with respect to the process (1.5.11) with small $(\kappa/k_0)^2$ in the vicinity of $\gamma/\omega_{\text{ex}}(k_0a)^2$. This instability brings about long period longitudinal ($\boldsymbol{\kappa}\|\boldsymbol{k}_0$) and transverse ($\boldsymbol{\kappa} \perp \boldsymbol{k}$) amplitude modulations of the travelling spin wave. The case of long spin waves in ferromagnets is more complicated because of the long-range magnetic dipole interaction leading to the non-analytical dependence of $T(\mathbf{12},\mathbf{34})$ on its arguments. Then (1.5.11) cannot be applied to spin waves in ferromagnets. Instead, it follows from (1.5.7, 8) (for more detail, see *Zakharov* et al. [4.1]):

$$\nu = -\gamma \pm \sqrt{[\hat{L}\kappa^2 + 2T(\boldsymbol{\kappa})|b|^2]^2 - 4|F(\boldsymbol{\kappa})b^2|^2} \ . \tag{4.1.15}$$

Here it is taken that $\kappa \ll k$ and the designations $T(\boldsymbol{\kappa})$ and $F(\boldsymbol{\kappa})$ are introduced for the following functions depending on the direction of $\boldsymbol{\kappa}$ only:

$$2T(\boldsymbol{\kappa}) = T(\boldsymbol{k}_0 + \boldsymbol{\kappa}, \boldsymbol{k}_0) + T(\boldsymbol{k}_0 - \boldsymbol{\kappa}, \boldsymbol{k}_0) - T(\boldsymbol{k}_0, \boldsymbol{k}_0) \ , \tag{4.1.16}$$

$$2F(\boldsymbol{\kappa}) = T(\boldsymbol{k}_0, \boldsymbol{k}_0, \boldsymbol{k}_0 + \boldsymbol{\kappa}, \boldsymbol{k}_0 - \boldsymbol{\kappa}) \ . \tag{4.1.17}$$

Note that if $T(\mathbf{12},\mathbf{34})$ analytically depends on the wave vectors, then $T(\boldsymbol{\kappa}) = F(\boldsymbol{\kappa}) = T(\mathbf{00})$ under $\kappa \to 0$ and the expression (4.1.15) is safely transformed into (1.5.11). Using (3.1.5,6), we can obtain

$$F(\boldsymbol{\kappa}) = T(\boldsymbol{\kappa}) = [g/(2\pi)^3 M][\varphi(\boldsymbol{\kappa})\omega_{\text{M}} \cos^2 \Theta - \omega_{\text{ex}}(ak_0^2)] \ , \tag{4.1.18}$$

$$\varphi(\boldsymbol{\kappa}) = (\Omega - \omega_{\text{M}} \sin^2 \Theta)/(\Omega + \omega_{\text{M}} \sin^2 \Theta) \ ,$$
$$\hat{L}\kappa^2 = \omega_{\text{ex}}(a\kappa)^2 + \kappa^2 \omega_{\text{M}} \sin^2 \Theta / 2k_0^2 \ , \tag{4.1.19}$$
$$\Omega = gH - \omega_{\text{M}} N_z \ , \quad \cos \Theta = \kappa_z/|\kappa| \ .$$

At $F = T$ the necessary condition of instability, as follows from (4.1.15, 19) is $TL^{-1} < 0$. Note that $\hat{L}\kappa^2$ is always positive, and $T(\boldsymbol{\kappa}) > 0$ only if $\omega_{\text{ex}}(ak_0)^2 < \omega_{\text{M}}$, therefore an external magnetic field larger than $4\pi M(N_z + 1)$ suppresses the modulation instability of the travelling wave with $\boldsymbol{k}\|\boldsymbol{M}$. If $H < 4\pi M(N_z + 1)$, then $\varphi(\boldsymbol{\kappa})$ will be negative for some directions which leads to an instability with small κ, first of all with those satisfying the equation $\hat{L}\kappa^2 = \gamma\kappa/k_0$. The threshold of this instability is given by the expression (4.1.14), where T is satisfied by (4.1.16).

When the spin wave propagates transversely ($\boldsymbol{k} \perp \boldsymbol{M}$) there can arise instabilities with large κ unlike in the case of the longitudinal propagation ($\boldsymbol{k}\|\boldsymbol{M}$). This happens because the decay of the wave with $\boldsymbol{k} \perp \boldsymbol{M}$ is accompanied by a decrease of its dipole–dipole energy which can compensate the decreased exchange energy under some κ. However, the instability threshold with large κ, has the same order of magnitude as the threshold for $\kappa \ll k$. For simplicity we shall consider only the latter case. The instability analysis under arbitrary ω_{M}, Ω and k_0 would be too cumbersome. Here only one interesting case will be presented: $\omega_{\text{M}} \gg \Omega, \omega_{\text{ex}}(ak_0)^2$. Other specific cases were considered by *Zakharov* et al. [4.1]. In our example we obtain from (3.1.5,6) and (3.1.22)

$$\hat{L}\kappa^2 = \frac{\omega(k_0)}{k_0^2} \frac{\kappa^2 \Omega}{64\pi^3 M[\Omega + \omega_{\text{ex}}(ak_0^2)]} \left[\frac{\omega_{\text{ex}} \sin^2 \Theta}{\Omega + \omega_{\text{ex}}(ak_0)^2} + \cos \Theta \right],$$

$$T(\boldsymbol{\kappa}) = \frac{g\omega_{\text{M}}^2 (\cos^2 \Theta - 2N_z)}{64\pi^3 M[\Omega + \omega_{\text{ex}}(ak_0^2)]}, \qquad (4.1.20)$$

$$F(\boldsymbol{\kappa}) = \frac{g\omega_{\text{M}}^2 (N_z - \cos^2 \Theta)}{64\pi^3 M[\Omega + \omega_{\text{ex}}(ak_0^2)]}.$$

Since the sign of $\hat{L}\kappa^2$ can be changed, spin waves are unstable whatever the sign of F and T. The threshold amplitude is

$$(\Delta M_z/M)_{\text{th}} = 8\gamma[\Omega + \omega_{\text{ex}}(ak_0^2)]/(\omega_{\text{M}}^2 |N_z - \cos^2 \Theta|). \qquad (4.1.21)$$

It is clear that for long spin waves ($\omega_{\text{ex}}(ak_0) \ll \omega_{\text{M}}$) propagating in a sample where the internal field is small ($\Omega \ll \omega_{\text{M}}$) the instability threshold decreases significantly.

In conclusion, let us estimate the spin wave intensity leading to modulation instability. The estimate for the energy flux in spin waves follows from (4.1.14)

$$P_{\text{th}} = \omega b_{\text{th}} \partial \omega / \partial k \simeq \gamma \omega_{\text{ex}} a^2 \, kM/g. \qquad (4.1.22)$$

For a monocrystal YIG $\gamma \simeq 10^6 \text{s}^{-1}$, $M \simeq 100$ Oe, $\omega_{\text{ex}} a^2 \simeq 0.1$ cm^{2}s^{-1}. If we put $k = 10$ cm, then $P_{\text{th}} \simeq 10^{-3}$ Wt·cm. It must be noted that the expression (4.1.23) sets the upper limit to the energy flux which can be transferred by a spin wave inside the crystal. This flux is usually much smaller that the maximum energy flux (0.1 Wt·cm) for a sound wave in ferromagnets (*Gurevitch*, [4.2]), therefore the critical values of the flux (4.1.23) can be reached as the sound is transformed to a spin wave in a nonuniform magnetic field (*Schlömann* et al. [4.3]).

4.2 Self-Focusing of Magnetoelastic Waves in Antiferromagnets

4.2.1 Structure of Basic Equations

The magnetic subsystem of a crystal is considerably nonlinear, which results in the nonlinearity of the system of other quasi-particles interacting with magnons. For example, in most solids under practically attainable deformations the nonlinear acoustic effects are small. This is due to the imperfections in the crystals leading to the failure of the sample under relative deformations u much less than unity ($u \simeq 10^{-4} - 10^{-6}$). In antiferromagnets because of the exchange amplification of the magnetostriction (Sect. 3.2) the nonlinearity and the dispersion of magnetoelastic waves may be considerable. The law of acoustic dispersion in this case has the following form:

$$\omega^2 = (V_S + Q_1 u + Q_2^2 u^2)^2 k_x + 2V_S^4 D k_x + 2V_S a k^2 . \qquad (4.2.1)$$

Here, the nonlinearity (which can be considered as the renormalization of the sound velocity) is taken into account, as well as the dispersion of the wave velocity and the diffraction (we assume the elastic wave to be weakly non-unidimensional $k^2 = k_x^2 + k_\perp^2$, $k_\perp \ll k_x$). In isotropic media $a = V_S/2$; in crystals because of the anisotropy of the elasticity and the magnetostriction the value of a can be either positive or negative (for more detail see *Turitsyn* and *Falkovich* [4.4]). At not too high excitation levels the evolution of magnetoelastic waves can be subdivided into quick and slow evolution. Quick evolution is the transfer of the initial excitation with the sound velocity and slow evolution is due to the small effects of nonlinearity, dispersion and diffraction. To study these effects the reference system moving with the sound velocity is convenient. To change over to this system we substitute in (4.2.1) $\omega \Rightarrow \Omega - V_S k_x$, $\Omega \ll V_S k_x$, and perform the inverse Fourier transform

$$u_t(\boldsymbol{r},t) = \int u(\boldsymbol{k},\Omega) \exp(i\boldsymbol{k}\boldsymbol{r} - i\Omega t)\, d\boldsymbol{k}d\Omega .$$

We obtain the following equation:

$$u_t + Q_1 u u_x + Q_2 u^2 u_x + D u_{xxx} + a \int_{-\infty}^{x} \Delta_\perp u\, dx' = 0 ,$$
$$u_t = \partial u/\partial t , \quad u_x = \partial u/\partial x , \quad u_{xxx} = \partial^3 u/\partial^3 x . \qquad (4.2.2)$$

The derivation of this equation is surely not precise and only serves as an illustration. The rigorous derivation of (4.2.2) from the Landau-Lifshitz equations and the theory of elasticity were obtained by *Turitsyn* and *Falkovich* in [4.4], where the expressions for the constants Q_j, D and a were calculated for antiferromagnets with rhombohedral structure. Note also that in some cases $Q = 0$, i.e. the first nonvanishing nonlinearity is the cubic one. The dispersion factor $D \simeq (V_S^2 - V_M^2)$ is proportional to the difference of the sound velocity and the limiting velocity of magnons. Therefore in antiferromagnets where $V_S > V_M$ (or, in other words, the Debye temperature is greater than the Néel temperature), $D > 0$, i.e. the dispersion is negative, and the velocity of magnetoelastic waves drops as the wave vector increases (*Ozhogin* et al. [4.5]).

Under unidimensionality (when the initial perturbation is uniform over the plane perpendicular to the line of propagation) (4.2.2) is reduced depending on the type of nonlinearity either to the Korteveg-de Vries (KdV) equation

$$u_t + Q u u_x + D u_{xxx} = 0 , \qquad (4.2.3)$$

or to the modified KdV (MKdV) equation

$$u_t + Qu^2 u_x + Du_{xxx} = 0 \, . \tag{4.2.4}$$

4.2.2 Properties of Unidimensional Equations

The one-dimensional equations (4.2.3-4) have the following remarkable property: If any solution of such an equation for $u(x,t)$ (soliton) is treated as the potential of the linear Schrödinger equation

$$\left[\frac{\partial^2}{\partial t^2} + u(x,t)\right] \Psi(x,t) = E\Psi(x,t) \, , \tag{4.2.5}$$

the energy spectrum E does not depend on time. In other words, the evolution $u(x,t)$ by virtue of the equations KdV (4.2.3) or MKdV (4.2.4) leads to isospectrum conversion of the Schrödinger equation potential. This fact, discovered by *Gardner* at al. [4.6], forms the basis of the method of the inverse problem of scattering and its further generalizations [4.7, 8]. The connection with (4.2.5) enables us to describe for (4.2.3-4) the evolution of the arbitrarily localized initial perturbation $u(x,0)$. To this end, the spectrum (4.2.5) must be found with the potential $u(x,0)$. The number of discrete levels in the spectrum is equal to the number of solitons resulting from the decay of the initial perturbation in the limit $t \to \infty$. The solitons, as in (1.5.24), under unidimensionality are asymptotic states. It is the study of (1.5.24) and (4.2.3) that revealed the basic role of solitons in nonlinear wave dynamics (see [4.7]). Such localized stationary pulses may appear beginning with the amplitudes under which the effects of the nonlinearity uu_x and the dispersion u_{xxx} are comparable. At such amplitudes plane waves are no longer well-defined objects since their interaction time becomes of the order of the scattering time of the wave packets due to the dispersion of the group velocities. Unfortunately, the soliton as a fundamental subject of description instead of the plane wave did not meet our expectations since solitons are very often unstable. As this instability increases, it often results in collapse (or, in other words, self-focusing) which therefore proves to be of no less importance in the nonlinear dynamics of waves than solitons.

4.2.3 Stability of Solitons and Self-Focusing Theorem

Let us consider the weakly non-unidimensional generalization of the MKdV equation (4.2.4):

$$\frac{\partial}{\partial x}(u_t + Du_{xxx} + Qu^2 u_x) = au_{yy} + cu_{zz} \, . \tag{4.2.6}$$

In the easy-plane antiferromagnets the case $Q > 0$ occurs most often (see [4.5]), so here we shall discuss only this case. The plane solitons

4.2 Self-Focusing of Magnetoelastic Waves in Antiferromagnets

$$u_0(x - Vt) = \frac{\sqrt{2V/Q}}{\cosh[(x - Vt)\sqrt{V/D}]}$$

travel with supersonic velocity ($V > 0$; it should be recalled that (4.2.4, 6) have been written in the reference system moving at the sound velocity). As can be seen from the expression for u_0, the solitons exist only at $D > 0$, i.e. for antiferromagnets where $V_S > V_M$ (when $D < 0$), the stationary solution has the form of a running "domain wall"

$$u_t = \frac{\sqrt{2V/Q}}{\coth[(x - Vt)\sqrt{V/D}]}.$$

The stability of unidimensional solitons with respect to transverse perturbation was first studied by *Kadomtsev* and *Petviashvili* [4.9]. They showed solitons to be stable only under $a < 0$, $c < 0$; if the sign of either one (or both) of these values is positive the plane solitons are unstable with respect to crimping. Qualitatively, this instability is due to the fact that a local increase of the soliton amplitude leads to its decreased velocity in the same place. This part of the soliton falls behind; the wave front is bent. This in turn leads to a concentration of energy and the further increase of the amplitude in the place of initial fluctuation etc. At $a > 0$, $c > 0$ the self-focusing of the elastic wave is possible. To illustrate this, let us represent (4.2.6) as

$$u_t = \frac{\partial}{\partial x}\frac{\delta \mathcal{H}}{\delta u}, \quad \mathcal{H} = \frac{1}{2}\int (Du_x^2 + aw_y^2 + cw_z^2 - 2u^4)\,d\boldsymbol{r}, \quad (4.2.7)$$

where $\mathcal{H}w_x = u$. As (1.5.32), (1.4.29), equations (4.2.7) are Hamiltonian and retain the Hamiltonian $\mathcal{H}$.

As in (1.5.3), we obtain the second derivative of the positive value proportional to the square of the characteristic transverse size of the beam with respect to time:

$$\frac{\partial^2}{\partial t^2}\int r^2 u^2\,d\boldsymbol{r} = 16a\mathcal{H} - 8aD\int u_x^2\,d\boldsymbol{r}. \quad (4.2.8)$$

Clearly, at $D > 0$ and $a = c > 0$ any distribution with the negative Hamiltonian $\mathcal{H}$ is self-focused. When the sign of even one of the D values is negative, (4.2.8) reveals the defocusing of the beam. Indeed, if $D > 0$, $a > 0$, a sufficient condition for the diffusion of the transverse imperfections is $\mathcal{H} < 0$. At $a < 0$

$$\frac{\partial}{\partial t^2}\int r^2 u^2\,d\boldsymbol{r} = 16a\left[\frac{1}{2}\int a(\nabla_\perp w)^2\,d\boldsymbol{r} - \int u^4\,d\boldsymbol{r}\right] > 0 \quad (4.2.9)$$

and, consequently, any initial distribution is diffuse.

4.2.4 Evolution of Magnetoelastic Waves in the Absence of a Linear Bond Between Magnons and Phonons

The system of equations describing the joint evolution of the stress tensor u and the angle of rotation of the antiferromagnetism vector φ ($\alpha = V_\text{M}/V_\text{S}$) in dimensionless variables has the form:

$$u_{tt} - u_{xx} = (\varphi^2)_{xx}, \quad \varphi_{tt} - \alpha^2 \varphi_{xx} + \varphi = 2u\varphi. \tag{4.2.10}$$

The further reductions (4.2.10) depend on the way the magnetic or elastic subsystem of the crystal are excited. Under parametric excitation of spin waves by a high-frequency electromagnetic field we may pass from $\varphi(x,t)$ to the smooth envelope $\Psi(x,t)$:

$$\varphi = \Psi \exp[ikx - it\sqrt{1+\alpha^2 k^2}] + \text{c.c.}$$

after which we get the system first obtained by *Zakharov* [4.10] for the interaction of the Langmuir and ion-sound waves in a plasma:

$$u_{tt} - u_{xx} = |\Psi|^2_{xx},$$

$$i(\Psi_t + \alpha^2 ik\Psi_x) + \frac{1}{2}\alpha^2 \Psi_{xx} = u\Psi. \tag{4.2.11}$$

Within this system there exist stable unidimensional solitons, and the non-unidimensional generalizations (4.2.11) describe the instability of the solitons with respect to the transverse crimping and (in the case of three-dimensions) collapse of waves.

The physical situation may be different, e.g. the slow sound wave motions of $u(x,t)$ are excited (say, by a piezotransducer). The system (4.2.10) has soliton solutions. In the initial variables

$$u_0 = -2\cosh^{-2}\left[\frac{x-Vt}{\sqrt{\alpha^2-V^2}}\right], \quad \varphi_0 = \sqrt{2(1-V^2)}\cosh^{-1}\left[\frac{x-Vt}{\sqrt{\alpha^2-V^2}}\right]$$

(unlike the *solitons of the envelopes* in (4.2.11)). These solitons, however, are unstable even in the one-dimensional geometry. In this case, too, we can prove the possibility of the collapse which by virtue of unidimensionality does not mean self-focusing, but wave overturn over finite time.

Obviously, the unlimited increase of the amplitude (or its derivatives) cannot be physically meaningful, since in this case we will fall outside the scope of applicability of the employed equations (1.5.23), (4.2.6) or (4.2.10). Here, the very possibility of the wave amplitude increasing to the point when nonlinearity is no longer small is important.

4.3 Methods of Parametric Excitation of Spin Waves

Theoretically, the parametric excitation of waves is simply a result of the developing decay instability of the first or second order as shown in Sects. 1.4.2 and 1.5.1. Therefore to obtain the threshold of the parametric excitation we have to calculate only the amplitude of the corresponding elementary process and the damping decrement of the waves in their final state.

4.3.1 Transverse Pumping of Spin Waves in FM

Transverse pumping of spin waves in ferromagnets, or *Suhl's instability of the first order* [4.11] are the usual notions for the decay of the homogeneous precession of the magnetization into a pair of spin waves (magnons) with the wave vectors $\boldsymbol{k}$ and $-\boldsymbol{k}$ and the frequencies $\omega(\boldsymbol{k}) = \omega(-\boldsymbol{k}) = \omega_p/2$. This process is described by the following terms of the Hamiltonian $\mathcal{H}_3$:

$$\mathcal{H}_3 = \frac{1}{2}\sum_{\boldsymbol{k}}[V(\boldsymbol{0},\boldsymbol{k},-\boldsymbol{k})b^*(0)b(\boldsymbol{k})b(-\boldsymbol{k}) + \text{c.c.}] \tag{4.3.1}$$

with the interaction amplitudes (3.1.19, 20):

$$\begin{aligned}V(\boldsymbol{0},\boldsymbol{k},-\boldsymbol{k}) &= 2\tilde{V}(\boldsymbol{k})u^2(\boldsymbol{k}) + 2u(\boldsymbol{k})\tilde{V}(\boldsymbol{k})V^*(\boldsymbol{k}) \\ &= \pi g\sqrt{2gM}c(\boldsymbol{k})\sin 2\Theta(\boldsymbol{k})\exp[i\varphi(\boldsymbol{k})]\;, \\ c(\boldsymbol{k}) &= 1 + \omega(\boldsymbol{k})/[A(\boldsymbol{k}) + |B(\boldsymbol{k})|]\;.\end{aligned} \tag{4.3.2}$$

It has been assumed here that the magnetization $\boldsymbol{M}$ is directed along the rotation axis of an ellipsoidal sample so that the polarization of the homogeneous precession is circular and accordingly $v_0 = 0$. In accordance with (1.4.21) the threshold amplitude of the homogeneous precession $b_{\text{th}}(0)$ with respect to its decay into a pair of magnons $\pm\boldsymbol{k}$ is given by

$$|V(\boldsymbol{0},\boldsymbol{k},-\boldsymbol{k})|b_{\text{th}}(0) = \gamma(\boldsymbol{k})\;. \tag{4.3.3}$$

Consequently the pair of magnons for which the ratio $\gamma(\boldsymbol{k})/|V(\boldsymbol{0},\boldsymbol{k},-\boldsymbol{k})|$ is a minimum are the first to be excited (as b increases). If $\gamma(\boldsymbol{k})$ does not depend on the angles $\Theta(\boldsymbol{k})$ and $\varphi(\boldsymbol{k})$ over the resonant surface $\omega(\boldsymbol{k}) = \omega_p/2$ (for the time being we shall not go into the details of this dependence), the first to be excited are the pairs for which $|V(\boldsymbol{0},\boldsymbol{k},-\boldsymbol{k})|$ is maximum. Their wave vectors lie on two circles (meridians) of the resonant surface and have $0 < \varphi(\boldsymbol{k}) < 2\pi$, $\Theta(\boldsymbol{k}) = \Theta_{\text{M}}, \pi - \Theta_{\text{M}}$, where Θ_{M} is a little less than $\pi/4$. This can be seen if we identically rewrite $|V(\boldsymbol{0},\boldsymbol{k},-\boldsymbol{k})|$ from (4.3.2) in the form

$$\begin{aligned}|V(\boldsymbol{0},\boldsymbol{k},-\boldsymbol{k})| &= V_1(\Theta) = (\pi g/\omega_p)\sqrt{2gM}\sin 2\Theta \\ &\quad \times \left(\omega_p + \omega_M\sin^2\Theta + \sqrt{\omega_p^2 + \omega_M^2}\sin\Theta\right)\;.\end{aligned} \tag{4.3.4}$$

This (when $\omega_p > \omega_M$) gives

$$b_{\text{th}} = \min_{\Theta(\mathbf{k})} b_{\text{th}}(\mathbf{k}) \simeq (\gamma/\omega_M)\sqrt{2M/g} \ . \tag{4.3.5}$$

To the canonical amplitude of the homogeneous precession b_{th} corresponds the critical precession angle Ψ_{th}:

$$\Psi_{\text{th}} = m_{\text{th},1}/M = \sqrt{2g/M}\, b_{\text{th}} \simeq 2\gamma/\omega_M \ . \tag{4.3.6}$$

For YIG (at $\omega_M = 1700$ Oe, $\gamma = 0.5$ Oe) $\Psi_{\text{th}} \simeq 2.5 \cdot 10^{-3}$.

Now the term "transverse pumping" must be explained. Usually the homogeneous precession is excited by a SHF magnetic field with a polarization $\mathbf{h}_\perp$ directed transverse to the constant magnetic field $\mathbf{H}$ (and accordingly to the stationary magnetization $\mathbf{M}$). The Hamiltonian of this interaction is due to the Zeeman energy $-\mathbf{h} \cdot \mathbf{M}(\mathbf{r},t) = -\mathbf{h}_\perp \cdot \mathbf{m}_\perp$ and has the form

$$\mathcal{H}_p^{(1)} = U[h_+(t) b_0^* + \text{c.c.}], \tag{4.3.7}$$

where $U = -\sqrt{gM_0/2}$, $h_+ = h_x + ih_y$. This Hamiltonian describes the well-known phenomenon of ferromagnetic resonance (FMR)

$$\frac{\partial b_0}{\partial t} + i\omega_0 b_0 + \gamma_0 b_0 = -i\frac{\delta \mathcal{H}_p^{(1)}}{\delta b_0^*} = -iuh_+(t) \ . \tag{4.3.8}$$

If the SHF field has right-hand polarization, then

$$h_+(t) = h_+ \exp(-i\omega t) \ ,$$

$$b_0(t) = \frac{\sqrt{gM_0/2}\, h_+ \exp(-i\omega t)}{\omega_0 - \omega - i\gamma_0} \ . \tag{4.3.9}$$

Under the linear polarization $h_y = 0$, $h_+ = h_x = 2h\cos\omega t$, and

$$\begin{aligned} b_0(t) &= b_0(\omega) \exp(-i\omega t) + b_0(-\omega) \exp(i\omega t), \\ b_0(\omega) &= h\sqrt{gM_0/2}(\omega - \omega_0 + i\gamma_0) \ . \end{aligned} \tag{4.3.10}$$

In the vicinity of the resonance when $|\omega - \omega_0| \ll \omega_0$ the term $b_0(\omega)$ describing the clockwise polarized part of b_0 is the principal one and the expression for the critical amplitude h_{th} corresponding to the instability threshold of the homogeneous precession (4.3.3) is the same for the clockwise circular and linear polarization of h:

$$h_{\text{th}} = b_{\text{th}}\sqrt{(\omega - \omega_0)^2 + \gamma_0^2}/U \simeq \gamma(\mathbf{k})\sqrt{(\omega - \omega_0)^2 + \gamma_0^2}/g\omega_M \ . \tag{4.3.11}$$

For exact resonance ($\omega = \omega_0$) h_{th} is very small. Taking for estimation $\gamma_0/g \simeq 1$ Oe, we have $h_{\text{th}} \simeq 0.002$ Oe. In experiments at the frequency 10^{10}s^{-1} the amplitude of the SHF field h reached values up of to 10 Oe. Therefore the amplitude of the precession b_{th} attains the critical values b_{th} not only under

FMR but also far from it. In such a case this instability is more naturally treated as a parametric instability of the external field with respect to decay into two waves with the wave vectors $\boldsymbol{k}$ and $-\boldsymbol{k}$. The equation (1.4.13) can therefore easily be rewritten in a form where the amplitude b_0 is expressed in terms of the amplitude of the pumping field h:

$$\begin{aligned}\partial b(\boldsymbol{k},t)/\partial t + [\gamma(\boldsymbol{k}) + i\omega(\boldsymbol{k})]b(\boldsymbol{k},t)\\
+ ih\exp(i\omega_\mathrm{p}t)\tilde{V}(\boldsymbol{k})b^*(-\boldsymbol{k},t) = 0\;,\\
\partial b^*(-\boldsymbol{k},t)/\partial t + [\gamma(\boldsymbol{k}) - i\omega(\boldsymbol{k})]b^*(-\boldsymbol{k},t)\\
- ih\exp(-i\omega_\mathrm{p}t)\tilde{V}(\boldsymbol{k})b(\boldsymbol{k},t) = 0\;.\end{aligned} \qquad (4.3.12)$$

Here the value $\tilde{V}(\boldsymbol{k})$ has been introduced which is the *effective amplitude of the magnon interaction with pumping*. In terms of this value the threshold amplitude h_th can be easily expressed:

$$h_\mathrm{th}|\tilde{V}(\boldsymbol{k})| = \gamma(\boldsymbol{k})\;. \qquad (4.3.13)$$

For the above treated case of circular polarization

$$\tilde{V}(\boldsymbol{k}) = \sqrt{\frac{gM}{2}}\frac{\tilde{V}(\boldsymbol{k})}{\omega_\mathrm{p} - \omega_0}\left[1 + \frac{A(\boldsymbol{k}) - |B(\boldsymbol{k})|}{\omega(\boldsymbol{k})}\right]\;. \qquad (4.3.14)$$

From (4.3.13, 14) the expression (4.3.11) for h_th can be obtained again. In the case of linear polarization far from the resonance the initial equations (1.4.13) must allow besides the term $V(0,\boldsymbol{k},-\boldsymbol{k})b_0(\omega)$ (which has now become non-resonant) also another non-resonant term $U(0,\boldsymbol{k},-\boldsymbol{k})b_0^*(-\omega)$. As a result (4.3.12) holds true, but the expression for $\tilde{V}(\boldsymbol{k})$ will change:

$$\begin{aligned}\tilde{V}(\boldsymbol{k}) &= \sqrt{\frac{gM_0}{2}}\left(\frac{V(0,\boldsymbol{k},-\boldsymbol{k})}{\omega_\mathrm{p} - \omega_0} - \frac{U(0,\boldsymbol{k},-\boldsymbol{k})}{\omega_\mathrm{p} + \omega_0}\right)\\
&= \sqrt{\frac{gM_0}{2}}\tilde{V}(\boldsymbol{k})\left(1 + \frac{2[A(\boldsymbol{k}) - |B(\boldsymbol{k})|]}{\omega_\mathrm{p}}\right)\\
&+ \left(\frac{1}{\omega_\mathrm{p} - \omega_0} + \frac{[\omega_\mathrm{p} + 2A(\boldsymbol{k})]\exp[2i\varphi(\boldsymbol{k})]}{\omega_\mathrm{p}[\omega_\mathrm{p} + \omega_0]}\right)\;.\end{aligned} \qquad (4.3.15)$$

It is significant that the first and second terms in this expression describing the contributions of the right-hand and left-hand polarization respectively to the linear pumping depend differently on the azimuthal angle of the spin waves $\varphi(\boldsymbol{k})$. Therefore the maximum of the modulo of the whole expression is achieved not over the whole meridian with the fixed $\Theta(\boldsymbol{k})$ and any $\varphi(\boldsymbol{k})$ as in the case of the circular polarization, but in two points $\varphi(\boldsymbol{k}) = 0, \pi$ corresponding to the excitation of the two pairs of the magnons $\boldsymbol{k}$ and $-\boldsymbol{k}$ on the plane of the pumping $\boldsymbol{h}$ and $\boldsymbol{M}$. This circumstance, i.e. the maximum interaction with the pumping not of a whole group of pairs with different $\varphi(\boldsymbol{k})$, but only of a finite number (in our case it is two) makes lateral pumping rather promising for future experimental studies of magnon behavior above threshold.

Concluding this section we shall give the expression for the minimum threshold of magnon excitation by transverse pumping with $\Theta(\boldsymbol{k}) = \pi/4$, $\varphi(\boldsymbol{k}) = 0, \pi$ in the case of $\gamma(\boldsymbol{k}) = \gamma(k)$ and $\omega_M \ll \omega_p$:

$$h_{\text{th}} = 2H\gamma(\boldsymbol{k})(\omega_p - \omega_0)/(g\omega_M) \ . \tag{4.3.16}$$

For monocrystals YIG with $\gamma(\boldsymbol{k}) = g \cdot 0.1$ Oe at the frequency $\omega_p = 2\pi \cdot 10^{10} \text{s}^{-1}$ we obtain $h_{\text{th}} \simeq 0.3$ Oe. This value can be easily achieved experimentally.

4.3.2 Parallel Pumping of Spin Waves in FM

This type of parametric instability appearing at $\boldsymbol{h} \parallel \boldsymbol{M}$ was predicted and observed by *Morgenthaler* [4.12] and *Schlomann* et al. [4.13]. In order to describe this instability let us discuss the longitudinal part of the Zeeman energy $-h_z m_z$ which results in the Hamiltonian

$$\mathcal{H}_{p2} = \mathcal{H}_{p1} + \mathcal{H}_p \ , \tag{4.3.17}$$

$$\mathcal{H}_p = \frac{1}{2}\sum_{\boldsymbol{k}}[h(t)V(\boldsymbol{k})b^*(-\boldsymbol{k})b^*(\boldsymbol{k}) + \text{c.c.}] \ . \tag{4.3.18}$$

$$\mathcal{H}_{p1} = \sum_{\boldsymbol{k}} U(\boldsymbol{k})b^*(\boldsymbol{k})b(\boldsymbol{k})[h^*(t) + h(t)] \ , \tag{4.3.19}$$

$$\begin{aligned}
V(\boldsymbol{k}) &= 2gu(\boldsymbol{k})v(\boldsymbol{k}) = gB(\boldsymbol{k})/\omega(\boldsymbol{k}) \\
&= [g\omega_M/2\omega(\boldsymbol{k})]\sin^2\Theta(\boldsymbol{k})\exp[2i\varphi(\boldsymbol{k})] \ , \\
U(\boldsymbol{k}) &= gu^2(\boldsymbol{k}) = g[A(\boldsymbol{k}) + \omega(\boldsymbol{k})]/2\omega(\boldsymbol{k}) \\
&= \frac{1}{2}g\left(1 + \sqrt{1 + \frac{\omega_M^2 \sin^2\Theta(\boldsymbol{k})}{4\omega^2(\boldsymbol{k})}}\right),
\end{aligned} \tag{4.3.20}$$

Here $h(t) = h\exp(-i\omega_p t)$, $h_z = h(t) + h^*(t)$. The Hamiltonian $\mathcal{H}_{p1}$ describes processes which do not change the total number of magnons. Due to the condition of time synchronism this term may prove to be important only at low frequencies Ω of $h(t)$. The frequency Ω approximates the spin-wave relaxation frequencies giving the accuracy of the magnetization oscillation frequency $\omega(\boldsymbol{k})$. The origin of the Hamiltonian $\mathcal{H}_p$ describing the parametric excitation of the pair of the spin waves by the longitudinal field can easily be illustrated by the following simple geometric consideration. Due to the magnetic dipole interaction and the crystallographic anisotropy the magnetization, at any point, precesses along the elliptic cone (formally it is revealed in the circular variables $a(\boldsymbol{k}) \sim m_x + im_y$ not being normal so that in order to diagonalize the quadratic term of the Hamiltonian the

u-v-transformation is necessary). Since the length of the vector $\boldsymbol{M}$ remains constant the base of the cone is not flat, which results in the appearance of a variable longitudinal component (z-component) of the vector $\boldsymbol{M}$ changing with the doubled precession frequency $2\omega(\boldsymbol{k})$. Clearly, those waves can be excited by a magnetic field with frequency $2\omega(\boldsymbol{k})$ polarized along z.

From the expressions (4.3.12) and (4.3.18) the threshold of the parametric pumping of magnons is readily obtained:

$$h_{\text{th}}|V(\boldsymbol{k})| = \gamma(\boldsymbol{k}) . \tag{4.3.21}$$

Here $V(\boldsymbol{k})$ is the amplitude of the $\pm \boldsymbol{k}$-magnon-pumping interaction (4.3.20). Evidently, $|V(\boldsymbol{k})|$ is maximum for magnon pairs with wave vectors on the equator of the resonance surface: $0 \leq \varphi(\boldsymbol{k}) < 2\pi$, $\Theta(\boldsymbol{k}) = \pi/2$. If $\gamma(\boldsymbol{k})$ depends slowly on $\Theta(\boldsymbol{k})$, these very pairs have the minimum excitation threshold

$$gh_{\text{th}} = \gamma\omega_{\text{p}}/\omega_{\text{M}} . \tag{4.3.22}$$

The order of magnitude of the parallel pumping field (4.3.22) is the same as that of the threshold field of the lateral pumping (4.3.13)-(4.3.15) far from FMR.

4.3.3 "Oblique" Pumping of Spin Waves in FM

"*Oblique pumping*" is a method of parametric excitation of spin waves in ferromagnets intermediate between the lateral and parallel pumping when the linearly polarized SHF field $\boldsymbol{h}(t)$ is at some angle Θ_0 to the magnetization. Therefore both lateral and longitudinal components of pumping must be simultaneously allowed for. The resulting effective amplitudes of the magnon interaction with the oblique pumping $V'(\boldsymbol{k}, \Theta_0)$ will have the form

$$V'(\boldsymbol{k}, \Theta_0) = \tilde{V}(\boldsymbol{k}) \sin \Theta_0 + V(\boldsymbol{k}) \cos \Theta_0 , \tag{4.3.23}$$

where $\tilde{V}(\boldsymbol{k})$ and $V(\boldsymbol{k})$ are given by (4.3.15) and (4.3.20) respectively. As it can be seen from these expressions, the threshold of parametric excitation of the spin waves is minimum only for the pair of waves with $\varphi(\boldsymbol{k}) = 0$ and some $\Theta(\boldsymbol{k})$ depending on Θ_0.

4.3.4 Suhl Instability of the Second Order in FM

If we decrease the pumping frequency ω_{p} or increase the constant magnetic field $\boldsymbol{H}$ and, accordingly, the gap in the magnon spectrum, it is easy to make the pumping frequency less than the double minimum (with respect to $\varphi(\boldsymbol{k})$ and $\Theta(\boldsymbol{k})$) gap in the magnon spectrum. Then the condition of the parametric resonance

$$\omega_p = \omega(\mathbf{k}) + \omega(-\mathbf{k}) \tag{4.3.24}$$

will be fulfilled for no $\mathbf{k}$ and, consequently, three-wave decay processes (4.3.24) of the external field into two waves will be forbidden. In this case we can observe the *second-order Suhl instability of the homogeneous precession of magnetization* [4.11]. The conditions for parametric resonance for this instability have the following form:

$$2\omega_p = \omega(\mathbf{k}) + \omega(-\mathbf{k}) \ . \tag{4.3.25}$$

Since the interaction Hamiltonian of the magnons with the external field (4.3.18) is proportional to the first power of h, the direct decay process of two quanta of the pumping into two magnons in ferromagnets will not take place. Only the indirect process involving the uniform precession is possible: the external field h spins it up in resonance $\omega_p \simeq \omega_0$ and then the following four-magnon parametric process takes place:

$$2\omega_0 = \omega(\mathbf{k}) + \omega(-\mathbf{k}) \ . \tag{4.3.26}$$

Clearly, the threshold amplitude of the precession $b_{0,\mathrm{th}}$ is given by the condition

$$|S(0, \mathbf{k})(b_{0,\mathrm{th}})^2| = \gamma(\mathbf{k}) \ , \tag{4.3.27}$$

where b_0 is connected with h by (4.3.9) or (4.3.10). Substituting $S(0, \mathbf{k})$ from (3.1.24) into (4.3.27) we have for the cubic ferromagnets:

$$(gh_{\mathrm{th}}/\omega_M)^2 = \gamma(\mathbf{k})/\omega(\mathbf{k}) \ . \tag{4.3.28}$$

If for the first-order parametric processes $gh_{\mathrm{th}}/\omega_M \simeq \gamma(\mathbf{k})/\omega_M$, for the second-order ones the threshold $gh_{\mathrm{th}}/\omega_M \simeq \sqrt{\gamma(\mathbf{k})/\omega_M}$, and the threshold amplitude h_{th} far away from ferromagnetic resonance appears significantly greater. When the resonance condition $\omega_p = \omega_0$ is fulfilled, however, it follows from (4.3.28) that $gh_{\mathrm{th}} \simeq \sqrt{\gamma_0 \gamma(\mathbf{k})}$, i.e. the amplitude $h_{\perp\mathrm{th}}$ of the process under consideration has the same order of magnitude as the amplitude $h_{\|\mathrm{th}}$ of the parallel pumping. It is important for the parametric process (4.3.26) that the instability threshold (4.3.28) is minimum for the magnons at the poles of the resonance surface $[\Theta(\mathbf{k}) = 0, \pi]$, i.e. at a single pair of points.

4.3.5 Parallel Pumping in "Easy-Plane" Antiferromagnets

As has already been mentioned in Chap. 2 when studying nonlinear properties of magnons, "easy-plane" antiferromagnets are most interesting since they have magnon branches with frequencies lying in the convenient range below 50 gHz. The mechanism of the parametric excitation of magnons by parallel pumping is similar here to the mechanism of transverse pumping by a linearly polarized field or of the oblique pumping of magnons in ferromagnets: both the direct magnon excitation by the field h in the lower branch $h \to a(\boldsymbol{k}) + a(-\boldsymbol{k})$ and their indirect excitation, the decay of magnons with $\boldsymbol{k} = 0$ (homogeneous precession) in the upper branch $b_0 \to a(\boldsymbol{k}) + a(-\boldsymbol{k})$, this precession b_0 being linearly excited by the nonresonant field $h^* \to b$, $h \to b^*$. Accordingly, the total Hamiltonian of the system including the interaction with the pumping has the form:

$$\mathcal{H} = \sum_{\boldsymbol{k}} [\omega(\boldsymbol{k}) a^*(\boldsymbol{k}) a(\boldsymbol{k}) + \Omega(\boldsymbol{k}) b^*(\boldsymbol{k}) b(\boldsymbol{k})] + 2hU(b_0 + b_0^*) \cos \omega_{\mathrm{p}} t$$

$$+ \frac{1}{2} \sum_{\boldsymbol{k}} [hV(\boldsymbol{k}) a^*(\boldsymbol{k}) a^*(-\boldsymbol{k}) \exp(-i\omega_{\mathrm{p}} t) + \text{c.c.}] + \mathcal{H}_3 ,$$

$$\mathcal{H}_3 = \sum_{2+3=1} [V^{(1)}_{1,23} b_1^* a_2 a_3 + V^{(2)}_{1,23} a_1^* b_2 a_3 + \text{c.c.}]$$

$$+ \frac{1}{2} \sum_{1+2+3=0} [U_{123} b_1^* a_2^* a_3^* + \text{c.c.}] .$$

(4.3.29)

In this equation $\omega(\boldsymbol{k})$ and $\Omega(\boldsymbol{k})$ are the frequencies of the quasi-ferromagnetic and quasi-antiferromagnetic branches of the spectrum (see (3.2)). The amplitudes U and $\tilde{V}(\boldsymbol{k})$ describe the linear and parametric interactions of the field h with antiferro- and ferromagnons, and, finally, the amplitudes $V^{(1)}(1; 2, 3)$, $V^{(2)}(1; 2, 3)$ and $U(1, 2, 3)$ describe three-magnon processes of special interest for us (for them see (3.2.10)). Calculating b_0 from the equations of motion (4.3.8), we obtain the expressions for the effective amplitude of the interaction between the magnons of the lower branch and the pumping:

$$\tilde{V}(\boldsymbol{k}) = V(\boldsymbol{k}) - U \left[\frac{V^{(1)}(0, \boldsymbol{k}, -\boldsymbol{k})}{\Omega_0 - \omega_{\mathrm{p}}} - \frac{U(0, \boldsymbol{k}, -\boldsymbol{k})}{\Omega_0 + \omega_{\mathrm{p}}} \right]$$

$$= \frac{g^2}{2\omega(\boldsymbol{k})} \left(H_D + \frac{2H \Omega_0^2}{\Omega_0^2 - \omega_{\mathrm{p}}^2} \right) ,$$

(4.3.30)

where H is the external constant field, H_D is the Dzyaloshinsky field. The result (4.3.30) was obtained by *Ozhogin* [4.14]. It differs essentially from ferromagnets in its spherical symmetry: $\tilde{V}(\boldsymbol{k})$ does not depend on the direction of $\boldsymbol{k}$ and therefore under the spherical symmetry of magnon damping in the above-threshold state pairs with all directions of $\boldsymbol{k}$ must be excited, i.e. "the whole resonance surface is excited". Taking into account that in easy-plane antiferromagnets $\omega_0^2 = g^2 H(H + H_D)$ (see (3.2.5)) and usually H and H_D are of the same order, clearly $\tilde{V}(\boldsymbol{k}) \simeq g$, i.e. has the same order of magnitude as $V(\boldsymbol{k})$ in the method of the parallel pumping in ferromagnets (4.3.21). Spin wave damping in good samples of antiferromagnetic $MnCO_3$ under helium temperatures is about 10^6 s^{-1}, so the threshold field h_{th} is of the order of 0.1 Oe. This value is readily obtained experimentally.

4.3.6 Parametric Pumping of Nuclear Magnons

It should be recalled that the concept of *nuclear magnons* in antiferromagnets has been briefly described in Sect. 3.2.3. Their dispersion law $\omega_n(\boldsymbol{k})$ is given by (3.2.14). In [4.5] the thresholds $h_{\text{th}}^{(nn)}$ and $h_{\text{th}}^{(en)}$ of the processes

$$\omega_p = \omega_n(\boldsymbol{k}) + \omega_n(-\boldsymbol{k}), \quad \omega_p = \omega_n(\boldsymbol{k}) + \omega_e(-\boldsymbol{k}). \qquad (4.3.31a,b)$$

were calculated. The first one is the parametric excitation of two nuclear magnons and the second one is the parametric excitation of one electron and one nuclear magnon in "easy–plane" antiferromagnets. In our notation

$$h_{\text{th}}^{(nn)} = \gamma_n(\boldsymbol{k})/|V^{(nn)}(\boldsymbol{k})|, \quad h_{\text{th}}^{(en)} = \sqrt{\gamma_e(\boldsymbol{k})\gamma_n(\boldsymbol{k})}/|V^{(en)}(\boldsymbol{k})|. \quad (4.3.32)$$

Here $\gamma_n(\boldsymbol{k})$ and $\gamma_e(\boldsymbol{k})$ are the damping decrements of the electron and nuclear magnons respectively: $V^{(nn)}(\boldsymbol{k})$ and $V^{(en)}(\boldsymbol{k})$ are the effective amplitudes of the magnon interaction with the pumping in the processes (4.3.31a) and (4.3.31b) respectively. At $\omega_p \ll \Omega_e(0)$ [4.5]:

$$\begin{aligned} V^{(nn)}(\boldsymbol{k}) &= \frac{g^2(2H + H_D)[\omega_n^2(0) - \omega_n^2(\boldsymbol{k})]}{2\omega_e^2(\boldsymbol{k})\omega_n(\boldsymbol{k})}, \\ V^{(en)}(\boldsymbol{k}) &= \frac{g^2(2H + H_D)\sqrt{\omega_n^2(0) - \omega_n^2(\boldsymbol{k})}}{2\omega_e(\boldsymbol{k})\sqrt{\omega_e(\boldsymbol{k})\omega_n(\boldsymbol{k})}}. \end{aligned} \qquad (4.3.33)$$

In the designation of the present section the interaction amplitude of the external field with two electron magnons (4.3.30) at $\omega_p \ll \Omega_e(\boldsymbol{k})$ has the form:

$$V(\boldsymbol{k}) = V^{(ee)}(\boldsymbol{k}) = g^2(2H + H_D)/[2\omega_e(\boldsymbol{k})]. \qquad (4.3.34)$$

Obviously, the relation

$$V_{\text{th}}^{(ee)}(\boldsymbol{k})V^{(nn)}(\boldsymbol{k}) = [\tilde{V}^{(en)}(\boldsymbol{k})]^2. \qquad (4.3.35)$$

is satisfied. A direct consequence of (4.3.35) is the relation for the threshold fields $h_{\text{th}}^{(ee)} h_{\text{th}}^{(nn)} = [h_{\text{th}}^{(en)}]^2$ corresponding to the excitation of the nuclear and electron magnons with the same values of the wave vector. This relation probably has no important physical meaning, but it can be useful for numerical estimations.

In conclusion, note that all the above processes have been observed and investigated experimentally. Thus, in MnCO$_3$ at $T \simeq 2$ K and frequency $\omega(\boldsymbol{k}) = 0.9$ gGz, $h_{\text{th}}^{(nn)} \simeq (0.08 - 0.15)$ Oe and $h_{\text{th}}^{en} \simeq (0.4 - 1.0)$ Oe depending on the value of the external field H.

5 Stationary Nonlinear Behavior of Parametrically Excited Waves. Basic S-Theory

5.1 History of the Problem

The development of the parametric instability in a continuous medium (if the system's size is big enough in comparison with the wavelength of the excited waves) results in a great number of waves simultaneously excited and intensively interacting. The state of the waves is determined mostly by some particular factors: the dispersion law of the waves and the nonlinear and dissipative properties of the medium. These factors can vary considerably in different cases.

It became clear in the early seventies that there is one simple and at the same time very important specific case when one can formulate a general nonlinear theory of the parametric excitation of waves. This is the case of pumping by a spatially homogeneous field with $k_p = 0$ or by a very long wave $k_p \ll k', k''$. Then the characteristics of the state above threshold may be assumed to be statistically homogeneous. This is the case in most experiments on the parametric excitation of spin waves in magnetically ordered dielectrics. Note that these are the most "pure" experiments in the physics of nonlinear waves because it is relatively simple to carry them out (in comparison, say, with experiments on plasma or the experimental study of nonlinear optics) and the high quality of the dielectrics. Experiments on the ferrimagnetic YIG - Yttrium Iron Garnet - are the most successful because of its many unique properties. Most of the experiments that are described in this book have been performed on YIG. Experiments on antiferromagnets $MnCO_3$, $CsMnF_3$ and $FeBO_3$ have also produced valuable results and are described here.

The first experimental data on the behavior of the spin waves above the threshold as well as the first models of their behavior above the threshold were obtained in the early 60s. The researchers who tried to suggest such models primarily aimed at obtaining the mechanism limiting the increasing amplitude of unstable spin waves. The first step in this direction was made by *Suhl* in [5.1]. He showed that under the pumping of spin waves by the homogeneous precession of the magnetization the principal limiting mechanism is the feedback effect on the pumping. This leads to "freezing" of the spin wave amplitude at the threshold level. However to consistently explain the nonlinear behavior of spin waves under parallel pumping was not so easy. The "traditional" mechanism limiting the parametric instability under parametric resonance with a small number of degrees of freedom, i.e. the nonlinear damping and nonlinear frequency shift, proven to be inadequate. The nonlinear damping in most cases is too weak and sensitive to the magnitude of the constant magnetic field to account for the observed level of the spin waves. The nonlinear frequency shift does not limit the parametric resonance in the continuous medium at all since at any amplitude the renormalized frequencies of some waves completely satisfy the resonance conditions. *Schlömann* in 1962 [5.2] made an important contribution to understanding the behavior of the spin waves above the threshold. He showed that the nonlinear interaction of the spin waves must be allowed for and presumed that the most important role in this interaction is played by the nonlinear processes satisfying the conditions

$$\omega(\boldsymbol{k}) + \omega(-\boldsymbol{k}) = \omega(\boldsymbol{k}_1) + \omega(-\boldsymbol{k}_1) \tag{5.1.1}$$

without taking the waves out of parametric resonance. Pioneering works by *Zakharov, L'vov* and *Starobinets* published in 1969–70 [5.3 – 6], opened up a new stage in the study of nonlinear processes under parametric excitation. The processes (5.1.1) were shown to retain phase correlation within each parametrically excited pair of waves with opposite wave vectors and to result in the self-consistent change of the total phase of waves in each pair. This decreases the energy flux from the pumping to the system of waves and leads to the limitation on their amplitudes. The waves whose renormalized frequencies fully satisfy the parametric resonance conditions in this case prove to be excited. This *phase mechanism* for limitation on the wave amplitude is typical in a continuous medium. It is essential in systems with the large sizes in comparison with the wavelength. Phase mechanism is of principal importance under the parallel pumping of spin waves in the non-decay part of spectrum. It is convenient to study the processes (5.1.1) (allowing for the necessary phase correlations) in the mean–field approximation where the role of the nonlinear wave interaction is reduced to the renormalizing coefficients of the linear equations (the wave frequency $\omega(\boldsymbol{k})$ and the pumping amplitude $hV(\boldsymbol{k})$), describing the parametric instability. The theory based on this approximation (*Zakharov, L'vov, Starobinets* [5.6]) was later called the S-*theory*. In their works published in 1970 – 74 these authors and their colleagues *Cherepanov, Musher, Zautkin, Rubentchik* and some others managed to develop considerably the study of spin wave behavior above the threshold within the S-theory, and its generalizations. A qualitative explanation has been given to numerous experimentally observed effects and good agreement between theory and experiment has been achieved [5.7–21]. For example, within the frame of the S-theory the giant auto-oscillations of the magnetization under parametric excitation of spin waves discovered by *Hartwick*, *Peressini* and *Weiss* [5.22] have been explained. A first presentation of the S-theory has been made in the review by *Zakharov*, *L'vov* and *Starobinets* published in [5.23] (based primary on their theoretical and experimental results).

In 1975–80 some interesting works developing the S-theory were published (see, for example, [5.24–30]). But the process of theoretical investigation of nonlinear phenomena under spin-wave parametric excitation slowed down. This was mainly due first to the fact that in 1974 the theory got ahead of experimental studies, and second, that it was not yet commonly accepted by physicists. At that time several publications appeared which either disproved the S-theory (see, e.g., [5.31]) or, on the contrary, obtained again its results using different methods (e.g. [5.30–32]). Incorrect results obtained in some theoretical publications of a transient character making use of the S-theory may be due to the insufficient understanding of the limits of its applicability (e.g., [5.33, 34]). At the same time *Melkov*, *Prozorova*, *Smirnov*, *Ozhogin*, *Zautkin* and their colleagues [5.35–68] obtained interesting and important experimental results on the nonlinear behavior of parametric spin waves. These works were based on the new understanding of the physical processes and phenomena beyond the threshold of the parametric excitation of waves. The basic deductions of the S-theory were confirmed, the details of the nonlinear behavior of the spin waves were clarified and qualitatively new nonlinear effects were discovered that would have been difficult to predict. Two comparatively short reviews devoted to the S-theory were published recently. The first one (*L'vov* [5.69]) deals mostly with theory, the second one (*L'vov* and *Prozorova* [5.70] is devoted primarily to experimental studies of the parametric excitation of magnons in ferro- and antiferromagnets. Chapters 5-7 of the present book give a detailed review of the current state of the S-theory describing the nonlinear behavior of parametrically excited waves in the mean–field approximation. A systematic presentation of some generalizations and modifications of the S-theory for more complicated conditions of parametric excitation (in the presence of the nonlinear wave damping, under violated space homogeneity, for incoherent pumping, under a swept pumping frequency, under simultaneously excited waves of two types etc.) is given for the first time. Much attention will also be given to the description and discussion of various experimental studies of the nonlinear behavior of parametrically excited magnons from the viewpoint of the S-theory. These results (theoretical as well as experimental) may be significant not only for a certain branch of the physics of magnetodielectrics. They are and most certainly will be of great importance for the development of the physics of nonlinear waves in other media. That is why we did our best to separate the results of the general nonlinear theory of the parametric wave excitation in our presentation from the results characteristic of the magnetics. Similarly, discussing the experimental studies of the nonlinear magnons I have tried to emphasize results generally significant for the physics of nonlinear waves. The introduction to the nonlinear theory of the parametric excitation of waves (Chaps. 5 and 7) proceeding from the Hamiltonian equations of motion is based on the same approach and gives no detailed treatment of the nature of specific waves or the medium in which they are excited. This chapter studies the stationary state of the parametrically excited waves.

5.2 Statement of a Problem of Nonlinear Wave Behavior

The following assumptions, simplifications and approximations were made in formulating the nonlinear theory of the parametric excitation of waves:

1. The medium in which waves are propagating (spin waves in magnetodielectrics, various waves in plasma, waves on a liquid surface etc.) is unbounded and spatially homogeneous. To this end the samples of magnetodielectrics, plasma etc. must be of sufficiently high quality, and their linear size L must significantly exceed the mean free path of the waves: $l \simeq \partial \omega / \gamma \partial k$. The effect of statistical random inhomogeneities on the propagation of the waves and their nonlinear behavior will be considered in Chap. 10.

2. Waves in the medium will be described within the frame of the classical Hamiltonian formalism described in Chap. 1, i.e. using the equations of motion (1.3.1) for the complex amplitudes of the travelling waves $b(\boldsymbol{k}), b^*(\boldsymbol{k})$, for which the quadratic Hamiltonian $\mathcal{H}_2$ is diagonal. This approach is in most cases indisputable. However, as mentioned in Chap. 3, it is not evident when the spin waves (magnons) are described at temperatures not small in comparison with the Curie temperature. The exact method for the description of magnons interacting under finite temperatures must be based on the spin diagram techniques for the non-equilibrium processes (*Belinicher, L'vov* [5.71]). However for simplicity we shall proceed here from the Bloch equations for the canonical variables assuming them to give considerable evidence for spin waves.

3. As in Chap. 1 the wave damping at this stage of investigation will be taken into account phenomenologically via the dissipative term $\gamma(\boldsymbol{k})b(\boldsymbol{k},t)$ in (1.3.3). A doubt may arise as to whether this procedure is justified in describing coherent wave systems where the phase relations are significant. Later in Chap. 10 the justification of this procedure will be presented and it will be shown that the decrement of the wave damping in (1.3.3) can be obtained by means of the ordinary kinetic equation. This, however, does not refer to the special case (considered in Chap. 10) when the damping of waves is due to scattering by random heterogeneities.

4. Simplifications of items 1–3 are not specific to the problem of the nonlinear behavior of parametrically excited waves; they are employed in the S-theory as in many other problems of the physics of nonlinear waves. Here we shall discuss a more specific problem: how to select the Hamiltonian of the wave interaction in the S-theory. The frequencies of all parametric waves are almost the same (in first-order processes of parametric excitation they equal half the pumping frequency) so only the four-wave scattering processes of the $2 \to 2$ type are in resonance with parametric waves. These processes were described by the Hamiltonian (1.1.32). The remaining part of $\mathcal{H}_{\text{int}}$ specifies the interaction of parametric waves with the thermal bath of the thermally excited waves and results in the damping of the parametric waves phenomenologically allowed for in (1.3.3). Running a little ahead it can be noted that this procedure has a large region of applicability with respect to the pumping amplitude if three-wave decay processes of one parametric wave into two thermal waves are forbidden. If these processes are allowed the region of applicability is narrower. Sections 5.6.2, 11.2.1 and 11.4.3 consider in detail the influence of the three-wave processes of the interaction of parametric waves with thermal waves on the nonlinear behavior of the parametric waves.

5. The external action on the medium – the parametric pumping – will be considered as monochromatic and spatially homogeneous. These assumptions prove to be correct for all methods of parametric excitation of spin waves considered in 5.4.3. Indeed, the characteristic values of the pumping frequencies ω_p of the spin waves are within the range from 10 to 36 gHz, which corresponds to the wavelengths of electromagnetic radiation from 3.0 to 0.8 cm. They considerably exceed the usual experimental values of the parametric spin wave lengths – $(10^{-1}\text{--}10^{-4})$ cm. As to the degree of pumping incoherence, the frequency linewidth of the experimentally used SHF radiation sources – magnetrons and klystrons – is at least by a factor of 10^2 less than the smallest attained linewidth of the spin waves $\gamma(\boldsymbol{k}) = 10^5$ Hz.

For first-order parametric processes (the decay of one "quantum" of the pumping field into two waves) the Hamiltonian of the interaction with the pumping $\mathcal{H}_\text{p}$ has the form (1.4.18):

$$\mathcal{H}_\text{p} = \frac{1}{2} \sum_{\boldsymbol{k}} \left[hV(\boldsymbol{k}) b^*(\boldsymbol{k}) b^*(-\boldsymbol{k}) \exp(i\omega t) + \text{c.c.} \right]. \tag{5.2.1}$$

For the Suhl second-order processes (the decay of two magnons with $\boldsymbol{k} = 0$ into two magnons with $\boldsymbol{k} \neq 0$) the Hamiltonian $\mathcal{H}_\text{p}$ can be derived from (5.2.1) by substituting $hV(\boldsymbol{k})\exp(i\omega t) \to S(0,\boldsymbol{k}) b_0^2$. We shall therefore take $\mathcal{H}_\text{p}$ to have the form (5.2.1) and if necessary this substitution will only in the final formula actually be carried out.

6. Here we shall write the equations of the wave motion that will be subsequently used to analyze the nonlinear behavior of the parametrically excited waves. Substituting into dynamic equations of motion (1.1.3) the

Hamiltonian of the waves $\mathcal{H}_{\text{int}} = \mathcal{H}_{\text{p}} + \mathcal{H}_4$ where $\mathcal{H}_{\text{p}}$ and $\mathcal{H}_4$ are given by (5.2.1) and (1.1.32) respectively, we finally get:

$$\left[\frac{\partial}{\partial t}+\gamma(\boldsymbol{k}) + i\omega(\boldsymbol{k})\right]b(\boldsymbol{k},t) + ihV(\boldsymbol{k})\exp(i\omega t)b^*(-\boldsymbol{k},t)$$
$$= -i\frac{\partial \mathcal{H}_4}{\partial b^*(\boldsymbol{k},t)} = -i\sum_{k+1=2+3} T(\boldsymbol{k},1;2,3)b_1^* b_2 b_3 \,. \qquad (5.2.2)$$

5.3 Phase Relations and Mechanisms for Amplitude Limitation

5.3.1 Analysis of Phase Relations

The initial equations of motion (5.2.2) in the linear approximation (i.e. at $\mathcal{H}_4 = 0$) split into independent pairs of equations for waves with equal and oppositely directed wave vectors $\pm \boldsymbol{k}$. On eliminating the fast time-dependence, i.e. in the "slow" variables

$$a(\boldsymbol{k},t) = b(\boldsymbol{k},t)\exp(i\omega t/2)\,, \qquad (5.3.1)$$

they assume the form

$$\begin{aligned}\{\partial/\partial t + \gamma(\boldsymbol{k}) + i[\omega(\boldsymbol{k}) - \omega_{\text{p}}/2]\}a(\boldsymbol{k},t) + ihV(\boldsymbol{k})a^*(-\boldsymbol{k},t) = 0\,,\\ \{\partial/\partial t + \gamma(\boldsymbol{k}) - i[\omega(\boldsymbol{k}) - \omega_{\text{p}}/2]\}a^*(-\boldsymbol{k},t) + ihV(\boldsymbol{k})a(\boldsymbol{k},t) = 0\,.\end{aligned} \qquad (5.3.2)$$

Taking $a(\boldsymbol{k},t)$, $a(-\boldsymbol{k},t) \propto \exp[\nu(\boldsymbol{k})t]$, we have

$$\nu(\boldsymbol{k}) = -\gamma(\boldsymbol{k}) \pm \sqrt{|hV(\boldsymbol{k})|^2 - [\omega(\boldsymbol{k}) - \omega_{\text{p}}/2]^2}\,. \qquad (5.3.3)$$

Obviously, this expression coincides with (1.4.19) which was obtained in Chap. 1, where the parametric instability had been considered as a special case of three-wave decay instability of the wave with $\boldsymbol{k}_{\text{p}} = 0$. The minimum threshold corresponding to the parametric resonance $2\omega(\boldsymbol{k}) = \omega_{\text{p}}$ is obtained from the condition $|hV(\boldsymbol{k})| = \gamma(\boldsymbol{k})$, which has a simple meaning of energy balance. Indeed, the energy flux W_+ from the pumping to the wave pair $\pm \boldsymbol{k}$ is

$$\begin{aligned}W_+ &= -\partial \mathcal{H}_{\text{p}}/\partial t = i\omega_{\text{p}}[hV(\boldsymbol{k})a^*(\boldsymbol{k})a^*(-\boldsymbol{k}) + \text{c.c.}]\\ &= 2|hV(\boldsymbol{k})|\omega_{\text{p}}|a(\boldsymbol{k})|^2 \sin[\Psi_{\text{p}}(\boldsymbol{k}) - \Psi(\boldsymbol{k})]\,.\end{aligned} \qquad (5.3.4)$$
$$a(\boldsymbol{k}) = |a(\boldsymbol{k})|\exp[-i\varphi(\boldsymbol{k})]\,,$$

where $\Psi(\boldsymbol{k}) = \varphi(\boldsymbol{k}) + \varphi(-\boldsymbol{k})$ is the common phase of the pair and $\Psi_{\text{p}}(\boldsymbol{k}) = \arg\{hV(\boldsymbol{k})\}$ is the pumping phase. On the other hand, the energy W_- dissipated by the pair per unit time is

$$W_- = 2\gamma(\boldsymbol{k})[\omega(\boldsymbol{k})|a(\boldsymbol{k})|^2 + \omega(-\boldsymbol{k})|a(-\boldsymbol{k})|^2] = 2\omega_{\rm p}\gamma(\boldsymbol{k})|a(\boldsymbol{k})|^2 \ . \quad (5.3.5)$$

At the threshold point $W_+ = W_-$. The greatest energy flux and the lowest instability threshold are characteristic of the pair with the most advantageous phase relation: $\Psi(\boldsymbol{k}) = \Psi_{\rm p}(\boldsymbol{k}) + \pi/2$. For the threshold in this case the relation $|hV(\boldsymbol{k})| = \gamma(\boldsymbol{k})$ holds true again. The condition for parametric resonance can evidently be simultaneously fulfilled for a great number of pairs whose wave vectors are on the resonance surface. The pairs with the minimum ratio $\gamma(\boldsymbol{k})/|hV(\boldsymbol{k})|$ have the minimum threshold of excitation: $h_{\rm th} = \min[\gamma(\boldsymbol{k})/|hV(\boldsymbol{k})|]$. At $h > h_{\rm th}$ the amplitudes of the pairs begin to increase exponentially:

$$\begin{aligned} a(\boldsymbol{k},t) &= a(\boldsymbol{k})\exp[\nu(\boldsymbol{k})t - i\Psi(\boldsymbol{k})/2], \\ a^*(-\boldsymbol{k},t) &= a^*(-\boldsymbol{k})\exp[\nu(\boldsymbol{k})t + i\Psi(\boldsymbol{k})/2] \end{aligned} \quad (5.3.6)$$

with the increment (5.3.3). It follows from (5.3.2) that

$$\cos[\Psi(\boldsymbol{k}) - \Psi_{\rm p}(\boldsymbol{k})] = [\omega(\boldsymbol{k}) - \omega_{\rm p}/2]/|hV(\boldsymbol{k})| \ . \quad (5.3.7)$$

This means that in the linear stage of the parametric instability a certain relation between the phases of the waves in pairs is established. The phase correlation of waves with equal and oppositely directed wave vectors may, by analogy with superconductivity, be called *coupling*. Unlike superconductivity, the physical cause for the wave coupling is that the pumping separates pairs of waves with the maximum instability increment from the thermal bath of waves with chaotic phases. Phase correlation will later be shown to be complete at the nonlinear stage of the development of the instability. This means that although the value $a(\boldsymbol{k},t)$ is random, the value $a(\boldsymbol{k},t)a(-\boldsymbol{k},t)$ will be dynamic. In this case

$$\langle a(\boldsymbol{k},t)a(-\boldsymbol{k},t)\rangle = a(\boldsymbol{k},t)a(-\boldsymbol{k},t), \quad \langle \Psi(\boldsymbol{k},t)\rangle = \Psi(\boldsymbol{k},t).$$

5.3.2 Nonlinear Mechanisms for Limiting Parametric Instability

Nonlinear damping, i.e. a dependence of $\gamma(\boldsymbol{k})$ on the squared amplitudes of parametric waves $|a(\boldsymbol{k})|^2$ (*Schlömann* [5.72], *Gottlib* and *Suhl* [5.73]) can serve as the simplest mechanism of this kind. The stationary amplitudes of waves are determined by the well-known condition of energy balance: $|hV(\boldsymbol{k})| = \gamma(\boldsymbol{k})$. The simplest dependence is chosen for the qualitative analysis: $\gamma = \gamma_0 + \eta \sum |a(\boldsymbol{k})|^2$. Then

$$\sum_{\boldsymbol{k}} |a(\boldsymbol{k})|^2 = (|hV| - \gamma_0)/\eta = V(h - h_{\rm th})/\gamma_0 \ . \quad (5.3.8)$$

The phases $\Psi(\boldsymbol{k})$ in this case are obtained from the conditions of the parametric resonances and are shifted by $\pi/2$ from the pumping phase.

One more evident limiting mechanism for parametric instability is connected with the feedback of parametrically excited waves on the pumping. The energy flux into the system of parametric waves required for maintaining their number at a finite level is taken off from the pumping and decreases its amplitude. In the simplest case when there are no other limiting mechanisms the pumping amplitude is "frozen" at the threshold level, and the number of excited parametric waves is obtained from the energy balance condition in the system of the pumping. In some cases when, say, the homogeneous precession of magnetization excited in resonance by the external SHF field serves as the pumping for spin waves in the ferromagnets, the *feedback mechanism* is really important and will be treated in detail in Sect. 5.6.3. In other cases, for example, under the parallel pumping when the spin waves in a small sample of a magnetodielectric are excited by the SHF field of a big cavity-resonator the feedback of the spin waves on the field in the resonant cavity, as a rule, can be neglected.

The third *phase limiting mechanism* suggested by *Zakharov, L'vov* and *Starobinets* [5.4] plays the key role in the parametric excitation of waves. The present chapter will deal with its detailed treatment. Now it will only be noted that the phase limiting mechanism is connected with the coupling which leads to the misphasing ($\sin \Psi(\mathbf{k}) < 1$) between the pairs and the external pumping, i.e. to the decreased energy flux into the system.

The fourth limiting mechanism due to the generation and the collapse of solitons is active when the amplitude of waves interacting with the pumping $V(\mathbf{k})$ in the Hamiltonian (5.2.1) is maximum at a single pair of points $\pm \mathbf{k}$ so that above the threshold a narrow wave packet is excited. This limiting mechanism will be described in Sect. 8.3.

5.4 Basic Equations of Motion in the S-Theory

5.4.1 Statistical Properties of a Non-Interacting Field

As known from statistical physics, in the absence of interaction the statistical properties of the non-interacting wave field are Gaussian. This means that all the odd-order correlators are zero:

$$\langle b(\mathbf{k},t) \rangle = 0, \quad \langle b^*(\mathbf{k},t)b(\mathbf{k}_1,t)b(\mathbf{k}_2,t) \rangle = 0, \tag{5.4.1}$$

and the even-order correlators are expressed in terms of various products of the double correlators (correlation functions)

$$\langle b(\mathbf{k},t)b^*(\mathbf{k}',t) \rangle = n(\mathbf{k},t)\delta(\mathbf{k}-\mathbf{k}') . \tag{5.4.2}$$

In particular, the splitting rule of the fourth-order correlators has the form

$$\langle b_1^* b_2^* b_3 b_4 \rangle = n_1 n_2 [\delta(\mathbf{k}_1 - \mathbf{k}_3)\delta(\mathbf{k}_2 - \mathbf{k}_4) + \delta(\mathbf{k}_1 - \mathbf{k}_4)\delta(\mathbf{k}_2 - \mathbf{k}_3)] . \tag{5.4.3}$$

Now and further on we shall use the short notation $b_j = b(\boldsymbol{k}_j,t)$, $n_j = n(\boldsymbol{k}_j,t)$. We shall also use the splitting rule of the sixth-order correlators

$$\begin{aligned}\langle b_1^* b_2^* b_3^* b_4 b_5 b_6 \rangle &= n_1 n_2 n_3 \\ &\times \{\delta(\boldsymbol{k}_1 - \boldsymbol{k}_4)[\delta(\boldsymbol{k}_2 - \boldsymbol{k}_5)\delta(\boldsymbol{k}_3 - \boldsymbol{k}_6) + \delta(\boldsymbol{k}_2 - \boldsymbol{k}_6)\delta(\boldsymbol{k}_3 - \boldsymbol{k}_5)] \\ &+ \delta(\boldsymbol{k}_1 - \boldsymbol{k}_5)[\delta(\boldsymbol{k}_2 - \boldsymbol{k}_4)\delta(\boldsymbol{k}_3 - \boldsymbol{k}_6) + \delta(\boldsymbol{k}_2 - \boldsymbol{k}_6)\delta(\boldsymbol{k}_3 - \boldsymbol{k}_4)] \\ &+ \delta(\boldsymbol{k}_1 - \boldsymbol{k}_6)[\delta(\boldsymbol{k}_2 - \boldsymbol{k}_4)\delta(\boldsymbol{k}_3 - \boldsymbol{k}_5) + \delta(\boldsymbol{k}_2 - \boldsymbol{k}_5)\delta(\boldsymbol{k}_3 - \boldsymbol{k}_4)]\}.\end{aligned} \quad (5.4.4)$$

Equations (5.3.4, 5) are the classical limit of the well-known Week theorem for the Bose-operators. It must also be noted that the δ-functions of the momenta differences in the right-hand sides of (5.4.2–4) are the consequence of the spatial homogeneity of the problem. And, finally, the correlators $n(\boldsymbol{k},t)$ have the dimensionality of the action (erg·s). They are the classical analogues of the dimensionless quantum-mechanical occupation numbers $n_{\text{qm}}(\boldsymbol{k},t)$. Within the limit $n_{\text{qm}} \gg 1$, $\hbar n_{\text{qm}}(\boldsymbol{k},t) = n(\boldsymbol{k},t)$.

5.4.2 Mean–Field Approximation

In this section the mean–field approximation will be formulated and statistical equations of motion for conjugate correlators describing the system of parametrically excited waves will be obtained in first-order perturbation theory with respect to the interaction Hamiltonian $\mathcal{H}_{\text{int}}$. Differentiating (5.4.2) for $n(\boldsymbol{k},t)$ with respect to time and using the equations of motion (5.2.2, 3), we obtain:

$$\left\{\left[\frac{\partial}{\partial t} + 2\gamma(\boldsymbol{k})\right]n(\boldsymbol{k},t) + 2\text{Im}\left[\hbar V^*(\boldsymbol{k})\sigma(\boldsymbol{k},t)\right]\right\}\delta(\boldsymbol{k} - \boldsymbol{k}_1) \\ = \text{Im}\left\{\sum_{k+2=3+4} T(\boldsymbol{k},2;3,4)\langle a_1^* a_2^* a_3 a_4\rangle\right\}. \quad (5.4.5)$$

where $a_j = a(\boldsymbol{k}_j,t)$ are slow variables (5.3.1). In (5.4.5) for the first time in our theory appeared a new object - the *anomalous double correlator* $\sigma(\boldsymbol{k},t)$, which will be given by the following formula

$$\langle a(\boldsymbol{k},t)a(\boldsymbol{k}_1,t)\rangle = \sigma(\boldsymbol{k},t)\delta(\boldsymbol{k} + \boldsymbol{k}_1) . \quad (5.4.6)$$

With such a definition of $\sigma(\boldsymbol{k},t)$ via slow variables, the explicit time dependence in (5.4.5) is absent. Recall that in the free wave field the anomalous correlator σ is zero because it vanishes under averaging over the wave phases. Indeed,

$$\sigma(\boldsymbol{k}) \propto \langle |a(\boldsymbol{k})||a(-\boldsymbol{k})|\rangle \exp\{i[\varphi(\boldsymbol{k}) + \varphi(-\boldsymbol{k})]\} . \quad (5.4.7)$$

In the presence of pumping, however, at the linear stage of the parametric instability the sum of phases $[\varphi(\boldsymbol{k}) + \varphi(-\boldsymbol{k})]$ according to (5.3.7) is rigidly fixed. Therefore at the linear stage $\sigma(\boldsymbol{k},t) \neq 0$ and just in case must be allowed for in the nonlinear equations, too.

Differentiating $\sigma(\boldsymbol{k},t)$ (5.4.6) with respect to time and employing the equations of motion (5.2.2) and (5.2.3), we obtain the equation of motion for $\sigma(\boldsymbol{k},t)$:

$$\left\{\left[\frac{\partial}{\partial t} + [\gamma(\boldsymbol{k}) + \gamma(-\boldsymbol{k})] + i[\omega(\boldsymbol{k}) + \omega(-\boldsymbol{k}) - \omega_{\mathrm{p}}]\right]\sigma(\boldsymbol{k},t) \right. $$
$$\left. + ihV(\boldsymbol{k})[n(\boldsymbol{k},t) + n(-\boldsymbol{k},t)]\right\}\delta(\boldsymbol{k}+\boldsymbol{k}_1) \qquad (5.4.8)$$
$$= \frac{1}{2}\sum_{2,3,4}\left[T(\boldsymbol{k},2;3,4)\langle a_1 a_2^* a_3 a_4\rangle + T(1,2;3,4)\langle a_k a_2^* a_3 a_4\rangle\right].$$

Equations (5.4.5, 8) are accurate but not constructive since they express double correlators $n(\boldsymbol{k})$ and $\sigma(\boldsymbol{k})$ via the fourth-order correlators. The equations for the fourth-order correlators, in turn, will contain the sixth-order correlators, etc.

At the first stage of the theory development this chain of equations should be closed in the simplest way taking the interaction Hamiltonian to be small in any required sense. This enables us to confine ourselves to first-order perturbation theory with respect to H_{int}. To formulate this approximation, note that the first parts of (5.4.5, 8) explicitly contain amplitudes of the interaction Hamiltonian $T(1,2;3,4)$. Therefore in the linear approximation with respect to $\mathcal{H}_4$ the fourth-order correlators must be calculated in the zeroth approximation with respect to $\mathcal{H}_4$, as for the free wave field in the presence of pumping. This implies that the fourth-order correlators can be expressed in terms of $n(\boldsymbol{k},t)$ and $\sigma(\boldsymbol{k},t)$ using the following splitting procedure generalizing the standard procedure (5.4.3):

$$\langle a_1^* a_2^* a_3 a_4\rangle = n_1 n_2 [\delta(\boldsymbol{k}_1-\boldsymbol{k}_3)\delta(\boldsymbol{k}_2-\boldsymbol{k}_4)$$
$$+ \delta(\boldsymbol{k}_1-\boldsymbol{k}_4)\delta(\boldsymbol{k}_2-\boldsymbol{k}_3)] \qquad (5.4.9)$$
$$+ \sigma_1^* \sigma_3 \delta(\boldsymbol{k}_1+\boldsymbol{k}_2)\delta(\boldsymbol{k}_3+\boldsymbol{k}_4),$$

$$\langle a_1^* a_2 a_3 a_4\rangle = n_1[\sigma_3\delta(\boldsymbol{k}_1-\boldsymbol{k}_2)\delta(\boldsymbol{k}_3+\boldsymbol{k}_4)$$
$$+ \sigma_2\delta(\boldsymbol{k}_1-\boldsymbol{k}_3)\delta(\boldsymbol{k}_2+\boldsymbol{k}_4)] \qquad (5.4.10)$$
$$+ \sigma_2\delta(\boldsymbol{k}_1-\boldsymbol{k}_4)\delta(\boldsymbol{k}_2+\boldsymbol{k}_3)]].$$

Similar relations are well known in theoretical physics, i.e. in the theory of superfluidity of a weakly non-ideal Bose gas [5.74]. Relations (5.4.9, 10) enable us to close the system of equations (5.4.5, 8):

$$\frac{\partial \sigma(\boldsymbol{k},t)}{\partial t} = \{-2\gamma(\boldsymbol{k}) + i[2\omega_{\mathrm{NL}}(\boldsymbol{k},t) - \omega_{\mathrm{p}}]\}\sigma(\boldsymbol{k},t)$$
$$- i[n(\boldsymbol{k},t) + n(-\boldsymbol{k},t)]P(\boldsymbol{k},t),$$
$$\frac{1}{2}\cdot\frac{\partial n(\boldsymbol{k},t)}{\partial t} = -\gamma(\boldsymbol{k})n(\boldsymbol{k},t) - \mathrm{Im}\{P^*(\boldsymbol{k},t)\sigma(\boldsymbol{k},t)\},$$
$$\frac{1}{2}\cdot\frac{\partial n(-\boldsymbol{k},t)}{\partial t} = -\gamma(\boldsymbol{k})n(-\boldsymbol{k},t) - \mathrm{Im}\{P^*(\boldsymbol{k},t)\sigma(\boldsymbol{k},t)\}, \qquad (5.4.11)$$
$$P(\boldsymbol{k},t) = hV(\boldsymbol{k}) + \sum_1 S(\boldsymbol{k},\boldsymbol{k}_1)\sigma(\boldsymbol{k}_1,t),$$
$$\omega_{\mathrm{NL}}(\boldsymbol{k},t) = \omega(\boldsymbol{k}) + 2\sum_1 T(\boldsymbol{k},\boldsymbol{k}_1)n(\boldsymbol{k}_1,t),$$

where $S(\boldsymbol{k},\boldsymbol{k}_1) = S^*(\boldsymbol{k}_1,\boldsymbol{k}) = T(\boldsymbol{k},-\boldsymbol{k},\boldsymbol{k}_1,-\boldsymbol{k}_1)/2$, and $T(\boldsymbol{k},\boldsymbol{k}_1) = T^*(\boldsymbol{k},\boldsymbol{k}_1) = T(\boldsymbol{k},\boldsymbol{k}_1,\boldsymbol{k},\boldsymbol{k}_1)/2$. The value $P(\boldsymbol{k},t)$ will be called the *complete pumping*. It differs from the *external pumping* $hV(\boldsymbol{k})$, describing the energy flux from the external fluid of the pumping h into the pair of waves with the wave vectors $\pm\boldsymbol{k}$ in the *self-consistent pumping* $\sum S(\boldsymbol{k},\boldsymbol{k}_1)\sigma(\boldsymbol{k}_1,t)$, specifying the energy exchange between the pairs due to their interaction under pairing (i.e. when $\sigma \neq 0$). The function $\omega_{\mathrm{NL}}(\boldsymbol{k},t)$ is the wave frequency renormalized due to the interaction. It differs from the frequency of the non-interacting wave field $\omega(\boldsymbol{k})$ in the nonlinear term proportional to $T(\boldsymbol{k},\boldsymbol{k}_1)$. We have already discussed such a renormalization when dealing with the four-wave processes in Sect. 1.5.

Equations (5.4.11) obtained in the present section will be referred to as the *basic equations of the S-theory*. The content of this theory presented in Chaps. 5–8 essentially consists of the analysis of different solutions of the equations (and also modifications for more complicated conditions of parametric excitation). The very name of "S-theory" reflects the decisive influence of the $S(\boldsymbol{k},\boldsymbol{k}_1)$ function (specifying the rate of energy exchange between $\pm\boldsymbol{k}$ and $\pm\boldsymbol{k}_1$ pairs) on the nonlinear behavior of the system of interacting parametric waves.

Recall that the basic equations of the S-theory (5.4.11) have been obtained in first-order perturbation theory with respect to $\mathcal{H}_4$ in the approximation not allowing for the correlation of the wave field fluctuations due to the interaction. In theoretical physics such an approximation is often called the *mean–field approximation*. Classical examples of such an approximation are the Curie-Weiss theory of the molecular field, the Landau theory of second-order phase transitions as well as the BCS-theory of superconductivity. Those highly meaningful examples refer to the physics of the second-order phase transitions. In the physics of nonlinear waves, however, the mean–field approximation is usually trivial since it does not account for non-trivial wave dynamics. It can be seen from (5.4.8) that in the absence of pumping and wave coupling (i.e. at $hV(\boldsymbol{k}) = 0$, $\sigma(\boldsymbol{k},t) = 0$) the kinetics of the number of waves is trivial: $\partial n(\boldsymbol{k},t)/\partial t = -2\gamma(\boldsymbol{k})n(\boldsymbol{k},t)$.

Non-trivial wave kinetics in the mean–field approximation can be a result not only of the coupling, but also of the spatial inhomogeneity or nonhermitian amplitude of the wave interaction $T(1,2;3,4)$. Theoretical treatment of the phenomena arising in this situation was presented in short e.g. in my "Lectures on the Physics of Nonlinear Phenomena" [5.75]. It is essential for explaining some facts of plasma physics, nonlinear optics and even environmental science. It has much in common with the S-theory.

5.4.3 General Analysis of Basic Equations of the S-Theory

Proceeding to the analysis of the basic equations of the S-theory (5.4.11), let us obtain from them the following relations

$$\begin{aligned}
\left[\frac{\partial}{\partial t} + 4\gamma(\boldsymbol{k})\right]\left[n(\boldsymbol{k},t)n(-\boldsymbol{k},t) - |\sigma(\boldsymbol{k},t)|^2\right] &= 0, \\
\left[\frac{\partial}{\partial t} + 2\gamma(\boldsymbol{k})\right]\left[n(\boldsymbol{k},t) - n(-\boldsymbol{k},t)\right] &= 0
\end{aligned} \quad (5.4.12)$$

showing that arbitrary initial distributions, over time of the order $1/\gamma(\boldsymbol{k})$, relax to the state (not necessarily stationary) in which $n(\boldsymbol{k},t) = n(-\boldsymbol{k},t)$ and $|\sigma(\boldsymbol{k},t)| = n(\boldsymbol{k},t)$. The conditions implying that the phases of wave pairs are fully correlated also at the nonlinear stage of the parametric instability make it possible to introduce instead of the variables $\sigma(\boldsymbol{k},t)$, $\sigma^*(\boldsymbol{k},t)$, $n(\boldsymbol{k},t)$ and $n(-\boldsymbol{k},t)$ only the two real variables $n(\boldsymbol{k},t)$ and $\Psi(\boldsymbol{k},t)$:

$$\begin{aligned}
\frac{\partial n(\boldsymbol{k},t)}{2\partial t} &= n(\boldsymbol{k},t)\left\{-\gamma(\boldsymbol{k}) + \mathrm{Im}\{P^*(\boldsymbol{k},t)\exp[-i\Psi(\boldsymbol{k},t)]\}\right\} \\
\frac{\partial \Psi(\boldsymbol{k},t)}{\partial t} &= \omega_{\mathrm{NL}}(\boldsymbol{k},t) - \frac{\omega_{\mathrm{p}}}{2} + \mathrm{Re}\{P^*(\boldsymbol{k},t)\exp[-i\Psi(\boldsymbol{k},t)]\}, \\
P(\boldsymbol{k},t) &= hV(\boldsymbol{k}) + \sum_1 S(\boldsymbol{k},\boldsymbol{k}_1)n(\boldsymbol{k}_1,t)\exp[-i\Psi(\boldsymbol{k},t)], \\
\omega_{\mathrm{NL}}(\boldsymbol{k},t) &= \omega(\boldsymbol{k}) + 2\sum_1 T(\boldsymbol{k},\boldsymbol{k}_1)n(\boldsymbol{k}_1,t) .
\end{aligned} \quad (5.4.13)$$

This may be written in a more compact form if instead of the two real variables, the wave number $n(\boldsymbol{k},t)$ and the phase of the pair $\Psi(\boldsymbol{k},t)$, we use one complex variable $c(\boldsymbol{k},t)$:

$$\begin{aligned}
n(\boldsymbol{k},t) = n(-\boldsymbol{k},t) &= |c(\boldsymbol{k},t)|^2, \qquad \sigma(\boldsymbol{k},t) = c^2(\boldsymbol{k},t), \\
c(\boldsymbol{k},t) &= \sqrt{n(\boldsymbol{k},t)}\exp[-i\Psi(\boldsymbol{k},t)/2] .
\end{aligned} \quad (5.4.14)$$

In those variables we have one complex equation instead of (5.4.11) for $n(\boldsymbol{k},t)$, $n(-\boldsymbol{k},t)$, $\sigma(\boldsymbol{k},t)$ and $\sigma^*(\boldsymbol{k},t)$ or (5.4.14) for $n(\boldsymbol{k},t)$ and $\Psi(\boldsymbol{k},t)$:

$$\frac{\partial c(\boldsymbol{k},t)}{\partial t} + \left\{\gamma(\boldsymbol{k}) + i\left[\omega_{\mathrm{NL}}(\boldsymbol{k},t) - \frac{\omega_{\mathrm{p}}}{2}\right]\right\}c(\boldsymbol{k},t) = -iP(\boldsymbol{k},t)c^*(\boldsymbol{k},t), \quad (5.4.15)$$

$$\omega_{\rm NL}(\bm{k},t) = \omega(\bm{k}) + 2\sum_{\bm{k}_1} T(\bm{k},\bm{k}_1)|c(\bm{k}_1,t)|^2 ,$$

$$P(\bm{k},t) = hV(\bm{k}) + \sum_{\bm{k}_1} S(\bm{k},\bm{k}_1)c^2(\bm{k},t) . \tag{5.4.16}$$

These equations can be obtained directly from the initial dynamic equations (5.2.2), if we assume $b(\bm{k})=b(-\bm{k})=c(\bm{k})$ and substitute the four-wave interaction Hamiltonian (1.1.32) for its diagonal part in pairs:

$$\tilde{\mathcal{H}}_{\rm S} = \sum_{\bm{k}_1,\bm{k}_2} T(\bm{k}_1,\bm{k}_2)|c_1|^2|c_2|^2 + \frac{1}{2}\sum_{\bm{k}_1,\bm{k}_2} S(\bm{k}_1,\bm{k}_2)c_1^{*2}c_2^2 . \tag{5.4.17}$$

This implies that the basic equations of the S-theory are in fact dynamic, although they are obtained through statistical averaging and correlation splitting by the (5.4.9,10) rule. Moreover, at $\gamma(\bm{k})=0$ the equations of the S-theory are Hamiltonian. In the variables c,c^* this fact is trivial, since the (5.4.15) coincide with (5.2.2). As to (5.4.13), obviously, they can be obtained by the variation of the Hamiltonian

$$\mathcal{H}_{\rm S} = \sum_1 \Big[\omega(\bm{k}_1) + \sum_2 T(\bm{k}_1,\bm{k}_2)n_2\Big]n_1 + \frac{1}{2}\sum_1 \Big[hV(\bm{k}_1)\cos\Psi_1 \\ + \frac{1}{2}\sum_2 S(\bm{k}_1,\bm{k}_2)n_2\cos(\Psi_1-\Psi_2)\Big] \tag{5.4.18}$$

by the rule

$$\frac{\partial n(\bm{k},t)}{\partial t} + \gamma(\bm{k})n(\bm{k},t) = -\frac{\delta \mathcal{H}_{\rm S}}{\delta n(\bm{k},t)},\quad \frac{\partial \Psi(\bm{k},t)}{\partial t} = \frac{\delta \mathcal{H}_{\rm S}}{\delta n(\bm{k},t)} . \tag{5.4.19}$$

The Hamiltonian $\tilde{\mathcal{H}}_{\rm S}$ (5.4.17) or $\mathcal{H}_{\rm S}$ (5.4.18) is called the *diagonal Hamiltonian of the S-theory*. In the Hamiltonians only such terms were retained that are either fully independent of the wave phases (the first sum in $\mathcal{H}_{\rm S}$), or depend only on the sum of the phases $\Psi(\bm{k})=\varphi(\bm{k})+\varphi(-\bm{k})$ in the pairs. All the other terms depending on the individual phases of waves (or, to be more exact, on the differences $\varphi(\bm{k})-\varphi(-\bm{k})$) are omitted. It is clear from the above that in first-order perturbation theory with respect to $\mathcal{H}_{\rm int}$ these terms are zero after averaging over wave phases of the Gaussian ensemble of the free wave field. The interaction itself, of course, leads to some correlations of the phases. Therefore, in the second (and higher) orders of perturbation theory with respect to $\mathcal{H}_{\rm int}$ additional terms will arise. Those terms will be proportional to $T^2(1,2;3,4)$, and will describe the four-wave scattering of particular parametric waves. This theory was named ST^2-*theory*; a short outline will be given in Sect. 11.2. The approximation of the S-theory will be shown to give a good description of the basic characteristics of parametric waves system up to the pumping amplitudes $h<h_{\rm S}$, where

$$h_{\rm S}V \simeq \sqrt{\gamma k \partial \omega(\bm{k})/\partial k} . \tag{5.4.20}$$

5.5 Ground State of System of Interacting Parametric Waves

5.5.1 Stationary States and Analysis of Instability

Now let us proceed to a discussion of the stationary states of the system of pairs in which all the numbers $n(\boldsymbol{k},t)$ and phases $\Psi(\boldsymbol{k},t)$ are time-independent. Taking in (5.4.13) derivatives with respect to time to be zero we immediately obtain for the point of $\boldsymbol{k}$-space, where $n(\boldsymbol{k}) \neq 0$, the condition

$$|P(\boldsymbol{k})|^2 = \gamma^2(\boldsymbol{k}) + [\omega_{\mathrm{NL}}(\boldsymbol{k}) - \omega_{\mathrm{p}}/2] \; . \qquad (5.5.1)$$

Before analyzing this result we shall make two remarks of a general character. First, it is clear that the amplitudes of the pairs differ from zero only over a thin layer near the resonance surface $2\omega(\boldsymbol{k}) = \omega_{\mathrm{p}}$. Because of this it is convenient to use the following coordinates in $\boldsymbol{k}$-space: κ is the deflection from this surface in the normal direction and Ω is the coordinate on the surface. Second, the coefficients of (5.4.13) with the dimensionality of frequency:

$$\gamma(\boldsymbol{k})\; , \quad hV(\boldsymbol{k})\; , \quad \sum_1 T(\boldsymbol{k},\boldsymbol{k}_1)n_1\; , \quad \sum_1 S(\boldsymbol{k},\boldsymbol{k}_1)n_1$$

are considerably less than the eigenfrequency $\omega(\boldsymbol{k})$. Therefore in theory it will be sufficient to allow dependence on κ only in the function $\omega(\boldsymbol{k}) - \omega_{\mathrm{p}}/2$ and in all other functions (namely $\gamma(\boldsymbol{k}), hV(\boldsymbol{k}), T(\boldsymbol{k},\boldsymbol{k}_1)$ and $S(\boldsymbol{k},\boldsymbol{k}_1)$) to substitute their values on the resonant surface ($\gamma(\Omega), hV(\Omega)$, etc). Making use of the above approximations one can easily obtain from (5.5.1) those κ for which $n(\kappa,\Omega)$ can be non-zero:

$$\omega_{\mathrm{NL}}(\kappa,\Omega) = \omega_{\mathrm{p}}/2 \pm \sqrt{|P(\Omega)|^2 - \gamma^2(\Omega)} \; . \qquad (5.5.2)$$

Thus, in the stationary state the distribution of the occupation numbers of the pairs is singular: $n(\kappa,\Omega)$ can differ from zero only on two surfaces (5.5.2). Note that there are numerous stationary states differing both in the function $P(\Omega)$ defining the surfaces (5.5.2) and in the distribution of $n(\Omega)$ over them. Indeed, one can arbitrarily set the direction of Ω where $n(\Omega)$ is equal to zero. In fact, of all the stationary states, only those states can be realized which are stable with respect to small perturbations. The requirement that the states should be stable considerably narrows the range of feasible stationary states. The study of stability of stationary states within the limits of the diagonal Hamiltonian can be shown to split into two independent problems: the investigation of *internal stability* with respect to the perturbation of amplitudes and the phases of pairs already existing and the study of the *external stability* with respect to the generation of new pairs.

The external stability is the easiest to treat. Let us write an equation for the wave pair of perturbations with the help of (5.4.15) $c(\boldsymbol{k},t)$ and $c^*(\boldsymbol{k},t)$ $\propto \exp[\nu(\boldsymbol{k}_1)t]$, analogous to (5.3.2):

$$\left\{\frac{\partial}{\partial t} + \gamma(\boldsymbol{k}_1) - i\left[\omega_{\mathrm{NL}}(\boldsymbol{k}_1) - \frac{\omega_{\mathrm{p}}}{2}\right]\right\}c(\boldsymbol{k}_1,t) + iP(\boldsymbol{k}_1)c^*(-\boldsymbol{k}_1,t) = 0 \ .$$

The expression for the external instability increment $\nu(\boldsymbol{k}_1)$ takes a form similar to (5.3.3) with the substitution $\gamma(\boldsymbol{k}) \to \gamma(\boldsymbol{k}_1)$, $\omega(\boldsymbol{k}) \to \omega_{\mathrm{NL}}(\boldsymbol{k}_1)$ and $hV(\boldsymbol{k}) \to P(\boldsymbol{k}_1)$:

$$\nu(\boldsymbol{k}) = -\gamma(\boldsymbol{k}_1) + \sqrt{|P(\boldsymbol{k}_1)|^2 - [\omega_{\mathrm{NL}}(\boldsymbol{k}_1) - \omega_{\mathrm{p}}/2]^2} \ .$$

The increment $\nu(\boldsymbol{k}_1)$ maximum with respect to k_1 (under fixed Ω_1) corresponds to $k_1 = k$, satisfying the equation $2\omega_{\mathrm{NL}}(\boldsymbol{k}) = \omega_{\mathrm{p}}$. This means that the most unstable waves have wave vectors $|\boldsymbol{k}_1| = k$ which are between the surfaces (5.5.2). The maximum $\nu(\Omega_1)$ of the increment $\nu(\boldsymbol{k}_1)$ equals:

$$\nu(\Omega_1) = \max_{k_1}\nu(\boldsymbol{k}_1) = |P(\Omega_1)| - \gamma(\Omega_1) \ .$$

The condition of external stability $\nu(\Omega) < 0$ can therefore be written in the form:

$$|P(\Omega)| \leq \gamma(\Omega) \ . \tag{5.5.3}$$

On the other hand, it follows from (5.5.2) that $|P(\Omega)| \geq \gamma(\Omega)$ for the directions Ω where $n(\kappa,\Omega) \neq 0$. Consequently, for those directions both inequalities are compatible only in the case $|P(\Omega)| = \gamma(\Omega)$, when two surfaces (5.5.2) converge into one

$$2\omega_{\mathrm{NL}}(\boldsymbol{k}) = \omega_{\mathrm{p}} \ . \tag{5.5.4}$$

Therefore the condition of external stability under the given angular distribution of the parametric waves completely eliminates the arbitrariness in choosing the surface on which $n(\boldsymbol{k}) \neq 0$. It will be called the *resonance surface* and the stationary state characterized by the external stability will be termed the *ground state*.

The above result has a simple physical meaning. As it is clear from (5.3.2), at the linear stage the waves with zero frequency shift are most closely connected to the pumping. The value $\omega_{\mathrm{NL}}(\boldsymbol{k}) - \omega_{\mathrm{p}}/2$ is a frequency shift allowing for the nonlinear terms. If two surfaces (5.5.2) on which $n(\boldsymbol{k}) \neq 0$ do not converge, the wave pair whose values fall into a spherical layer between these surfaces prove to be more closely connected with the pumping than the waves already excited. As a consequence, the waves in that spherical layer will increase.

It must be noted that the above-described ambiguity of the solution of stationary equations and the elimination of this arbitrariness (complete or partial) via the condition of stability is specific not only for the S-theory but is a common property of the mean–field approximation in the theory of nonlinear waves. Another method for eliminating that arbitrariness (technically a little bit more complicated but more natural for many people) consists in introducing the thermal noise into the S-theory by substituting the difference $\gamma(\boldsymbol{k})[n(\boldsymbol{k},t) - n_0(\boldsymbol{k})]$ instead of $\gamma(\boldsymbol{k})n(\boldsymbol{k},t)$ into equation (5.4.13) for correlator $n(\boldsymbol{k},t)$. This substitution in the absence of pumping (at $h = 0$) ensures that the relaxation $n(\boldsymbol{k},t)$ tends not towards zero, but to the thermodynamic equilibrium at temperature T value of $n_0(\boldsymbol{k}) = T/\omega(\boldsymbol{k})$. On the other hand, under $hV(\boldsymbol{k}) > \gamma(\boldsymbol{k})$ the term $\gamma(\boldsymbol{k})n_0(\boldsymbol{k})$ maintains the development of all instabilities and, as a consequence, ensures the uniqueness of the stationary solutions (5.4.2). This question will be treated in more detail in Sect. 6.4.

It would be also interesting to find out how the thermal noise $n_0(\boldsymbol{k})$ in the course of development of the parametric instability finally brings about the wave state $n(\boldsymbol{k}) \simeq \delta[\omega_{\mathrm{NL}}(\boldsymbol{k}) - \omega_{\mathrm{p}}/2]$ coherent in $\boldsymbol{k}$. This question will be discussed in Sect. 7.5. We shall give a theoretical and computer proof, in particular, of the fact that after a certain time after the pumping is turned on, the distribution of waves in $\boldsymbol{k}$ has a Gaussian form, with the distribution width asymptotically tending to zero as $1/\sqrt{t}$.

Let us now turn to the question of the pair distribution over the resonance surface. Let us introduce the distribution function $n(\Omega)$, "the number" of pairs per unit of the solid angle, defining it in the following way:

$$N = \sum_{\boldsymbol{k}} n(\boldsymbol{k}) = \int n(\Omega) d\Omega \ . \tag{5.5.5}$$

The stationary equation for $n(\Omega)$ and $\Psi(\Omega)$ following from (5.4.13) and (5.5.4) will be rewritten as

$$\{P(\Omega)\exp[i\Psi(\Omega)] - i\gamma(\Omega)\}n(\Omega) = 0 \ ,$$

$$P(\Omega) = hV(\Omega) + \int S(\Omega,\Omega_1)n(\Omega_1)\exp[-i\Psi(\Omega_1)]d\Omega_1 \ . \tag{5.5.6}$$

These equations do not yet determine unambiguously the distribution $n(\Omega)$ and $\Psi(\Omega)$ since the areas on the surface where $n(\Omega) = 0$ can be given arbitrarily. As will be shown in the following section, the condition of external stability with respect to the generation of new pairs on the resonance surface considerably reduces the class of possible solutions and as a result in some cases only one stable distribution remains.

It is useful geometrically to interpret the condition of external stability in the following way. The expressions $\gamma = \gamma(\Omega)$ and $|P| = |P(\Omega)|$ are the equations of a surface in $\boldsymbol{k}$-space. The condition (5.5.3) implies that the surface $|P(\Omega)|$ is wholly contained within the surface $\gamma(\Omega)$ and is tangent to it at the points $\tilde{\Omega}$ where the solution is concentrated, i.e. $n(\tilde{\Omega}) \neq 0$. By virtue of the relations $V(\mathbf{k}) = V(-\mathbf{k})$ and $S(\mathbf{k}, \mathbf{k}_1) = S(-\mathbf{k}, \mathbf{k}_1)$ both the surfaces have a symmetry center. The tangency of the surfaces $|P(\Omega)|$ and $\gamma(\Omega)$ can take place either over a discrete set of points or over a continuum, i.e. a line or even part of the surface. In the first case a finite number of monochromatic wave pairs is excited in the ground state; in the second case the distribution of $n(\Omega)$ is continuous. The situation can be intermediate when the surfaces are tangent at an isolated pair of points and or over some line. In this case a monochromatic pair and a continuous background are simultaneously present in the system. Note that the field of applicability of the S-theory in the case when there is a small number of discrete pairs requires special justification including the investigation of their stability within an exact Hamiltonian.

5.5.2 Ground State Under Low Supercriticality

The simplest ground state in the S-theory refers to the case of spherical symmetry which is realized, for instance, in antiferromagnets (when the exceptionally small magnetic dipole interaction is neglected). At $V(\Omega) = V$, $\gamma(\Omega) = \gamma$ and $S(\Omega, \Omega_1) = S$ the equations (5.5.6) have an isotropic solution $n(\Omega) = N/4\pi$, under which

$$N = \sqrt{h^2 V^2 - \gamma^2}/|S|, \quad hV \sin \Psi = \gamma. \tag{5.5.7}$$

These equations have the same solution if $S(\Omega, \Omega_1)$ depends only on the angle of $\mathbf{k}$ with respect to $\mathbf{k}_1$. Then in (5.5.7) $4\pi S = \int S(\Omega, \Omega_1) d\Omega_1$. The case of axial symmetry, for which

$$V(\Omega) = V(\Theta, \varphi) = V(\Theta) \exp(im\varphi),$$
$$S(\Omega, \Omega_1) = S(\Theta, \Theta_1, \varphi - \varphi_1), \quad T(\Omega, \Omega_1) = T(\Theta, \Theta_1, \varphi - \varphi_1).$$

is also of great interest.

By way of example, it should be recalled that under parametric excitation of spin waves in ferromagnets by parallel pumping (see (4.3.21)) $V(\Theta) = V \sin^2 \Theta$, $m = 2$, under parallel pumping (see (4.3.2)) $V(\Theta) \simeq V \sin 2\Theta$, $m = 1$. Let us show that in the case of axial symmetry the dependence on the azimuthal angle φ can be excluded from the basic equations of the S-theory for the class of axially symmetric solutions $n(\Theta, \varphi) = 2\pi n(\Theta)$ by substituting new variables. From (5.5.6) for $n(\Omega)$ it is clear that $P(\Omega) \exp[i\Psi(\Omega)]$ must not depend on φ. Then (5.5.6) for $P(\Omega)$ gives

$$\Psi(\Omega) = \Psi(\Theta, \varphi) = \Psi_{\text{inv}}(\Theta) + m\varphi. \tag{5.5.8}$$

This is the equation for the *invariant phase* of the pairs $\Psi_{\text{inv}}(\Theta)$ which is independent of φ. With the help of (5.5.8) Eqs. (5.5.6) are represented as:

$$P_{\text{inv}}(\Theta) = hV(\Theta) + \int S_{\text{inv}}(\Theta, \Theta_1) n(\Theta_1) \exp[-i\Psi_{\text{inv}}(\Theta_1)] \sin \Theta_1 d\Theta_1,$$

5 Stationary Behavior of Parametric Waves

$$\{P_{\text{inv}}(\Theta)\exp[i\Psi_{\text{inv}}(\Theta)] - i\gamma(\Theta)\}n(\Theta) = 0;$$

$$2\pi S_{\text{inv}}(\Theta,\Theta_1) = \int S(\Theta,\Theta_1,\varphi-\varphi_1)\exp[im(\varphi-\varphi_1)]d\Theta_1, \quad (5.5.9)$$

$$P_{\text{inv}}(\Theta) = P(\Theta,\varphi)\exp(-im\varphi).$$

If desired, these equations can be additionally simplified by means of the following substitution after which the amplitudes of the wave interaction with the pumping will no longer depend on x ($x = \cos\Theta$):

$$N(x) = n(x)|f(x)|, \quad f(x) = V(x)/V_1, \quad V_1 = \max\{V(x)\},$$

$$\tilde{P}(x) = P(x)/f(x), \quad \Gamma(x) = \gamma(x)/|f(x)|, \quad (5.5.10)$$

$$\tilde{S}(x,x_1) = S(x,x_1)/f(x)f(x_1).$$

For the functions $N(x)$, $\Psi_{\text{inv}}(x)$ the equations of motion retain the form (5.5.9) with the substitution $\gamma \to \Gamma$, $P_{\text{inv}} \to \tilde{P}$. In the expression for P the amplitudes of the wave-pumping interaction prove to be constant:

$$\tilde{P}(x) = hV_1 + \int \tilde{S}(x,x_1)N(x_1)\exp[-i\Psi_{\text{inv}}(x_1)]dx_1,$$

$$\{\tilde{P}(x)\exp[i\Psi_{\text{inv}}(x)] - i\Gamma(x)\}N(x) = 0. \quad (5.5.11)$$

To solve the problem of the distribution $N(x)$ under small supercriticality (when the pumping amplitude h is only a little above the threshold of the parametric excitation) we shall use the above-mentioned geometric interpretation of the external stability condition (5.5.3). Under very small supercriticality when the wave amplitudes are small, the line $|\tilde{P}(x)|$ differs insignificantly from a straight line $hV_1 = \text{const}$. The curvature of the line $|\tilde{P}(x)|$ is also small. It is clear that the lines $|\tilde{P}(x)|$ and $\Gamma(x)$ are tangent only at the point $x = x_1$, where the function $\Gamma(x)$ is minimum. This means that under small supercriticality the distribution $N(x)$ differs from zero only at $x = x_1$, and in this case for the total number of waves N and phase $\Psi_1 = \Psi_{\text{inv}}(x_1)$ we can readily obtain from (5.5.11)

$$N(x) = N_1\delta(x-x_1), \quad N_1 = \sqrt{h^2V_1^2 - \Gamma_1^2}/|S_{11}|,$$

$$hV_1\sin\Psi_1 = \Gamma_1, \quad \Gamma_1 = \Gamma(x_1), \quad S_{11} = \tilde{S}(x_1,x_1). \quad (5.5.12)$$

By way of a third example of solving the basic equations of the S-theory under small supercriticality, consider the case when the function $\Gamma(\Omega) = \gamma(\Omega)V_1/|V(\Omega)|$ has a maximum at one pair of points. This is the case in particular with spin waves in ferromagnets parametrically excited by "oblique pumping" (see (4.3.23)) or under the second-order Suhl instability. In the second case the problem has axial symmetry; the amplitude of the spin wave interaction with the pumping is maximum at the pole of resonance surface at $\Theta = 0$ (in this case in (5.5.9) $m = 0$). Clearly, the solution (5.5.12) holds true also in this case, only by S_{11} we should understand $S(\Omega, \Omega_1)$ at $\Theta = \Theta_1 = 0$.

Thus, we consider three cases of distributions of parametric waves with different dimensionality d. In the first case (under spherical symmetry) the function $n(\Omega)$ is nonzero on the hole resonance surface, and, consequently, $d = 2$. In the second case (when the symmetry is axial) $n(\Omega)$ differs from zero on the lines (on the two parallels of the resonance surface $\Theta = \Theta_1$ and $\Theta = \pi - \Theta_1$ or on the equator $\Theta_1 = \pi/2$) and therefore $d = 1$. In the third case (when, for instance, there is no symmetry) $n(\Omega)$ is non-zero at one or several pairs of points and $d = 0$. It is essential that in all these cases expressions (5.5.7, 8) for the total number of parametric waves N_1 and invariant phase Ψ_1 (which is the same for all the excited parametric waves) can be similarly represented

$$N_1 = \int n(\Omega) d\Omega = \frac{\sqrt{h^2 V_1^2 - \gamma_1^2}}{|S_{11}|}, \quad hV_1 \sin \Psi_1 = \gamma_1 . \qquad (5.5.13a)$$

Here V_1 and γ_1 are the amplitudes of the interaction with the pumping and damping of parametric waves in the area where $n(\Omega) \neq 0$ (it shall be recalled that in this area $|V|$ and γ do not depend on Ω),

$$S_{11} = \int S(\Omega, \Omega') n(\Omega) n(\Omega') d\Omega d\Omega' / N_1^2 \qquad (5.5.13b)$$

is the average value $S(\Omega, \Omega')$ in the area where $n(\Omega) \neq 0$. Therefore the total number of excited waves does not drastically depend on the dimensionality of their distribution d. This fact is typical not only of the S-theory; it is characteristic of many different versions of mean–field theories. Further the character and magnitude of corrections to the S-theory due to the influence of the thermal bath (Sect. 6.4), scattering of parametric waves by each other (Sect. 10.2), etc. depend cardinally on the dimensionality of the ground state and of the medium. This circumstance is also typical of mean–field theories. One of the best known examples of this is the physics of second-order phase transitions.

Finally we shall account for the physical cause of the limiting of the amplitude of parametric waves in this version of the S-theory. It is very simple. Each wave pair "does not know" anything about the external pumping hV and "feels" only the total pumping $P = hV + SN_1 \exp(-i\Psi_1)$. "Trying" to receive as much energy from the external pumping as possible it "turns" its phase Ψ_1 at a right angle from the phase of total pumping $\Psi_p = \arg\{P\}$ (note that this energy flux is proportional to $\sin(\Psi_1 - \Psi_p)$). Therefore, the difference between the phase of pair Ψ_1 and the phase of the external pumping $\arg\{hV\}$ (which we assumed to be equal to zero) falls off from the optimum value $\pi/2$, which results in a decreased energy flux from the external pumping $W_+ = \omega_p hV_1 N_1 \sin \Psi_1$ and, consequently, in the limitation on the total number of parametric waves N_1. To calculate N_1 and Ψ_1, remember the expression for the dissipation rate $W_- = \omega_p \gamma_1 N_1$. Therefore, the condition of energy balance $W_+ = W_-$ has the form $hV_1 = \gamma \sin \Psi_1$. This relation coincides with (5.5.13b) and specifies the dependence of the phases of pairs on the supercriticality. On the other hand, the energy flux from total pumping

$$W_+ = \omega_p |P| N_1 \sin(\Psi_1 - \Psi_p) = \omega_p |P| N_1 .$$

This results in another relation $|P| = \gamma_1$ (compare with (5.5.3)). Thus, we have a triangle a three vectors hV_1, $SN_1 \exp(i\Psi_1)$ and

$$P = hV + SN_1 \exp(i\Psi_1) = \gamma_1 \exp(i\Psi_p) ,$$

with the two last vectors being perpendicular. By Pythagoras theorem $(SN_1)^2 + \gamma_1^2 = (hV_1)^2$, which coincides with (5.5.13a) and specifies the dependence of N on the supercriticality.

Finally it must be noted that the fundamentally important conclusion of the S-theory about the dependence of the phase of the pair Ψ_1 on the supercriticality (5.5.13b) has been verified by direct experiments measuring the phase of the pair Ψ_1 under parametric excitation of magnons in an antiferromagnet $MnCO_3$ and in a ferromagnet YIG. These experiments were carried out by *Prozorova* and *Smirnov* [5.61] and then by *Melkov* and *Krutsenko* [5.52] in YIG. These will be described in Chap. 9. They corroborate the conclusion of the S-theory about the essential features of the above-threshold behavior of the parametric spin waves.

5.5.3 Threshold of Generation of Second Group of Pairs

Let us proceed to the study of the parametric wave distribution $n(\Omega)$ over the resonance surface under an increasing amplitude of the external field h. Note first that geometrical considerations employed in 5.4.3 can easily be generalized to the case of arbitrary dependence $V(\Omega)$. A general theorem can be proven [5.6] on the fact that under sufficiently small supercriticality, $N(\Omega)$ is nonzero only at those points of the resonance surface where $|V(\Omega)|/\gamma(\Omega)$ is maximum. For spherical symmetry these are all the points of the surface, under axial symmetry these are points of one (at $\Theta_1 = \pi/2$) or two lines. Under lower symmetry it is one or several equivalent pairs of points.

Let us return to the case of axial symmetry and assume for simplicity that the first group of pairs has been generated at the equator: $x_1 = \cos \Theta_1 = 0$. This is so, for instance, when the spin waves are parametrically excited in ferromagnets by parallel pumping (see 4.3.2). Let us show that the distribution of the pairs (5.5.12) concentrated at the equator remains stable with respect to the generation of pairs at other latitudes up to sufficiently great supercriticality. To this end, consider the function $|\tilde{P}(x)|$. From (5.5.10-12) we obtain

$$|\tilde{P}(x)|^2 = N_1^2 [S_{11} - \tilde{S}(x, x_1)]^2 + \Gamma_1^2 . \qquad (5.5.14)$$

It is evident that the state (5.5.12) will remain stable until $|\tilde{P}(x)| < \Gamma(x)$ for all x, except $x = x_1 = 0$. The *second threshold* $p_2 = h_2^2/h_{\text{th}}^2$ corresponds to the minimum value of $p = h^2/h_{\text{th}}^2$, under which the lines $|\tilde{P}(x)|$ and $\Gamma(x)$ are tangent under a given $x = x_2 \neq x_1$. The value p_2 is determined from the condition $p_2 = \min\{p(x)\}$, where

$$p(x) = 1 + S_{11}^2[\Gamma(x) - \Gamma_1]/\Gamma_1^2[S_{11} - \tilde{S}(x,x_1)]^2$$
$$= 1 + \frac{S_{11}^2[\gamma^2(x)V_1^2 - \gamma_1^2 V^2(x)]}{\gamma_1^2[S_{11}V(x) - S(x,x_1)V_1]^2} \ . \tag{5.5.15}$$

Let us make the simplest assumption on the functions in this formula $\gamma(x) = \gamma$, $\tilde{S}(x,x_1) = S_{11}$ and take $V(x) = V_1[1 - x^2]$. Then expression for $p(x)$ assumes the form:

$$p(x) = 1 + (2 - x^2)/x^2 \ . \tag{5.5.16}$$

The minimum of this function is attained at $x = x_2 = 1$ and equals 2. Therefore in this simplest case the second pair of waves is generated at the poles under the supercriticality 3 dB ($p_2 = 2$).

Bearing in mind the experimental situation in ferromagnets consider a more meaningful example making the dependence of $S(x,x_1)$ on x more complicated:

$$\tilde{S}(x,x_1) = S_{11}(1 - x^2)(1 + bx^2) \tag{5.5.17}$$

and the same x-dependences for $V(x) = V[1 - x^2]$ and $\gamma(x) = \gamma$. In this case the expression (5.5.15) for $p(x)$ is transformed to:

$$p(x) = 1 + (2 - x^2)/b^2 x^2 (1 - x^2)^2 \ . \tag{5.5.18}$$

The minimum of this function p is realized at $x = x_2$, where

$$x_2 = (3 - \sqrt{5})/2 \simeq 0.38 \ , \tag{5.5.19}$$

$$p_2 = 1 + (11 + 5\sqrt{5})/2b^2 \simeq 1 + 11.1/b^2 \ . \tag{5.5.20}$$

The large numeric factor before b^{-2} in (5.5.20) is quite remarkable. If, for instance, $b = 1$, then $p_2 = 12$, which makes the stability of the first group of pairs (and accordingly, the applicability of (5.5.12)) equal to 13 dB. If $b < 1$, the region of the first-group stability increases. Qualitatively, we can conclude from the example that in the S-theory the applicability of the simple formulae (5.5.7), (5.5.11) and (5.5.13) describing the state with a single group of pairs and initially obtained at small supercriticalities appears to be anomalously great $p < 10$. This theoretical conclusion has been verified by a number of experiments described in Chap. 9.

Note again that in the first example the generation angle of the second group pairs is given by $\Theta_2 = 0$. In the second example (see (5.5.20a)),

$\Theta_2 \simeq 51°$. Consideration of more complicated cases shows that the angle Θ_2 can be arbitrary and can also be close to Θ_1 (see 6.1.2).

The analytical solution of the basic equations of the S-theory for two or more groups of pairs is very cumbersome and is performed elsewhere. Some results of this kind have been obtained in [5.76]. The problem is considerably simpler in the region of high supercriticalities. Now we shall proceed to this task.

5.5.4 Ground State Under High Supercriticality

The behavior of a parametric wave system with a continuous distribution can be described most conveniently in the limiting case of very strong pumping ($hV \gg \gamma$). In the zeroth approximation in the parameter γ/hV, $P(\Omega) = 0$ whence according to (5.5.6) at real $S(\Omega, \Omega')$ it follows that

$$hV(\Omega) + \int S(\Omega, \Omega')X(\Omega')d\Omega' = 0 , \qquad (5.5.21a)$$

$$\int S(\Omega, \Omega')Y(\Omega')d\Omega' = 0 , \qquad (5.5.21b)$$

$$X(\Omega) = n(\Omega)\cos\Psi(\Omega) , \quad Y(\Omega) = n(\Omega)\sin\Psi(\Omega) . \qquad (5.5.21c)$$

The problem is therefore reduced to the solution of first-order linear Fredholm equations. For simplicity assume that in the absence of pumping the medium is isotropic, i.e. $V(\Omega)$ depends only on the polar angle Θ and $S(\Omega, \Omega')$ depends only on the angle α of the direction Ω with the direction Ω'. Then, expanding all the functions in (5.5.21a, b) in terms of Legendre polynomials $P_n(\cos\alpha)$ and making use of the theorem of addition for $P_n(\cos\alpha)$, we obtain after elementary integration over Θ' and Ψ' the formal solution

$$4\pi X(\Theta) = -\sum_{n=0}(2n+1)hV_n P_n(\cos\Theta)/S_n ,$$

$$S(\cos\alpha) = \sum_{n=0} S_n P_n(\cos\alpha) . \qquad (5.5.22)$$

Here V_n and S_n are the expansion factors of the functions $V(\Theta)$ and $S(\alpha)$ into the series of Legendre polynomials.

A general consideration of the solutions of (5.5.21) includes three possible cases:

1. Equation (5.5.21b) has nontrivial solutions not orthogonal to $V(\Omega)$. Then (5.5.21a) has no solutions, and the theory proceeding from the diagonal Hamiltonian imposes no limitation on the amplitude. This case will be called *singular*. For the isotropic model the singular case occurs, for example, if $S_n = 0$ and $V_n \neq 0$ for some particular n. It has been shown in the previous section that under $V \neq$ const and not too large a supercriticality a solution in the form of one or more pairs exists also in the singular case. Thus, there is a certain critical amplitude h_c of the pumping and when it is attained the stationary state is disturbed. Above this amplitude the limitation takes place due to weaker nonlinear mechanisms. It follows from (5.5.22) that in most singular case when $S_0 = 0$ and $V = $ const, the amplitude is not limited, under any h, i.e. $h_c = h_{\text{th}}$. In less singular cases $h_c > h_{\text{th}}$.

Consider in detail the case when $S = $ const and $V(\Theta)$ decreases monotonically from V_1 to V_2. In this case the surface $|P(\Theta)|$ has not more than two maxima. Therefore not more than two pairs can be in a stationary state. Seeking the solution of (5.5.6) we obtain the following expression for the total number of waves:

$$N_1 + N_2 = \frac{\gamma}{|S|} \frac{h(V_1 + V_2)}{\sqrt{4\gamma^2 - h^2(V_1 - V_2)^2}} . \qquad (5.5.23)$$

It is clear from the above that $h_c = 2\gamma/(V_1 - V_2)$ at $S = $ const.

2. Equations (5.5.21b) have non-trivial solutions orthogonal to $V(\Omega)$. Then (5.5.21a) has solutions calculated to an accuracy of the solutions of (5.5.21b) and the theory based on the diagonal Hamiltonian gives no unambiguous definition of the stationary state. For the isotropic models the indeterminate case is realized if $S_m = 0$ and $V_m = 0$ for one m at least. Under $V \neq $ const and $h - h_{\text{th}} \ll h_{\text{th}}$ there is the unique solution (describing the set of similar pairs) as in the singular case. Therefore there are some critical pumping amplitudes h_u such that under $h > h_u$ the solution loses its uniqueness. The indeterminacy in the solution increases with increasing number of harmonics with $S_m = V_m = 0$. If $S_m = V_m = 0$ for any m except $m = 0$ (i.e. $S = $ const and $V = $ const) the most indeterminate case applies. Only the 0-th harmonic of the $n(\Omega)$-distribution is determined by the basic equation of the S-theory (e.g., (5.4.15)) in this case, while all the other harmonics can be chosen at will in the S-theory approximation. For the 0-th harmonic n_0 the following expression can easily be obtained

$$4\pi n_0 = \int n(\Omega) d\Omega = \sqrt{h^2 V^2 - \gamma^2}/|S| . \qquad (5.5.24)$$

coinciding with (5.5.7).

3. Equation (5.5.21b) has no non-trivial solution. This case will be called *regular*. In the regular case all $S_n = 0$. Then (5.5.21a) has the unique solution (5.5.22) which, however, is physically meaningful only under some limitations. First, there must be a fast enough convergence of the sequence V_n/S_n to zero under $n \to \infty$. This condition, however, is not sufficient. Indeed, one must allow for the finiteness of the damping γ in (5.5.21a, b). To this end make use of (5.5.6)

$$-\gamma \sin \Psi(\Omega) = hV(\Omega) + \int S(\Omega, \Omega_1) n(\Omega_1) \cos \Psi(\Omega_1) d\Omega_1 ,$$

$$-\gamma \cos \Psi(\Omega) = \int S(\Omega, \Omega_1) n(\Omega_1) \sin \Psi(\Omega_1) d\Omega_1 \tag{5.5.25}$$

for the isotropic model. To obtain the solution (5.5.22) it was assumed that $\gamma/h\nu$ equals zero. To obtain corrections to that solution we linearize (5.5.6) to get for small $\gamma/h\nu$

$$-\gamma \delta\Psi(\Omega) = \int S(\cos\alpha) \delta n(\Omega_1) d\Omega_1 ,$$
$$-\gamma = \int S(\cos\alpha) n(\Omega_1) \delta\Psi(\Omega_1) d\Omega_1 , \tag{5.5.26}$$

whence

$$\delta\Psi(\Theta) = -\gamma/4\pi S_0 N(\Theta) , \quad N(\Theta) = \sum_m N_m P_m(\cos\Theta) ,$$
$$N_m = -\gamma(2m+1)\delta\Psi/4\pi S_m . \tag{5.5.27}$$

Therefore, for the region of applicability of (5.5.22) to exist the distribution $n(\Omega)$ must never become zero. Recall that under $V = \text{const}$ an accurate solution of equations with damping can be obtained for any pumping amplitude. This solution has the form (5.5.24). In the regular case, moreover, $N = \text{const}$, i.e. $n(\Omega) = N/4\pi$.

Qualitatively we can conclude from the above that the distribution of pairs $n(\Omega)$ under large amplitude h is highly sensitive to the fine structure of the functions $V(\Omega)$ and $S(\Omega, \Omega_1)$: in some cases continuous distributions of pairs over the resonance surface are established, in other cases the limitations are violated and the picture becomes essentially nonstationary.

5.5.5 Nonlinear Susceptibilities of Parametric Waves

One of the methods most commonly used for the experimental study of the parametric excitation of waves is based on measuring the energy flux W_+ from the pumping into the system of parametric waves. In statistical physics [5.77] the value W_+ is expressed in terms of the generalized susceptibility $\chi(\omega)$:

$$W_+ = \frac{1}{2} \text{Im}\{\chi(\omega)\} |f(\omega)|^2 . \tag{5.5.28}$$

In the general case $f(\omega)$ is the ω-Fourier transform of the generalized force $f(t)$, $\chi(\omega)$ is the proportionality coefficient of the generalized coordinate $x(\omega)$ to $f(\omega)$:

$$x(\omega) = \chi(\omega) f(\omega) . \tag{5.5.29}$$

The generalized coordinate $x(t)$ is selected so that the interaction energy of the external force with the system has the form

$$\mathcal{H}_{\text{int}} = -x(t)f(t) \; . \tag{5.5.30}$$

For magnetodielectrics $f(t)$ is the magnetic field of the pumping $h(t)$, x is the magnetization m, the energy (5.5.30) is the Zeeman energy. Consequently, in magnetodielectrics χ is the magnetic susceptibility. Within the general theory of the parametric excitation of waves (not specifying their nature) proceeding from the "standard" Hamiltonian of the pumping (5.2.1) $\text{Im}\{\chi(\omega)\}$ can be calculated most readily if we compare (5.5.28) and (5.3.4) for W_+:

$$\chi'' = \text{Im}\{\chi(\omega_{\text{p}})\} = -\frac{2}{h}\sum_{\bm{k}} \text{Im}\{V^*(\bm{k})\sigma(\bm{k})\} \; . \tag{5.5.31a}$$

A similar expression can also be obtained for the real part of this susceptibility:

$$\chi' = \text{Re}\,\{\chi(\omega_{\text{p}})\} = -\frac{2}{h}\sum_{\bm{k}} \text{Re}\,\{V^*(\bm{k})\sigma(\bm{k})\} \; . \tag{5.5.31b}$$

These formulae can readily be combined

$$\chi = -\frac{2}{h}\sum_{\bm{k}} V^*(\bm{k})\sigma(\bm{k}) \; . \tag{5.5.32}$$

By means of the angular distribution (5.5.5) the latter formula can be represented in a more convenient form:

$$\chi = -\frac{2}{h}\int V^*(\Omega)n(\Omega)\exp[-i\Psi(\Omega)]d\Omega \; . \tag{5.5.33}$$

The behavior of susceptibilities χ' and χ'' above the threshold is largely determined by the mechanism limiting the amplitude. Thus, it follows for the mechanism of nonlinear damping from (5.5.33) and (5.3.8) that

$$\chi' = 0 \; , \quad \chi'' = 2V^2(h - h_{\text{th}})/\eta h \; , \tag{5.5.34}$$

and for the phase limiting mechanism in the isotropic case it follows from (5.5.7):

$$\chi' = 2V^2(h^2 - h_{\text{th}}^2)/Sh^2 \; , \quad \chi'' = 2V^2 h_{\text{th}}\sqrt{h^2 - h_{\text{th}}}/|S|h^2 \; . \tag{5.5.35}$$

Obviously, these formulae hold true also in the absence of spherical symmetry in a narrower range of supercriticalities below the generation threshold of the second group of pairs: ($h_{\text{th}} < h < h_2$). Equations (5.5.34, 35) show that the fundamental difference of the dissipative and phase mechanisms manifests itself in the behavior of the value χ'; $\chi' = 0$ for the dissipative and $\chi' \simeq \chi''$ for the phase mechanisms. Experimental results to be discussed in detail in Chap. 9 indicate that the real susceptibility χ' differs from zero and may be of the order of or even greater than χ''. Those facts also provide evidence of the phase mechanism for limiting of the parametric wave amplitudes.

6 Advanced S-Theory: Supplementary Sections

The previous chapter treated the main concepts of the S-theory of the parametric excitation of waves in terms of spatially homogeneous pumping. Yet, a variety of interesting situations has not been treated.

In particular, nonlinear behavior of the system of parametric waves (with the exception of their general characteristics) was found to be very sensitive to the fine details of the functions $V(\boldsymbol{k})$ and $S(\boldsymbol{k},\boldsymbol{k}')$, specifying the interaction Hamiltonian. Only the simplest assumptions concerning these functions have been made, but these are not always valid. New details of the nonlinear behavior of the system of parametric waves caused by a complex form of the interaction Hamiltonian will be described in Sect. 6.1.

In the basic S-theory the damping of parametric waves was assumed to be independent of their amplitudes. If this assumption does not hold the damping of waves is nonlinear. Some mechanisms of the nonlinear damping will be considered in Sect. 10.2. The nonlinear damping of parametric waves can easily be allowed for within the S-theory. This is done in Sect. 6.2.

We also believe that the amplitude of the pumping field is the external parameter of the theory given by the experimentator. This is really so, if the role of pumping is played by an external field (electric or magnetic) whose energy is substantially higher than the energy of parametrically excited waves. If pumping is one of the oscillatory modes of the sample itself (e.g. the uniform precession of the magnetization under the parametric excitation of spin waves), its energy can as a rule be compared with the energy of the system of parametrically excited waves. In this case the feedback effect on the pumping must be taken into account. To this end, the equation for the pumping amplitude must be added to the basic equations of the basic S-theory and their simultaneous solutions must be obtained. This will be done in Sect. 6.3 for the processes of the parametric instability of the first and second order.

In Sect. 6.4 the fine effects will be studied caused by the interaction of parametrically excited waves with a thermal bath, which is a medium in thermal equilibrium with the temperature T. The basic S-theory described in Chap. 5 may be called a limiting case of the S-theory treating a thermal bath temperature tending to zero.

In Chap. 5 the pumping was assumed to be spatially homogeneous and the sample in which waves are excited parametrically was taken to be unbounded. Clearly this is not always correct and it would be interesting to study how the spatial inhomogeneity influences the nonlinear behavior of the system of parametrically excited waves. Some steps in this direction have been made in Sect. 6.5. The corrections to the basic S-theory obtained in this section are small if the characteristic size of the inhomogeneity L is great compared with the mean free path of parametric waves $|\partial\omega(\boldsymbol{k})/\partial\boldsymbol{k}|/\gamma(\boldsymbol{k})$.

The basic S-theory can easily be generalized to the "asymmetrical case" when the waves from different spectrum branches enter into the parametric pairs, e.g. electronic and nuclear spin waves in magnets, Langmuir and ionic-sonic waves in nonisothermic plasma, etc. An introduction into the asymmetrical S-theory describing such situations is given in Sect. 6.6. I consider the predicted (but not yet experimentally observed) phenomenon of the *correlation instability of waves* to be of special interest.

One more interesting problem is to study how the incoherence of the pumping influences the nonlinear behavior of the system of parametric waves. This problem is discussed in Sect. 6.7. The basic S-theory in the case of coherent pumping developed in Chap. 5 holds true in the limit where the linewidth of the pumping generator $\Delta\omega_p$ is small compared with the damping decrement of parametric waves γ. Special attention is devoted to the opposite case $\Delta\omega_p > \gamma$. It has been shown in particular that the coupling of waves and the phase mechanism of amplitude limitation is retained also in this case, although it becomes $\gamma/\Delta\omega_p$ times less effective (under the same supercriticality).

The physical situations treated and generalized in the above mentioned seven subsections by no means exhaust the diversity of the general picture. The S-theory can be developed further and further, as I hope it will be. In particular, the nonlinear theory of the parametric excitation of waves by the plane wave of pumping determined on the boundary of the sample must be developed. This is possible if the ideas of Sects. 6.3, 6.5 and, perhaps, 6.6 are combined. The theory must be developed further also for the cases which will be studied in Sects. 6.5–7.

And, finally, the results outlined in the sections of this chapter are to a great extent independent of each other and have only one very important point in common: they are treated on the basis of the basic S-theory presented in Chap. 5. Therefore it is not necessary to read and study Chap. 6 section by section, the reader can choose the problems interesting to him and skip the others or leave them till later.

6.1 Ground State Evolution of System with Increasing Pumping Amplitude

Studying the basic S-theory we tacitly made some simplifying assumptions about the form of the functions $T(\boldsymbol{k}_1, \boldsymbol{k}_2; \boldsymbol{k}_3, \boldsymbol{k}_4)$, describing the four-wave interaction. It was done to avoid digression from the presentation of the main concept of the S-theory. Now we shall return to this problem.

The first assumption. The formula (5.4.12) gives the definition of the functions $T(\boldsymbol{k}, \boldsymbol{k}_1)$ and $S(\boldsymbol{k}, \boldsymbol{k}_1)$, which is given below

$$T(\boldsymbol{k},\boldsymbol{k}_1) = T(\boldsymbol{k},\boldsymbol{k}_1;\boldsymbol{k},\boldsymbol{k}_1)/2 \;, \quad S(\boldsymbol{k},\boldsymbol{k}_1) = T(\boldsymbol{k},-\boldsymbol{k};\boldsymbol{k}_1,-\boldsymbol{k}_1)/2 \;. \quad (6.1.1)$$

The interaction amplitudes $T(\boldsymbol{k}_1, \boldsymbol{k}_2; \boldsymbol{k}_3, \boldsymbol{k}_4)$, are the functions of four vector arguments $\boldsymbol{k}_1, \boldsymbol{k}_2, \boldsymbol{k}_3$ and $\boldsymbol{k}_4$. However, in a spatially homogeneous medium (we shall confine ourselves to this case throughout the book) they are defined over the hypersurface $\boldsymbol{k}_1 + \boldsymbol{k}_2 = \boldsymbol{k}_3 + \boldsymbol{k}_4$. Therefore these amplitudes are the functions of three vector arguments. The functions of two vector arguments $T(\boldsymbol{k}, \boldsymbol{k}_1)$ and $S(\boldsymbol{k}, \boldsymbol{k}_2)$ are defined by the additional reduction (6.1.1) of these functions. Trying to use the future theory (the untrusting reader is free to think that we dreamt it up) one can suggest the following definitions of the functions T and S:

$$\begin{aligned} T(\boldsymbol{k},\boldsymbol{k}_1,\boldsymbol{\kappa}) &= T(\boldsymbol{k}+\boldsymbol{\kappa},\boldsymbol{k}_1-\boldsymbol{\kappa},\boldsymbol{k}-\boldsymbol{\kappa},\boldsymbol{k}+\boldsymbol{\kappa})/2 \;, \\ S(\boldsymbol{k},\boldsymbol{k}_1,\boldsymbol{\kappa}) &= T(\boldsymbol{k}+\boldsymbol{\kappa},-\boldsymbol{k}+\boldsymbol{\kappa},\boldsymbol{k}_1+\boldsymbol{\kappa},-\boldsymbol{k}+\boldsymbol{\kappa})/2 \;, \end{aligned} \quad (6.1.2)$$

such that

$$T(\boldsymbol{k},\boldsymbol{k}_1) = T(\boldsymbol{k},\boldsymbol{k}_1,0) \;, \quad S(\boldsymbol{k},\boldsymbol{k}_1) = S(\boldsymbol{k},\boldsymbol{k}_1,0) \;. \quad (6.1.3)$$

It must be noted here that the functions $T(\boldsymbol{k}_1, \boldsymbol{k}_2; \boldsymbol{k}_3, \boldsymbol{k}_4)$ are not necessarily continuous functions of their arguments. For example, the calculations of the Hamiltonian of the spin wave interaction in ferromagnets performed in Chap. 3 (see (3.1.22)) for $c(\boldsymbol{k})$ and $c(0)$) show that the limits

$$\lim T(\boldsymbol{k},\boldsymbol{k}_1,\boldsymbol{\kappa}), \quad \lim S(\boldsymbol{k},\boldsymbol{k}_1,\boldsymbol{\kappa}) \quad \text{at} \quad \boldsymbol{\kappa} \to 0 \quad (6.1.4)$$

may fail to exist at all (may depend on the direction of the vector $\boldsymbol{\kappa}$) and, second, may fail to coincide with the values of these functions at $\kappa = 0$. Therefore the definition of the functions $S(\boldsymbol{k}, \boldsymbol{k}_1)$ and $T(\boldsymbol{k}, \boldsymbol{k}_1)$ is ambiguous. Either the definition (6.1.3) is correct, or $S(\boldsymbol{k}, \boldsymbol{k}_1)$ and $T(\boldsymbol{k}, \boldsymbol{k}_1)$ denote the limits (6.1.4), but then how is the vector $\boldsymbol{\kappa}$ directed? We have no answers to these questions (in Chap. 5 and in general) and we can affirm only that the basic S-theory holds true if the limits of (6.1.4) exist and equal to the values of (6.1.3).

For other cases the S-theory must be correspondingly generalized. This future theory must be spatially non-homogeneous and allow for the shape of the sample. One of its basic elements will take into account the long-range dipole-dipole interaction of the parametrically excited waves with the eigenmodes (electrostatic or magnetostatic) of the sample. Since at present there is no such theory, the four-wave amplitude of interaction $T(\boldsymbol{k}_1, \boldsymbol{k}_2; \boldsymbol{k}_3, \boldsymbol{k}_4)$ throughout this book will be assumed to be a continuous function of its arguments. Then the limits (6.1.4) exist and it is easy to define the functions $T(\boldsymbol{k}, \boldsymbol{k}_1)$ and $S(\boldsymbol{k}, \boldsymbol{k}_1)$.

The second assumption. In the analysis of the ground state of the system of parametric waves the function $S(\Omega, \Omega_1)$ was assumed in Sect. 5.5 to be real. Usually this is so, though not always. In Sect. 6.1.1 the behavior of the system of parametric waves will be studied in the case when the function $S(\Omega, \Omega_1)$ is complex.

The third assumption. Studying the step-by-step excitation of parametric waves in Sect. 5.5.4 we basically assumed that $S(\Omega, \Omega_1)$ is not only continuous but everywhere a differentiable function of its arguments. However, the calculation of this function performed in Sect. 3.1.4 showed (the details will be given later) that in cubic ferromagnets this is not the case. The non-analytical contribution to the function $S(\Omega, \Omega_1)$ can qualitatively change the nonlinear behavior of parametric waves. This will be considered in detail in Sects. 6.1.2, 3.

6.1.1 Ground State of Parametric Waves for Complex Pair Interaction Amplitudes

It must be recalled that by using the invariant phase (5.5.8) and the substitution (5.5.10) the equation of the basic S-theory can be reduced to the form of (5.5.11) where the amplitude of the pumping wave interaction $V(\Omega)$=const and real. This can be done with unrestricted generalization. It can naturally be expected that under small supercriticality only one group of pairs will be excited in the ground state. Under the axial symmetry of the problem these pairs are located in parallels on the resonant surface Θ_1 and $\pi - \Theta_1$, at arbitrary φ. Their polar angle Θ_1 can be defined by the geometrical interpretation of the condition of the external stability discussed at the end of Sect. 5.5.1. In the terms of (5.5.11) this is formulated in the following way: the whole line $\Gamma(x)$ ($x = \cos \Theta$) is below the line $|\tilde{P}(x)|$ and is tangent to it at the points where $N(x)=0$. This means that the location of the first pair (i.e. the value $x_1 = \pm \cos \Theta_1$) is determined from the following condition:

$$d[\Gamma(x) - |\tilde{P}(x)|]/dx = 0 \quad \text{at} \quad x = x_1. \tag{6.1.4}$$

The imaginary part of the function $\tilde{S}(x, x_1)$ ($\text{Im}\{\tilde{S}(x, x_1)\}$) results in the changed expression for $\tilde{P}(x)$. Instead of (5.5.14) one can obtain:

$$|\tilde{P}(x)|^2 = N_1[S(x_1,x_1) - \text{Re}\{\tilde{S}(x_1,x)\}]^2 + [\Gamma_1 - N_1\text{Im}\{\tilde{S}(x_1,x)\}]^2 \ . \quad (6.1.5)$$

From the condition (6.1.4) we have:

$$N_1 d[\text{Im}\{\tilde{S}(x_1,x)\}]/dx = d\Gamma(x)/dx \ . \quad (6.1.6)$$

Thus, unlike before, $d\Gamma(x)/dx=0$ only at the threshold point. Afterwards the pair "leaves" the point x_1 (x_1 is the coordinate of the pair at $p=0$) and the *departure* $\Delta x = x - x_1$, being proportional to N_1. This results in a further weakening of the wave interaction with the pumping and is, in addition to the phase mismatch, one more cause of the limitation of the pair amplitude. Simultaneously solving (5.5.11) and (6.1.6) we obtain at small N_1 (for more detail, see [6.1]):

$$N_1^2 = [h^2 V_1^2 - \Gamma^2]\left\{S_{11}^2 + \frac{[d\text{Im}\{\tilde{S}(x_1,x)\}/dx]^2 h^2 V_1^2}{[d\Gamma(x)/dx]^2}\right\}^{-1} . \quad (6.1.7)$$

Comparing this result with (5.5.12) we can see that this departure leads to the appearance of an additional term in the denominator of (6.1.7), generally speaking, of the same order of magnitude. It must be noted that under the excitation of the magnons by parallel pumping in ferromagnets, the first group of pairs appears at symmetrical position of the resonant surface, i.e. at its equator. As the supercriticality increases, the pairs do not leave the equator because they "do not know" where to go: northwards or southwards. Formally this is reflected in the fact that in (6.1.6) $d\,\text{Im}\{S(x,x_1)\}/dx = 0$ at $x=x_1=0$ ($\cos\Theta_1=0$). On the contrary, under the transverse pumping of magnons the first group of the pairs is excited at $\Theta = \Theta_1$ and $\pi - \Theta_1$, $\Theta_1 \simeq \pi/4$. Because of absence of symmetry in this case it is quite possible for magnons "to go" to the equator or to poles. The dependence of the location of the pairs on the supercriticality has not yet been studied experimentally or theoretically for particular cases. Here we only point out the possibility of such a situation.

6.1.2 The Second and Intermediate Thresholds

In order to compare later the theoretical results with the experimental data we shall study a concrete example of the parametric excitation of magnons in ferromagnets by parallel pumping. First of all, the particular form of the function $\tilde{S}(x,y)$ for this case must be obtained. From the definition of the function $S(\boldsymbol{k},\boldsymbol{k}')$ (5.4.12) and (5.5.9, 10) it follows that the function $\tilde{S}(x,y)$ is the even function of each of the arguments

$$\tilde{S}(x,y) = \tilde{S}(-x,y) = \tilde{S}(x,-y) = \tilde{S}(-x,-y) \quad (6.1.8a)$$

and its real part $\text{Re}\{\tilde{S}(x,y)\}$ is symmetrical about the permutation x,y. The calculations for ferromagnets show that $\text{Im}\{\tilde{S}(x,y)\}=0$. Thus

$$\text{Re}\{\tilde{S}(x,y)\} = \tilde{S}(x,y) = \tilde{S}(y,x) \ . \tag{6.1.8b}$$

Since the problem is axially symmetrical, $S(\Theta,\Theta',\varphi-\varphi')$ is independent of $(\varphi-\varphi')$ if $\Theta=0$ or $\Theta'=0$. Therefore the integral (5.5.9) must become zero if $x \to 1$ or $y \to 1$ and function $\tilde{S}(x,y)$ accordingly remains finite at $x,y \to 1$, in spite of the singular factor $f(x) = 1-x$ in the denominator of (5.5.10). Thus, $\tilde{S}(x,y) = \tilde{S}(y,x)$ are finite.

The behavior of $\tilde{S}(x,y)$ under small x, y will be studied on the basis of the explicit form of (3.1.24, 26) for $S(\boldsymbol{k},\boldsymbol{k}')$. Some terms of these formulae are proportional to $\cos^2[\Theta(\boldsymbol{k}+\boldsymbol{k}')]$ and $\cos^2[\Theta(\boldsymbol{k}-\boldsymbol{k}')]$. Each of these terms as a result of the integration over φ according to (5.5.9) gives the non-analytical contribution into $\tilde{S}(x,y)$, proportional to the powers of $|x \pm y|$. Taking this fact and the above discussed properties of symmetry (6.1.8) into account the expansion of $\tilde{S}(x,y)$ under small x and y to the accuracy of the terms $\propto x^2, y^2$ will take the form

$$\tilde{S}(x,y) = S[1 + (1/2)a(|x+y| + |x-y|) + b(x^2+y^2)] \ . \tag{6.1.9}$$

Computer calculations for YIG at room temperature according to (3.1.24, 26) show that within the range of magnetic fields $H - H_{\text{th}}$ from 100 Oe to 500 Oe the coefficient a changes within the limits 2.6 – 2.1 and the coefficient b changes from –1.8 to 1.3 becoming zero at $H = H_{\text{th}} - 180$ Oe (note that under $H = H_{\text{th}}$, $\omega(\boldsymbol{k},H) = \omega_{\text{p}}/2$ at $\boldsymbol{k} \perp \boldsymbol{M}$ and $k \to 0$).

Now we can further analyze the problem of the generation of the second group of pairs we started to consider in Sect. 5.5.3 on the basic S-theory. Substituting (6.9) into the expression (5.5.16) for $p(x)$ and expanding it in terms of x, we obtain

$$\begin{aligned} p(x) &= p(0) + |x|p' + x^2 p''/2 \ , \quad p(0) = 1 + \frac{\Gamma''}{a^2\Gamma} \ , \\ p' &= \frac{-b\Gamma''}{a^3\Gamma} \ , \quad p'' = \frac{3\Gamma''^2 + \Gamma\Gamma^{(4)}}{6a^2\Gamma^2} - \frac{2\Gamma''b^2}{\Gamma a^4} \ . \end{aligned} \tag{6.1.10}$$

Here $\Gamma'' = \Gamma^{(2)}$ and $\Gamma^{(4)}$ denote the second and fourth derivatives of $\Gamma(x)$ at $x = x_1 = 0$. It is clear that the fact is that $\tilde{S}(x,y)$ is non-analytical (a=0) leads to the finiteness of $p(0)$. The case when at x=0 the minimum of the function $p(x)$ is realized will subsequently be considered. For the time being we shall assume $p < 0$. This happens in YIG in the fields $H > H_{\text{th}} + 180\text{Oe}$ when the coefficient a is positive. It can be seen from (6.1.10) that

$$|x_2| = -p'/p'' \ , \qquad p_2 = p(0) - 2p'^2/p'' \ . \tag{6.1.11}$$

This expression holds true under small b when $|x_2| \ll 1$ and we can use the result of the expansion of (5.5.16) in the form of (6.1.10).

Now let us turn to the range of magnetic fields with negative coefficient a in the expansion (6.1.9) for $\tilde{S}(x,y)$. We shall subsequently see that in this case with increasing supercriticality the distribution function of parametric waves over the resonant surface could be quite different and not result in the generation of the second group of pairs. Under some $p = p_{\text{th}}$ (this value of the supercriticality is called the *intermediate threshold*) the regular part appears in $N(x)$ in the vicinity of the equator: $N(x) \neq 0$ in the range $|x| < \delta$ the width of the distribution $N(x)$ being zero at $p = p_{\text{th}}$ and further increasing as the supercriticality increases.

To find p_{th} let us substitute the expression (6.1.9) for $\tilde{S}(x,y)$ in the expression (5.5.14) for the self–consistent pumping. As a result, we obtain:

$$|\tilde{P}(x)|^2 = \Gamma^2 \{1 + x^2[a^2 + 2ab|x| + b^2 x^2](p-1)\} . \qquad (6.1.12a)$$

Expanding $\Gamma(x)$ into a series we obtain:

$$\Gamma^2(x) = \Gamma^2 + \Gamma\Gamma'' x^2 + (3\Gamma''^2 + \Gamma\Gamma^{(4)})x^4/12 . \qquad (6.1.12b)$$

The comparison of the (6.1.12a) and (6.1.12b) shows that at $p_2 = p(0)$ given by (6.1.10) the coefficients of x^2 of these dependences become equal: at the point $x=0$ second order tangency of the lines $|\tilde{P}(x)|$ and $\Gamma(x)$ occurs – the values of these functions coincide as well as the values of their first and second derivatives. Therefore $p_{\text{th}} = p(0)$. At $p > p(0)$ this results in the appearance of the instability region at $|x| \leq \delta$ (where $|\tilde{P}(x)| > \Gamma(x)$).

The boundary $\delta_1(p)$ of the instability region is obtained from the comparison of the two formulae (6.1.12):

$$p - p(0) = \frac{2b(p-1)|\delta_1|}{a\Gamma^2} + \left[\frac{b^2(p-1)}{a^2} - \frac{2\Gamma''^2 + \Gamma\Gamma^{(4)}}{12\Gamma^2} \right] \delta_1^2 . \qquad (6.1.13a)$$

As the smallness of δ_1 is $p - p(0)$ we can substitute p in the right-hand side of (6.1.13a) for $p(0)$. Then (6.1.13a) can be represented in the form similar to (6.1.10):

$$p - p(0) = p'\delta_1 + p''\delta_1^2/2 . \qquad (6.1.13b)$$

As is clear from this equation under $p' > 0$ the instability region is limited and $\delta_1 \propto [p - p(0)]$. If $p'=0$, then for δ_1 being limited it is necessary that $p'' > 0$, then $\delta_1 \propto \sqrt{p - p(0)}$.

The subsequent section, which is based on the study of *Zautkin, L'vov* and *Podivilov* [6.2], shows that the development of the instability results in a region term in the distribution function of parametric waves with the width over x equal to $\delta(p)$, and under the small supercriticality above the intermediate threshold $p(0)$ the dependence $\delta(p)$ can be obtained from the equation similar to (6.1.13):

$$p - p(0) = 6[p'|\delta| + p''\delta^2/2] . \qquad (6.1.14)$$

It will also be shown that an considerable change in the distribution function above the intermediate threshold (i.e. when $p > p(0)$) does not seriously affect the total characteristics of the system of parametric waves such as $\chi' \chi'', N$, etc.

6.1.3 Nonlinear Behavior of Non-Analytic Pair Interaction Amplitudes

First of all we obtain the nonlinear equations for the regular term of the distribution function of parametric waves $N(x)$ and their phases $\Psi(x)$. To this end, we calculate $\tilde{P}(x)$ by using (5.5.11) and (6.1.9). Under $0 \leq x \leq \delta$ (δ is the distribution width):

$$\tilde{P}(x) = A + Bx^2 + 2aS[x \int_0^x \sigma(y)\,dy + \int_x^\delta y\sigma(y)\,dy], \qquad (6.1.15a)$$

$$A = hV + S(\Sigma_0 + b\Sigma_2), \quad B = bS\Sigma_0, \quad \Sigma_0 = \int_{-\delta}^\delta \sigma(y)\,dy, \qquad (6.1.15b)$$

$$\Sigma_2 = \int_{-\delta}^\delta y^2 \sigma(y)\,dy, \quad \sigma(y) = N(y)\exp[-i\Psi(y)]. \qquad (6.1.15c)$$

At $x > \delta$ the expression for $\tilde{P}(x)$ has another form, i.e.:

$$\tilde{P}(x) = A + Bx + aS\Sigma_0 x. \qquad (6.1.15d)$$

In combination with (5.5.11) this results in a closed system of integral equations for $N(x)$ and $\Psi(x)$. Proceeding from the above we can obtain the integrodifferential equations for these values which are easier to analyze. It follows from (6.1.15) that the functions $\tilde{P}(x)$ and $d\tilde{P}(x)/dx$ are continuous at $x = \delta$, whereas:

$$\tilde{P}(\delta) = hV + S\Sigma_0(1 + a\delta + b\delta^2) + bS\Sigma_2], \qquad (6.1.16a)$$

$$d\tilde{P}(x)/dx = S\Sigma_0(a + 2b\delta) \quad \text{at} \quad x = \delta. \qquad (6.1.16b)$$

Subsequent differentiation yields

$$d^2\tilde{P}(x)/dx = 2bS\Sigma_0, \quad \text{at } x > \delta; \qquad (6.1.16c)$$

$$d\tilde{P}(x)/dx = 2bS\Sigma_0 + 2aS\sigma(x), \quad \text{at } 0 < x < \delta. \qquad (6.1.16d)$$

At $x < \delta$ the equations of the S-theory (5.5.11) in combination with (6.1.15, 16) yield the required system of equations:

$$2aSN(x) = -2bS \int_{-\delta}^\delta N(y)\exp\{-i[\Psi(y) - \Psi(x)]\}dy$$

$$+ \Gamma(x)\frac{d^2\Psi}{dx^2} + \frac{d\Gamma}{dx}\frac{d\Psi}{dx} + i\left[\frac{d^2\Gamma}{dx^2} - \Gamma(x)\left(\frac{d\Psi}{dx}\right)^2\right]. \qquad (6.1.17)$$

These equations can most easiest be solved for the case $b=0$ when the integrated term is absent. Then the x-dependence of the phase Ψ has the form:

$$\frac{d\Psi(x)}{dx} = \Psi(0) + \int_0^x \sqrt{\frac{d^2\Gamma(y)}{\Gamma(y)dy^2}}\, dy. \qquad (6.1.18a)$$

The distribution $N(x)$ contains the singular term due to the salient point of the first derivative of $\Psi(x)$ at $x=0$:

$$N(x) = N_1\delta(x) + N_2(x),$$
$$N_1 = N_{1\mathrm{cr}} = \sqrt{\Gamma\Gamma''}/aS, \quad \Gamma'' = d^2\Gamma(x)/dx^2 \text{ at } x = 0, \qquad (6.1.18b)$$
$$N_2(x) = \left[\frac{d^3\Gamma(x)}{dx^3}\Gamma(x) + \frac{d\Gamma(x)}{dx}\frac{d^2\Gamma(x)}{dx^2}\right]\Big/2aS\sqrt{\frac{\Gamma(x)d^2\Gamma(x)}{dx^2}}.$$

This solution is realized under the supercriticalities p above the intermediate threshold $p(0)$. Under $p < p(0)$ the ordinary singular solution of the basic S-theory (5.5.13a) is realized, which in the case under consideration is increased as p increases according to the law

$$N_1(p) = \Gamma(0)\sqrt{p-1}/|S|. \qquad (6.6.19)$$

Using (6.1.10b) for $p(0)$ one can see that the value of the function $N_1(p)$ under $p=p(0)$ coincides with the value $N_{1,\mathrm{cr}}$ in (6.1.18). Therefore in our solution the number of parametric waves $N_1(p)$ under $p < p(0)$ increases according to (6.1.19) and at $p > p(0)$ stops at the threshold level $N_{1,\mathrm{cr}}$ (6.1.18c). According to (6.1.18) $N_2(x)=0$ at $x=0$ and linearly increases with $|x|$ at small x

$$N_2(x) = |x|[\Gamma\Gamma^{(4)} + (\Gamma^{(2)})^2]/4aS\sqrt{\Gamma\Gamma^{(2)}}, \qquad (6.1.20)$$

where $\Gamma^{(n)}$ is the derivative of $\Gamma(x)$ with respect to x of the order n at $x=0$.

When $b \neq 0$ the integrodifferential equation (6.1.17) can be solved under small supercriticalities above the intermediate threshold, i.e. under $p - p(0) \ll 1$. To this end, the theory of perturbations with respect to δ (the width of the continuous part of the distribution function) should be employed assuming that the value δ must be small. We shall present the result of the corresponding calculations

$$\frac{d\Psi}{dx} = \mathrm{sign}x\sqrt{\frac{d^2\Gamma(x)}{\Gamma(x)dx^2}}\left[1 + (2\delta|x| - \delta^2)\frac{b^2}{a^2}\right],$$
$$N_1 = N_{1\mathrm{cr}}[1 - \delta^2 b^2/a^2], \qquad N_2(x) = N_{21} + N_{22}x, \qquad (6.1.21)$$
$$N_{21} = N_{1\mathrm{cr}}[-b/a + 3\delta^2 b^2/a^2] - N_{22}\delta^2 b,$$
$$N_{22} = N_{1\mathrm{cr}}[(\Gamma\Gamma^{(4)} + (\Gamma^{(2)})^2/4\Gamma\Gamma^{(2)} + \delta^2 b^2/a^2].$$

This solutions differ from (6.1.18) obtained under $b=0$ in two significant characteristics. First, the singular part of the N-distribution in (6.1.21) decreases as the supercriticality increases and apparently must become zero under some high p; second, the regular part of the $N(x)$-distribution at $x=0$ is nonzero and increases with the increasing p.

Now we shall find how the width of the distribution δ depend on the supercriticality $p - p(0)$. To this end, we shall make use of the fact that $|\tilde{P}(x)| = \Gamma(x)$ at $x \leq \delta$. Therefore

$$|\tilde{P}(\delta)|^2 = \Gamma^2(\delta), \tag{6.1.22a}$$

$$d|\tilde{P}(\delta)|/dx = d\Gamma^2(x)/dx \quad \text{at} \quad x = \delta \tag{6.1.22b}$$

$$d^2|\tilde{P}(\delta)|^2/d^2x = d^2\Gamma^2(x)/dx^2 \quad \text{at} \quad x = \delta - 0 \tag{6.1.22c}$$

At $x > \delta$ the inequality $|\tilde{P}(x)|^2 \leq \Gamma^2(x)$ holds. Taking into account the relations (6.1.22a, b) we have

$$d^2|\tilde{P}|^2/dx^2 \leq d^2\Gamma^2/dx \quad \text{at} \quad x = \delta + 0.$$

A much more detailed analysis based on the condition of the stability of the obtained solution in the wave package narrowing (decrease of δ) shows that the equality sign must appear in the last formula

$$d^2|\tilde{P}|^2/dx^2 = d^2\Gamma^2/dx^2 \quad \text{at} \quad x = \delta + 0. \tag{6.1.22d}$$

Substituting into (6.1.22a, b and c) the expressions for $\tilde{P}(\delta)$, $\tilde{P}'(\delta)$ and $\tilde{P}''(\delta + 0)$ from (6.1.16a, b and c) after simple transformations we obtain

$$Xf^2 + 2Yf = \Gamma^2 - h^2V^2, \quad Xff_1 + Yf_1 = \Gamma\Gamma', \tag{6.1.23a,b}$$

$$X[f_1^2 + 2ff_2] + 2Yf_2 = \Gamma'^2 + \Gamma\Gamma'', \tag{6.1.23c}$$

$$X = |S\Sigma_0|^2, \quad Y = hVS\text{Re}\{\Sigma_0\}, \tag{6.1.23d}$$

$$f(x) = S(0,x)/S(0,0) = 1 + ax + bx^2, \tag{6.1.23e}$$

$$f_1(x) = f'(x) = a + 2bx \quad f_2(x) = f''/2 = b. \tag{6.1.23f}$$

The values of all functions f, f_1, f_2 and $\Gamma, \Gamma', \Gamma''$ are taken at the point $x = \delta$. In these equations the term $b\Sigma_2$ in comparison with Σ_0 was neglected. Using the solution (6.1.21) we can estimate $\Sigma_2 \simeq -b\delta^3 \Sigma_0/3a$. Therefore our approximation holds true if $b\delta^3 \ll 3a$ and can be employed either in the case of small b and arbitrary δ or in the case of arbitrary b and small δ. This approximation considerably simplifies the situation: the equations (6.1.23) become closed and specify the dependences of δ and Σ_0 on hV. This enables one to obtain not only the width of the package of parametric waves δ, but also the total characteristic of the system, such as the nonlinear susceptibilities χ' and χ'' without the explicit solution of the original integrodifferential equations (6.1.17) and subsequent integration according to the formulae

$$\chi' = -\frac{2}{h} \int V(x) \operatorname{Re}\sigma(x) dx , \quad \chi'' = -\frac{2}{h} \int V(x) \operatorname{Im}\sigma(x) dx , \qquad (6.1.24a)$$

Indeed, in our notation (6.1.24a) are reduced to the form

$$\chi' = -2V \operatorname{Re}\Sigma_0/h , \quad \chi'' = 2V \operatorname{Im}\Sigma_0/h \qquad (6.1.24b)$$

i.e. are expressed in terms of the value Σ_0, whose dependence on the hV is given by (6.1.23). In order to find the dependence of the package width δ on the supercriticality let us eliminate X and Y from (6.1.23). This results in:

$$f_1^3(h^2V^2 - \Gamma^2) = f[ff_1(\Gamma'^2 + \Gamma\Gamma'') - 2\Gamma\Gamma'(f_1^2 + ff_2)] . \qquad (6.1.25a)$$

Substituting in the above the explicit form of the functions f, f_1 and f_2 from (6.1.23 e and f) we can obtain accurately to the terms not higher than δ^3:

$$p - p(0) = 6[p'(1 + a\delta^3/3 - 5b\delta/a) + p''^2/2] . \qquad (6.1.26a)$$

The coefficients $p(0)$, p' and p'' are given by (6.1.10b); a and b are the expansion coefficients of the function $\tilde{S}(x,y)$ in (6.1.9). Under $b=0$, when $p'=0$ in (6.1.26a) we can limit ourselves to the terms linear in δ. Then

$$\delta = [p - p(0)]/6p' . \qquad (6.1.26b)$$

If $b=0$, then $p'=0$ and

$$\delta^2 = [p - p(0)]/3p'' . \qquad (6.1.26c)$$

Both these cases are described by the interpolation formula (6.1.14), whose form is simpler than (6.1.26a). Comparison of (6.1.26) with (6.1.13b) enables us to draw a qualitative conclusion about the width of the nonlinear package of the parametric waves δ being smaller by a factor from 6 to $\sqrt{6}$ than the width of the instability region δ_1. The S-theory generally shows the trend towards the narrowing of the parametric wave package at the nonlinear stage of its evolution and this is one of its manifestations.

Seeking the solution of (6.1.23) with respect to X and Y we can easily obtain

$$|S\Sigma_0|^2 = X = [h^2V^2 - \Gamma^2]/f^2 + 2\Gamma\Gamma'/ff , \qquad (6.1.27a)$$

$$-hVS\operatorname{Re}\Sigma_0 = -Y = [h^2V^2 - \Gamma^2]/f - \Gamma\Gamma'/f_1 . \qquad (6.1.27b)$$

The right-hand parts of these equations are the functions of δ which depends on the supercriticality according to (6.1.25a) or (6.1.26a). At $p - p(0) \ll 1$ (6.1.27) can be reduced to the accuracy of the second order of δ to the following form:

$$\begin{aligned}|S\Sigma_0|^2 &= \Gamma^2(0)[p - 1 - ap'\delta^2] , \\ -hVS\operatorname{Re}\Sigma_0 &= \Gamma^2(0)[p - 1 - a\delta^2/3]\end{aligned} \qquad (6.1.28)$$

It should be recalled that at $p < p(0)$ when $N(x) \propto \delta(x)$, these values are described by the following simple formulae of the basic S-theory:

$$|S\Sigma_0|^2 = -hVS\,\mathrm{Re}\Sigma_0 = \Gamma^2(0)[p-1] \qquad (6.1.29)$$

It can easily be seen from the comparison of the above formulae with (6.1.28) that the significant qualitative rearrangement of the wave distribution function $N(x)$ above the intermediate threshold (i.e. at $p - p(0) \ll 1$) produces a small effect on the total characteristic of the wave system (Σ_0, χ' and χ'') if $p - p(0)$ is small. Indeed, at $b \neq 0$ the difference of (6.1.28) and (6.1.29) is proportional to $[p - p(0)]^2$. If $b=0$, then $p'=0$ and the difference between (6.1.28) and (6.1.29) is proportional to $[p - p(0)]^{3/2}$. Therefore this rearrangement of the distribution function above the threshold $p(0)$ is not easily detected in experiments aimed at obtaining the total characteristics of the system of the parametric waves (Σ_0, χ' and χ''). Nevertheless, the above-described rearrangements of the parametric wave distribution function under increasing supercriticality have been detected and studied by *Zautkin* et al. in [6.2] under parallel pumping of magnons in YIG (see also Sect. 9.6.2. and Fig. 9.26).

6.2 Influence of Nonlinear Damping on Parametric Excitation

It is assumed in the basic S-theory that the damping of the parametric waves $\gamma(\boldsymbol{k})$ is independent of their number. In Sect. 1.3.3 it has already been mentioned that this assumption is an ideal case, generally speaking $\gamma(\boldsymbol{k})$ is the functional of $n(\boldsymbol{k}')$. According to (1.3.10) under small $n(\boldsymbol{k}')$ it can be assumed that

$$\gamma(\boldsymbol{k}) = \gamma_0(\boldsymbol{k}) + \int \eta(\boldsymbol{k},\boldsymbol{k}')n(\boldsymbol{k}')\,d\boldsymbol{k}' \;. \qquad (6.2.1)$$

Since we are mainly interested in the qualitative influence of the nonlinear damping on the behavior of parametric waves, in this section we shall confine ourselves mostly to the linear dependence (6.2.1). The subsequent results can readily be generalized to the more complex models of the nonlinear damping.

6.2.1 Simple Theory

For simplicity, the function $\eta(\boldsymbol{k},\boldsymbol{k}')$ will be assumed in this subsection to be a continuous function of its arguments [6.3]. Its values over the resonant surface will enter the fundamental equations of the basic S-theory (5.5.6):

$$\gamma(\Omega) = \gamma_0(\Omega) + \int \eta(\Omega, \Omega') n(\Omega') d\Omega' . \qquad (6.2.2)$$

As is known, under small supercriticality in the ground state there is only one group of pairs (located in one pair of points on the line or over the entire resonant surface depending on the symmetry of the problem). In this case

$$\eta(N) = \gamma_0 + \eta N ,$$
$$N = \int n(\Omega) d\Omega , \quad \eta = \int \eta(\Omega, \Omega') n(\Omega') d\Omega' / N . \qquad (6.2.3)$$

The total number of pairs N and their phase Ψ is obtained from the following equations generalizing (5.5.7):

$$hV \sin \Psi = \gamma(N) , \quad -hV \cos \Psi = SN . \qquad (6.2.4)$$

From this equation and from (6.2.3) the dependence of N on the supercriticality p (p is the ratio of the power of the pumping to its threshold value $p = h^2 V^2 / \gamma^2$) can readily be obtained:

$$N = \frac{\gamma_0}{|S|} \frac{-c \pm \sqrt{p(c^2+1) - 1}}{c^2 + 1} , \quad c = \frac{\eta}{|S|} . \qquad (6.2.5)$$

In the absence of the nonlinear damping (at $c=0$) this formula goes over to (5.5.7). Under large positive c (6.2.5) changes into

$$N = \gamma_0 [\sqrt{p} - 1]/\eta = (hV - \gamma_0)/\eta . \qquad (6.2.6)$$

This expression trivially follows from (6.2.4) at $S = 0$ ($\cos \Psi = 0$, $\sin \Psi = 1$, $hV = \gamma(N) = \gamma_0 + \eta N$) and is a condition of the energy balance in the absence of the phase mechanism of the amplitude limitation. In this case the stationary value of N is entirely due to the increase of damping as the N increases.

Under negative nonlinear damping (i.e. under $\eta < 0$) the dependence $N(p)$ becomes ambiguous (see Fig. 6.1). *Hard excitation* of waves arises: at the instability threshold (at $p=1$) the wave number increases abruptly from zero to N, where

$$N_+ = N(p = 1) = 2c\gamma_0 / |S|(c^2 + 1) . \qquad (6.2.7)$$

The further increase of p results in the increasing number of waves according to (6.2.5). As p decreases from the values of $p > 1$ to $p < 1$ (but $p > p_-$, where $p_- = 1/(c^2 + 1)$) the number of waves decreases according to (6.2.5) to $N_- = N_-/2$ and then abruptly falls to zero. Therefore the hard excitation of the waves inevitably is accompanied by the hysteresis of the dependence $N(p)$.

With $N(p)$ known, the susceptibility χ of the parametric wave system can readily be calculated. Its imaginary part χ'' is proportional to the power

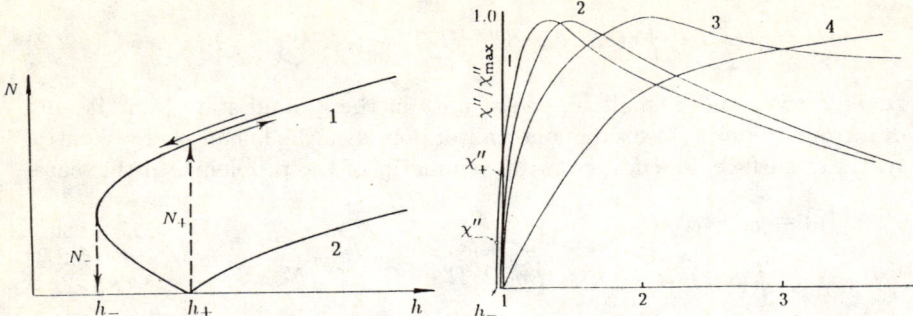

Fig. 6.1. (left) Hard excitation of parametric waves under negative nonlinear damping: theoretical dependences (6.2.5) of the total number of parametric waves N on the pumping amplitude h. The line (2) is unstable

Fig. 6.2. (right) Theoretical dependences (6.2.8) of imaginary part of nonlinear susceptibility χ'' on dimensionless pumping amplitudes h/h_{th} at different values of nonlinear damping: (1) negative nonlinear damping ($\eta = -0.25|S|$); (2) linear damping ($\eta = 0$); (3 and 4) positive nonlinear damping at $\eta/|S|$ equals 0.25 and 1.0, respectively

absorbed by the waves $W_+ = \omega_p \chi'' h^2/2$ and is connected with their total number N by the following formula:

$$\chi'' = (2/h^2) \int \gamma(x) N(x)\, dx$$
$$= \frac{2V^2}{|S|} \frac{\sqrt{p(c^2+1)-1}}{p(c^2+1)} \sqrt{p(c^2+1)^2 - [\sqrt{p(c^2+1)-1}-c]^2}. \qquad (6.2.8)$$

The dependences $\chi''(p)$ for different $c = \nu/|S|$ are plotted in Fig. 6.2. At $c > 0$ the finite slope $\chi''(p)$ appears under $p=1$, and the finite value of $\chi'(\infty) = V^2\eta/(\eta^2 + S^2)$. It is interesting to note that at $c < 1$ ($\eta < S$) the maximum value $\chi''_{\max} = V^2/|S|$, it is independent of η and coincides with the maximum value of χ'' calculated by the basic S-theory without the nonlinear damping (at $c=0$). As η increases the position of the maximum p is shifted to the greater p: $p_m = 2S^2/(|S|-\eta)^2$. At $\eta > |S|$ the susceptibility χ'' increases monotonically with the increase of p.

For $\eta < 0$, as already said, the hard excitation of the parametric waves takes place. Naturally, it is accompanied by the hysteresis of the dependence $\chi''(p)$. When p assumes the threshold value of $p = 1$ from the side of the smaller p the susceptibility abruptly changes at $p=1$ from zero to the following value

$$\chi_+ = -\frac{4V^2}{S} \frac{c\sqrt{(c^2+1)^2 - 4c^2}}{(c^2+1)^2}. \qquad (6.2.9)$$

As the amplitude of the pumping decreases to some values $p_- \simeq 1/(c^2+1)$ (which is below the threshold p_+) than the reverse abrupt change of the susceptibility takes place. In the case of $|c| < 1$ this reverse change is half as great as the direct change

$$\chi_- = \chi_+/2 = 2V^2\eta/S^2 \ . \tag{6.2.10}$$

This hard excitation phenomenon and the hysteresis of χ'' was discovered by *Le Gall, Lemaire* and *Sere* in the YIG under parametric excitation of the magnons by parallel pumping at the frequency of 9.8 hHz [6.4].

In conclusion, we shall briefly discuss the region of applicability of the obtained results. Obviously, (6.2.3) for damping holds true if $\eta N \ll \gamma_0$. Under $0 < c \ll 1$ this condition is satisfied within the wide range of the supercriticality values $p \ll 1/\sqrt{c}$, and at $c \gg 1$ is satisfied only when the supercriticality values are only insignificantly above the threshold, i.e. at $p - 1 \ll 1$. In the case of negative nonlinear damping the applicability criterion of the results obtained is still more stringent. The condition $\eta N \ll \gamma_0$ brings about the requirement $|c| \ll 1$. In this case the above formulae hold true at $p \ll 1/|c|$.

The applicability of the simple S-theory with the negative nonlinear damping can be significantly extended if instead of (6.2.3a) with $\eta < 0$ the more realistic model dependence $\eta(N)$ is used:

$$\gamma(N) = \gamma_1 + \gamma_2^2/(\gamma_2 + \eta N) \ . \tag{6.2.11}$$

It holds true qualitatively for the common mechanisms of the negative nonlinear damping of magnons even when $\eta N > \gamma_2$ [6.5].

6.2.2 Influence of Non-Analyticity on Nonlinear Damping

The subsequent study of different mechanisms of nonlinear damping (Sect. 11.2) will show that in some cases the function $\eta(\boldsymbol{k}, \boldsymbol{k}')$ is non-analytical under $\boldsymbol{k} \to \boldsymbol{k}'$. In the region $|\boldsymbol{k} - \boldsymbol{k}'| = \kappa \ll k, k'$ in some approximation it resembles $1/\kappa$, i.e. it has an integrated singularity.

By way of example, let us consider the problem of the parametric excitation of the spin waves in ferromagnets (for more detail, see [6.6]). As shown in Sect. 5.5.2 in this case under $\eta=0$ (or when the dependence $\eta(\boldsymbol{k}, \boldsymbol{k}')$ is analytical) and when the supercriticalities are not too high, the waves are excited only on the equator of the resonant surface (i.e. $N(x) \propto \delta(x)$). Therefore it should be expected that at $\eta N \ll \gamma_0$ the distribution function $n(\boldsymbol{k}) = n(k, x, \kappa)$ will be non-zero only under small x and k, approaching the radius of the equator k_0. The calculations of the nonlinear damping performed in Sect. 11.2 showed that in this range, i.e. at $x_1, x_2 \ll 1$ and $|\boldsymbol{k}_{1,2} - \boldsymbol{k}_0| \ll k_0$ the function $\eta(\boldsymbol{k}_1, \boldsymbol{k}_2)$ can be represented in the following form

$$\eta(\boldsymbol{k}_1, \boldsymbol{k}_2) = \eta\{\sin^2(\frac{\psi_1 - \psi_2}{2}) + (\frac{\alpha_1 - \alpha_2}{4}) + (\frac{k_1 - k_2}{2k_1})^2\}^{-1/2} \ . \tag{6.2.12}$$

Here $\alpha = \pi/2 - \Theta$ and η is the analytical function of the magnetic field, k and other parameters of the problem. The entire non-analytical part of the function $\eta(\boldsymbol{k}_1, \boldsymbol{k}_2)$ is enclosed in braces.

In order to simplify the problem further note that the expected width of the packet $n(\boldsymbol{k})$ is significantly less in module k than in polar angle Θ. It can be assumed that $(k - k_0)[\partial \omega(\boldsymbol{k})/\gamma(\boldsymbol{k})\partial k] \simeq \alpha$. Then the last term in braces in (6.2.12) can be neglected, since it is by a factor of $[k_0 \gamma(\boldsymbol{k})]/[\partial \omega(\boldsymbol{k})/\partial k]$ smaller than the last but one. Let us also take into account the axial symmetry of the problem and average (6.2.12) over the difference of the azimuthal $\varphi_1 - \varphi_2$. This yields

$$\gamma(\alpha) = \gamma_0 + \eta \int \ln|\alpha - \beta| N(\beta)\, d\beta\,. \tag{6.2.13}$$

In this formula the numerical factor of the integral of the order of unity has been dropped as unimportant. The integral equations of the basic S-theory (5.5.6) we shall represent as

$$[P(\alpha)\exp[i\Psi(\alpha)] - i\gamma(\alpha)]N = 0\,, \tag{6.2.14a}$$

$$P(\alpha) = hV(\alpha) + \int S(\alpha,\beta) N(\beta) \exp[-i\Psi(\beta)]\, d\beta\,. \tag{6.2.14b}$$

Our task therefore consists in the simultaneous solution of the integral equations (6.2.13, 14). We shall be interested mainly in the qualitatively new results due to the non-analyticity of the function of the nonlinear damping in (6.2.13), and therefore the problems associated with (6.2.14b) will be significantly simplified. To this end, let us choose a function $S(\alpha, \beta)$ in the factorized form

$$S(\alpha, \beta) = S f(\alpha) f(\beta)\,, \quad f(\alpha) = V(\alpha)/V\,, \quad V = V(0)\,. \tag{6.2.15}$$

Then the dependence $P(\alpha)$ will easily be found

$$P(\alpha) = P f(\alpha)\,, \quad P = hV + S\Sigma\,,$$
$$\Sigma = \int f(\alpha) N(\alpha) \exp[-i\Psi(\alpha)]\, d\alpha\,. \tag{6.2.16}$$

This enables us to obtain from (6.2.13, 14a) a closed equation for $N(\alpha)$:

$$\gamma_0 + \eta \int \ln|\alpha - \beta| N(\beta)\, d\beta = |P| f(\alpha)\,. \tag{6.2.17}$$

From (6.2.14a) it also follows that

$$\Psi(\alpha) = \Psi\,, \qquad \mathrm{Re}\{P \exp \Psi\} = 0\,. \tag{6.2.18}$$

Let us first consider the case of **positive nonlinear damping**: $\eta > 0$. The general solution of this equation localized in some range $-a < \alpha < a$ and equal to zero when $|\alpha| > a$ has the following form

$$N(\alpha) = \frac{1}{\pi\eta} \int_{-a}^{a} \sqrt{\frac{a^2 - t^2}{a^2 - \alpha^2}} \frac{P'(t)dt}{t - \alpha} + \frac{A}{\sqrt{a^2 - \alpha^2}} \qquad (6.2.19)$$

where $P' = Pf'$ is the derivative of the pumping with respect to the angle, A denotes an arbitrary constant and the integral has the meaning of the Cauchy's principal value. As can easily be shown, the solution limited for the both end points of the interval exists only if the equality

$$\int_{-a}^{a} \frac{P'(t)\, dt}{\sqrt{a^2 - t^2}} = 0$$

is satisfied, which is ensured by the evenness of the function $P(t)$. This solution has the following form:

$$N(\alpha) = \frac{1}{\pi\eta} \int_{-a}^{a} \sqrt{\frac{a^2 - \alpha^2}{a^2 - t^2}} \frac{P'(t)dt}{t - \alpha}. \qquad (6.2.20)$$

It must be recalled that our consideration holds true when the excited packet is narrow ($a \ll 1$). Therefore the integral in (6.2.20) must be calculated expanding $P(t)$ under small t:

$$P(t) = P[1 - (1/2)f''t^2]\,, \quad f'' = d^2f/d\alpha^2 \text{ at } \alpha = 0\,, \quad P'(t) = -Pf''t\,.$$

The form of the packet in this case is as follows:

$$N(\alpha) = (|P|f''/\pi\eta)\sqrt{a^2 - \alpha^2}\,. \qquad (6.2.21a)$$

This expression can be conveniently represented as

$$N(\alpha) = (2N/\pi a)\sqrt{1 - (\alpha/a)^2}\,, \qquad (6.2.21b)$$

$$N = \int N(\alpha)d\alpha = |P|f''a^2/2\eta \qquad (6.2.21c)$$

where N is the total number of waves in the packet. Substituting (6.2.21b) in the initial equation (6.2.17) we obtain the integrated relation

$$|P| = \gamma_0 + \eta N \ln(2/a)\,. \qquad (6.2.22)$$

From (6.2.16, 18) we can readily obtain in the approximation of narrow packets the usual relations of the basic S-theory

$$\Sigma = N \exp[-i\Psi], \quad SN = -hV\cos\Phi, \quad |P|^2 = (hV)^2 - (SN)^2\,. \quad (6.2.23)$$

The relations (6.2.21c, 22, 23) close the problem of the self-consistency and make it possible to determine the dependences of $|P|$, N, Ψ and a on the supercriticality. This can be done as follows. At the first stage the weak (logarithmic) dependence of $|P|$ on a in (6.2.22) can be neglected if we assume

$$|P| = \gamma_0 + \tilde{\eta} N , \quad \tilde{\eta} = \eta \ln(2/a) = \text{const} . \tag{6.2.24}$$

Substituting this expression into (6.2.23), we obtain the formulae of the simple theory developed in the previous section. For the supercriticality dependences N and χ'' this yields (6.2.5, 8), η being replaced by $\tilde{\eta}$. By substituting the dependence $N(p)$ into (6.2.21c), we can obtain the dependence of the packet width a on the supercriticality p.

Now let us consider the case of **negative nonlinear damping**. Under $\eta < 0$ equation (6.2.17) (corresponding to the mean–field approximation) admits only the solutions with the integrated singularity, i.e.

$$N(\alpha) = A/\sqrt{a^2 - \alpha^2} + P f'' \sqrt{a^2 - \alpha^2}/\pi \eta .$$

By substituting this solution into the equation of the energy balance we find the following connection between P, a and A:

$$-\frac{P^2 a^2}{\eta} \ln\left(\frac{2}{a}\right) - A \ln\left(\frac{2}{a}\right) = -\left(\frac{P f''}{\eta}\right) \frac{P - \gamma}{P} . \tag{6.2.25}$$

From the condition $\int N(\alpha)\, d\alpha = N$ we can obtain the second connection between these three parameters. Therefore we have a one-parameter set of solutions. Which of them is actually realized? Note that for each of these solutions the renormalized damping and pumping coincide for all α and not only for $-a \leq \alpha \leq a$. This implies that any of the solutions is indifferently stable with respect to the appearance of the waves outside the integral $[-a, a]$. Consequently, the solution with the maximum (under the given supercriticality) width of a will be realized. Allowing for the condition $N(\alpha) \geq 0$ we find that the maximum width corresponds to the choice $A = -2 f'' a P/\pi \eta$ and is equal to

$$a^2 = -\eta f'' N / 2P , \tag{6.2.26}$$

i.e. it coincides with the expression (2.2.21c) for the case $\eta > 0$ (it must only be assumed that $c < 0$ and the general sign must be changed); and the dependence of the total number of waves N on the dimensionless pumping power p is still given by (6.2.5) but now with $c < 0$.

In conclusion, compare the influences of the non-analyticity of the main functions of the S-theory $S(\boldsymbol{k}_1, \boldsymbol{k}_2)$ and $\eta(\boldsymbol{k}_1, \boldsymbol{k}_2)$ that determine the nonlinear renormalization of the self-consistent pumping $P[n(\boldsymbol{k}), \Psi(\boldsymbol{k})]$ and the self-consistent damping $\eta[n(\boldsymbol{k})]$. In both cases the non-analyticity results in the broadening of the parametric wave packet over the polar angle. However, the non-analyticity of the function S proved to be much weaker than the non-analyticity of the function η (under the axial symmetry S is characterized by a discontinuity of the derivatives while η has a logarithmic singularity). Therefore the broadening of the packet when S is non-analytic is substantially weaker: it occurs only when the supercriticality p is above the intermediate threshold $p(0)$. At the same time the width of the packet δ increases rather slowly, i.e. proportionally to $[p-p(0)]^n$, $n=2$ or $3/2$. When η is non-analytic the packet broadening starts immediately above the threshold of the parametric excitation, i.e. under $p > 1$ and is faster, proportionally to $(p-1)$ under the positive nonlinear damping and is discontinuous when the nonlinear damping is negative. On the other hand, the non-analyticity of the two functions S and η is so weak that they practically do not influence the general characteristics of the system of parametric waves such as N, χ'', χ', etc. It must be added that the weak dependence of the general characteristics of nonlinear systems on the fine properties of the functions describing the interaction is not characteristic of the problem of parametric wave excitation, but is typical of the systems that can be described within approximations like the mean-field.

6.3 Parametric Excitation Under the Feedback Effect on Pumping

6.3.1 Hamiltonian of the Problem

For definiteness, we shall consider this problem for the case of the nonlinear theory of the ferromagnetic resonance [6.7]. This theory must describe the amplitude dependence of the uniform precession of the magnetization (UP) c_0 on the frequency ω_p and the amplitude $h_\perp$ of the external magnetic field $\mathbf{h}_\perp$ (oriented transverse to the magnetization $\mathbf{h}_\perp \perp \mathbf{M}$) when the amplitude $h_\perp$ is so large that the nonlinear effects must be taken into consideration. In addition to the nonlinear frequency shift of the UP proportional to $|c_0|^2$, the most important of these effects are the processes of parametric excitation of magnons by the uniform precession (see the description in Sect. 4.3.1, 4) and the feedback effect of the parametric magnons on the uniform precession. All the above processes are described by the Hamiltonian

$$\mathcal{H} = \omega_0 c_0 c_0^* + \mathcal{H}_{00} + \mathcal{H}_\perp + \sum_{\mathbf{k}} \mathcal{H}_{0\mathbf{k}} + \sum_{\mathbf{k}} \omega(\mathbf{k}) c(\mathbf{k}) c^*(\mathbf{k}) + \mathcal{H}_\mathrm{S} \,, \quad (6.3.1a)$$

$$\mathcal{H}_{00} = T_{00} c_0^2 c_0^*\,, \quad \mathcal{H}_\perp = [h_\perp \exp(-i\omega_\mathrm{p} t) U c_0^* + \text{c.c.}]\,, \quad (6.3.1b)$$

$$\mathcal{H}_{0\mathbf{k}} = (1/2)[V^*(0,\mathbf{k},-\mathbf{k}) c^*(\mathbf{k}) c^*(-\mathbf{k}) + \text{c.c.}]\,,$$
$$+ [S(0,\mathbf{k}) c_0^* c_0^* c(\mathbf{k}) c(-\mathbf{k}) + \text{c.c.}] + T(0,\mathbf{k}) c_0^* c_0 c^*(\mathbf{k}) c(\mathbf{k})\,, \quad (6.3.1c)$$

$$\mathcal{H}_\mathrm{S} = \sum_{1,2} T(\mathbf{k}_1, \mathbf{k}_2) c^*(\mathbf{k}_1) c^*(\mathbf{k}_2) c(\mathbf{k}_2)$$
$$+ \frac{1}{2} \sum_{1,2} S(\mathbf{k}_1, \mathbf{k}_2) c^*(\mathbf{k}_1) c^*(-\mathbf{k}_1) c(\mathbf{k}_2) c(-\mathbf{k}_2)\,. \quad (6.3.1d)$$

Here $\mathcal{H}_{00}$ describes the nonlinear eigen shift of the UP frequency, $\mathcal{H}_\perp$ gives the UP interaction with the uniform magnetic UHF field with the frequency ω_p (for simplicity, we assume this field to be circularly polarized in the plane perpendicular to M) and, finally, $\mathcal{H}_{0k}$ describes the interaction of the uniform precession with the magnons. The first two terms describe the parametric interaction of the first and second order respectively and the last term describes the nonlinear frequency shift. $\mathcal{H}_S$ is the diagonal in pairs Hamiltonian of the spin wave (magnon) interaction in the basic S-theory (5.4.17). The nonlinear theory of the ferromagnetic resonance based on the Hamiltonian (6.3.1) was first developed in 1971 by *Starobinets* and *L'vov* [6.7]. This theory was based on the earlier theory of the ferromagnetic resonance suggested by *Suhl* [6.8]. The Suhl theory, however, allowed only for the magnon-UP interaction and neglected the interaction between the magnons (magnon-magnon interaction), though the latter is considerable under large wave amplitudes and cannot be neglected. *Schlömann* [6.9] wrote: "This approximation (the exclusion of the magnon-magnon interaction) was intended for the mathematical simplification of the problem and, generally speaking, cannot be justified. Assuming this approximation we, probably, lose the major part of the important physical information".

From the theoretical viewpoint it is clear that as the number of the parametric magnons increases above the threshold the energy of their interaction $\mathcal{H}_S$ at first becomes equal and then exceeds the energy of the magnon interaction with UP. The power of the pumping under which the Hamiltonian $\mathcal{H}_S$ becomes important depends on the properties of the spin system and the type of the nonlinear processes and, as a rule, may be exceeded in an experiment. At the same time it was commonly accepted that the amplitude limitation at the parametric excitation is mostly due to the reverse effect of UP on magnons (described by the Hamiltonian $\mathcal{H}_{0k}$). This opinion is based on the "freezing" UP above the threshold which has been predicted by the "magnon-UP-theory" and was applied to the ferrite power limiters.

As shown by *Schlömann* et al. [6.10] the increase of the UP amplitude almost stops under the ferromagnetic resonance for the first-order processes. On the other hand, some facts cannot be explained only by the magnon-UP interaction. In particular, the actual behavior of the nonlinear susceptibilities χ', χ'' far from the resonance differs from the theoretically predicted one when the pumping power considerably exceeds the threshold value (*Damon* [6.11]). It must be specially emphasized that the real part of the susceptibility χ' is practically unchanged above the threshold whereas according to the theory of the magnon-UP interaction it must decrease as $1/h_\perp^2$ ($h_\perp$ is the amplitude of the microwave magnetic field). We must also mention the experiments performed by *Gurevich* and *Starobinets* [6.12] on "saturation" of the ferromagnetic resonance (in the case of second order processes) testifying to the "increased UP amplitude above the threshold".

The above-mentioned and some other facts are readily explained in the subsequently presented theory of *L'vov* and *Starobinets* [6.7] simultaneously taking into account the magnon-UP interaction H_{0k} as well as the magnon-magnon interaction in the approximation of the diagonal Hamiltonian $\mathcal{H}_S$ in the basic S-theory.

It must also be noted that the Hamiltonian (6.3.1) refers not only to the nonlinear theory of the ferromagnetic resonance. It can equally describe also other cases of parametric excitation of waves where the pumping field (or some other eigenmode of the medium oscillations acting as pumping) is linearly excited by an external inducing force. Thus, under parallel pumping of magnons in ferromagnets the variable c_0 is proportional to the amplitude of magnetic field $h_\perp$ in the resonator; ω_0 is the eigenfrequency of the resonator; $T_{00}=0$, because the eigen nonlinearity of the resonator usually is vanishingly small ($T_{00} \ll T(\boldsymbol{k}_1, \boldsymbol{k}_2)$); $S_{0k}=0$, if $\boldsymbol{h} \parallel \boldsymbol{M}$; $V(0, \boldsymbol{k}, -\boldsymbol{k}) \propto Q$, where Q is the filling factor of the resonator by the sample, the amplitude of magnetic field in the waveguide serves as h , the U-factor describes the coupling between the fields in the waveguide and the resonator.

6.3.2 General Analysis of the Equations of Motion

The equations of motion with the Hamiltonian (6.3.1)

$$\left[\frac{\partial}{\partial t} + \gamma_0\right] c_0 = -i\frac{\delta \mathcal{H}}{\delta c_0^*}, \quad \left[\frac{\partial}{\partial t} + \gamma(\boldsymbol{k})\right] c(\boldsymbol{k}) = -\frac{\delta \mathcal{H}}{\delta c^*(\boldsymbol{k})} \qquad (6.3.2a,b)$$

can be represented in the following form usual for the basic S-theory (5.4.16)

$$\left\{\frac{\partial}{\partial t} + \gamma_0 + i\bigl[\omega_{\mathrm{NL}}(0) - \omega_\mathrm{p}\bigr]\right\} c_0(t) + iP_0 c_0^*(t) = 0 , \qquad (6.3.3a)$$

$$\left\{\frac{\partial}{\partial t} + \gamma(\boldsymbol{k}) + i\left[\omega_{\mathrm{NL}}(\boldsymbol{k}) - \frac{\ell\omega_\mathrm{p}}{2}\right]\right\} c(\boldsymbol{k}),t) + iP(\boldsymbol{k})c^*(-\boldsymbol{k},t) = 0, \quad (6.3.3b)$$

$$\omega_{\mathrm{NL}}(0) = \omega_0 + 2T_{00}|c_0|^2 + 2\sum_{\boldsymbol{k}} T(0,\boldsymbol{k})|c(\boldsymbol{k})|^2 , \qquad (6.3.4a)$$

$$\omega_{\mathrm{NL}}(\boldsymbol{k}) = \omega(\boldsymbol{k}) + 2\sum_{1} T(\boldsymbol{k},\boldsymbol{k}_1)|c(\boldsymbol{k}_1)|^2 + 2T(0,\boldsymbol{k})|c_0|^2 . \qquad (6.3.4b)$$

Here c_0 is the amplitude of the spatially homogeneous pumping whose role in the problem of the nonlinear ferromagnetic resonance is played by the uniform precession of the magnetization, $\omega_{\mathrm{NL}}(0)$ and $\omega_{\mathrm{NL}}(\boldsymbol{k})$ describe the nonlinear frequency of the uniform precession and the parametric magnons, $\ell=1,2$ respectively for the processes of the first and second order. The expressions for the values P_0, $P(\boldsymbol{k})$ denoting the "complex energy fluxes" into the pumping and the $\boldsymbol{k}$-pair of magnons depend on the type of the process under consideration. For the first-order processes

$$P_0 = h_\perp U + \frac{1}{2} \sum_k V(0, \boldsymbol{k}, -\boldsymbol{k}) c(\boldsymbol{k}) c(-\boldsymbol{k}) , \qquad (6.3.5a)$$

$$P(\boldsymbol{k}) = V(0, \boldsymbol{k}, -\boldsymbol{k}) c_0 + \sum_{\boldsymbol{k}'} S(\boldsymbol{k}, \boldsymbol{k}') c(\boldsymbol{k}') c(-\boldsymbol{k}') . \qquad (6.3.5b)$$

The second-order instability threshold is much higher than the first-order one. However, if the frequency ($\omega_p/2$) is outside the magnon spectrum, first-order processes are forbidden and second-order instability can be observed with $\omega(k) = \omega_p$. For it

$$\begin{aligned} P_0 &= \frac{hU}{c_0^*} + \sum_k S(0, \boldsymbol{k}) c(\boldsymbol{k}) c(-\boldsymbol{k}) , \\ P(\boldsymbol{k}) &= S(0, \boldsymbol{k}) c_0^2 + \sum_{\boldsymbol{k}'} S(\boldsymbol{k}, \boldsymbol{k}) c(\boldsymbol{k}') c(-\boldsymbol{k}') . \end{aligned} \qquad (6.3.6)$$

Our task is now reduced to the study of solutions of the coupled equations (6.3.3-6) depending on the amplitude of the external field $h_\perp$. It is clear that the equations (6.3.3b) for parametric magnons assuming c_0=const. fully coincide with the equations describing the parametric instability described in detail in Chap. 5. By substituting the solution of these equations into (6.3.3a) we obtain one complex equation specifying the dependence $c_0(h_\perp)$. This is a rather complicated task requiring computer processing and actually to carry it out is worthwhile only for some important cases (e.g. for YIG monocrystals). This, however, has not yet been done. Here we shall only qualitatively analyze the behavior of the parametric magnons and the pumping above the instability threshold of the nonlinear pumping resonance. Taking into account that the total characteristics of the parametric wave system are not very sensitive to the particular forms of the functions $V(0, k, -k)$, $S(0, k)$ and $S(k_1, k_2)$ we shall assume that $V(0, k, -k)$ and $S(0, k)$=const. Then for the first-order processes from (5.5.7) we obtain

$$N_1 = \sqrt{V_{01}^2 n_0 - \gamma_1^2}/|S_{11}|, \quad \sin(\Psi_1 - \Psi_0/2) = \gamma_1/\sqrt{V_{01}^2 n_0} . \qquad (6.3.7)$$

Similarly, for second-order processes we have

$$N_1 = \sqrt{S_{01}^2 n_{01}^2 - \gamma_1^2}/S_{11}, \quad \sin(\Psi_1 - \Psi_0) = \gamma_1/|S_{01}| n_0 . \qquad (6.3.8)$$

Here N_1 is the total number of parametric waves, $V_{01} = V(0, k_1, -k_1)$, $S_{01} = S(0, k)$, S_{11} is the mean value of $S(k_1, k_2)$ over part of the resonant surface occupied by parametric waves.

6.3.3 First-Order Processes

The stationary solution of (6.3.2a, 7) for the uniform precession has the following form:

$$c\bigl[\omega_{\mathrm{NL}}(0) - \omega_{\mathrm{p}} + i\gamma_{\mathrm{NL}}(0)\bigr] = hU , \qquad (6.3.9)$$

where $\omega_{\mathrm{NL}}(0)$ and $\gamma_{\mathrm{NL}}(0)$ are the frequency and damping of the uniform precession renormalized due to the interaction with the parametric waves:

$$\omega_{\mathrm{NL}}(0) = \omega_0 + 2T_{00}|n_0|^2 + 2T_{01}N_1 - S_{11}N_1^2/2|n_0| ,$$

$$\gamma_{\mathrm{NL}}(0) = \gamma_0 + \gamma_1 N_1/2n_0 , \quad n_0 = |c_0|^2 . \qquad (6.3.10)$$

Here T_{01} is the mean value of $T(0, \boldsymbol{k})$. Now we have the complete system of equations (5.5.7, 9) required for calculating the stationary state. Subsequently, these equations should be rendered dimensionless which will facilitate the estimation of the relative value of the different terms. Introducing to this end the following dimensionless values

$$x_0 = \frac{n_0}{n_{\mathrm{th}}} , \quad x_1 = \frac{N_1}{n_{\mathrm{th}}} , \quad n_{\mathrm{th}} = \frac{\Gamma_1}{V_{01}^2} , \quad p = \left(\frac{h}{h_{\mathrm{th}}}\right)^2 ,$$

$$h_{\mathrm{th}} = \frac{\gamma_0 \gamma_1}{U V_{01}} , \quad \delta = \frac{(\omega_0 - \omega_{\mathrm{p}})}{\gamma_0} , \quad d = \frac{2\gamma_0 S_{11}}{V_{01}^2} , \qquad (6.3.11)$$

$$a = \frac{T_{00}\gamma_1^2}{S_{11}\gamma_0^2} , \quad b = \frac{2T_{01}\gamma_1}{S_{11}\gamma_0} ,$$

we reduce the initial equations to the form

$$\gamma_1 |d| x_1 = 2\gamma_0 \sqrt{x_0 - 1} , \quad (\delta_{\mathrm{NL}}^2 + \gamma_{\mathrm{NL}}^2) x_0 = p\gamma_0^2 , \qquad (6.3.12)$$

$$\gamma_{\mathrm{NL}}^2 = \gamma_0^2 [1 + \sqrt{x_0 - 1}/|d|x_0] ,$$

$$\delta_{\mathrm{NL}} = \delta - (x_0 - 1)/dx_0 + b\sqrt{x_0 - 1} + dax_0 . \qquad (6.3.13)$$

These equations contain the small parameter $|d| \simeq \gamma/\omega_M \simeq 10^{-3} - 10^{-4}$, which in the theory not allowing for the interaction of parametric waves was assumed to be zero.

1 Behavior of Uniform Precession in Resonance. As follows from (6.3.12, 13), the amplitude x_0 is reaches its maximum under $\delta_{\mathrm{NL}} = 0$ and is given by the following equation

$$\sqrt{x_0 - 1} + |d|x_0 = d\sqrt{px} . \qquad (6.3.14)$$

The smallness of d leads to the appearance of two characteristic regions of the solution of this equation: the region of small supercriticality $|d|\sqrt{p} \ll 1$, where

$$x = 1/(1 + d^2 p) , \quad x_1 = 2(\sqrt{p} - 1)\gamma_0/\gamma_1 , \qquad (6.3.15)$$

and the region of large supercriticality $|d|\sqrt{p} \simeq 1$, where

$$x_0 = 1/(1-d^2 p), \qquad x_1 = 2\gamma_0(\sqrt{p}-1)/\gamma_1 - d^2 p). \qquad (6.3.16)$$

These relations hold true as long as $x < |d|^{-2/3}$. Under other supercriticalities, as follows from (6.3.14), the amplitude of the uniform pumping must linearly increase $x_0 = p$ (Fig. 6.3), but in this case we enter the region $x_0 \simeq |d|^{-1}$ where the initial equations (6.3.2) are no longer valid.

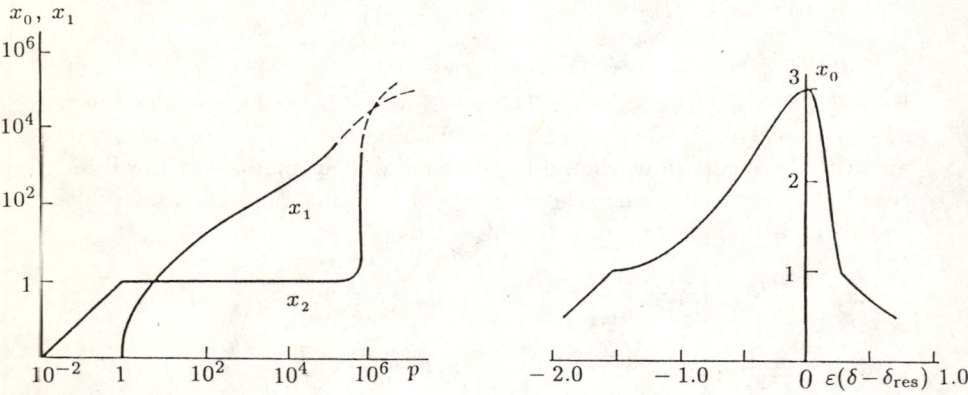

Fig. 6.3. (left) Theoretical dependences (6.3.16) of the square of the homogeneous precession amplitude $x_0 = |c_0|^2/n_{\text{th}}$ and number of magnons N_1/n_{th} on the dimensionless power of the pumping $p = (h/h_{\text{th}})^2$ for first-order processes. Dashed lines denote the region where theory is not valid

Fig. 6.4. (right) Theoretical form (6.3.18) of the nonlinear resonant curve (x_0 versus p) at large pumping power for first-order processes

2 Form of Resonant Curve. When the exact resonance is impossible or is absent, the threshold condition has the following form

$$p = p_{\text{th}} = (1+\delta^2)$$

At $p \to p_{\text{th}} > |d|^2$ let us obtain from (6.3.12, 13) the following relation

$$x_0 = 1 + d^2(p-\delta^2)/(1-\delta d)^2, \qquad (6.3.17)$$

which holds as long as $x_0 < 1/|d|^{2/3}$. The relation (6.3.17) shows that within the limit of large p the amplitude of the homogeneous pumping increases linearly

$$x_0 = \frac{|d|^2(1+\delta^2)}{(1-d\delta)^2} \frac{p}{p_{\text{th}}}.$$

Far from resonance, when $(1-d\delta) \simeq 1$, the proportionality factor approaches unity, i.e. its order of magnitude is the same as in the linear mode. The form of the resonant curve of the nonlinear resonance is appreciated after performing an identical transformation of (6.3.17) yielding:

$$x_0 = \{1 - d^2[(dp - \delta)/(1 - d\delta)]^2\}/(1 - d^2\delta) . \qquad (6.3.18)$$

It is readily seen that the resonant value x_0 is given by (6.3.16) and is attained at $\delta_{\text{res}} = dp$ which corresponds to the condition of the nonlinear resonance above the threshold. The resonant curve (6.3.16) shown in Fig. 6.4 is significantly different from the usual Lorentzian curve of the linear resonance. The half-width of the curves at a half-height (on the left and on the right of the resonance) equals

$$\delta_\pm = \delta_{\text{res}} \pm \delta_{1/2} = \pm [1 - d^2 p]/[d(\sqrt{2} \mp 1)] .$$

so that the ratio of $\delta_-/\delta_+ = 6$.

3 Nonlinear susceptibility. From (6.3.5b, 9) one can obtain

$$\chi = 2Uc_0/h = 2U^2/[(\omega_{\text{NL}} - \omega_{\text{p}}) - i\gamma_{\text{NL}}] . \qquad (6.3.19)$$

In the dimensionless form

$$\chi' = \chi_0 \frac{(\omega_{\text{NL}} - \omega_{\text{p}})x_0}{\gamma p} , \quad \chi'' = \chi_0 \frac{\gamma_{\text{NL}} x_0}{\gamma p} , \qquad (6.3.20)$$

where $\chi_0 = 2U^2/\gamma_0$ denotes the resonant susceptibility of the homogeneous pumping to the transverse SHF field in the linear mode. Now let us consider the behavior of the nonlinear susceptibilities far from the resonance ($\delta \gg 1$), i.e. in the region of the *additional absorption*. It follows from (6.3.17, 20) that:

$$\chi'' = \chi_0 \sqrt{p - \delta^2}/(p|1 - d\delta|), \quad \text{at} \quad p - p_{\text{th}} > |d| . \qquad (6.3.21)$$

Hence, in particular, it follows that above the threshold χ'' linearly increases and attains the maximum

$$\chi''_{\text{max}} = \chi_0/(2\delta|1 - d\delta|) \qquad (6.3.22)$$

under $p = 2\delta \simeq 2p_{\text{th}}$. The decrease χ'' under large supercriticality according to (6.3.22) obeys the law $1/\sqrt{p}$. The real part of the nonlinear susceptibility changes within a much narrower range. From (6.3.16, 20) we obtain

$$\chi' = (\chi_0/\delta)[1 + (1 - \delta^2/p)/(d\delta - 1)], \quad \text{at} \quad p - p_{\text{th}} \gg 1 . \qquad (6.3.23)$$

In the theory not allowing for parametric wave interaction, $d = 0$ and χ' according to (6.3.23) falls off to zero as $1/p$. If this interaction is allowed for, this result changes drastically, i.e. under large supercriticality $\chi' \to$

$\chi_0 d/|1-d\delta|$, that is the order of magnitude of the susceptibility is the same as below the threshold. This accounts for the fact that the phase constants of the ferrite SHF equipment are practically independent of the power level [6.13].

6.3.4 Second-Order Processes

By using (6.3.2) we can obtain (6.3.9) for the amplitude of the homogeneous pumping. However, the expressions for ω_{NL} and γ_{NL} for second-order processes will differ from the corresponding expressions (6.3.10) obtained for the first-order processes, i.e.

$$\omega_{\text{NL}} = \omega_0 + 2T_{00}n_0 + 2T_{01}N_1 - S_{11}N_1^2/n_0, \quad \gamma_{\text{NL}} = \gamma_0 + \gamma_1 N_1/n_0 \,. (6.3.24)$$

These relations together with (6.3.8, 9) set the complete system of equations for the definition of the stationary mode of second-order processes. It is convenient to represent it in the dimensionless form:

$$x_0^2 = 1 + \left(\frac{r^2\gamma_1^2 x_1^2}{\gamma_0^2}\right)(\Omega_{\text{NL}}^2 + \gamma_{\text{NL}}^2)\, x_0 = \gamma_0^2 p,$$

$$\gamma_{\text{NL}} = \gamma_0 \left\{1 + \sqrt{x_0^2 - 1}/rx_0\right\}, \qquad (6.3.25)$$

$$\Omega_{\text{NL}} = \delta\gamma_0 + \gamma_1 \left[\frac{2x_0 T_{00}}{S_{01}} + 2\sqrt{x_0^2 - 1}\frac{T_{01}}{S_{11}} - \frac{(x_0^2 - 1)S_{11}}{S_{01}x_0}\right].$$

Here x_0 and x_1 as before denote the dimensionless amplitudes n/n_{th} and N/n_{th}, but now the threshold amplitude $n_{\text{th}} = \gamma_1/|S_{01}|$ and the smallest parameter d is replaced by the coefficient $r = \gamma_0 S_{11}/\gamma_1 S_{01} \simeq 1$. The latter determines the drastic difference between first-order and second-order processes.

The dependence of the resonance amplitude x_0 on p is given by the following equation

$$rx_0 + \sqrt{x_0^2 - 1} = r\sqrt{px_0} \,. \qquad (6.3.26)$$

If the interaction of parametric waves is not taken into account, then $r = 0$ and $x_0^2 = 1$. Actually, no "freezing" of the amplitude of the homogeneous pumping takes place and even at small $(p-1)$ the amplitude x_0 significantly differs from unity:

$$x_0^2 = 1 + r^2(\sqrt{p} - 1)/2 \quad \text{at} \quad \sqrt{p} - 1 < 1 \,. \qquad (6.3.27)$$

Hence it is clear that we can avoid taking into account the interaction of parametric waves only near the threshold itself when $(\sqrt{p} - 1) \ll 1/r \simeq 1$. For the first-order process the corresponding condition is satisfied in a much wider pumping range $(\sqrt{p} - 1) \ll 1/d \simeq 10^3 - 10^4$. Under large supercriticalities $(p \gg 1)$

$$x_0 = [r\sqrt{p}/(r+1)]^2 + (r+1)/(r^2 p) \ . \tag{6.3.28}$$

The dependences $x_0(p)$, $x_1(p)$ are shown in Fig. 6.5. Figure 6.6 shows nonlinear resonant curves $x_0(\delta)$ computer processed according to (6.3.25) for various powers p of the external pumping field. Subsequently, in Sect. 9.2.4 we shall compare this theory with the experimental data of *Gurevich* and *Starobinets* [6.12] on the nonlinear ferromagnetic resonance.

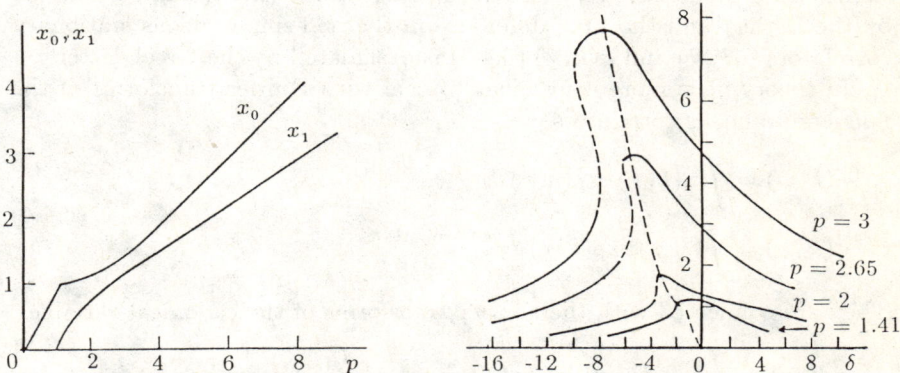

Fig. 6.5. (left) Solution of (6.3.25, 26) for the square of the homogeneous precession amplitude $x_0 = |c_0|^2/n_{\rm th}$ and number of magnons $N_1/n_{\rm th}$ on the dimensionless power of the pumping $p = (h/h_{\rm th})^2$ for second-order processes

Fig. 6.6. (right) Nonlinear resonant curve at various pumping power $p = (h/h_{\rm th})^2$ for second-order processes which is the solution of (6.3.25)

6.4 Nonlinear Theory of Parametric Wave Excitation at Finite Temperatures

6.4.1 Different Time Correlators and Frequency Spectrum

It will be shown later that the interaction of the parametric wave system with the thermal bath leads to a non-trivial behavior of its non-simultaneous correlators $n(\boldsymbol{k}, t, \tau)$ and $\sigma(\boldsymbol{k}, t, \tau)$. The definition of these functions generalizes (5.4.2, 7) for the non-simultaneous (different time) correlators:

$$n(\boldsymbol{k}, t) = n(\boldsymbol{k}, t, 0) \, , \qquad \sigma(\boldsymbol{k}, t) = \sigma(\boldsymbol{k}, t, 0) \ . \tag{6.4.1}$$

That is

$$n(\boldsymbol{k},t,\tau)\delta(\boldsymbol{k}-\boldsymbol{k}_1) = \langle b(\boldsymbol{k},t+\tau/2)b^*(\boldsymbol{k}_1,t-\tau/2)\rangle,$$
$$\sigma(\boldsymbol{k},t,\tau)\delta(\boldsymbol{k}+\boldsymbol{k}_1) = \langle b(\boldsymbol{k},t+\tau/2)b(\boldsymbol{k}_1,t-\tau/2)\rangle \exp(i\omega_\mathrm{p} t). \quad (6.4.2)$$

In the present section we shall be interested only in the stationary state of the parametric wave system. Therefore the argument t of the correlators will be dropped, and we shall assume

$$n(\boldsymbol{k},t,\tau) = n(\boldsymbol{k},\tau), \qquad \sigma(\boldsymbol{k},t,\tau) = \sigma(\boldsymbol{k},\tau). \quad (6.4.3)$$

In order to avoid ambiguity the time argument difference in the non-simultaneous non-stationary correlators $n(\boldsymbol{k},t)$ and $\sigma(\boldsymbol{k},t)$ (6.4.1) is denoted by the Latin t and the time difference in the non-simultaneous stationary correlators $n(\boldsymbol{k},\tau)$ and $\sigma(\boldsymbol{k},\tau)$ (6.4.3) is designated by the Greek letter τ.

In theory, it is more convenient to deal with Fourier transforms of the non-simultaneous correlators

$$n(\boldsymbol{k},\omega) = \int n(\boldsymbol{k},\tau)\exp(i\omega\tau)\,d\tau,$$
$$\sigma(\boldsymbol{k},\omega) = \int (\boldsymbol{k},\tau)\exp(i\omega\tau)\,d\tau. \quad (6.4.4)$$

They are connected with the Fourier transforms of the canonical variables

$$b(\boldsymbol{k},\omega) = \int b(\boldsymbol{k},t)\exp(i\omega t)\,dt, \; b(\boldsymbol{k},\omega) = \int b(\boldsymbol{k},t)\exp(-i\omega t)\,dt. \quad (6.4.5)$$

in the following way

$$n(\boldsymbol{k},\omega)\delta(\boldsymbol{k}-\boldsymbol{k}_1)\delta(\omega-\omega_1) = \langle b(\boldsymbol{k},\omega)b^*(\boldsymbol{k}_1,\omega_1)\rangle, \quad (6.4.6a)$$
$$\sigma(\boldsymbol{k},\omega)\delta(\boldsymbol{k}+\boldsymbol{k}_1)\delta(\omega+\omega_1-\omega_\mathrm{p}) = \langle b(\boldsymbol{k},\omega)b(k_1,\omega_1)\rangle. \quad (6.4.6b)$$

These relations can be verified by the direct substitution of the definitions (6.4.5) in (6.4.2). The relations (6.4.6) illustrate the physical meaning of the functions $n(\boldsymbol{k},\omega)$ and $\sigma(\boldsymbol{k},\omega)$ – these are the power spectra of the normal and abnormal correlators describing the spectral density of the wave energy at various frequencies.

6.4.2 Basic Equations of Temperature S-Theory

According to the results obtained in Sect.1.3.3 let us take into account the interaction of the system of parametric waves with the thermal bath, adding the Langevin random force $f(\boldsymbol{k},t)$ (1.3.6) into (5.2.2). This yields

$$[\partial/\partial t + \gamma(\boldsymbol{k}) + i\omega(\boldsymbol{k})]b(\boldsymbol{k},t) + ihV(\boldsymbol{k})b^*(-\boldsymbol{k},t)\exp(-i\omega_\mathrm{p} t)$$
$$= \frac{-i}{2}\sum_{k+1=2+3} T(\boldsymbol{k},1;2,3)b_1^* b_2 b_3 + f(\boldsymbol{k},t). \quad (6.4.7)$$

Here, as usual, $\bm{j} = \bm{k}_j$, $b_j = b(\bm{k}_j, t)$. In Fourier representation the equations assume the following form:

$$\{i[\omega(\bm{k}) - \omega] + \gamma(\bm{k})\} b(\bm{k}, \omega) + ihV(\bm{k}) b^*(-\bm{k}, \omega_p - \omega)$$
$$= -\frac{i}{2} \sum_{k+1=2+3} \int [T(\bm{k}, 1; 2, 3) b_1^* b_2 b_3 \qquad (6.4.8)$$
$$\times \delta(\omega + \omega_1 - \omega_2 - \omega_3)] \, d\omega_1 \, d\omega_2 \, d\omega_3 + f(\bm{k}, \omega) \, .$$

Here and in the following $b_j = b(\bm{k}_j, \omega)$ which differs from the above assumed denotation $b_j = b(\bm{k}_j, t)$. This ambiguity will not lead to confusion since the context always shows the function of which additional argument (t or ω) is b_j.

Deriving the equations of the S-theory for the correlations $n(\bm{k}, \omega)$ and $\sigma(\bm{k}, \omega)$ we shall use almost the same procedure as in Sect. 5.4.2 when we derived the equations for the simultaneous correlators $n(\bm{k}, t)$ and $\sigma(\bm{k}, t)$. We shall first multiply (6.4.8) by $b^*(\bm{k}', \omega')$, then by $b(\bm{k}', \omega)$ and each time we shall average the obtained equations splitting the fourth-order correlators through the product of pair correlators. This yields the following equations:

$$\gamma(\bm{k}) n(\bm{k}, \omega) + \mathrm{Im}\{P(\bm{k}), \sigma^+(\bm{k}, \omega)\} = \mathrm{Im}\{G(\bm{k}, \omega)\} f^2(\bm{k}, \omega),$$
$$\{-i[\omega_p - \omega_{\mathrm{NL}}(\bm{k}) - \omega_{\mathrm{NL}}(-\bm{k})] + [\gamma(\bm{k}) + \gamma(-\bm{k})]\} \sigma^+(\bm{k}, \omega) \qquad (6.4.9)$$
$$- iP^+(\bm{k})[n(\bm{k}, \omega) + n(-\bm{k}, \omega_p - \omega)]$$
$$= [L^*(\bm{k}, \omega) + L^*(-\bm{k}, \omega_p - \omega)] f^2(\bm{k}, \omega) \, .$$

Here, as in the basic S-theory, $\omega_{\mathrm{NL}}(\bm{k})$ and $P(\bm{k})$ are the frequency and pumping renormalized by the interaction (in the first-order theory of perturbation with respect to the interaction Hamiltonian $\mathcal{H}_4$):

$$\omega_{\mathrm{NL}}(\bm{k}) = \omega(\bm{k}) + 2 \int T(\bm{k}, \bm{k}_1) n(\bm{k}_1, \omega_1) \, d\bm{k}_1 d\omega_1 / 2\pi \, ,$$
$$P(\bm{k}) = hV(\bm{k}) + \int S(\bm{k}, \bm{k}_1) \sigma(\bm{k}_1, \omega_1) \, d\bm{k}_1 d\omega_1 / 2\pi \, . \qquad (6.4.10)$$

Two new very important functions $G(\bm{k}, \omega)$ and $L(\bm{k}, \omega)$ emerged in (6.4.9). They are called the *normal* and *abnormal Green's functions* and they are defined by the following equations:

$$G(\bm{k}, \omega) \delta(\bm{k} - \bm{k}_1) \delta(\omega - \omega_1) = \langle b(\bm{k}, \omega) f^*(\bm{k}_1, \omega_1) \rangle / f^2(\bm{k}, \omega) ,$$
$$L(\bm{k}, \omega) \delta(\bm{k} + \bm{k}_1) \delta(\omega + \omega_1 - \omega_p) = \langle b(\bm{k}, \omega) f(\bm{k}_1, \omega_1) \rangle / f^2(\bm{k}, \omega) \, . \qquad (6.4.11)$$

Here $f^2(\bm{k}, \omega)$ is the Fourier transform of the non-simultaneous correlator of the Langevin random force. According to (1.3.6) it is independent of frequency and equals to

$$f^2(\bm{k}, \omega) = f^2(\bm{k}) = 2\gamma(\bm{k}) n_0(\bm{k}) \, . \qquad (6.4.12)$$

In (6.4.9) and henceforth the following designations are used:

$$G^+(\boldsymbol{k},\omega) = G^*(-\boldsymbol{k},\omega_\mathrm{p}-\omega), \quad L^+(\boldsymbol{k},\omega) = L(-\boldsymbol{k}\omega_\mathrm{p}-\omega),\ldots \quad (6.4.13)$$

Here the symbol + denotes the complex conjugation and the substitution $\boldsymbol{k}\to-\boldsymbol{k}, \omega\to\omega_\mathrm{p}-\omega$.

The equations for the Green's functions G and L can be derived from (6.4.8) in the same way as (6.4.9) for the correlators n and σ. Namely, (6.4.8) must be multiplied by $f^*(\boldsymbol{k}',\omega')$ and $f(\boldsymbol{k}',\omega')$ and the averaging must be performed by splitting the fourth-order correlators into various products of the pair ones. This yields:

$$\begin{aligned}[\omega-\omega_\mathrm{NL}(\boldsymbol{k})+i\gamma(\boldsymbol{k})]G(\boldsymbol{k},\omega) - P(\boldsymbol{k})L^+(\boldsymbol{k},\omega) &= 1, \\ -P^+(\boldsymbol{k})G(\boldsymbol{k},\omega) + [\omega_\mathrm{p}-\omega-\omega_\mathrm{NL}(\boldsymbol{k})-i\gamma(\boldsymbol{k})]L^+(\boldsymbol{k},\omega) &= 0.\end{aligned} \quad (6.4.14)$$

These equations can be solved for the Green's functions

$$G(\boldsymbol{k},\omega) = \frac{\omega_\mathrm{p}-\omega-\omega_\mathrm{NL}(\boldsymbol{k})-i\gamma(\boldsymbol{k})}{\Delta(\boldsymbol{k},\omega)}, \quad L(\boldsymbol{k},\omega) = \frac{P(\boldsymbol{k})}{\Delta(\boldsymbol{k},\omega)}, \quad (6.4.15)$$

$$\Delta(\boldsymbol{k},\omega) = [\omega-\omega_\mathrm{NL}(\boldsymbol{k})+i\gamma(\boldsymbol{k})][\omega_\mathrm{p}-\omega-\omega_\mathrm{NL}(\boldsymbol{k})-i\gamma(\boldsymbol{k})] - |P(\boldsymbol{k})|^2.$$

The solution of (6.4.9) for the pair correlators (allowing for (6.4.12, 15)) can be represented as

$$\begin{aligned}n(\boldsymbol{k},\omega) &= n_0(\boldsymbol{k})\frac{2\gamma(\boldsymbol{k})\{[\omega_\mathrm{p}-\omega-\omega_\mathrm{NL}(\boldsymbol{k})]^2+\gamma^2(\boldsymbol{k})+|P(\boldsymbol{k})|^2\}}{|\Delta(\boldsymbol{k},\omega)|^2}, \\ \sigma(\boldsymbol{k},\omega) &= n_0(\boldsymbol{k})\frac{2P(\boldsymbol{k})\gamma(\boldsymbol{k})[\omega_\mathrm{p}-2\omega_\mathrm{NL}(\boldsymbol{k})+2i\gamma(\boldsymbol{k})]}{|\Delta(\boldsymbol{k},\omega)|^2}.\end{aligned} \quad (6.4.16)$$

These equations together with (6.4.10) expressing $P(\boldsymbol{k})$ and $\omega_\mathrm{NL}(\boldsymbol{k})$ in terms of $n(\boldsymbol{k},\omega)$ and $\sigma(\boldsymbol{k},\omega)$ are the basic equations for the so called *temperature S-theory* which takes into account the finite temperature of thermal bath.

6.4.3 Separation of Waves into Parametric and Thermal

Equations (6.4.16) describe the waves over the entire $\boldsymbol{k},\omega$-space. In the absence of the pumping they describe the state of the thermodynamic equilibrium at the temperature T. Indeed, under $hV=0$ it follows from (6.4.16) that $\sigma(\boldsymbol{k},\omega)=0$ and $n(\boldsymbol{k},\omega)=n_0(\boldsymbol{k},\omega)$, where

$$n_0(\boldsymbol{k},\omega) = \frac{2\gamma(\boldsymbol{k})n_0(\boldsymbol{k})}{[\omega(\boldsymbol{k})-\omega]^2+\gamma^2(\boldsymbol{k})}. \quad (6.4.17)$$

Here $n_0(\boldsymbol{k}) = T/\omega(\boldsymbol{k})$ are the Raleigh-Jeans distributions. This formula describes the occupation numbers of the *thermal waves*, i.e. waves excited by the thermal bath with the temperature T. Obviously,

$$\int n(\boldsymbol{k},\omega)\,d\omega/2\pi = n_0(\boldsymbol{k})\,. \tag{6.4.18}$$

The pumping changes the occupation numbers $n(\boldsymbol{k},\omega)$ not only in the region of parametric resonance but also over the entire $\boldsymbol{k}$-space. Therefore the following question arises: What waves must be considered parametrically excited and which waves must be as before referred to as thermal waves? The qualitative answer is: Parametric waves are the pairs whose phases are correlated with the pumping phase, other waves could be called thermal if the $n(\boldsymbol{k})$ for them does not differ greatly from the equilibrium level. The formula for the numbers of parametric waves n_{p} can be written as follows

$$n_{\mathrm{p}}(\boldsymbol{k},\omega) = n(\boldsymbol{k},\omega) - n_{\mathrm{T}}(\boldsymbol{k},\omega)\,, \quad n_{\mathrm{T}}(\boldsymbol{k},\omega) = \frac{2\gamma(\boldsymbol{k})n_0(\boldsymbol{k})}{[\omega(\boldsymbol{k}) - \omega]^2 + \gamma^2(\boldsymbol{k})}\,. \tag{6.4.19}$$

Here $n_{\mathrm{T}}(\boldsymbol{k},\omega)$ is defined by analogy with (6.4.17). But unlike in (6.4.19) $n_{\mathrm{T}}(\boldsymbol{k},\omega)$ it is defined not by means of the thermodynamic equilibrium spectrum $\omega(\boldsymbol{k})$, but using the real wave frequency $\omega_{\mathrm{NL}}(\boldsymbol{k})$, calculated in the presence of pumping. In such a case $n(\boldsymbol{k},\omega)$ asymptotically tends to $n_{\mathrm{T}}(\boldsymbol{k},\omega)$ as it recedes from the resonant surface.

6.4.4 Two-Dimensional Reduction of Basic Equations

The obtained equations (6.4.16) may seem rather complicated. This is a system of essentially nonlinear integral equations in the four-dimensional $\boldsymbol{k},\omega$-space. However, because the packets $n_{\mathrm{p}}(\boldsymbol{k},\omega)$ and $\sigma(\boldsymbol{k},\omega)$ are narrow with respect to $\omega(\boldsymbol{k})$ and ω, (6.4.16) can be reduced to the two-dimensional integral equations over the resonant surface and effectively analyzed afterwards. Indeed, the main dependence of $n_{\mathrm{p}}(\boldsymbol{k},\omega)$ and $\sigma(\boldsymbol{k},\omega)$ on ω and $\omega(\boldsymbol{k})$ in (6.4.16) is explicitly specified, therefore the dependences $\gamma(\boldsymbol{k})$, $P(\boldsymbol{k})$ and $n_0(\boldsymbol{k})$ on the module of k may be neglected replacing those functions by their values on the resonant surface $\gamma(\Omega)$, $P(\Omega)$ and $n_0(\Omega)$. This enables one explicitly to integrate (6.4.16) with respect to ω and $(\omega\boldsymbol{k})$ and to obtain the closed two-dimensional equations

$$\begin{aligned} n_{\mathrm{p}}(\Omega) &= \pi k^2(\Omega) n_0(\Omega) |P(\Omega)|^2 / v(\Omega)\nu(\Omega)\,, \\ \sigma(\Omega) &= i\pi k^2(\Omega) n_0(\Omega)\gamma(\Omega) P(\Omega)/v(\Omega)\nu(\Omega)\,, \\ \nu^2(\Omega) &= \gamma^2(\Omega) - |P(\Omega)|^2 \end{aligned} \tag{6.4.20}$$

for the values $n_{\mathrm{p}}(\Omega)$ and $\sigma(\Omega)$, integrated in ω and k and depending only on the angular coordinates $\Omega = \Theta, \varphi$ on the resonant surface:

$$\begin{aligned} n_{\mathrm{p}}(\Omega) &= [k^2(\Omega)/2\pi v(\Omega)] \int n_{\mathrm{p}}(\boldsymbol{k},\omega)\,d\omega(\boldsymbol{k})\,d\omega/2\pi\,, \\ \sigma(\Omega) &= [k^2(\Omega)/2\pi v(\Omega)] \int \sigma(\boldsymbol{k},\omega)\,d\omega(\boldsymbol{k})\,d\omega/2\pi\,. \end{aligned} \tag{6.4.21}$$

Here $k(\Omega)$ and $v(\Omega)$ are the wave vector and the group velocity of the waves at the point of resonant surface with the angular coordinate Ω. Substituting $\sigma(\Omega)$ from (6.4.21) into the usual expression of the basic S-theory for the self-consistent pumping $P(\Omega)$ we shall obtain the nonlinear integral equation for $P(\Omega)$ on the resonant surface:

$$P(\Omega) = hV(\Omega) + i\pi n_0(\Omega) \int \frac{S(\Omega, \Omega')\gamma(\Omega')P(\Omega')k^2(\Omega')\,d\Omega'}{\sqrt{[\gamma^2(\Omega') - |P(\Omega')|^2]}v(\Omega')}. \qquad (6.4.22)$$

On solving this equation we can determine from (6.4.20) the integrated characteristics of the parametric waves $n_p(\Omega)$ and $\sigma(\Omega)$ on the resonant surface, and then using (6.4.16) we can study the distribution structure of $n_p(\boldsymbol{k},\omega)$ and $\sigma(\boldsymbol{k},\omega)$ with respect to ω and $\omega_{\mathrm{NL}}(\boldsymbol{k})$ near it.

6.4.5 Distribution of Parametric Waves in $\boldsymbol{k}$

On integrating (6.4.16) only with respect to ω and taking into account (6.4.19) we obtain

$$n(\boldsymbol{k}) = n_0(\boldsymbol{k})\frac{[\omega_{\mathrm{NL}}(\boldsymbol{k}) - \omega_{\mathrm{p}}/2]^2 + \gamma^2(\Omega)}{\Delta(\boldsymbol{k})}, \quad n_p(\boldsymbol{k}) = n_0(\boldsymbol{k})\frac{|P(\Omega)|^2}{\Delta(\boldsymbol{k})},$$

$$\sigma(\boldsymbol{k}) = n_0(\boldsymbol{k})\frac{P(\Omega)[\omega_{\mathrm{NL}}(\boldsymbol{k}) - \frac{\omega_{\mathrm{p}}}{2} + i\gamma(\Omega)]}{\Delta(\boldsymbol{k})},$$
$$\Delta(\boldsymbol{k}) = [\omega_{\mathrm{NL}}(\boldsymbol{k}) - \frac{\omega_{\mathrm{p}}}{2}]^2 + \nu^2(\Omega). \qquad (6.4.23)$$

It can be seen from (6.4.23) that the distribution of the $n_p(\boldsymbol{k})$ and $\sigma(\boldsymbol{k})$ in their eigen frequencies $\omega_{\mathrm{NL}}(\boldsymbol{k})$ (or in module k) has the form of the Lorentzian function with the maximum on the resonant surface and the halfwidth $\nu(\Omega)$. This value will be calculated a bit later (see (6.4.30, 31, 33)). The equations (6.4.23) can be identically transformed to the following form:

$$\gamma(\Omega)n(\boldsymbol{k}) + \mathrm{Im}\{P^*(\Omega)\sigma(\boldsymbol{k})\} = \gamma(\Omega)n_0(\boldsymbol{k}),$$
$$\{\gamma(\Omega) + i[\omega_{\mathrm{NL}}(\boldsymbol{k}) - \omega_{\mathrm{p}}/2]\}\sigma(\boldsymbol{k}) + iP(\omega)n(\boldsymbol{k}) = 0. \qquad (6.4.24)$$

These equations were first intuitively written and treated in detail by Zakharov and myself in 1971 [6.14]. The equations (6.4.24) differ from the basic equations of the S-theory (5.4.11) only in the inhomogeneous term $\gamma(\Omega)n_0(\boldsymbol{k})$ describing the influence of the thermal bath with the non-zero temperature on the system of parametric waves.

6.4.6 Spectrum of Parametric Waves

The function of parametric wave distribution in actual frequencies, i.e. the quantity

$$n_p(\Omega, \omega) = [k(\Omega)/v(\Omega)] \int n(\boldsymbol{k}, \omega) d\omega(\boldsymbol{k}) \tag{6.4.25}$$

can also be readily obtained. To this end, (6.4.16a) must be integrated with respect to $\omega_{NL}(\boldsymbol{k})$ neglecting the dependence of the function $\gamma(\boldsymbol{k}), P(\boldsymbol{k})$ and $n_0(\boldsymbol{k})$ on the modulus k and substituting in (6.4.16a) their values on the resonant surface. This yields:

$$\begin{aligned} n(\Omega, \omega) &= \sqrt{2}\gamma^3(\Omega) n_0(\Omega) k^2(\Omega)/\delta^3(\Omega, \omega), \\ \delta^6(\Omega, \omega) &= \left\{ \nu^2(\Omega) + \sqrt{\nu^4(\Omega) + 2\gamma^2(\Omega)(\omega - \omega_p/2)^2} \right\} \\ &\quad \times \left\{ \nu^4(\Omega) + 4\gamma^2(\Omega)(\omega - \omega_p/2)^2 \right\}. \end{aligned} \tag{6.4.26}$$

The width of this function in ω is $\nu^2(\Omega)/2\gamma(\Omega)$. The line shape differs from the Lorentzian shape which has wider wings. The dependence of the function $\nu(\Omega)$ on Ω and the supercriticality p will be obtained in Sect. 6.4.8 (see (6.4.30, 31, 33)).

6.4.7 Heating Below Threshold

Below the threshold of the parametric instability (i.e. under $hV(\Omega) < \gamma(\Omega)$) the wave interaction can be neglected and in all formulae we can assume $P(\Omega) = hV(\Omega)$. Then formulae (6.4.20a, c) will describe the heating below the threshold and the phase correlation of waves by the pumping

$$n_p(\Omega) = \frac{\pi k^2(\Omega) n_0(\Omega) |hV(\Omega)|^2}{v(\Omega)\sqrt{\gamma^2(\Omega) - |hV(\Omega)|^2}}, \quad \sigma(\Omega) = i n_p(\Omega) \frac{hV(\Omega)}{\gamma(\Omega)}. \tag{6.4.27}$$

Clearly, below the threshold the phase correlation is not complete (i.e. $|\sigma| < n$), the number of parametric waves n at $hV \ll \sigma$ increases proportionally to the pumping power. Then (approaching the threshold) the correlation becomes complete ($|\sigma| \to n_p$), n_p tends to infinity and the wave interaction resulting in the limitation of their amplitude must be taken into account.

6.4.8 Influence of Thermal Bath on Total Characteristics

This problem can be best studied for the case of the spherical symmetry when (6.4.22) becomes algebraic:

$$P(1 - ir\gamma/\nu) = hV, \quad \nu^2 = \gamma^2 - |P|^2. \tag{6.4.28a,b}$$

Here r is the small parameter characterizing the influence of the thermal bath:

$$r = SN_{\rm T}/kv, \quad N_{\rm T} = 4\pi^2 n_0(\boldsymbol{k})k^3 \ . \tag{6.4.28c}$$

In this formula $N_{\rm T}$ is approximately equal to the number of thermal waves inside the resonant surface. In typical experiments on the ferromagnetic YIG $r = 10^{-2} - 10^{-3}$. The parameter r for the Heisenberg ferromagnet can be theoretically estimated as

$$r \simeq (ak)^2 T/T_{\rm c}, \tag{6.4.29}$$

where T denotes the thermal bath temperature, $T_{\rm c}$ is the Curie temperature, a designates the lattice constant and k is the radius of the resonant surface.

The solution of (6.4.27) $P(p)$ is substituted in (6.4.26a) and for $p = (\hbar V/\gamma) > 1$ we obtain

$$SN = \sqrt{p-1}\left[1 + \frac{r^2}{2(p-1)}\right], \quad \nu = \frac{r\gamma}{\sqrt{p-1}} \quad \text{at} \quad \frac{r^2}{(p-1)} \ll 1 . \tag{6.4.30}$$

It is clear that within the applicability of these formulae the dependence of $N(p)$ does not differ much from the dependence predicted by the basic S-theory. But the basic difference consists in the appearance of the finite width of the packet $n(\boldsymbol{k},\omega)$ in the eigenfrequency $\omega_{\rm NL}(\boldsymbol{k})$ ($\delta\omega_{\rm NL}(\boldsymbol{k}) = \nu$) and in the actual frequency ($\delta\omega = \nu^2/2\gamma$). The width ν below the threshold increases as $\gamma\sqrt{p-1}$ and then decreases according to (6.4.30) as $1/\sqrt{p-1}$. The maximum value $\nu_{\max} = \sqrt{r}$ is attained on the threshold (at $p = 1$).

The situation is quite different when the number of parametric waves is limited by the nonlinear damping. In this case $P = \hbar V = \gamma + \eta N$. On substituting this relation in (6.4.20a) we obtain

$$\nu = r\gamma p\eta/S[\sqrt{p} - 1] \ . \tag{6.4.31}$$

Hence

$$\delta\omega(\boldsymbol{k}) = \frac{r\gamma p\eta}{S[\sqrt{p}-1]} \quad \delta\omega = \frac{r^2\gamma p^{3/2}\eta^2}{S^2[\sqrt{p}-1]^2} \ . \tag{6.4.32}$$

We must, however, draw attention to the limited applicability of these expressions because in their derivation the nonlinear behavior of the heated waves far from the resonant surface was not allowed for.

In conclusion, the results will be presented for the case of axial symmetry when, according to the basic S-theory, only one group of equivalent pairs (on the equator of the resonant surface) is excited. In this case above the threshold [6.14]:

$$SN = \gamma\sqrt{p-1}\left[1 + p^2 \exp\left(-\frac{\sqrt{p-1}}{r}\right)\right], \quad \nu^2(x) = \nu_0^2 + c\gamma^2 x^2 ,$$

$$\nu_0 = \gamma\exp\left(-\frac{\sqrt{p-1}}{r}\right) \quad x = \cos\Theta, \quad c \simeq 1 \tag{6.4.33}$$

at $p < \exp(\sqrt{p-1}/2r)$. Clearly, in this case the influence of the thermal bath is exponentially small.

Finally, a general remark. In the basic S-theory the solution of the stationary equations was highly ambiguous. This ambiguity was eliminated by the condition of the external stability, from which it followed, in particular, that parametric waves are excited on the equator of the resonant surface $\omega_{\mathrm{NL}}(\boldsymbol{k}) = \omega_{\mathrm{p}}/2$. This ambiguity can be removed by taking into account the wave interaction with the thermal bath: the solution of (6.4.16) is unique (and, naturally, is concentrated over the resonant surface). As a result of the thermal bath influence all the possible instabilities obtain the "initial impact" and develop removing the ambiguity of parametric waves' state.

6.5 Introduction to Spatially Inhomogeneous S-Theory

In describing the nonlinear behavior of parametric waves the statistical properties of the wave field have been assumed to be spatially homogeneous. In terms of the correlation functions this implies that

$$\langle c(\boldsymbol{k})c^*(\boldsymbol{k}')\rangle \propto \delta(\boldsymbol{k}-\boldsymbol{k}'), \quad \langle c(\boldsymbol{k})c(-\boldsymbol{k}')\rangle \propto \delta(\boldsymbol{k}-\boldsymbol{k}').$$

If we abandon this assumption, and take the space inhomogeneity to be smooth compared with the wavelength, then the values

$$n(\boldsymbol{k},\boldsymbol{k}') = \langle c(\boldsymbol{k})c^*(\boldsymbol{k}')\rangle, \quad \sigma(\boldsymbol{k},\boldsymbol{k}') = \langle c(\boldsymbol{k})c(-\boldsymbol{k}')\rangle . \tag{6.5.1}$$

will no longer be proportional to $\delta(\boldsymbol{k}-\boldsymbol{k}')$: with respect to $(\boldsymbol{k}-\boldsymbol{k}')$ they will be concentrated in a narrow layer with the width of $1/L$, where L is the characteristic size of the inhomogeneity. In the present section the S-theory equations for $n(\boldsymbol{k},\boldsymbol{k}',t)$ and $\sigma(\boldsymbol{k},\boldsymbol{k}',t)$ will be obtained and analyzed for the case of the smooth inhomogeneity.

6.5.1 Basic Equations

In order to obtain the equations of the S-theory for the correlators (6.5.1), let us obtain the derivatives

$$\frac{\partial n(\boldsymbol{k},\boldsymbol{k}',t)}{\partial t} = \left\langle c(\boldsymbol{k},t)\frac{\partial c^*(\boldsymbol{k}',t)}{\partial t}\right\rangle + \left\langle \frac{\partial c(\boldsymbol{k},t)}{\partial t}c^*(\boldsymbol{k}')\right\rangle . \tag{6.5.2}$$

As in the derivation of the basic equations of the spatially homogeneous S-theory (5.4.11) the dynamical equations of motion (5.2) will be used and we shall "split" the fourth order correlators into the paired correlators according to the rule generalizing (5.4.10):

$$\langle c_1^* c_2^* c_3 c_4 \rangle = n_{31} n_{42} + n_{41} n_{32} + \sigma_{12}^* \sigma_{34},$$

$$\langle c_1^* c_2 c_3 c_4 \rangle = n_{21} \sigma_{3,-4} + n_{31} \sigma_{2,-4} + n_{41} \sigma_{2,-3},$$

$$\langle c_1^* c_2^* c_3^* c_4 \rangle = n_{41} \sigma_{2,-3} + n_{42} \sigma_{1,-3}^* + n_{43} \sigma_{1,-2}^*, \qquad (6.5.3)$$

$$n_{12} = n(\boldsymbol{k}_1, \boldsymbol{k}_2), \quad \sigma_{12} = \sigma(\boldsymbol{k}_1, \boldsymbol{k}_2).$$

As a result, it follows from (5.2.2, 3) and (6.5.2)

$$\left\{ \frac{\partial}{\partial t} + \gamma(\boldsymbol{k}) + \gamma(\boldsymbol{k}') + i[\omega(\boldsymbol{k}) - \omega(\boldsymbol{k}')] \right\} n(\boldsymbol{k}, \boldsymbol{k}', t)$$

$$= -i \sum_{1,2,3} \left\{ T_{k1,23} (n_{2k'} n_{31} + n_{3k'} n_{21} + \sigma_{1,-k'}^* \sigma_{2,-3}) \delta(\boldsymbol{k} + 1 - 2 - 3) \right.$$

$$\left. - T_{23,k'1} (n_{k2} n_{13} + n_{k3} n_{12} + \sigma_{k,-1}^* \sigma_{2,-3}) \delta(\boldsymbol{k}' + 1 - 2 - 3) \right\}, \qquad (6.5.4)$$

$$\left\{ \frac{\partial}{\partial t} + \gamma(\boldsymbol{k}) + \gamma(\boldsymbol{k}') + i[\omega(\boldsymbol{k}) + \omega(\boldsymbol{k}')] - i\omega_{\mathrm{p}} \right\} \sigma(\boldsymbol{k}, \boldsymbol{k}', t)$$

$$= -i \sum_{1,2,3} \left\{ T_{k1,23} (n_{k'1} \sigma_{23} + n_{21} \sigma_{3k'} + n_{2,3} \sigma_{2,k'}) \delta(\boldsymbol{k} + 1 - 2 - 3) \right.$$

$$\left. - T_{k'1,23} (n_{k1} \sigma_{23} + n_{k1} \sigma_{k3} + n_{3,1} \sigma_{k,2}) \delta(-\boldsymbol{k}' + 1 - 2 - 3) \right\}.$$

Subsequently, for simplicity we shall assume the spatial homogeneity to be unidimensional (along the z-axis). Then

$$n(\boldsymbol{k}, \boldsymbol{k}', t) = n(\boldsymbol{k}_\perp, k_z, k_z', t) \delta(\boldsymbol{k}_\perp - \boldsymbol{k}'_\perp),$$
$$\sigma(\boldsymbol{k}, \boldsymbol{k}', t) = \sigma(\boldsymbol{k}_\perp, k_z, k_z', t) \delta(\boldsymbol{k}_\perp - \boldsymbol{k}'_\perp). \qquad (6.5.5)$$

Taking into account that under the fixed $\boldsymbol{k}_\perp$ the packets $n(\boldsymbol{k}_\perp, k_z, k_z', t)$ and $\sigma(\boldsymbol{k}_\perp, k_z, k_z', t)$ are concentrated in the narrow layer $\delta k \ll k$, it is possible to expand $\omega(\boldsymbol{k}_\perp, k_z)$ in (6.5.4) into the series of $k_z - k_z^0$ (k_z^0 is the center of the packet: $k_z^0 = f(\boldsymbol{k}_\perp)$), and the dependence T on $k_z - k_z^0$ can be neglected. The obtained equations can be reduced after the transition to the $\boldsymbol{r}$-representation with respect to the z-coordinate.

$$\left\{ \frac{\partial}{\partial t} + 2\gamma(\boldsymbol{k}) + v(\boldsymbol{k}) \left(\frac{\partial}{\partial z} + \frac{\partial}{\partial z'} \right) + i \left[\omega_{\mathrm{NL}}(\boldsymbol{k}, z) - \omega_{\mathrm{NL}}(\boldsymbol{k}, z') \right] \right\} n(\boldsymbol{k}, z, z', t)$$

$$+ i[P(\boldsymbol{k}, z', z')\sigma(-\boldsymbol{k}, z, z', t) - P^*(\boldsymbol{k}, z, z)\sigma(\boldsymbol{k}, z', z, t)] = 0,$$

$$\left\{ \frac{\partial}{\partial t} + 2\gamma(\boldsymbol{k}) + v(\boldsymbol{k}) \left(\frac{\partial}{\partial z} + \frac{\partial}{\partial z'} \right) \right. \qquad (6.5.6)$$

$$\left. + i \left[\omega_{\mathrm{NL}}(\boldsymbol{k}, z) + \omega_{\mathrm{NL}}(\boldsymbol{k}, z') - \omega_{\mathrm{p}} \right] \right\} \sigma(\boldsymbol{k}, z', z, t)$$

$$+ i[P(\boldsymbol{k}, z', z')n(-\boldsymbol{k}, z', z, t) + P(-\boldsymbol{k}, z, z, t)n(\boldsymbol{k}, z, z', t)] = 0$$

Here

$$(2\pi)^3 n(\boldsymbol{k}, z, z') =$$
$$= \int n(\boldsymbol{k}_\perp, k_z, k'_z) \exp\{i[k_z z - k'_z z' - k^0_z(z-z')]\} \, dk_z dk'_z,$$
$$(2\pi)^3 \sigma(\boldsymbol{k}, z', z) = (2\pi)^3 \sigma(-\boldsymbol{k}, z, z')$$
$$= \int \sigma(\boldsymbol{k}_\perp, k_z, k'_z) \exp\{i[k_z z - k'_z z' - k^0_z(z-z')]\} \, dk_z dk'_z, \qquad (6.5.7)$$
$$\omega_{\text{NL}}(\boldsymbol{k}, z) = \omega(\boldsymbol{k}) + 2\int T(\boldsymbol{k}, \boldsymbol{k}') n(\boldsymbol{k}', z, z) \, d\boldsymbol{k}',$$
$$P(\boldsymbol{k}, z, z') = hV(\boldsymbol{k}) + \int S(\boldsymbol{k}, \boldsymbol{k}') \sigma(\boldsymbol{k}', z', z) d\boldsymbol{k}' \, .$$

For simplicity we dropped the terms with the anti-Hermitian parts S and T and the diffraction proportional to ω'' since these terms are usually insignificant.

In the case of the space homogeneity (6.5.6) go over into the basic equations of the S-theory. If the pumping amplitude h changes slowly (in comparison with the wavelength $1/k$) in space then in (6.5.6) h depends on z. It is important that the equations considered above admit the factorized solution of the form

$$n(\boldsymbol{k}, z, z') = A^*(\boldsymbol{k}, z) A(\boldsymbol{k}, z'), \quad \sigma(\boldsymbol{k}, z, z') = A(\boldsymbol{k}, z') A(-\boldsymbol{k}, z), \qquad (6.5.8)$$

where $A(\boldsymbol{k}, z)$ satisfy the equation generalizing (5.4.15):

$$\left\{ \frac{\partial}{\partial t} + \gamma(\boldsymbol{k}) + v(\boldsymbol{k}) \frac{\partial}{\partial z} \right.$$
$$\left. + i \left[\omega_{\text{NL}}(\boldsymbol{k}, z, t) - \frac{\omega_{\text{p}}}{2} \right] \right\} A(\boldsymbol{k}, z, t) + i P(\boldsymbol{k}, z, t) A^*(-\boldsymbol{k}, z, t) = 0,$$
$$P(\boldsymbol{k}, z, t) = h(z) V(\boldsymbol{k}) + \int S(\boldsymbol{k}, \boldsymbol{k}') A(\boldsymbol{k}', z, t) A(-\boldsymbol{k}', z, t) d\boldsymbol{k}', \qquad (6.5.9)$$
$$\omega_{\text{NL}}(\boldsymbol{k}, z, t) = \omega(\boldsymbol{k}) + 2 \int T(\boldsymbol{k}, \boldsymbol{k}') |A(\boldsymbol{k}', z, t)| d\boldsymbol{k}' \, .$$

6.5.2 Parametric Threshold in Inhomogeneous Media

In studying the problem, one should take into account the energy flux from the range of the positive increment. Therefore unlike for the case of space homogeneity, even the problem of obtaining the threshold of the parametric instability proves to be sufficiently meaningful.

Evidently, in order to calculate the threshold we can confine ourselves in (6.5.9) to the terms linear in the wave amplitude, i.e.

$$\left\{ \frac{\partial}{\partial t} + \gamma(\boldsymbol{k}) + v(\boldsymbol{k}) \frac{\partial}{\partial z} + i \left[\omega_{\text{NL}}(\boldsymbol{k}, z, t) - \frac{\omega_{\text{p}}}{2} \right] \right\} A(\boldsymbol{k}, z, t) \qquad (6.5.10)$$
$$+ ihV(\boldsymbol{k}, z, t) A^*(-\boldsymbol{k}, z, t) = 0 \, .$$

Obviously, there is no point in studying these equations under the arbitrary dependence $h(z)$. The problem of the decay instability of the homogeneous wave ($h(z) = h$) in the non-homogeneous medium has been thoroughly investigated in connection with the laser problems [6.15-16]. We are interested in a different case of inhomogeneous pumping in the homogeneous medium. This situation is often observed under the parametric excitation of spin waves and waves on the water surface [6.17]. It would be natural to consider two opposite cases, i.e. fast and slowly decreasing field of the pumping. In the first case the dependence $h(z)$ can be approximated by the rectangle

$$h(z) = h \quad \text{under } 0 < z < L \quad h(z) = 0 \quad \text{at } z < 0 \text{ or } z > L . \qquad (6.5.11)$$

In the second case

$$h(z) = h/[1 + z^2/2L^2] . \qquad (6.5.12)$$

1 Threshold of Parametric Instability in a Plate. A more general presentation of the problem (6.5.11) about the parametric instability in the plane layer has been given in some studies (e.g., [6.18]). For the subsequent study of the nonlinear stage of the parametric wave excitation in the plane layer we shall briefly outline some results of the linear theory. Obviously, the boundary conditions have the following form:

$$A(\boldsymbol{k}, 0) = A(-\boldsymbol{k}, L) = 0 . \qquad (6.5.13)$$

It can be assumed without loss of generality that in (6.5.10) $\omega(\boldsymbol{k}) = \omega_{\text{p}}/2$. Then

$$A(\boldsymbol{k}, z) = A \sin(kz) \quad A(-\boldsymbol{k}, z) = A \sin[\kappa(L - z)], \qquad (6.5.14)$$

κ and the threshold value of $h = h_{\text{th}}$ being given by the following equations

$$h_{\text{th}} V(\boldsymbol{k}) \cos(\kappa L) = -\gamma(\boldsymbol{k}), h_{\text{th}} V(\boldsymbol{k}) \sin(\kappa) L = \kappa v(\boldsymbol{k}) . \qquad (6.5.15)$$

Hence

$$[h_{\text{th}} V(\boldsymbol{k})]^2 = \gamma(\boldsymbol{k})^2 + [\pi b v(\boldsymbol{k})/L]^2 . \qquad (6.5.16)$$

The numerical coefficient b depends on the ratio of the mean free path v/γ to the size of the layer L and varies in the range from unity (at $v \ll \gamma L$) to the half of the quantity in the opposite limiting case. Note that in the threshold formula (6.5.16) the squares of the eigen damping of waves $\gamma(\boldsymbol{k})$ and the effective damping $(\pi b v/L)$ caused by the energy flux from the pumping region (or by the absorption on the boundary of the sample) are added.

2 Threshold Under Smoothly Non-Homogeneous Pumping (6.5.12).

To calculate this threshold let us take in (6.5.10) $\partial/\partial t = 0$ and drop $A(-\boldsymbol{k}, z)$. Then

$$\left[\gamma^2 - h^2(z)V^2 - v^2\frac{\partial^2}{\partial z^2} + \frac{\partial h(z)}{h(z)\partial z^2}\left(\gamma + v\frac{\partial}{\partial z}\right)\right]A(\boldsymbol{k}, z) = 0 \ . \quad (6.5.17)$$

Here and in the following for brevity the argument $\boldsymbol{k}$ of the functions γ, V and v is dropped. The boundary conditions for this equation have the form $A(\boldsymbol{k}, \infty) = A(\boldsymbol{k}, -\infty) = 0$. Taking into account, first, that for the smoothly decreasing pumping $L \gg v/\gamma$ and, second, that the solution is concentrated in the region of the size l satisfying the following inequality

$$L \gg l \gg v/\gamma, \quad (6.5.18)$$

(6.5.17) can be reduced to the form

$$\left[\gamma^2 + h^2V^2 + \frac{(hV)^2}{L^2}\left(z - \frac{\gamma v}{2n^2V^2}\right) - v^2\frac{\partial^2}{\partial z^2}\right]A(\boldsymbol{k}, z) = 0 \ . \quad (6.5.19)$$

The solution of the equation we are interested in has the form

$$A(\boldsymbol{k}, z) = A\exp[-(\gamma/2vL)(z - v/2\gamma)^2] \quad (6.5.20)$$

under the threshold value of the pumping amplitude

$$h_{\text{th}}V = \gamma + v/2L \ . \quad (6.5.21)$$

The characteristic size of the solution is

$$l = \sqrt{Lv/\gamma} \quad (6.5.22)$$

and the inequality (6.5.18) necessarily holds true.

Note that the maximum of the packet $A(\boldsymbol{k}, z)$ (6.5.20) is shifted to the right by the distance equal to the mean free path. The maximum of the packet $A(-\boldsymbol{k}, z)$ is shifted to the left:

$$A(-\boldsymbol{k}, z) = A\exp[-(/2vL)(z + v/2)] \ . \quad (6.5.23)$$

The maximum of their product obviously remains at the point $z = 0$ where the pumping amplitude is at maximum (6.5.12). That is

$$A(\boldsymbol{k}, z)A(-\boldsymbol{k}, z) = A^2 \exp[z^2 + v^2/4\gamma^2] \ .$$

Interestingly, the obtained expression for the threshold (6.5.21) under the continuous inhomogeneity differs greatly for the corresponding expression (6.5.16) for the rectangular profile $h(z)$. To explain this difference, let us obtain the expressions for the profiles from the following simple considerations. As seen from (6.5.10), the characteristic size of the region, l, where the parametrically excited waves are concentrated, is given by the expression

$$(v/l)^2 = (hV)^2 - \gamma^2 \ . \tag{6.5.24}$$

In its turn, l is of the order of magnitude of a size of the region, where $h(z) > \gamma$. For (6.5.11) this is, naturally, the pumping scale L given by (6.5.11) and then from (6.5.24) follows the estimate of h_{th} coinciding with (6.5.16). If the pumping amplitude continuously decreases, $h(z)V = hV/(1+z/L)^n$, then $\gamma^2(l/L)^n = (hV)^2 - \gamma^2$. Combining this relation with (6.5.24), we obtain

$$l \simeq L(v/\gamma L)^{2/(n+2)}, \quad (h_{\text{th}}V)^2 - \gamma^2 \simeq (v/L)^2 (\gamma L/v)^{4/(n+2)}, \tag{6.5.25}$$

hence at $n = 2$ we get the estimate coinciding with (6.5.21) and at $n \gg 1$ we get the result (6.5.16) for the "rectangular" pumping (6.5.11).

3 Excitation Threshold of Oblique Waves in a Plate. In order to calculate the excitation threshold of the waves propagating at an angle Θ with the direction of the non-homogeneity, it is sufficient to allow for the dependences $V(\Theta)$, $v(\Theta) = v\cos\Theta$ and $\gamma(\Theta)$ in (6.5.16). By minimizing the expression (6.5.16) for the threshold pumping amplitude we obtain the threshold of generation and location Θ_1 of the wave pair first to be excited. Here are some simple examples.

A. For the threshold to be minimum for the waves propagating along the non-homogeneity, the maximum $V(\Theta)$ at $\Theta = 0$ must be sufficiency sharp:

$$\left[\frac{\partial^2 V}{V \partial \Theta^2}\right]_{\Theta=0} > 2\left[\frac{h_{\text{th}}^2 V^2 - \gamma^2}{h_{\text{th}}^2 V^2}\right]_{\Theta=0} = 2\frac{(\pi bv/L)^2}{\gamma^2 + (\pi bv/L)^2} \ . \tag{6.5.26}$$

B. If the wave-pumping interaction amplitude is isotropic, i.e $V(\Theta) = V$, then the waves propagating across the non-homogeneity are the first to be excited. Their distribution is homogeneous in space and it is therefore described by the usual formulae of the basic S-theory. It must be noted that under $\Theta = \pi/2$ the term with the group velocity in (6.5.10) becomes zero. Thus, we must allow for the diffraction effects described by the term $\omega'' \partial^2/\partial z^2$ and which we had failed to take into account in (6.5.10). All this results in the following for a plate of thickness L at $V(\Theta) = V$:

$$h_{\text{th}}^2 V^2 = \gamma^2 + (\pi^2 \omega''/2L) \ . \tag{6.5.27}$$

6.5.3 Stationary State in Non-Homogeneous Media

The instability treated in the previous section is absolute; thus, under $h > h_{\text{th}}$ the wave amplitudes are limited by their nonlinear interaction. The resulting distribution is described by (6.5.9) with the corresponding boundary conditions. Let us take some interesting examples confining ourselves to the consideration of the parametric excitation of waves in the plane (6.5.11).

1 Stationary conditions under low supercriticality. Let the pumping amplitude be not too great and let only one set of pairs be excited of the parallels Θ and $\pi - \Theta$. Then the stationary equations (6.5.9) are reduced to

$$\left\{v\frac{d}{dz} + \gamma + i\left[\omega_{\mathrm{NL}}(\boldsymbol{k}) - \frac{\omega_{\mathrm{p}}}{2}\right]\right\} A(\boldsymbol{k},z) = -iP(\boldsymbol{k})A^*(-\boldsymbol{k},z) \ . \quad (6.5.28)$$

By substituting here the expressions for $\omega_{\mathrm{NL}}(\boldsymbol{k})$ and $P(\boldsymbol{k})$ and selecting (without loss of generality) $\omega(\boldsymbol{k}) = \omega_{\mathrm{p}}/2$ we obtain

$$\begin{aligned}\left\{v\frac{d}{dz} + \gamma + 2i[T_1|A(\boldsymbol{k},z)|^2 + T_2|A(-\boldsymbol{k},z)|^2]\right\} A(\boldsymbol{k},z) \\ = -i[hV + 2S_1 A(\boldsymbol{k},z)A(-\boldsymbol{k},z)]A^*(-\boldsymbol{k},z) \ .\end{aligned} \quad (6.5.29)$$

Let us pass to the amplitude phase variables:

$$A(\boldsymbol{k},z) = a(z)\exp[-i\varphi(z)], \quad A(-\boldsymbol{k},z) = b(z)\exp[-i\psi(z)] \ . \quad (6.5.30)$$

The equation for the phase difference $(\varphi - \psi)$ in this case is split out and we come to the following closed system of equations:

$$\begin{aligned} v\frac{da}{dz} + \gamma a &= hVb\sin\Phi, \quad -v\frac{db}{dz} + \gamma b = hVa\sin\Phi, \\ v\frac{d\Phi}{dz} + 2S(a^2 - b^2) &+ \frac{hV(a^2 - b^2)}{ab}\cos\Phi = 0, \\ \Phi &= \varphi + \psi, \quad S = T_2 - T_1 + S_1 \ . \end{aligned} \quad (6.5.31)$$

Equations (6.5.31) can be immediately verified to have the following integral of motion

$$I = ab[Sab + hV\cos\Phi] = \mathrm{const} \ . \quad (6.5.32)$$

On the boundaries of the sample $ab = 0$, therefore $I = 0$ and in the volume of the plane

$$Sab + hV\cos\Phi = 0 \ . \quad (6.5.33)$$

2 Amplitude and profile of distributions $a(z)$ and $b(z)$ under low supercriticality. These values can be obtained by making use of the perturbation theory with respect to the parameter $\delta h/h_{\mathrm{th}} \ll 1$ ($\delta h = h - h_{\mathrm{th}}$). According to (6.5.14) at $h = h_{\mathrm{th}}$, $\Phi = \Phi_0 = \pi/2$ and:

$$a = a_0 = \sqrt{N/2}\sin[\kappa z], \quad b = b_0 = \sqrt{N/2}\sin[\kappa(L-z)] \ . \quad (6.5.34)$$

Here N is the number of parametric waves which in the linear theory obviously remains undetermined. Under $0 < \delta h/h_{\mathrm{th}} \ll 1$ we assume

$$\Phi = \pi/2 + \Phi_1, \quad a = a_0 + a_1, \quad b = b_0 + b_1 . \tag{6.5.35}$$

From the integral of motion (6.5.33) we obtain $\Phi_1(z)$:

$$\Phi_1(z) = (SN/2h_{\text{th}}V) \sin(\kappa z) \sin[\kappa(L-z)] . \tag{6.5.36}$$

Now for a_1 and b_1 from (6.5.31) we can obtain

$$\hat{L}_0 \begin{bmatrix} a_1 \\ b_1 \end{bmatrix} = \begin{bmatrix} (v\frac{d}{dz}+\gamma)a_1 & -hVb_1 \\ hVa_1 & +(v\frac{d}{dz}+\gamma)b_1 \end{bmatrix} = Y(z) \begin{bmatrix} a_0 \\ b_0 \end{bmatrix} \tag{6.5.37}$$

$$Y(z) = hV - (Sa_0b_0)^2/2h_{\text{th}}V .$$

Clearly, the zeroth eigenfunctions of the operator $\hat{L}_0^*$ are $(b,-a)$. The condition of the right-hand side of (6.5.37) being orthogonal to $(b,-a)$ gives the dispersion equation $\int_0^L (a_0^2 + b_0^2) Y \, dz = 0$, specifying the total number of parametric waves

$$(SN)^2 = 2h\delta hV^2 d , \tag{6.5.38}$$

where d is the ration of integrals

$$d = \frac{2N \int_0^L (a_0^2 + b_0^2) \, dz}{\int_0^L (a_0^2 + b_0^2) a_0^2 b_0^2 \, dz} .$$

In changing over from the thin sample to the thick one d will change from 12.8 to 32. Note that in the basic S-theory for the unbounded medium SN is given by (5.5.7), which can be reduced to the form of (6.5.38) with the factor $d = 1$.

3 Profile $a(z)$ in a thin plate. In thin samples where even when for low supercriticalities the damping can be neglected, the equations (6.5.31) have one more integral of motion

$$N = a^2(z) + b^2(z). \tag{6.5.39}$$

This enables one to solve (6.5.31) by quadratures. The wave distribution is qualitatively similar to the distribution considered in Subsect. 1: $a(z)$ monotonically increases deep into the sample. The transcendental equation for the determination of the total number of waves N has the form

$$hVL = v \int \frac{dy}{\sqrt{[1-(SN/2hV)^2 y_2][1-y^2]}} . \tag{6.5.40}$$

Under low supercriticalities we can obtain from the above (6.5.16) for the instability threshold (at $\gamma = 0$ and $b = 1/2$) and (6.5.38) for N (at $d = 32$). Under high supercriticalities (i.e. at $hVL \gg v$), SN quickly tends to $2hV$:

$$SN = hV[2 - \exp(-hVL/v)] . \tag{6.5.41}$$

In this case, too, the solution profile can easily be calculated

$$2ab = N\tanh(2hVz/v)\text{, at }\quad z < L/2\;.\tag{6.5.42}$$

Hence it is clear that at the depths of the order v/hV the solution becomes practically homogeneous.

4 On a Solution profile in an arbitrary case.

The amplitudes for the depth of the sample $a(\infty) = b(\infty)$ (in the case of a single set of pairs) are determined by the value of the integral (6.5.32) $I = 0$, obtained on the boundary of the sample. But according to (5.5.7) the integral

$$I = -SN/2 = -\sqrt{(hV)^2 - \gamma^2}/2$$

and is non-zero. Therefore the values of $a(\infty)$, $b(\infty)$ in the depth of the sample obtained as solutions of (6.5.31) will differ from the values (5.5.7)

$$a_S = b_S = [(hV^2 - \gamma^2]^{1/4}/\sqrt{2S}\,,\tag{6.5.43}$$

obtained in the basic S-theory for the unbounded sample. As is known, under homogeneity all the solutions different from (5.5.7) are unstable. Therefore it must be expected that under $SN > v/L$ the above obtained solution is unstable, and, consequently, can be realized only within a narrow range with the thickness of the order of v/SN near the boundary. In order to describe the range transient to the values of a, b (6.5.43) of the basic S-theory, (6.5.29) must allow for the time derivatives and dispersion terms proportional to ω''.

How do the amplitudes $a(z), b(z)$ become the asymptotic values a_S, b_S given by the basic S-theory? In order to answer this question, let us linearize (6.5.29) in small deviations $A(\boldsymbol{k}, z)$ and $A(-\boldsymbol{k}, z)$ from the solution of the basic S-theory and let us assume that all these deviations are proportional to $\exp(-\kappa z)$. The value of κ can be determined from the condition of the determinant of the obtained homogeneous system of equations being equal to zero. This yields

$$(\kappa v)^2 = 4(T_2 - T_1)(S_1 + T_1 + T_2)N = \Delta^2\;.\tag{6.5.44}$$

Therefore at $\Delta^2 > 0$ the amplitudes exponentially become the asymptotic value of the basic S-theory with the characteristic length $L_1 = 1/\kappa = v/\Delta$. If, on the contrary, $\Delta < 0$ then, as it will be seen, the stationary solution of the basic S-theory a, b proves to be unstable which results in spontaneous auto-oscillations.

On the basis of the above-considered examples we shall try to describe qualitatively the profile $a(z)$ for the plate of the arbitrary thickness. First let us consider a narrow plate ($L \ll v/\gamma = l$). Then up to the supercriticality of about unity the solution of Subsect. 3 will hold true. Under high supercriticalities the wave propagating from the boundary increases to the value approximating the value of a_S (6.5.43) at the distance $L_2 \simeq v/hV$. The wave incident on the boundary does not change significantly after the homogeneous solution a is exponentially attained with characteristic size $L_1 \simeq v/\Delta$ (6.5.44).

In a thick sample ($L \gg v/\gamma = l$) the profile $a(z)$ is similar to the one above considered. Under very small supercriticalities when $SN < L/v$ the solution of the Subsect. 2 is realized. In the intermediate case when $hV \gg SN > v/L$ the solution is quite remarkable. Then the intermediate region consists of two parts. First, at the distance from the boundary of about $l = v/\gamma$ the wave increases up to the value close to a and then over a rather long distance (with the size of about value $L_1 = v/\Delta \simeq l(\gamma/SN)$ the homogeneous solution of the basic S-theory is being exponentially attained.

In conclusion I should like to emphasize that Sect. 6.5 is only the introduction to the inhomogeneous S-theory. We obtained the equations of the S-theory for the case of the smooth unidimensional non-homogeneity and analyzed them in the simplest situations. This almost exhausted the modern scientific data on this question. A host of interesting but unsolved problems have been left out.

6.6 Nonlinear Behavior of Parametric Waves from Various Branches. Asymmetrical S-Theory

6.6.1 Derivation of Basic Equations

In this section it will be assumed as before that the field $h(r,t) = h\exp(-i\omega t)$, homogeneous in space, serves as pumping so that the wave vectors $k, -k$ and wave frequencies $\omega(k)$ and $\Omega(-k)$, belonging to various branches of the spectrum, are obtained from the relation [6.20]

$$\omega(k) + \Omega(-k) = \omega_{\mathrm{p}} . \qquad (6.6.1)$$

The Hamiltonian of the problem

$$\mathcal{H} = \sum_{k}[\omega(k)a(k)a^*(k) + \Omega(k)b(k)b^*(k)] + \mathcal{H}_{\mathrm{p}} + \mathcal{H}_{\mathrm{int}} \qquad (6.6.2a)$$

comprises the squared Hamiltonian of the two-wave type, the Hamiltonian of their interaction with the pumping

$$\mathcal{H}_{\mathrm{p}} = \sum_{k}[h(t)V(k)a^*(k)b^*(-k) + \text{c.c.}] \qquad (6.6.2b)$$

and the Hamiltonian of the wave interaction

$$\mathcal{H}_{\mathrm{int}} = \sum_{1+2=3+4} \left\{ \frac{1}{4}T_{\mathrm{LL}}(1,2;3,4)a^*(k_1)a^*(k_2)a(k_3)a(k_4) \right.$$
$$+ \frac{1}{4}T_{\mathrm{SS}}(1,2;3,4)b^*(k_1)b^*(k_2)b(k_3)b(k_4) \qquad (6.6.2c)$$
$$\left. + T_{\mathrm{SL}}(1,2;3,4)a^*(k_1)b^*(k_2)a(k_3)b(k_4) \right\} .$$

The dynamical equations of motion allowing for the wave damping are represented in the following form:

$$\begin{aligned}\partial a(k,t)/\partial t + \gamma(k)a(k,t) &= -i\delta\mathcal{H}/\delta a^*(k,t), \\ \partial b(k,t)/\partial t + \Gamma(k)b(k,t) &= -i\delta\mathcal{H}/\delta b^*(k,t) .\end{aligned} \qquad (6.6.3)$$

As in the basic S-theory the energy flux from the pumping with Hamiltonian $\mathcal{H}_{\mathrm{p}}$ (6.6.2b) can be shown to be proportional to $\sin[\varphi_{\mathrm{S}}(k) + \varphi_{\mathrm{L}}(-k) - \varphi_{\mathrm{p}}]$, where φ_{S} and φ_{L} are the phases of the waves $a(k)$ and $b(k)$ with the respective dispersion laws $\omega(k)$ and $\Omega(k)$, φ_{p} denotes the pumping phase. This implies that pumping $\mathcal{H}_{\mathrm{p}}$ leads to the correlation of the sum of phases $\Psi(k)$ in the pair $a(k), b(-k)$:

$$\Psi(k) = \varphi_{\mathrm{L}}(k) + \varphi_{\mathrm{S}}(-k) . \qquad (6.6.4)$$

Therefore at the nonlinear stage of the parametric instability the anomalous correlator

$$\sigma(\boldsymbol{k})\delta(\boldsymbol{k}-\boldsymbol{k}_1) = \langle a(\boldsymbol{k})b(-\boldsymbol{k})\exp(i\omega_\mathrm{p}t)\rangle \tag{6.6.5}$$

must be allowed for. Let us also determine the normal correlators

$$n_\mathrm{L}(\boldsymbol{k})\delta(\boldsymbol{k}-\boldsymbol{k}_1) = \langle a(\boldsymbol{k})a^*(\boldsymbol{k}_1)\rangle\,,\quad n_\mathrm{S}(\boldsymbol{k})\delta\boldsymbol{k}-\boldsymbol{k}_1) = \langle b(\boldsymbol{k})b^*(\boldsymbol{k}_1)\rangle \tag{6.6.6a}$$

It will subsequently be shown that the state described by these correlators as a result of the nonlinear interaction of waves may prove to be unstable with respect to the wave pair inside each branch. This results in new anomalous correlators

$$\begin{aligned}\sigma_\mathrm{L}(\boldsymbol{k})\delta(\boldsymbol{k}-\boldsymbol{k}_1) &= \langle a(\boldsymbol{k})a(-\boldsymbol{k})\exp[2i\omega_1(\boldsymbol{k})t]\rangle\,,\\ \sigma_\mathrm{S}(\boldsymbol{k})\delta(\boldsymbol{k}-\boldsymbol{k}_1) &= \langle b(\boldsymbol{k})b(-\boldsymbol{k})\exp[2i\Omega_1(\boldsymbol{k})t]\rangle\,,\\ n_\mathrm{SL}(\boldsymbol{k})\delta(\boldsymbol{k}-\boldsymbol{k}_1) &= \langle a(\boldsymbol{k})b^*(\boldsymbol{k}_1)\exp i[\omega(\boldsymbol{k})-\Omega(\boldsymbol{k})]\rangle\,,\\ n_\mathrm{SL}(\boldsymbol{k}) &= n_\mathrm{SL}^*(\boldsymbol{k})\,,\quad \omega_1(\boldsymbol{k})\simeq\omega(\boldsymbol{k})\,,\quad \Omega_1(\boldsymbol{k})\simeq\Omega(\boldsymbol{k})\,.\end{aligned} \tag{6.6.6b}$$

By differentiating these relations with respect to time and making use of (6.6.3) with the Hamiltonian (6.6.2) one can obtain the equations of motion for all correlators (6.6.6). By splitting the fourth-order correlators of the wave amplitudes through various products of the paired correlators similar to (6.6.6) we obtain in the first order of the perturbation theory with respect to $\mathcal{H}_\mathrm{int}$:

$$\begin{aligned}\frac{\partial n_\mathrm{L}(\boldsymbol{k},t)}{2\partial t} &+ \gamma(\boldsymbol{k})n_\mathrm{L}(\boldsymbol{k},t) = \mathrm{Im}\{P(\boldsymbol{k})\sigma^*(\boldsymbol{k},t)\\ &+ P_\mathrm{L}(\boldsymbol{k})\sigma_\mathrm{L}^*(\boldsymbol{k},t) + G(\boldsymbol{k})n_\mathrm{SL}^*(\boldsymbol{k},t)\}\,,\\ \frac{\partial n_\mathrm{S}(\boldsymbol{k},t)}{2\partial t} &+ \Gamma(\boldsymbol{k})n_\mathrm{S}(\boldsymbol{k},t) = \mathrm{Im}\{P(-\boldsymbol{k})\sigma^*(-\boldsymbol{k},t)\\ &+ P_\mathrm{S}(\boldsymbol{k})\sigma_\mathrm{S}(\boldsymbol{k},t) + G^*(\boldsymbol{k})n_\mathrm{LS}^*(\boldsymbol{k},t)\}\,,\\ \frac{\partial \sigma(\boldsymbol{k},t)}{\partial t} &+ \{\gamma(\boldsymbol{k})+\Gamma(\boldsymbol{k})+i[\omega_\mathrm{NL}(\boldsymbol{k})+\Omega_\mathrm{NL}(\boldsymbol{k})-\omega_\mathrm{p}]\}\sigma(\boldsymbol{k},t)\\ &= -iP(\boldsymbol{k})[n_\mathrm{L}(\boldsymbol{k},t)+n_\mathrm{S}(\boldsymbol{k},t)] - iP_\mathrm{L}(\boldsymbol{k})n_\mathrm{SL}(\boldsymbol{k},t)\\ &\quad - iP_\mathrm{S}(\boldsymbol{k})n_\mathrm{LS}(\boldsymbol{k},t) - iG(\boldsymbol{k})\sigma_\mathrm{S}(\boldsymbol{k},t) - iG^*(\boldsymbol{k})\sigma_\mathrm{L}(\boldsymbol{k},t)\,,\end{aligned} \tag{6.6.7}$$

Introduction to Asymmetrical S-Theory

$$\frac{\partial \sigma_{\mathrm{L}}(\boldsymbol{k},t)}{2\partial t} + \{\gamma(\boldsymbol{k}) + i[\omega_{\mathrm{NL}}(\boldsymbol{k}) - \omega_1(\boldsymbol{k})]\}\sigma_{\mathrm{L}}(\boldsymbol{k},t)$$
$$+ iP(\boldsymbol{k})n_{\mathrm{LS}}(\boldsymbol{k},t) + iP_{\mathrm{L}}(\boldsymbol{k})n_{\mathrm{L}}(\boldsymbol{k},t) = -iG(\boldsymbol{k})\sigma(\boldsymbol{k},t),$$
$$\frac{\partial \sigma_{\mathrm{S}}(\boldsymbol{k},t)}{2\partial t} + \{\varGamma(\boldsymbol{k}) + i[\varOmega_{\mathrm{NL}}(\boldsymbol{k}) - \varOmega_1(\boldsymbol{k})]\}\sigma_{\mathrm{L}}(\boldsymbol{k},t)$$
$$+ iP(\boldsymbol{k})n_{\mathrm{LS}}(\boldsymbol{k},t) + iP_{\mathrm{L}}(\boldsymbol{k})n_{\mathrm{S}}(\boldsymbol{k},t) = -iG^*(\boldsymbol{k})\sigma(-\boldsymbol{k},t), \quad (6.6.8)$$
$$\frac{\partial n_{\mathrm{LS}}(\boldsymbol{k},t)}{\partial t} + [\gamma(\boldsymbol{k}) + \varGamma(\boldsymbol{k})]n_{\mathrm{LS}}(\boldsymbol{k},t)$$
$$+ i[\omega_{\mathrm{NL}}(\boldsymbol{k}) - \omega_1(\boldsymbol{k}) + \varOmega_{\mathrm{NL}}(\boldsymbol{k}) - \varOmega_1(\boldsymbol{k})]n_{\mathrm{LS}}(\boldsymbol{k},t)$$
$$+ iP(\boldsymbol{k})\sigma_{\mathrm{S}}(\boldsymbol{k},t) - iP(\boldsymbol{k})\sigma_{\mathrm{L}}(\boldsymbol{k},t) + iP_{\mathrm{L}}(\boldsymbol{k})\sigma^*(\boldsymbol{k},t)$$
$$- iP_{\mathrm{S}}(\boldsymbol{k})\sigma(-\boldsymbol{k},t) + iG(\boldsymbol{k})[n_{\mathrm{S}}(\boldsymbol{k},t) - n_{\mathrm{L}}(\boldsymbol{k},t)] = 0$$

In these equations we selected the values of the frequencies $\omega_1(\boldsymbol{k})$ and $\varOmega_1(\boldsymbol{k})$ entering into the definitions of correlators σ_{L} and σ_{S} in such a way:

$$\omega_1(\boldsymbol{k}) + \varOmega_1(\boldsymbol{k}) = \omega_{\mathrm{p}} \qquad (6.6.9)$$

and introduced the following designations for the frequencies and self-consistent pumpings renormalized to the interaction:

$$\omega_{\mathrm{NL}}(\boldsymbol{k}) = \omega(\boldsymbol{k}) + 2\sum_1 [T_{\mathrm{L}}(\boldsymbol{k},\boldsymbol{k}_1)n_{\mathrm{L}}(\boldsymbol{k}) + T(\boldsymbol{k},\boldsymbol{k}_1)n_{\mathrm{S}}(\boldsymbol{k}_1)],$$
$$\varOmega_{\mathrm{NL}}(\boldsymbol{k}) = \varOmega(\boldsymbol{k}) + 2\sum_1 [T_{\mathrm{S}}(\boldsymbol{k},\boldsymbol{k}_1)n_{\mathrm{S}}(\boldsymbol{k}) + T(\boldsymbol{k}_1,\boldsymbol{k})n_{\mathrm{L}}(\boldsymbol{k}_1)],$$
$$P(\boldsymbol{k}) = hV(\boldsymbol{k}) + \sum_1 S(\boldsymbol{k},\boldsymbol{k}_1)\sigma(\boldsymbol{k}_1), \qquad (6.6.11)$$
$$P_{\mathrm{L}}(\boldsymbol{k}) = \sum_1 S_{\mathrm{L}}(\boldsymbol{k},\boldsymbol{k}_1)\sigma_{\mathrm{L}}(\boldsymbol{k}_1), \quad P_{\mathrm{S}}(\boldsymbol{k}) = \sum_1 S_{\mathrm{S}}(\boldsymbol{k},\boldsymbol{k}_1)\sigma_{\mathrm{S}}(\boldsymbol{k}_1),$$
$$G(\boldsymbol{k}) = \sum_1 F(\boldsymbol{k},\boldsymbol{k}_1)n_{\mathrm{LS}}(\boldsymbol{k}_1), \quad F(\boldsymbol{k},\boldsymbol{k}_1) = T_{\mathrm{LS}}(\boldsymbol{k},\boldsymbol{k}_1;\boldsymbol{k}_1,\boldsymbol{k}).$$

$$T_{\mathrm{L}}(\boldsymbol{k},\boldsymbol{k}_1) = T_{\mathrm{LL}}(\boldsymbol{k},\boldsymbol{k}_1;\boldsymbol{k},\boldsymbol{k}_1)/2, \qquad T_{\mathrm{S}}(\boldsymbol{k},\boldsymbol{k}_1) = T_{\mathrm{SS}}(\boldsymbol{k},\boldsymbol{k}_1;\boldsymbol{k},\boldsymbol{k}_1)/2,$$
$$S_{\mathrm{L}}(\boldsymbol{k},\boldsymbol{k}_1) = T_{\mathrm{LL}}(\boldsymbol{k},-\boldsymbol{k},\boldsymbol{k}_1,-\boldsymbol{k}_1)/2, \quad S_{\mathrm{S}}(\boldsymbol{k},\boldsymbol{k}_1) = T_{\mathrm{SS}}(\boldsymbol{k},-\boldsymbol{k},\boldsymbol{k}_1,-\boldsymbol{k}_1)/2,$$
$$T(\boldsymbol{k},\boldsymbol{k}_1) = T(\boldsymbol{k},\boldsymbol{k}_1;\boldsymbol{k},\boldsymbol{k}_1), \qquad S(\boldsymbol{k},\boldsymbol{k}_1) = T_{\mathrm{LS}}(\boldsymbol{k},-\boldsymbol{k};\boldsymbol{k}_1,-\boldsymbol{k}_1)/2.$$

6.6.2 Stationary States in Isotropic Case

The equations of the asymmetric S-theory (6.6.7–11) are rather cumbersome. Therefore we shall confine ourselves to the consideration of the isotropic case which is realized, for instance, under parametric excitation of magnons of different branches in antiferromagnets. In the isotropic case $\gamma(\boldsymbol{k}) = \gamma(k)$, $\varGamma(\boldsymbol{k}) = \varGamma(k)$ and $V(\boldsymbol{k}) = V(k)$ and the interaction amplitudes T in $\mathcal{H}_{\mathrm{int}}$ depend only on k, k_1 and $(\boldsymbol{k}\boldsymbol{k}_1)$. Then the solution of the equations will be isotropic and $P(\boldsymbol{k}) = P(k)$, $P_{\rm S}(\boldsymbol{k}) = P_{\rm S}(k)$, $P_{\rm L}(\boldsymbol{k}) = P_{\rm L}(k)$ and $G(\boldsymbol{k}) = G(k)$. As in the basic S-theory we can neglect all the dependences of all the functions except the frequencies $\omega(\boldsymbol{k})$ and $\Omega(\boldsymbol{k})$ on the module k, substituting into the equations their values on the resonance surface (6.6.1). Thus it can be assumed that

$$\gamma(\boldsymbol{k}) = \gamma\,,\ \Gamma(\boldsymbol{k}) = \Gamma\,,\ V(\boldsymbol{k}) = V\,,\ P(\boldsymbol{k}) = P\,,$$
$$P_{\rm L}(\boldsymbol{k}) = P_{\rm L}\,,\ P_{\rm S}(\boldsymbol{k}) = P_{\rm S}\,,\ G(\boldsymbol{k}) = G \qquad (6.6.12)$$

and in the expressions (6.6.10) we can replace the following functions $T_{\rm L}$, $T_{\rm S}$, T, $S_{\rm L}$, $S_{\rm S}$, S and F of the argument $x = \cos\Theta_1 = (\boldsymbol{k}\boldsymbol{k}_1)/kk_1$ for their mean values

$$T_{\rm L} = \frac{1}{2}\int_{-1}^{1} T_{\rm L}(x)\,dx\,, \qquad T_{\rm S} = \frac{1}{2}\int_{-1}^{1} T_{\rm S}(x)\,dx \text{ and so on.} \qquad (6.6.13a)$$

As can easily be seen, (6.6.7–11) in the stationary case are a system of linear equations with coefficients γ, Γ, $\omega_{\rm NL}(\boldsymbol{k})$, $\Omega_{\rm NL}(\boldsymbol{k})$, P, $P_{\rm S}$, $P_{\rm L}$ and G. These equations have non-zero solutions only at such values $\omega_{\rm NL}(\boldsymbol{k})$ and $\Omega_{\rm NL}(\boldsymbol{k})$, under which the determinant of the system becomes zero. Therefore the stationary solutions are concentrated in the $\boldsymbol{k}$-space on the spherical surfaces. Because the determinant of the system is a second-order polynomial with respect to $\omega_{\rm NL}(\boldsymbol{k})$ there can be no more than two such surfaces.

To analyze the stationary states and the stability of the solutions with $k = \text{const}$ we obtain from (6.6.7–11) the equations for the following integral values:

$$N_{\rm L} = \sum_{\boldsymbol{k}} n_{\rm L}(\boldsymbol{k})\,, \quad N_{\rm S} = \sum_{\boldsymbol{k}} n_{\rm S}(\boldsymbol{k})\,, \quad \Sigma = \sum_{\boldsymbol{k}} \sigma(\boldsymbol{k})\,,$$
$$\Sigma_{\rm L} = \sum_{\boldsymbol{k}} \sigma_{\rm L}(\boldsymbol{k})\,, \quad \Sigma_{\rm S} = \sum_{\boldsymbol{k}} \sigma_{\rm S}(\boldsymbol{k})\,, \quad N_{\rm LS} = \sum_{\boldsymbol{k}} n_{\rm LS}(\boldsymbol{k})\,. \qquad (6.6.13b)$$

To this end, the equations (6.6.7, 8) must be summed with respect to $\boldsymbol{k}$, the relations (6.6.12, 13) must be allowed for and one must keep in mind that the wave amplitudes are non-zero only on the sphere. Then we have:

$$\frac{dN_{\rm L}}{2dt} + \gamma N_{\rm L} + \text{Im}\{hV\Sigma^*\} = 0\,,$$
$$\frac{dN_{\rm S}}{2dt} + \Gamma N_{\rm S} + \text{Im}\{hV\Sigma^*\} = 0\,, \qquad (6.6.14)$$
$$\frac{d\Sigma}{dt} + \{\gamma + \Gamma + i[\omega_{\rm NL}(\boldsymbol{k}) + \Omega_{\rm NL}(\boldsymbol{k}) - \omega_{\rm p} + S(N_{\rm L} + N_{\rm S})]\}\Sigma$$
$$+ ihV(N_{\rm L} + N_{\rm S}) + i(F + S_{\rm S})\Sigma_{\rm L}N_{\rm LS} = 0\,,$$

$$\frac{dN_{\text{LS}}}{dt} + \{\gamma + \Gamma + i[\omega_{\text{NL}}(\boldsymbol{k}) - \Omega_{\text{NL}}(\boldsymbol{k})$$
$$- \omega_1(\boldsymbol{k}) + \Omega_1(\boldsymbol{k}) + iF(N_{\text{S}} - N_{\text{L}})]\}N_{\text{LS}}$$
$$+ i[hV + (S - S_{\text{S}})\Sigma]\Sigma_{\text{S}}^* - i[hV + (S - S_{\text{L}})\Sigma^*]\Sigma_{\text{L}} = 0,$$
$$\frac{d\Sigma_{\text{L}}}{2dt} + \{\gamma + i[\omega_{\text{NL}}(\boldsymbol{k}) - \omega_1(\boldsymbol{k}) + S_{\text{L}}N_{\text{L}}]\}\Sigma_{\text{L}} \qquad (6.6.15)$$
$$+ i[hV + (S + F)\Sigma]N_{\text{LS}} = 0,$$
$$\frac{d\Sigma_{\text{S}}^*}{2dt} + \{\Gamma - i[\Omega_{\text{NL}}(\boldsymbol{k}) - \Omega_1(\boldsymbol{k}) + S_{\text{S}}N_{\text{S}}]\}\Sigma_{\text{S}}^*$$
$$- i[hV + (S + F)\Sigma^*]N_{\text{LS}} = 0,$$

Proceeding to the analysis of the stationary solutions of these equations note that the group of equations (6.6.15) can be treated as the system of the linear algebraic equations homogeneous in N_{LS}, Σ_{L} and Σ_{S}^*. Therefore the solutions of the entire system of equations (6.6.14, 15) can be of two types, i.e. state **A** where $N_{\text{LS}} = \Sigma_{\text{L}} = \Sigma_{\text{S}}^* = 0$, and the state **B** where these correlators differ from zero. In the state **A**

$$\Sigma = M\exp(-i\Psi), \quad M^2 = N_{\text{L}}N_{\text{S}},$$
$$\gamma N_{\text{L}} = \Gamma N_{\text{S}} = hVM\sin\Psi, \qquad (6.6.16a)$$

$$[hV\cos\Psi + SM][\sqrt{\gamma/\Gamma} + \sqrt{\Gamma/\gamma}] = \omega_{\text{p}} - \omega_{\text{NL}}(\boldsymbol{k}) - \Omega_{\text{NL}}(\boldsymbol{k}),$$
$$N_{\text{LS}} = \Sigma_{\text{L}} = \Sigma_{\text{S}} = 0. \qquad (6.6.16b)$$

In the state **B**

$$\Sigma = M\exp(-i\Psi), \quad M^2 = N_{\text{L}}N_{\text{S}},$$
$$\Sigma_{\text{S}} = N_{\text{S}}\exp(-i\varphi_{\text{S}}), \quad \Sigma_{\text{L}} = N_{\text{L}}\exp(-i\varphi_{\text{L}}), \qquad (6.6.17a)$$
$$N_{\text{LS}} = M\exp(-i\varphi), \quad \Psi = \varphi_{\text{L}} - \varphi = \varphi_{\text{S}} + \varphi$$

$$[hV\cos\Psi + SM][\sqrt{\gamma/\Gamma} + \sqrt{\Gamma/\gamma}]$$
$$= \omega_{\text{p}} - \omega_{\text{NL}}(\boldsymbol{k}) - \Omega_{\text{NL}}(\boldsymbol{k}) - \Delta_{\text{S}} - \Delta_{\text{L}},$$
$$\gamma N_{\text{L}} = \Gamma N_{\text{S}} = hVM\sin\Psi, \qquad (6.6.17b)$$
$$\Delta_{\text{S}} = S_{\text{S}}N_{\text{S}} + FN_{\text{L}} \quad , \Delta_{\text{L}} = S_{\text{L}}N_{\text{L}} + FN_{\text{S}}.$$

In both cases (states **A** and **B**) the radius of the resonant surface k is still arbitrary. Like in the basic S-theory it is determined from the condition of the external stability. This will be studied below.

1 Stationary state A. Making use of (6.6.3) with the Hamiltonian (6.6.2) for the pair of the perturbation waves $a(\boldsymbol{k}_1), b(-\boldsymbol{k}_1)$ and assuming that $\omega + \Omega = \omega_{\text{p}}$ and:

$$a(\boldsymbol{k}) \propto \exp(i\omega t + 2\nu(\boldsymbol{k}_1)t], \quad b^*(-\boldsymbol{k}) \propto \exp(-i\Omega t + 2\nu(\boldsymbol{k}_1)t], \quad (6.6.18)$$

we obtain for the increment $\nu(\boldsymbol{k}_1)$:

$$\{\nu(k_1)+\gamma+i[\omega_{\text{NL}}(\boldsymbol{k}_1)-\omega]\}\{\nu(k_1)+\varGamma-i[\varOmega_{\text{NL}}(\boldsymbol{k}_1)-\varOmega]\} = |P|^2. \quad (6.6.19)$$

The maximum (with respect to k_1) increment $\nu_{\max}(\boldsymbol{k}_1) = \nu$ corresponds to the wave vector k_1 satisfying the relations

$$\varOmega_{\text{NL}}(\boldsymbol{k}_1) + \omega_{\text{NL}}(\boldsymbol{k}_1) = \omega_{\text{p}}/2 \quad (6.6.20)$$

and is determined by the equation:

$$\nu^2 + \nu(\gamma + \varGamma) + \gamma\varGamma - |P|^2 = 0. \quad (6.6.21)$$

Therefore the condition of the external stability $\nu \le 0$ has the following form

$$|P|^2 \le \gamma\varGamma. \quad (6.6.22)$$

Allowing for this equation, (6.6.16b) enables us to find the radius of the resonant surface for the stable condition (6.6.19). In addition, these relations give:

$$M^2 = \gamma N_{\text{L}}^2/\varGamma = \varGamma/N_{\text{S}}^2/\gamma = \sqrt{h^2 V^2 - \gamma\varGamma}, \quad hV\sin\varPsi = \sqrt{\gamma\varGamma}. \quad (6.6.23)$$

These relations explicitly generalize the corresponding results of the basic S-theory (see (5.5.7)). To study the stability of this state with respect to the emergence of new anomalous correlators (6.6.6b) is of great interest. Taking Σ_{L}, Σ_{S} and N_{NL} to be proportional to $\exp[2\mu t]$, we obtain from (6.6.15) the equations for the increment of the *correlation instability* μ:

$$\begin{aligned}
&2\mu^3 + \mu^2[3(\gamma+\varGamma)+i(\Delta_{\text{L}}+S_{\text{L}}N_{\text{L}}-\Delta_{\text{S}}-S_{\text{S}}N_{\text{S}})] \\
&\quad + \mu[(\gamma+\varGamma)^2 + \Delta_{\text{L}}S_{\text{S}}N_{\text{S}} + \Delta_{\text{S}}S_{\text{L}}N_{\text{L}} \\
&\quad + 3i(\varGamma\Delta_{\text{L}}-\gamma\Delta_{\text{S}}) + 2i\sqrt{\gamma\varGamma N_{\text{L}}N_{\text{S}}}(S_{\text{L}}-S_{\text{S}})] \\
&\quad + (\gamma+\varGamma)[\Delta_{\text{L}}(S_{\text{S}}N_{\text{S}}+i\varGamma)+\Delta_{\text{S}}(S_{\text{L}}N_{\text{L}}-i\gamma)] = 0.
\end{aligned} \quad (6.6.24)$$

Hence under low supercriticalities we can obtain

$$(\gamma+\varGamma)^3 \operatorname{Re}\mu = -2\gamma\varGamma(S_{\text{L}}N_{\text{L}}+S_{\text{S}}N_{\text{S}})^2 - (\gamma+\varGamma)^2 F(S_{\text{L}}N_{\text{L}}^2+S_{\text{S}}N_{\text{S}}^2). \quad (6.6.25)$$

Hence and from (6.6.16a) we can readily obtain the conditions of correlation stability under low supercriticality

$$\gamma\varGamma(\varGamma S_{\text{L}}+\gamma S_{\text{S}})^2 + (\gamma+\varGamma)^2 F(\varGamma^2 S_{\text{L}}+\gamma^2 S_{\text{S}}) > 0 \quad (6.6.26a)$$

Similarly, we can find the condition of the correlation stability under high supercriticality

$$4(\varGamma S_{\text{L}}+\gamma S_{\text{S}})[(\varGamma S_{\text{L}}+\gamma S_{\text{S}})+F(\gamma+\varGamma)]+F^2(\gamma-\varGamma)^2 > 0. \quad (6.6.26b)$$

Interestingly, at $\gamma = \varGamma$ the condition of the correlation stability under high and low supercriticalities coincide

$$(S_L + S_S)(S_L + S_S + 2F) > 0. \tag{6.6.26c}$$

Of fundamental importance is the fact that under some relations between the coefficients this relation can fail to be observed. This means that the stationary state **A** is unstable with respect to the pairing inside the branches of the spectrum. It will be natural to assume that the development of this instability results in the transition of the parametric wave system to the state **B** with the complete phase correlation. Let us study the peculiarities of this state

2 Stationary state B. In order to obtain the radius of the resonance surface we must employ the condition of the external stability. Studying it we obtain a coupled system of equations for the following four variables:

$$a(\boldsymbol{k}), \quad b^*(-\boldsymbol{k}), \quad a^*(-\boldsymbol{k}), \quad b(\boldsymbol{k}). \tag{6.6.27}$$

The increment $\nu(\boldsymbol{k}_1)$ of the external instability is found from the condition that the determinant of the system of equations equals zero. The determinant is fourth-order polynomial with respect to $\nu(\boldsymbol{k}_1)$. It has four roots. One of them (namely $\nu(\boldsymbol{k}_1)$ at $k_1 = k$) leads to the condition

$$\gamma[\omega_{\mathrm{NL}}(\boldsymbol{k}) + \Delta_\mathrm{L} - \omega] = \Gamma[\Omega_{\mathrm{NL}}(\boldsymbol{k}) + \Delta_\mathrm{S} - \Omega], \quad \omega + \Omega = \omega_\mathrm{p}. \tag{6.6.28}$$

In addition, for the external instability $\partial = \mathrm{Re}\{\nu(\boldsymbol{k}_1)\}/\partial k$ must be zero at $k = k_1$. This can be shown to result in the following equation

$$v_\mathrm{L} D_\mathrm{L} + v_\mathrm{S} D_\mathrm{S} = 0, \tag{6.6.29}$$

where $v_\mathrm{L} = \partial \omega(k)/\partial k$, $v_\mathrm{S} = \partial \Omega(k)/\partial k$ and

$$\begin{aligned} D_\mathrm{L} = &\Delta_\mathrm{L}[\Omega_{\mathrm{NL}}(k) + \Delta_\mathrm{S} - \Omega][\Omega_{\mathrm{NL}}(k) + \Delta_\mathrm{L} - \Omega] \\ &+ \Delta_\mathrm{S}[\Omega_{\mathrm{NL}}(k) + \Delta_\mathrm{S} - \Omega][\omega_{\mathrm{NL}}(k) + \Delta_\mathrm{L} - \omega] \\ &+ \Gamma\{\gamma[\Omega_{\mathrm{NL}}(k) + \Delta_\mathrm{S} - \Omega] - \Gamma[\omega_{\mathrm{NL}}(k) + \Delta_\mathrm{L} - \omega]\} \\ &+ FN_\mathrm{L}(S_\mathrm{L} N_\mathrm{L} \Delta_\mathrm{S} + S_\mathrm{S} N_\mathrm{S} \Delta_\mathrm{L}). \end{aligned} \tag{6.6.30}$$

We can obtain hence the expression for D_S by replacing the values of $\Gamma \leftrightarrow \gamma$, $\Omega \leftrightarrow \omega$ and indices $L \leftrightarrow S$. Equations (6.6.28–30) specify two values $\Omega_{\mathrm{NL}}(k)$, $\Omega_{\mathrm{NL},1}(k)$ and $\Omega_{\mathrm{NL},2}(k)$ at which the condition of external stability can be satisfied. Accordingly, there are two types of stationary states **B1** and **B2**. The study of the remaining roots ν_2, ν_3 and ν_4 (carried out in my doctorate thesis) for the case of $\gamma = \Gamma$ and $v_\mathrm{L} = v_\mathrm{S}$ showed that the stability of the states **B1** and **B2** is determined by the relations between the coefficients F, S_L and S_S. For example, at $F = 0$ and $S_\mathrm{L} S_\mathrm{S} > 0$ the state **B1** is stable and **B2** is unstable, and when $F = 0$ and $S_\mathrm{L} S_\mathrm{S} < 0$ the opposite case takes place; at $F \gg S_\mathrm{L}, S_\mathrm{S}$ the state **B1** is stable and **B2** is unstable. Both states cannot be stable at the same time. Sometimes (e.g. at $2F > S_\mathrm{L} = S_\mathrm{S} > \sqrt{2}F > 0$) they are both unstable.

In conclusion we shall present the expressions for the numbers of parametric waves in the states **B1,2** at $\gamma = \Gamma$ and $v_\text{L} = v_\text{S}$. In the state **B1**:

$$N = \sqrt{h^2 V^2 - \gamma^2}/|S + F|. \tag{6.6.31}$$

In the state **B2**:

$$N = \frac{\sqrt{h^2 V^2 - \gamma^2}|2F + S_\text{L} + S_\text{S}|}{|F(S_\text{L} + S_\text{S} + 2S) + S(S_\text{L} + S_\text{S}) + 2S_\text{L} S_\text{S}|}. \tag{6.6.32}$$

Both these dependences differ from the function $N(hV)$ for the state **A** in the numeric factor. A more detailed study carried out in my doctorate thesis shows that the states **B1,2** can prove to be unstable with respect to decrease of the anomalous correlators N_LS, Σ_L and Σ_S. The development of this instability can bring the wave system into the state **A** when these correlators are zero. In its turn, the state **A** can prove to be unstable with respect to the increase of these correlators. Since we considered all possible stationary states and showed that they can all (in some range of the parameters F, S_L and S_S) be unstable, the only remaining possibility for the wave system is to perform *the correlation auto-oscillations*, under which the correlators of the wave system (N_LS, Σ_L, Σ_S, etc.) are non-stationary. We discovered the correlation auto-oscillations in the computer simulation of (6.6.15). It would be very interesting to observe this phenomenon experimentally.

It must be noted that the very possibility of the appearance of the correlation auto-oscillations is basically connected with the non-equilibrium of the system. In the thermodynamic equilibrium one of the states of the system (the *ground state*) must be absolutely stable and auto-oscillations are impossible. Thus in the problem of the superconductivity at $T > T_\text{th}$ the normal state of electrons (without pairing) is stable, and at $T < T_\text{th}$ stable is the superconducting state with anomalous correlators.

One more remark to the whole section 6.6. Everything above must be treated as the introduction to the problem of the parametric excitation of waves from the different spectrum waves. This theory may be developed in more detail as the experimental data are accumulated.

6.7 Parametric Excitation of Waves by Noise Pumping

It is well-known that not only monochromatic, but also noise pumping can lead to the parametric instability. The threshold of this instability depends on the frequency width of the pumping Δ. In the order of magnitude

$$\langle h^2 V^2 \rangle \simeq \gamma(\gamma + \Delta). \tag{6.7.1}$$

It is often erroneously assumed that under incoherent pumping the "phase" mechanism for the limitation of the parametric wave amplitude does not function. However, it can easily be seen that the amplitude of parametric waves and phase relations calculated according to the S-theory are independent of the width of the pumping spectrum Δ at $\Delta < \gamma$. Therefore it must be expected that at $\Delta \simeq \gamma$ these values will not change qualitatively, i.e. the anomalous correlators $\sigma(\mathbf{k})$ in the pairs of parametric waves are not small in comparison with the normal correlators $n(\mathbf{k})$. Consequently, also under $\Delta > \gamma$ the anomalous correlators behave like some function of Δ/γ which tends to zero when $\Delta/\gamma \to \infty$, and if this function decreases slowly enough as Δ/γ increases, there exists a range $\Delta > \gamma$ where the self-consistent interaction is the most important.

6.7.1 Equations of S-Theory Under Noise Pumping

Nonlinear behavior of parametric waves under incoherent (noise) pumping was studied by *Cherepanov* in the approximation of the S-theory [6.21]. This *"Noise S-theory"* is presented below.

The pumping $h(t)$ will be assumed to be $h(t) = \tilde{h}(t) \exp(-i\omega_p t)$, where $\tilde{h}(t)$ is the random stationary function with the spectrum width Δ such that $\omega_p \gg \Delta \gg \gamma$. The equations of the S-theory for the slow amplitudes of parametric waves in this case can readily be obtained from (5.4.15, 16) by replacing $h \to \tilde{h}(t)$. From them it follows, in particular, that

$$\partial |a(\mathbf{k},t)|^2 / \partial t + 2\gamma(\mathbf{k})|a(\mathbf{k},t)|^2 = A(t),$$
$$A(t) = 2\mathrm{Im}\{P(\mathbf{k},t)a^*(\mathbf{k},t)a^*(-\mathbf{k},t)\}. \tag{6.7.2}$$

The solution of this equation has the following form:

$$|a(\mathbf{k},t)|^2 = \int_0^\infty A(t-t_1)\exp[-2\gamma(\mathbf{k})t]\,dt. \tag{6.7.3}$$

The function $A(t)$ may be divided into two parts $A(t) = \langle A(t) \rangle + \delta A(t)$ where the angular brackets $\langle\ \rangle$ designate averaging over the random phases of the pumping. The statistical properties of the pumping are stationary, and therefore the function $\langle A \rangle$ is independent of time and in (6.7.3) it can be factored outside the integral sign. Then

$$\mathrm{Im}\langle A \rangle = \gamma(\mathbf{k})n(\mathbf{k}), \quad n(\mathbf{k}) = \langle |a(\mathbf{k},t)|^2 \rangle. \tag{6.7.4}$$

Here the contribution to the integral (6.7.3) of the quickly fluctuating part of $A(t)$ has been neglected. Indeed, the characteristic frequency of these fluctuations is $\Delta \gg \gamma$. Therefore the contribution of the $\delta A(t)$ to the integral (6.7.3) is (γ/Δ) times as little as the corresponding contribution of $\langle A \rangle$. Thus the wave number $n(\mathbf{k})$ as well as the renormalized frequency $\omega_{\mathrm{NL}}(\mathbf{k})$ do not fluctuate under noise the pumping. For the correlators

$$\sigma(\mathbf{k},t) = \langle a(\mathbf{k},t)a(-\mathbf{k},t) \rangle \exp(i\omega_p t) \tag{6.7.5}$$

we can similarly obtain

$$\partial \sigma(\boldsymbol{k},t)/\partial t + \{2\gamma(\boldsymbol{k}) + i[2\omega_{\text{NL}}(\boldsymbol{k}) - \omega_p]\}\sigma(\boldsymbol{k},t_1) = -2iP(\boldsymbol{k},t)n(\boldsymbol{k}). \quad (6.7.6)$$

Seeking the solution of the equation we have

$$\sigma(\boldsymbol{k},t) = -\int_0^\infty \sigma(t-t_1)\exp\{[-2\gamma(k) - i(2\omega_{\text{NL}}(k) - \omega_p)]t\}\,, \quad (6.7.7)$$
$$\sigma(t) = n(\boldsymbol{k})P(\boldsymbol{k},t).$$

The functions $\sigma(\boldsymbol{k},t)$ and $P(\boldsymbol{k},t)$ contain the same fast time dependence with the characteristic time $1/\Delta$. The function $P^*\sigma$ in our approximation does not depend on time. In order to calculate this function, (6.7.7) must be multiplied by $P^*(\boldsymbol{k},t)$ and the integral must be calculated taking into account only the fast time dependence. This yields

$$\text{Im}\langle P^*(\boldsymbol{k},t)\sigma(\boldsymbol{k},t)\rangle = -2\pi n(\boldsymbol{k})P^2[k,\omega_{\text{NL}}(\boldsymbol{k})],$$
$$P(\boldsymbol{k},t) = \int_\infty^{-\infty} P(\boldsymbol{k},)\omega)\exp(-i\omega t)\,d\omega\,, \quad (6.7.8)$$
$$\langle P(\boldsymbol{k},\omega)P^*(\boldsymbol{k},\omega)\rangle = P^2(\boldsymbol{k},\omega)\delta(\omega - \omega_1).$$

The combination of (6.7.2b, 4, 5) and (6.6.8) yields

$$\gamma(\boldsymbol{k})n(\boldsymbol{k}) = 2\pi n(\boldsymbol{k})P^2[k,\omega_{\text{NL}}(\boldsymbol{k})].$$

The non-trivial stationary solution of this equation exists if

$$\gamma(\boldsymbol{k}) = 2\pi P^2[k,\omega_{\text{NL}}(\boldsymbol{k})]. \quad (6.7.9a)$$

If for some $\boldsymbol{k}$ the following inequality

$$\gamma(\boldsymbol{k}) < 2\pi P^2[k,\omega_{\text{NL}}(\boldsymbol{k})], \quad (6.7.9b)$$

is satisfied, then the amplitudes of waves with corresponding $\boldsymbol{k}$ exponentially increase, which contradicts the assumptions about the stationarity of the state. Therefore, for all $\boldsymbol{k}$ the condition of the external stability

$$\gamma(\boldsymbol{k}) \leq 2\pi P^2[k,\omega_{\text{NL}}(\boldsymbol{k})], \quad (6.7.10)$$

must be satisfied, whereas $n(\boldsymbol{k})$ is non-zero only when there is an equality sign in (6.7.10). This condition is analogous to the condition of the external stability arising under monochromatic pumping $P(\boldsymbol{k}) \leq \gamma(\boldsymbol{k})$. From the usual expression for the renormalization of the pumping (e.g. (5.4.12)) and (6.7.6) we obtain

$$h^2(\omega)V(\boldsymbol{k})V^*(\boldsymbol{k}') = P^2(\boldsymbol{k},\boldsymbol{k}',\omega)$$
$$-2\int\left\{\frac{S(\boldsymbol{k},\boldsymbol{k}')P(\boldsymbol{k},\boldsymbol{k}',\omega)n(\boldsymbol{k}_1)}{\omega-2\omega_{\mathrm{NL}}(\boldsymbol{k})+2i\gamma(\boldsymbol{k})}\right.$$
$$\left.+\frac{S(\boldsymbol{k}',\boldsymbol{k}_1)P(\boldsymbol{k},\boldsymbol{k}_1,\omega)n(\boldsymbol{k}_1)}{\omega-2\omega_{\mathrm{NL}}(\boldsymbol{k})-2i\gamma(\boldsymbol{k}_1)}\right\}d\boldsymbol{k}_1 \quad (6.7.11)$$
$$+\int\frac{n(\boldsymbol{k}_1)n(\boldsymbol{k}_1)S(\boldsymbol{k},\boldsymbol{k}_1)S^*(\boldsymbol{k}',\boldsymbol{k}_2)P^2(\boldsymbol{k}_1,\boldsymbol{k}_2,\omega)}{[\omega-2\omega_{\mathrm{NL}}(\boldsymbol{k}_1)+2i\gamma(\boldsymbol{k}_1)][\omega-2\omega_{\mathrm{NL}}(\boldsymbol{k}_1)+2i\gamma(\boldsymbol{k}_2)]}d\boldsymbol{k}_1 d\boldsymbol{k}_2\ .$$

Solving this equation simultaneously with the condition of the external stability (6.7.10) we can obtain the distribution of parametric waves $n(\boldsymbol{k})$.

As it is known, under parametric excitation of waves by coherent pumping $n(\boldsymbol{k})$ differs from zero in a small region of the $\boldsymbol{k}$-space. It will subsequently be shown that at $\delta < kv$ (v is the group velocity of parametric waves) the region occupied by parametric waves is small, too. This enables us to assume that all the coefficients in (6.7.11) are independent of k. In some important cases, e.g. under spherical and axial distribution symmetry of parametric waves we can eliminate the angular dependence of the interaction amplitudes in the equation. This makes it possible to simplify (6.7.11), passing from the variable $\boldsymbol{k}$ to the variable $\omega_1 = \omega_{\mathrm{NL}}(\boldsymbol{k})$:

$$h^2(2\omega)V^2 = P^2(2\omega)\left\{1 - S\int\frac{n(\omega_1)\,d\omega_1}{(\omega_1-\omega+i0)}\right.$$
$$\left.+ S^2\left[\int\frac{n(\omega_1)\,d\omega_1}{(\omega_1-\omega+i0)}\right]^2 + [\pi Sn(\omega)]^2\right\}, \quad (6.7.12)$$
$$n[\omega(\boldsymbol{k})] = \int d\Omega k^2(\Omega)n(k,\Omega)/v(\Omega)\ .$$

6.7.2 Distribution of Parametric Waves Above Threshold

Cherepanov [6.21] analyzed (6.7.12) for the specific form of the spectral density of the noise pumping:

$$2\pi h^2(2\omega)V^2 = \gamma p(1-\omega^2/\Delta^2), \qquad \Delta \gg \gamma. \quad (6.7.13)$$

In the region of low supercriticalities ($p-1 < 1$) he obtained

$$3\pi Sn(\omega) = (p-1)\sqrt{(\sqrt{3}\omega+\mu)^3(3\mu-\sqrt{3}\omega)}/\Delta^2$$
$$\mu = \Delta\sqrt{p-1}\cdot\mathrm{sign}\,S, \quad (6.7.14)$$

It must be noted that this solution is non-zero in the asymmetric range $(-\mu/\sqrt{3},3\mu)$ and is not symmetric in spite of the symmetrical profile of the pumping (6.7.13). In the limiting case of high supercriticalities it follows from (6.8.10, 12) that

$$3\pi Sn(\omega) = \sqrt{(p-1)(3-4\omega^2/\Delta^2)^3}. \qquad (6.7.15)$$

This solution differs from zero in the symmetric region $|\omega| < \omega_1 = \sqrt{3}\Delta/2$, which is narrower than the instability region $|\omega| < \Delta$, where $h(2\omega) > 0$. Such a narrowing is characteristic of the S-theory. Because of this the behavior of $h^2(2\omega)$ at $|\omega| > \omega_1$ is not significant for our problem and the solution of (6.7.10) qualitatively holds true whatever the shape of the spectral plane of the pumping.

Integrating (6.7.14, 15) with respect to ω we can readily obtain the dependence of the total number of parametric waves on the supercriticality

$$N = 4\Delta(p-1)^{3/2}/9\pi\sqrt{3}|S|, \text{ at } p-1 < 1, \qquad (6.7.16)$$

$$N = \sqrt{3(p-1)}/2\pi|S|, \text{ at } p \gg 1. \qquad (6.7.17)$$

At $\Delta \simeq \gamma$ the expression (6.7.17) is in agreement with (5.5.7) for the case of the coherent pumping. As for the agreement of (6.7.15) and (5.5.7) for low supercriticalities, under $\Delta \simeq \gamma$ the first formula holds true because in this case there is not rigid correlation of the wave phases in the pairs. If the ratio γ/Δ becomes smaller than 1 then the term $\text{Re}\{\langle P^*(\boldsymbol{k})\sigma(\boldsymbol{k})\rangle\}$ is about γ/Δ. Therefore under small γ/Δ the term quadratic in σ must be allowed for.

Now let us obtain the limit of applicability of the mean-field approximation under the parametric wave excitation by the noise pumping. To this end, let us compare the term $\text{Im}\{\langle P^*(\boldsymbol{k})\sigma(\boldsymbol{k})\rangle\}$ with the terms of the kinetic equation (of the order of magnitude $(SN)^2 n(\boldsymbol{k})/(kv)$ omitted in (6.7.11). As a result we find that our approach holds true under

$$\Delta^2(p-1) < \gamma kv. \qquad (6.7.18)$$

In solids $kv/\gamma \simeq 10^2 - 10^6$ which provides wide enough applicability scope for the S-theory approximation. Note also that the opposite limiting case when the scattering of parametric waves exceeds their self-consistent interaction has been studied by *Levinson* [6.22].

Now let us present the estimation of the angular size of the excited packets. As in the case of the monochromatic pumping the angular sizes of the packet of parametric waves are zero. This can easily be checked taking the example of the simplest model

$$V(\boldsymbol{k}) = Vf(\Theta), S(\boldsymbol{k},\boldsymbol{k}') = Sf(\Theta)f(\Theta'), \gamma(\boldsymbol{k}) = \gamma,$$

qualitatively valid for ferromagnets. In this case

$$P(\boldsymbol{k},\boldsymbol{k}',\omega) = P^2(\omega)f(\Theta)f(\Theta')$$

and in (6.7.11) the dependence on the angles can be eliminated, and it is possible to show that parametric waves are concentrated in the angle Θ_0, where $|f(\Theta_0)|$ is at maximum. The weak scattering of parametric waves can be shown to lead to the packet broadening by the angle $\Delta\Theta$:

$$\Delta\Theta \simeq (SN/kv) \simeq \sqrt{p-1}/kv \,. \tag{6.7.19}$$

In conclusion note that the above-developed theory is in good agreement with the specially designed experiment on the parametric excitation of magnons in the ferrimagnetic YIG performed by *Zautkin, Orel* and *Cherepanov* [6.23] (see Sect. 9.10.2). It enables us to assume that the amplitude of parametric waves under noise pumping is limited, as under coherent excitation, by the S-theory phase mechanism.

7 Non-Stationary Behavior of Parametrically Excited Waves

7.1 Spectrum of Collective Oscillations (CO)

Chapters 5, 6 described the stationary state of the system of parametrically excited waves. The present chapter treats the non-stationary processes due to the time dependence of the experimental conditions. First we shall consider small *collective oscillations* (CO) with respect to the stationary state.

7.1.1 Spectrum of Spatially Homogeneous CO in the Non-Dissipation Limit

If the damping of the parametrically excited waves γ is zero, the Hamiltonian of the system $\mathcal{H}_0$ is the integral of motion. Let the value of $\mathcal{H}$ in the perturbed state differ from its value $\mathcal{H}_0$ in the ground state. This implies that the system will never attain the ground state and, since it has no other stable stationary states, its behavior will be essentially non-stationary. To study it, we must obtain the frequencies (spectrum) of the collective oscillations of the parametric wave system. For definiteness, consider the cubical ferromagnet under the supercriticality below the second threshold when parametric waves in the ground state are excited in the equator plane. Let $d(\boldsymbol{k},t)$ denote the deviations from the ground state. Then we shall extract the part $\mathcal{H}_0$ of the S-theory Hamiltonian corresponding to the ground states and the parts $\mathcal{H}_1$ and $\mathcal{H}_2$ containing the linear terms and the terms quadratic in $d(\boldsymbol{k},t)$. Taking into account the equations of motion we can see that the ground state energy is extreme: $\mathcal{H}_1 = 0$. The part of the Hamiltonian quadratic in small perturbations after we pass from summation to integration assumes the following form:

$$\mathcal{H}_2 = \frac{N_1}{4\pi^2}\Big\{[\exp(i\Psi)\int T_{12}d_1d_2\exp[-i(\varphi_1+\varphi_2)]+\text{c.c.}]\,d1d2$$

$$+ 2\int \{S_{12}\exp[-i(\varphi_1-\varphi_2)]+T_{12}\exp[i(\varphi_1-\varphi_2)]d_1d_2^*\}\,d1d2\Big\},$$

$$d_j = d(\varphi_j, t)\,,\quad dj = d\varphi_j\,,\quad T_{12} = T(\varphi_1-\varphi_2) = T(\boldsymbol{k}_1,\boldsymbol{k}_2)\,. \quad (7.1.1)$$

Here $\Theta = \Theta_1 = \pi/2$, $|\boldsymbol{k}_1| = |\boldsymbol{k}_2| = |\boldsymbol{k}_0|$. Changing over to Fourier components and using the axial symmetry of the problem we obtain

$$\mathcal{H}_2 = N_1 \sum_{m=-\infty}^{\infty} [2(T_m + S_m) d_m d_m^* + (T_m d_m d_{-m} + \text{c.c.})] , \qquad (7.1.2)$$

$$2\pi T_m = \int_0^{2\pi} T(\varphi - \varphi_1) \exp[-im(\varphi - \varphi_1)] \, d(\varphi - \varphi_1) ,$$

$$2\pi S_m = \int_0^{2\pi} S(\varphi - \varphi_1) \exp[-i(m-2)(\varphi - \varphi_1)] \, d(\varphi - \varphi_1). \qquad (7.1.3)$$

The Hamiltonian (7.1.2) can be diagonalized by the linear canonical transformation which yields

$$\mathcal{H}_2 = \sum_m \Omega_m d_m^* d_m , \qquad (7.1.4a)$$

$$\Omega_m = \left\{ (S_m - S_{-m}) + \sqrt{(S_m + S_{-m})(S_m + S_{-m} + 4T_m)} \right\} N_1 . \qquad (7.1.4b)$$

This transformation is possible if the frequency of the collective oscillations Ω_m in the system of parametric waves is real. At $(S_m + S_{-m})(S_m + S_{-m} + 4T_m) < 0$ this frequency is imaginary which indicates to the instability of the ground state with respect to the excitation of the exponentially increasing oscillations (the *internal instability* mentioned in Chap.5).

The canonical transformation unambiguously determines the sign before the radical in the expression (7.1.4b) for the frequency of the collective oscillations. But again, this sign can be easily obtained from simple considerations. Under small T_m the sign of Ω_m obviously coincides with the sign of $(S_m + T_m)$. Therefore the positive sign before the radical must be chosen for (7.1.4b). As T_m continuously increases the sign before the radical remains the same because the canonical transformation is continuous. Note that the frequency Ω_m may be negative. This means that the excitation of the collective excitation has resulted in the decreased energy of the system of parametric waves. Their relaxation is accompanied by the increase of the system energy. This does not violate the law of conservation of energy since the system of parametric waves acquires energy from the pumping.

In cubic ferromagnets under symmetric directions of magnetization ($\boldsymbol{M} \| [111]$) or [100]) the values S_m and T_m are non-zero only at $m = 0, \pm 1, \pm 2$. The expressions for coefficients S_m, T_m, corresponding to these modes can be obtained from (3.1.25, 26) and (7.1.3):

$$S_0 = 2\pi g^2 (\omega_M/\omega_p)^2 \left\{ \sqrt{1 + (\omega_p/\omega_M)^2} + N_{z0,\text{ef}} - 1 \right\} ,$$

$$T_0 = S_0 + 2\pi g^2 (N_{z0,\text{ef}} - 1) , \qquad N_{z0,\text{ef}} = N_z + \beta \omega_p / \omega_M ,$$

$$\beta = \begin{cases} -8 , & \text{for } \boldsymbol{M} \| [111] , \\ +9 , & \text{for } \boldsymbol{M} \| [100] , \end{cases}$$

$$S_1 = S_{-1} = 0 \,, \quad T_1 = T_{-1} = \frac{1}{2}(T_2 + T_{-2} + T_0 + S_0 - S_2 - S_{-2}) \,,$$

$$T_{\pm 2} = \frac{\pi g^2 \omega_M^2}{2\omega_p^2} \left[(N_{z2,\mathrm{ef}} - 1)u_\pm^2 + \sqrt{1 + (\omega_p/\omega_M)^2} \right] \,,$$

$$S_{\pm 2} = 2\pi g^2 \left[(N_{z2,\mathrm{ef}} - 1)u_\pm^2 + u_\pm \omega_M / 2\omega_p \right] \,,$$

$$u_\pm = (\sqrt{1 + (\omega_M/\omega_p)^2} \mp 1)/2 \,, \quad N_{z2,\mathrm{ef}} = \omega_{\mathrm{ex}}(ka)^2 + \omega_a/\omega_p \,. \quad (7.1.5)$$

The formulae (7.1.5) show the dependence of the CO frequency on the experimental conditions, i.e. the supercriticality, pumping frequency, magnetization, external magnetic field, the shape of the sample and crystallographic anisotropy. For the easy-plane ferromagnet it is clear from (3.2.13) that $S_0 = T_0$, and all the rest S_m and T_m are zero. This leads to the stability of the ground state experimentally observed by *Kotuzhansky* and *Prosorova* [7.1], *Prosorova* and *Smirnov* [7.2] up to very high supercriticality.

7.1.2 Influence of Wave Damping on the CO Spectrum

How does the damping of waves influence the spectrum of collective oscillations? This question is very important especially because the damping rate of parametrically excited waves $\gamma(\boldsymbol{k})$ can be of the same order as the frequency of collective oscillations Ω_m. By linearizing (5.4.15) with respect to the deviations from the ground state (5.5.7) and assuming $d, d^* \simeq \exp(-i\Omega t)$ we obtain the system of algebraic equations homogeneous in d, d^*. The condition of their solvability determines the frequency and damping of collective oscillations:

$$\begin{aligned}\Omega_\mathrm{m} = &- i\gamma + (S_\mathrm{m} - S_{-\mathrm{m}})N_1 \\ &+ \sqrt{(S_\mathrm{m} + S_{-\mathrm{m}})(S_\mathrm{m} + S_{-\mathrm{m}} + 4T_\mathrm{m})N_1^2 - \gamma^2} \,.\end{aligned} \quad (7.1.6)$$

Hence we can draw an important conclusion that the criterion on the emergence of the inner instability of the collective modes is independent of the damping rate and is, as in the conservative case, given by the inequality

$$(S_\mathrm{m} + S_{-\mathrm{m}})(S_\mathrm{m} + S_{-\mathrm{m}} + 4T_\mathrm{m}) < 0 \,. \quad (7.1.7)$$

Within the frame of the elementary theory collective oscillations of the parametric wave system are spatially homogeneous. When the spatial dispersion is allowed for a whole branch $\Omega_\mathrm{m}(\boldsymbol{\kappa})$ corresponds to each normal mode, (7.1.6) specifying the gap of this branch. The dependence $\Omega_\mathrm{m}(\boldsymbol{\kappa})$ will be calculated in the following section.

7.1.3 Spectrum of Spatially Non-Homogeneous CO

In order to obtain the frequency of non-homogeneous collective oscillations $\Omega_m(\kappa)$ (*spectrum of CO*) (6.5.9) must be linearized in small deviations from the ground state $A(\boldsymbol{k}, z, t) = A(\boldsymbol{k}, z) + d(\boldsymbol{k}, z, t)$ and the following form of their solution must be sought $d(\boldsymbol{k}, z, t)$ and $d^*(\boldsymbol{k}, z, t) \propto \exp[i(\kappa z - \Omega t)]$. The results for two simple, but important cases are given below (for more details, see *L'vov* and *Rubenchik* [7.3]).

1 Isotropic model: $S(\boldsymbol{k}\boldsymbol{k}') = S$, $T(\boldsymbol{k}\boldsymbol{k}') = T$, $V(\boldsymbol{k}) = V$. This case is characteristic of the easy-plane antiferromagnets. The dependence of the frequency of collective oscillations on their wave vector is given by their dispersion equation

$$\Delta^2(\Omega + 2i\gamma)\Omega \int_0^{2\pi} \left[(kv)^2 \cos^2\varphi + \Omega(\Omega + 2i\gamma)\right] d\varphi = 4\pi^2, \qquad (7.1.8)$$

where $\Delta = \Omega_0$, Ω_0 is given by (7.1.6) at $\gamma = 0$. For $(kv) \ll L$ the solution of (7.1.8) has the form

$$\Omega(\kappa) = -i\gamma \pm \sqrt{\Delta^2 + (kv)^2 - \gamma^2}. \qquad (7.1.9)$$

Under $\kappa = 0$ (7.1.9) passes to the expression for the frequency of the spatially-homogeneous collective oscillations discussed in the previous section. The negative Δ^2 results in the development of instability. The oscillations are the most unstable under $\kappa = 0$, the instability region extends to $\kappa v = \Delta/2$. At $\Delta^2 > 0$ the stationary solution is stable under any κ.

2 Axially symmetrical model: Let us study the spectrum of collective oscillations $\Omega(\boldsymbol{\kappa})$ in the case very important for experiments on ferromagnets when the wave amplitude is non-zero at the latitudes $\Theta = \Theta_0$ and $\Theta = \pi/2 - \Theta_0$. In the cubic ferromagnets under parallel pumping with the supercriticality up to 6 – 8 dB $\Theta_0 = \pi/2$ (equator). The transverse pumping often leads to the realization of the case $\Theta_0 \simeq \pi/4$. The dependence $\Omega(\kappa)$ has the simplest form when $\boldsymbol{\kappa}$ is parallel to the axis of symmetry. For each harmonic with the number m we can obtain

$$\Omega_{m1+}(\kappa) = -i\gamma \pm \sqrt{\delta_{m1}^2 - \gamma^2}, \quad \Omega_{m2+}(k) = -i\gamma \pm \sqrt{\delta_{m2}^2 - \gamma^2}.$$

$$\delta_{m1,2}^2 = (\kappa v)^2 + [\Delta_{m1}^2 + \Delta_{m2}^2]/2$$
$$\pm \sqrt{(\Delta_{m1}^2 - \Delta_{m2}^2)^2/4 + (\kappa v)^2(\Delta_{m1}^2 + \Delta_{m2}^2 + \Delta_m^2)}.$$

$$\Delta_m^2 = 4(T_m^+ - T_m^-)(T_m^+ + T_m^- + S_m)N^2,$$
$$\Delta_{m1}^2 = [\omega''\kappa^2/2 + (2S_m + T_m^+ + T_m^-)N]^2 - (T_m^+ + T_m^-)^2 N^2,$$
$$\Delta_{m2}^2 = [\omega''k^2/2 + (T_m^+ - T_m^-)N]^2 - (T_m^+ - T_m^-)^2 N^2. \qquad (7.1.10)$$

Here $T_m^\pm$, S_m - are the Fourier harmonics of the coefficients $T(k_z, k_z', (\varphi - \varphi'))$, $S(k_z, k_z'(\varphi - \varphi'))$ (6.5.9). For simplicity, in (7.1.10) we assumed $S_m = S_{-m}$. At $\delta_{m1,2}^2(\kappa) > 0$ the expressions (7.1.10) correspond to collective oscillations with the frequency $\delta_{m1,2}$, damping decrement γ_2 equal to the damping decrement of parametric waves. At $\delta_{m1,2}^2(\kappa) < 0$: Im$\Omega > 0$. This corresponds to the exponential increase of the spatially inhomogeneous collective oscillations, i.e. to the instability of the spatially homogeneous distribution of parametric waves.

Let us analyze the expressions obtained (7.1.10) for the frequencies of collective oscillations. At $\kappa = 0$

$$\delta_{m1}^2(0) = \Delta_{m1}^2(0) = 4S_m(T_m^+ + T_m^- + S_m)N^2 . \tag{7.1.11}$$

The oscillation $\Omega_{m1}(0)$ is stable if $S_m(T_M^+ + T_M^- + S_m) > 0$. The oscillation Ω_{m2} is neutrally stable. At small κv we have

$$\delta_{m1}^2(\kappa) = \delta_{m1}^2(0) + (\kappa v)^2(2S_m + T_m^+ - T_m^-)/S_m ,$$

$$\delta_{m1}^2(\kappa) = (\kappa v)^2(T_m^+ - T_m^-)/S_m . \tag{7.1.12}$$

Therefore if $S_m(T_m^+ - T_m^-) < 0$ the branch of the collective oscillations $\delta_{m2}(\kappa)$ at $\kappa \neq 0$ becomes unstable. At the same time as κ increases the margin of stability of the branch δ_{m2} also increases. Assuming $\Delta_{m1}^2 > 0$, $\Delta_m^2 > 0$ we find that the branch δ_{m1} is stable at any κ, and the instability region of the branch δ_{m2} is between $\kappa = 0$ and $\kappa = \kappa_0$, where $(\kappa_0 v)^2 = \Delta_m$. In the cubic ferromagnets κv may prove to be zero. This happens if the waves are excited on the equator and the perturbation is perpendicular to its plane and if because of the anisotropic dispersion law group velocity of the excited waves $v(\Theta)$ turns into zero for $\Theta = \Theta_0$. In this case (7.1.10) is reduced and assumes the following form:

$$\Omega_{m+}(\kappa) = -i\gamma \pm \sqrt{[2(T_m + S_m)N + \omega''\kappa^2/2]^2 - 4T_m^2 N^2 - \gamma^2} . \tag{7.1.13}$$

For simplicity, we took $T^+ = T^- = T$. This relation is always satisfied, for example for waves excited on the equator. At the same time the second branch of oscillations δ_{m2} is always stable. From (7.1.13) it is clear that even the $\delta_{m1}^2(0) > 0$ instability can emerge in the region $\omega''\kappa^2 \simeq 4(T_m + S_m)N^2$ if $\omega''(T_m + S_m) > 0$. As will be shown subsequently, the development of the instability with $\kappa = 0$ leads to the emergence of self-oscillations of the total number of waves. Under parametric excitation of spin waves in ferromagnets they manifest themselves experimentally as magnetization auto-oscillations. The sign of ω'' in cubic ferromagnets is determined by the value of the external field, and we can easily provide the condition when the spatially-homogeneous oscillations are stable and the instability is localized within the range $\omega''\kappa^2 \simeq SN$. The development of this instability can result in a stationary, spatially-homogeneous picture $A(\boldsymbol{k}, z)$ with the scale $\sqrt{SN/\omega''}$, modulation depth of the order of unity and the average number of the order

N. This raises the interesting problem of finding such stationary states and studying their stability within (6.5.9) which has not been solved yet. To observe such phenomenon experimentally would also be interesting.

7.2 Linear Theory of CO Resonance Excitation

Collective oscillations (CO) of the system of parametric waves discussed in the previous section can be excited (like all other types of oscillations) in different ways: by resonance external action, parametrically or by impact. All these ways have been employed in experiments on cubic ferromagnets and easy-plane antiferromagnets. To formulate the theory and to interpret experimental data would be easier for the resonance excitation of collective oscillations. This way was employed in 1972 by *Zautkin, L'vov* and *Starobinets* [7.14] who discovered and studied COs. In these experiments in addition to parallel pumping, another SHF signal with the same polarization $\boldsymbol{h}_z \| \boldsymbol{h} \| \boldsymbol{M}$, $h_z(t) \simeq \exp(-i\omega t)$ was fed to the sample. Its frequency ω differed from the pumping frequency ω_p by the frequency of collective oscillations Ω. The beatings between two SHF-signals at the resonance frequency served in this case as the resonance external force. Later (in 1975) *Orel* and *Starobinets* [7.5] employed the direct resonance excitation by an alternating magnetic field at the frequency of the COs which is within the single-frequency range (about 1 MHz). Collective oscillations could be excited also by a sound whose frequency and wave vector coincide with the frequency and wave vector of collective oscillations. All these methods of excitation of collective oscillations will be discussed in this section.

7.2.1 Basic Equations and Their Solution

As we are interested only in the behavior of spatially-homogeneous collective oscillations of the zeroth mode ($\kappa = 0$, $m = 0$) we shall average only the equations of motion of the basic S-theory (5.4.15) over all $\boldsymbol{k}$ directions in the package of parametric waves:

$$\{\partial/\partial t + \gamma + i[\omega(\boldsymbol{k}) - \omega_p/2 + 2T_0 c^* c]\}c + i(hV_0 + S_0 c^2)c^* = -if(t), \quad (7.2.1)$$

$$f(t) = f(\Omega)\exp(-i\Omega t) + f(-\Omega)\exp(i\Omega t) . \quad (7.2.2)$$

Here $|c(t)|^2 = N(t)$ is the total number of parametric waves in the package V, T and S are the coefficients of the Hamiltonian of the problem averaged over the package. For cubic ferromagnets they are given by (7.1.5); for easy-plane ferromagnets they are specified by (3.2.13). The periodic force $f(t)$ is added to the first right-hand side of (7.2.1), its nature will be discussed in detail.

A. When an additional weak signal acts on the system of parametric waves at the frequency ω (see Problem 5.3) then

$$f(\Omega) = h_z V_0 c_0 , \quad f(-\Omega) = 0 , \quad \Omega = \omega - \omega_p . \tag{7.2.3}$$

B. If the frequency of the weak signal Ω is within the radio frequency range, then (see Problem 5.3)

$$f(\Omega) = f(-\Omega) = U_0 c_0 h_z , \quad U_0 = g[1 + \sqrt{1 + \omega_M^2/\omega_p^2}]/2 . \tag{7.2.4}$$

The expression for U_0 corresponds to the cubic ferromagnet [(see (4.3.20)] when parametric waves are concentrated on the equator.

C. If collective oscillations in the easy-plane ferromagnets are excited by the sound, then (see Problem 5.4)

$$\begin{aligned} f(\Omega,\boldsymbol{\kappa}) &= V_{sm}(\boldsymbol{\kappa},\boldsymbol{k}) c_0 [\beta(\Omega,\boldsymbol{\kappa}) + \beta^*(-\Omega,-\boldsymbol{\kappa})] , \\ f(-\Omega,\boldsymbol{\kappa}) &= V_{sm}(-\boldsymbol{\kappa},\boldsymbol{k}) c_0 [\beta(\Omega,-\boldsymbol{\kappa}) + \beta^*(-\Omega,\boldsymbol{\kappa})] . \end{aligned} \tag{7.2.5}$$

Here $\beta(\Omega,\boldsymbol{\kappa})$ denotes the Fourier transform of the canonical amplitude of the sound $\beta(\Omega,t)$, and V_{sm} designates the amplitudes of the Hamiltonian of the sound interaction with magnons. This Hamiltonian has been calculated by *Lutovinov* [7.6]:

$$\mathcal{H}_{sm} = \sum_{\boldsymbol{\kappa},\boldsymbol{k}} V_{sm}(\boldsymbol{\kappa},\boldsymbol{k})[\beta(\boldsymbol{\kappa},t) c(\boldsymbol{k},t) c^*(\boldsymbol{k}+\boldsymbol{\kappa},t) + \text{c.c.}] , \tag{7.2.6}$$

where $V_{sm} \simeq \sqrt{\Omega(\boldsymbol{\kappa})/[\omega(\boldsymbol{k})\omega(\boldsymbol{k}+\boldsymbol{\omega})]}$. Equations (7.2.2) suggest how to seek the solution of (7.2.1) in the approximation linear in f:

$$\begin{aligned} c(t) &= c_0 + d(\Omega) \exp(-i\Omega t) + d(-\Omega) \exp(i\Omega t) , \\ d(\Omega) &= d_+(\Omega) f(\Omega) + d_-(\Omega) f^*(-\Omega) . \end{aligned} \tag{7.2.7}$$

Simple calculations here yield

$$\begin{aligned} d_+(\Omega) &= -[\Omega + i\gamma + 2(T_0 + S_0)N]/\Delta(\Omega) , \\ c_0 d_+ &= c_0^* d_- , \quad N = |c_0|^2 . \\ \Delta(\Omega) &= \Delta_0^2 - \Omega^2 - 2i\gamma\Omega , \quad \Delta_0^2 = 4S_0(2T_0 + S_0)N^2 . \end{aligned} \tag{7.2.8}$$

As it should be expected, the susceptibilities d are maximum when the frequency of the external force is close to the eigenfrequency of the collective resonance.

7.2.2 CO Excitation by a Microwave Field

The effectiveness of the excitation of collective oscillations by the magnetic field h_z can be characterized by the susceptibility $\chi(\omega_p + \Omega)$:

$$m_z(\omega_p + \Omega) = \chi(\omega_p + \Omega) h_z(\omega_p + \Omega) . \tag{7.2.9}$$

Here $m_z(\omega_p + \Omega)$ denotes the oscillation amplitude at the frequency $(\omega_p + \Omega)$ of the magnetization component $m_z(t)$ which depends on the number of the parametric magnons and, consequently, changes when collective oscillations are excited. In our approximation and using the notation of this section we can easily obtain $m_z((\omega_p+\Omega) = V_0 c_0 d(\Omega)$. Employing (7.2.7, 8) for c_0, $d(\Omega)$ and $f(\Omega)$ and the definitions of (7.2.6) we obtain the expressions for the imaginary part of the susceptibility to the weak SHF-signal, characterizing the absorption of its energy by the system of parametric magnons:

$$\chi''(\omega_p + \Omega) = \chi''\left(\frac{h}{h_{th}}\right)\frac{2\gamma^2[\Omega_0^2 + \Omega^2 + 4\Omega(T_0 + S_0)N}{(\Omega_0^2 - \Omega^2)^2 + 4\gamma^2\Omega^2} . \qquad (7.2.10)$$

Here χ'' is the imaginary part of the susceptibility to the pumping field h, given by (5.5.31a). Under high supercriticality when $\Delta_0^2 > \gamma^2$ the line shape (7.2.10) is close to the Lorentz shape and the line width is equal to the damping of magnons γ. At the resonance points the susceptibility is equal to

$$\chi''(\omega_p \pm \Omega) = \chi'' \frac{h^2}{h_{th}^2}\left\{1 \pm \sqrt{1 + \frac{T_0}{S_0(2T_0 + S_0)}}\right\} . \qquad (7.2.11)$$

The susceptibility may become negative. This fact is of cardinal importance, and corresponds not to the absorption but to the amplification of the weak signal. It follows from (7.2.11) that the absorption takes place at the frequency $\omega_p + \Omega$ and the amplification occurs at the image frequency $\omega_p - \Omega$. The amplification can be interpreted as the result of the instability with respect to decay of the ground state (with the "slow" frequency equal to zero) into the electromagnetic irradiation (with the slow frequency Ω) and collective oscillations with the eigenfrequency Δ_0. The law of conservation of energy in this process is $\delta_0 + \Omega = 0$. Therefore the amplification occurs at the frequency $\Omega = -\delta_0$ which corresponds to (7.2.11). The absorption in these terms is the consequence of the decay of the weak signal into the ground state and collective oscillations with the conservation law $\Omega = \Omega_0 + 0$.

The collective resonance has been experimentally observed by *Zautkin* et al. [7.4] on the YIG monocrystals. A good quantitative agreement with the theory presented here has been observed. In particular, the experimental dependences of the susceptibility to the weak signal and of collective oscillation frequency on the pumping intensity are in good agreement with the dependences given by (7.2.8, 11) up to the second threshold $h/h_2 = 8\,\text{dB}$. Note also that the collective resonance can be used as a convenient and instructive method for changing the relaxation times of magnons. Measurements show that in accordance with the theory the line width of resonance absorption is practically independent of the pumping intensity and is in good agreement with the value obtained from the threshold value of parallel pumping. A more detailed comparison of theory with experiment will be given in Sect. 9.5.

7.2.3 Direct CO Excitation by a Radio Frequency Field

The pumping method of collective oscillations at the frequency Ω by means of the longitudinal SHF field h_2 at the frequency $\omega_p + \Omega$ described in the previous subsection seems somewhat artificial. A more natural way would be to employ the radio frequency magnetic field at the frequency Ω. However, the first attempts to observe this effect were unsuccessful because of the low susceptibility in the radio frequency range and only in 1975 did *Orel* and *Starobinets* [7.5] succeed in detecting this effect by means of their own sensitive methods.

In order to calculate the susceptibility $\chi(\Omega)$ to the longitudinal field $h_z(\Omega)$ note that in the notation of this section the expression for the alternating part m_z with the frequency Ω has the form

$$m_z(\Omega) = U_0[c_0^* d(\Omega) + c_0 d^*(-\Omega)] = \chi(\Omega) h_z(\Omega) . \tag{7.2.12}$$

From this equation and (7.2.4, 7, 8), we readily obtain

$$\chi = 2U_0^2 S_0 N^2 / \Delta^*(\Omega) , \qquad \Delta(\Omega) = \Delta_0^2 - \Omega^2 - 2i\gamma\Omega . \tag{7.2.13}$$

The maximum value $\chi''(\Omega)$ corresponds to the resonance frequency Ω_{res} which, generally speaking, does not coincide with Δ_0 and under $\Delta_0 > 2\gamma$ is given to a good accuracy by

$$\Omega_{\text{res}}^2 = \Delta_0^2 - \gamma^2 . \tag{7.2.14}$$

In this approximation the half-width of collective resonance curve measured at the level 1/2 equals to the relaxation of magnons. Let us give the value of the susceptibility in resonance

$$\chi''_{\text{res}} = \frac{g^2 \sqrt{\Omega_0^2 - \gamma^2} S_0 N_0^2}{\gamma[\Omega_0^2 - (3/4)\gamma^2]} . \tag{7.2.15}$$

Under high supercriticality $\Delta_0^2 \gg \gamma^2$ and (7.2.15) is also reduced. Substituting the dependence of Ω_0 and N_0 on $p = (h/h_{\text{th}})^2$ we obtain

$$\chi''_{\text{res}} = \frac{S_0 g^2 \sqrt{p-1}}{|S_0| 2 S_0 \sqrt{T_0 + S_0}} . \tag{7.2.16}$$

From this formula an important conclusion can be drawn: at $S_0 > 0$, (the case of ferromagnets) $\chi'' > 0$ and radio-frequency radiation absorbs in the sample; at $S_0 < 0$ (the case of easy-plane antiferromagnets), $\chi'' < 0$, which corresponds not to the absorption but to the amplification of the radio-frequency signal. If this amplification exceeds the damping in the oscillatory circuit, it will lead to instability with respect to the excitation at the frequency Ω – a laser effect of a kind.

The theory presented in this section is in good agreement with the experimental data for the ferromagnetic YIG (see Chap. 9).

7.2.4 Coupled Motions of Collective Excitations of Parametric Waves and Sound

Theoretically, this question does not differ from the problem of CO interaction with the radio frequency field, only the expression for the force $f(\Omega)$ (7.2.4) must be substituted for expression (7.2.5). However, to setup an experiment on measuring the susceptibility of the collective oscillations to the sound is rather difficult. Therefore, in studying the interaction of collective oscillations with the sound attention should be rather paid to another statement of the problem aimed at the study of their coupled motions.

Let us calculate the frequencies of these motions [7.6]. From the expressions for the Hamiltonian of the magnetoelastic interaction $\mathcal{H}_{sm}$ (7.2.6) follow the equations of motion for the complex amplitude of sound $\beta(\Omega,\boldsymbol{\kappa})$:

$$[\Omega - \Omega_S(\boldsymbol{\kappa}) + i\Gamma_S(\boldsymbol{\kappa})]\beta(\Omega,\boldsymbol{\kappa}) = V^*(\boldsymbol{k},\boldsymbol{\kappa})[c_0^* d(\Omega) + c_0 d^*(-\Omega)],$$
$$[\Omega + \Omega_S(\boldsymbol{\kappa}) + i\Gamma_S(\boldsymbol{\kappa})]\beta(-\Omega,-\boldsymbol{\kappa}) = -V(\boldsymbol{k},\boldsymbol{\kappa})[c_0^* d(\Omega) + c_0 d^*(-\Omega)]. \quad (7.2.17)$$

Here $\Omega_S(\boldsymbol{\kappa})$ and $\Gamma_S(\boldsymbol{\kappa})$ are the frequency and damping of the sound with the wave vector $\boldsymbol{\kappa}$ in the absence of parametric magnons. Considering these equations simultaneously with (7.2.5, 7, 8), we obtain from the condition of zero determinant of the complete set of equations for $\beta(\Omega,\boldsymbol{\kappa})$, $\beta(-\Omega,-\boldsymbol{\kappa})$, $d(\Omega)$ and $d^*(-\Omega)$ the following:

$$\{[\Omega + i\Gamma_S(\boldsymbol{\kappa})]^2 - \Omega_S^2(\boldsymbol{\kappa})\}[\Omega(\Omega + i\gamma) - \Delta_0^2] = 8\Omega_S(\boldsymbol{\kappa})|V(\boldsymbol{k},\boldsymbol{\kappa})|^2 . \quad (7.2.18)$$

The simplest case for study is when the damping of both magnons and sound can be neglected. Then

$$2\Omega_\pm^2(\boldsymbol{\kappa}) = \Omega_S^2(\boldsymbol{\kappa}) + \Omega_0^2 + [(\Omega_S^2(\boldsymbol{\kappa}) - \Omega_0^2) + 32\Omega_S^2(\boldsymbol{\kappa})SN^2|V(\boldsymbol{k},\boldsymbol{\kappa})|^2] . \quad (7.2.19)$$

The interaction amplitude S is negative (see (3.2.13)), therefore near the resonance the coupled oscillation of parametric magnons and the elastic subsystem are unstable. If one takes into account damping of parametric magnons and sound then instability of the coupled oscillations (7.2.18) takes place under the number of parametric magnons N exceeds the threshold number $N_{cr}(\boldsymbol{k})$. The minimum value of N_{cr} is attained at resonance under $\Omega_0 = \Omega_S(\boldsymbol{\kappa})$:

$$2|SN_{cr}|^2 = -\gamma\Gamma\Omega_0^2|S|/|V(\boldsymbol{k},\boldsymbol{\kappa})|^2\Omega_S(\boldsymbol{\kappa}) . \quad (7.2.20)$$

The interaction of collective oscillations with the sound is significant also far from the resonance. In this case it follows from (7.2.18) that

$$\Omega_1(\boldsymbol{\kappa}) = \Omega_S(\boldsymbol{\kappa}) + 4SN^2 \frac{|V(\boldsymbol{k},\boldsymbol{\kappa})|^2}{\Omega_S^2(\boldsymbol{\kappa}) - \Omega_0^2}$$

$$- i\left\{\Gamma_S(\boldsymbol{\kappa}) + 8\gamma(\boldsymbol{k})\frac{\Omega_S(\boldsymbol{\kappa})SN^2|V(\boldsymbol{k},\boldsymbol{\kappa})|^2}{[\Omega_S^2(\boldsymbol{\kappa}) - \Omega_0^2]^2}\right\},$$

$$\Omega_2(\boldsymbol{\kappa}) = \Omega_0 + 4SN^2 \frac{|V(\boldsymbol{k},\boldsymbol{\kappa})|^2 \Omega_S(\boldsymbol{\kappa})}{\Omega_S^2(\boldsymbol{\kappa}) - \Omega_0^2}\Omega_0$$

$$- i\left\{\gamma(\boldsymbol{k}) + 8\Gamma_S(\boldsymbol{\kappa})\frac{\Omega_S(\boldsymbol{\kappa})SN^2|V(\boldsymbol{k},\boldsymbol{\kappa})|^2}{[\Omega_0^2 - \Omega_S^2]^2}\right\},$$

(7.2.21)

In the absence of the magneto-elastic interaction the oscillations with the frequency Ω_1 are purely acoustic and with the frequency Ω_2 they are collective oscillations of parametric magnons. The damping of the long-wave sound $\Gamma(\boldsymbol{\kappa})$ is as a rule small in comparison with the damping of parametric magnons $\gamma(\boldsymbol{k})$. Therefore under negative S the damping of the sound-type oscillations $\Omega_1(\kappa)$ can become negative. This implies instability. Its threshold is attained at a certain number of parametric magnons $N_{\mathrm{cr}}(\kappa)$:

$$N_{\mathrm{cr}} = \Gamma(\boldsymbol{\kappa})[\Omega_S^2(\boldsymbol{\Omega}) - \Omega_0^2]/64|SV^2(\boldsymbol{k},\boldsymbol{\kappa})|\Omega_S(\boldsymbol{\kappa})\gamma(\boldsymbol{k}). \tag{7.2.22}$$

In 1983 *Smirnov* [7.7] discovered the excitation of the acoustic mode of the sample at a certain pumping intensity. As the sample was finite the frequency of the acoustic oscillations $\Omega_S(\boldsymbol{\kappa})$ significantly exceeded the frequency of collective oscillations Ω_0. In this case the excitation threshold of the sound oscillations equals N_{cr} (7.2.22). According to the S-theory the corresponding value $p_{\mathrm{cr}} = (h_{\mathrm{cr}}/h_{\mathrm{th}})^2$ for Smirnov's experiment is $p = 25$. Experimentally this quantity has the value 50 – 100. Taking into account that we have no exact values of $V(\boldsymbol{k},\boldsymbol{\kappa})|S|$ and the damping $\gamma(\boldsymbol{k})$, such an agreement of theory and experiment must be considered satisfactory.

7.3 Threshold Under Periodic Modulation of Dispersion Law

Now we must study the large-amplitude collective oscillations. They can be excited in the ferromagnets also by the strong periodic modulation of the magnetic field at the frequency approximating the eigenfrequency of collective oscillations. Clearly, before investigating the above-threshold state of parametric magnons in the presence of a strong RF-field we must study the influence of this field on the threshold of the parametric excitation. This problem was first discussed by *Suhl* [7.8], who showed that under the periodic modulation of the biasing field (or of the pumping frequency) the threshold of the parametric instability increases because the condition of the parametric resonance is violated. According to Suhl such frequency and amplitude of the modulation can be chosen in such a way that the instability disappears as its threshold becomes infinitely high. Suhl's idea has been experimentally checked by *Hartwick, Peressini* and *Weiss* [7.9]. Their conclusions, however were not in full agreement with this theory, in particular they could not increase the threshold by more than 10dB. This lack of agreement will be theoretically explained later. Detailed experimental data corroborating our theory will be given in Chap. 9.

Let us consider the parametric excitation of spin waves in the following magnetic field [7.10, 11]:

$$H = H_M(t) + H_0 + h\cos\omega_p t . \qquad (7.3.1)$$

Here $\mathcal{H}_0$ is the constant magnetic field, $H_M(t)$ stands for the modulating field, $h\cos\omega t$ denotes the SHF pumping at the frequency ω_p. It is rather difficult to calculate the threshold of parallel pumping in the field (7.3.1) with an arbitrary modulation law $H_M(t)$. Therefore first the simplest case of the modulation will be considered, i.e. the modulation by rectangular repetitive pulses (meander-type mode with the period 2τ). The equations of motion for the normal amplitudes of spin waves taking into account the Hamiltonian of their interaction with the pumping $\mathcal{H}_p$ (4.3.18) and with the modulating field $\mathcal{H}_{p1}$ (4.3.19) have the following form:

$$\left\{\frac{\partial}{\partial t} + \gamma(\boldsymbol{k}) + i[\Delta(\boldsymbol{k}) + U(\boldsymbol{k})H_M]\right\} c(\boldsymbol{k},t) = -ihV(\boldsymbol{k})c^*(-\boldsymbol{k},t)$$

at $0 < t < \tau$, $2\tau < t < 3\tau$, .. and so on . $\Delta(\boldsymbol{k}) = \omega(\boldsymbol{k}) - \omega_p/2$,

$$\left\{\frac{\partial}{\partial t} + \gamma(\boldsymbol{k}) + i[\Delta(\boldsymbol{k}) - U(\boldsymbol{k})H_M]\right\} c(\boldsymbol{k},t) = -hV(\boldsymbol{k})c^*(-\boldsymbol{k},t) ,$$

at $\tau < t < 2\tau$, $3\tau < t < 4\tau$, .. and so on. $\qquad (7.3.2)$

Taking $c(\boldsymbol{k},t), c^*(-\boldsymbol{k},t) \propto \exp(\nu t)$ we find that the increment ν over different times is equal to:

$$\nu_1 = \nu(0,-\tau) = -\gamma + \sqrt{|hV(\boldsymbol{k})|^2 - [\Delta(\boldsymbol{k}) + U(\boldsymbol{k})H_M]^2} ,$$

$$\nu_2 = \nu(\tau,-2\tau) = -\gamma + \sqrt{|hV(\boldsymbol{k})|^2 - [\Delta(\boldsymbol{k}) - U(\boldsymbol{k})H_M]^2} . \qquad (7.3.3)$$

Let us consider the case $\gamma\tau > 1$ when the damping solutions corresponding to the negative sign before the square root in (7.3.3) can be neglected. Then the total increment ν during the total pulse period is $\nu_1 + \nu_2$. From the condition $\text{Re}\,\nu = 0$ we obtain two expressions for the instability threshold:

$$|h_{\text{th1}}V| = \gamma^2(\boldsymbol{k}) + (U(\boldsymbol{k})H_M)^2 + \Delta^2(\boldsymbol{k})[1 + U^2(\boldsymbol{k})H_M^2/\gamma^2] , \qquad (7.3.4)$$

$$|h_{\text{th2}}V| = 4\gamma^2(\boldsymbol{k}) + [\Delta(\boldsymbol{k}) \mp U(\boldsymbol{k})H_M]^2 . \qquad (7.3.5)$$

The expression (7.3.4) corresponds to the small amplitudes of modulations when $|hV(\boldsymbol{k})| > |U(\boldsymbol{k})H \pm \Delta(\boldsymbol{k})|$ and minimum threshold of instability is reached at $\Delta(\boldsymbol{k}) = 0$, i.e. for the spin waves over the resonance surface $\omega(\boldsymbol{k}) = \omega_\mathrm{p}/2$. The corresponding minimum threshold is

$$|h_\mathrm{th}V|^2 = \gamma^2(\boldsymbol{k}) + U^2(\boldsymbol{k})H_\mathrm{M}^2 , \qquad \Delta(\boldsymbol{k}) = 0 . \tag{7.3.6}$$

The large modulation amplitudes result in an additional local minimum of the threshold at $\Delta(\boldsymbol{k}) = \pm U(\boldsymbol{k})H_\mathrm{M}$. The value of the threshold on these surfaces is constant and equals $2\gamma(\boldsymbol{k})$. Therefore, the critical amplitude of the modulation $\mathcal{H}_\mathrm{m,cr} = \sqrt{3\gamma/U(\boldsymbol{k})}$ separates two modes of parallel pumping, i.e. the mode of weak modulation when the minimum threshold is specified by (7.3.6) and the mode of strong modulation when its increase stops and the minimum threshold is equal to 2γ. In the last mode the modulating field leads to a detuning whose value increases proportional to the amplitude of modulation. This characteristic effect of the strong modulation, i.e. the "freezing" of the threshold under $H_\mathrm{M} > H_\mathrm{m.cr}$ is, generally speaking inherent only in the modulation of the rectangular type. Thus it can be readily shown (*Zautkin* and *Orel* [7.11]) that for the sinusoidal and saw-tooth modulation the quantity value h_th increases, slowly but without bounds, with the increase of H_M. Under $H_\mathrm{M} \gg H_\mathrm{m,cr}$ for the sinusoidal and saw-tooth modulations respectively they have [7.11]:

$$2|h_\mathrm{th1,sin}V| = \gamma^{2/3}(U(\boldsymbol{k})H_\mathrm{M})^{1/3} , \qquad |h_\mathrm{th,saw}V| = \sqrt{\gamma U(\boldsymbol{k})H_\mathrm{M}} . \tag{7.3.7}$$

Until now we considered the low-frequency modulation with the period less than the relaxation time of spin waves. Now we shall drop this limitation and consider parallel pumping of spin waves in the magnetic field modulated by a sinusoid of arbitrary frequency Ω : $H_\mathrm{M}(t) = H_\mathrm{M}\cos\Omega t$ [7.10]. The linearized equations of motion have the following form:

$$\left\{\frac{\partial}{\partial t}+\gamma(\boldsymbol{k})+i[\Delta(\boldsymbol{k})+U(\boldsymbol{k})H_\mathrm{M}\cos(\Omega t)]\right\}c(\boldsymbol{k},t) = -ihV(\boldsymbol{k})c^*(-\boldsymbol{k},t) . \tag{7.3.8}$$

Confining ourselves to the small amplitudes of H_M let us write

$$c(\boldsymbol{k},t) = c_0(\boldsymbol{k}) + d(\Omega,\boldsymbol{k})\exp(-i\Omega t) + d(-\Omega,\boldsymbol{k})\exp(i\Omega t) . \tag{7.3.9}$$

The stationary solution of (7.3.8) corresponds to the instability threshold. Substitution of (7.3.9) in (7.3.8) yields the system of equations for $c_0(\boldsymbol{k})$, $d(\pm\Omega,\boldsymbol{k})$. Hence for the approximation quadratic in d we obtain the relation specifying the threshold

$$|h_\mathrm{th}V(\boldsymbol{k})|^2 = [\gamma^2(\boldsymbol{k}) + \Delta^2(\boldsymbol{k})]\left[1 + \frac{U^2(\boldsymbol{k})H_\mathrm{M}^2}{\Omega^2 + 4\gamma^2}\right]^2 . \tag{7.3.10}$$

As should be expected the minimum threshold for the mode of weak modulation is reached at $\Delta(\boldsymbol{k}) = 0$. At low frequencies (7.3.10) at $\Delta(\boldsymbol{k}) = 0$ is equivalent to the exact formulae (7.3.6) obtained above for the rectangular modulation. This apparently means that (7.3.10) derived within perturbation theory is applicable with a good accuracy up to the values $H_M = H_{m,\mathrm{cr}}$. This can be accounted for by the fact that at high frequencies of modulation the small parameter of the theory is $U(\mathbf{k})H_M/\Omega$ and at low frequencies it is $\Omega/\gamma(\mathbf{k})$.

In the general case of the modulation with arbitrary frequency and arbitrary amplitude we can obtain a comparatively simple expression for the instability threshold only for the rectangular modulation. Let us present without deviation the equation specifying h_{th} (see [7.10]):

$$\cos h[2\gamma(\mathbf{k})\tau] = \cos h(r_+\tau)\cos h(r_-\tau)$$
$$+ \sin h(r_+\tau)\sin h(r_-\tau)[|h_{\mathrm{th}}V(\mathbf{k})|^2 - \Delta_+\Delta_-]/d_+d_- \;,$$

$$\Delta_\pm = \Delta(\mathbf{k}) + U(\mathbf{k})H_M\;, \qquad d_\pm^2 = |h_{\mathrm{th}}V(\mathbf{k})|^2 - \Delta_\pm^2\;. \qquad (7.3.11)$$

When the frequencies are low (τ are large) we can easily obtain (7.3.2). At high frequencies ($\tau \to 0$) we have

$$|h_{\mathrm{th}}V(\mathbf{k})| = \gamma(\mathbf{k})\sqrt{1 + U(\mathbf{k})H_M\tau)^2/3}\;. \qquad (7.3.12)$$

Finally, it must be noted that the conclusion about the monotonic increase of the threshold by (7.3.7) for the periodic modulation refers to the excitation of waves with the limiting frequency shift $\Delta(\mathbf{k}) = \pm U(\mathbf{k})H$. It is shown by *Frishman* in his interesting study [7.12] that under large H_M the waves with the shift $\Delta_n(\mathbf{k}) = n\Omega/2$ (Ω is the frequency of the RF-field) are excited, if the pumping field exceeds the value $h_{\mathrm{th}}/|J_n(z)|$. Here $J_n(z)$-is the Bessel function of the argument $z = -U(\mathbf{k})H_M/\Omega$. Hence the threshold of the parametric instability $h_{\mathrm{th}}(H_M,\Omega) = h_{\mathrm{th}}(0)[\max_n |J_n(z)|]^{-1}$ is a complex non-monotonic function of H_M and Ω. This non-trivial conclusion of the theory is in good quantitative agreement with the experimental data of *Ozhogin* et al. [7.13] on the parametric excitation of spin waves in ferromagnets.

In Chap. 9 the experimental data on the parametric excitation of magnons under conditions of the periodic RF-modulation of the magnetic field H_M supporting the above-developed theory will be discussed. Now we shall only note that this theory is applicable not only to spin waves. It describes also the excitation of waves of different nature whose frequency is periodically changed with the amplitude $A_M = U(\mathbf{k})H_M$.

7.4 Large-Amplitude Collective Oscillations and Double Parametric Resonance

7.4.1 Stationary State Under Periodic Modulation

It would be natural to assume that the periodic modulation of the magnetic field does not qualitatively change the processes resulting in the limitation of the parametric instability. They are, however, accompanied by some additional effects that can significantly change the stationary values of the total number N and phase Ψ of parametric wave pairs. Under the influence of the modulating RF-field collective oscillations of parametric waves are excited. They nonlinearly interact with the initial parametric waves and with the RF-field. This results in the increased damping of parametric waves and in the change in their interaction. The interaction processes of the type $H_\mathrm{M} c d^*$, (which are actually the confluence of parametric waves with the RF-photon resulting in the excitation of the collective mode d) are responsible for the additional damping. Obviously, these processes lead to the addition to γ proportional to the square of the amplitude of the modulating field. Other interaction processes similar to $c^* c^* dd$ provide an additional coupling between parametric waves via the collective oscillations. These processes can be interpreted as the renormalization of the amplitudes of the four-magnon interaction $S(\boldsymbol{k}, \boldsymbol{k}')$. In order to allow for all these effects let us write an equation of motion of parametric waves taking into account their interaction with each other and the RF-field

$$\{\frac{\partial}{\partial t}+\gamma(\boldsymbol{k})+i[\Delta(\boldsymbol{k})+2T|c(t)|^2+UH_\mathrm{M}(t)]\}c(t) = -i[hV+Sc^2(t)]c^*(t). \quad (7.4.1)$$

It is the result of the averaging of the equations of motion for the amplitudes $c(\boldsymbol{k},t)$ and $c^*(\boldsymbol{k},t)$ over all the directions of the wave vector $\boldsymbol{k}$ for which in the ground state $c_0(\boldsymbol{k}) \neq 0$. It can be done without loss of generality since we are interested in the reaction of parametric waves to the homogeneous RF-field and we do not consider here the non-homogeneous modes of the collective oscillations. Assuming $H_\mathrm{M}(t) = H_\mathrm{M} \cos \Omega t$ we obtain from (7.2.4, 7, 8) the linear reaction of the collective oscillations $d(\pm\Omega)$ to the field H_M; substituting the expression for this reaction into (7.4.1) we obtain the equation for the stationary amplitude c_0:

$$\gamma\{1+[UH_\mathrm{M}\Omega/\Delta(\Omega)]^2\} + i\{\Delta(\boldsymbol{k}) + [2T+S+(S\Delta_0^2+2T\Omega^2 \\ +3S\Omega^2)U^2 H_\mathrm{M}^2/2\Delta^2(\Omega)]N + hV\exp(i\Psi) = 0 \; . \quad (7.4.2)$$

Here $\Delta(\Omega)$ and Δ_0 are given by (7.2.8) and $\Delta(\boldsymbol{k})$ denotes frequency shift. The position of the surface in $\boldsymbol{k}$-space where parametric waves in stationary state are excited, i.e. the value $\Delta(\boldsymbol{k})$ must be obtained from the condition of the stability of the solution (7.4.2) with respect to small perturbations δ_c outside the stationary surface, i.e. from the condition of the *external stability*. The reader can find it independently solving the Problem 7.5. Substituting the answer (7.4.11) into (7.4.2) it is possible to represent the latter in the following form:

$$hV\exp(i\Psi) + \tilde{S}N = i\Gamma, \quad \Gamma = \gamma[1 + (UH_\mathrm{M})|/\Delta(\Omega)|^2], \tag{7.4.3}$$

$$\tilde{S} = S\left\{1 + 2\frac{(UH_\mathrm{M})^2(3\Omega^2 + 4\gamma^2)(\Omega^2 - 4S^2N^2)}{2(\Omega^2 + 4\gamma^2)|\Delta(\Omega)|^2}\right\}. \tag{7.4.4}$$

The form of (7.4.3) coincides with the form of the equation for the stationary state of the basic S-theory (5.5.6) if in the latter equation we substitute $\gamma \to \Gamma$, $S \to \tilde{S}$ according to (7.4.4). Therefore in the presence of the RF-field we must write instead of (5.5.7)

$$|\tilde{S}|N = \sqrt{|hV|^2 - \Gamma}, \quad hV\sin\Psi = \Gamma, \quad hV\cos\Psi + \tilde{S}N = 0. \tag{7.4.5}$$

These equations enable us to calculate the dependences of the nonlinear susceptibility χ' and χ'' on the supercriticality p and the magnitude of the RF-field H_M. These dependences, as shown by *Zautkin* et al. [7.10] are in qualitative agreement with the experimental results on the YIG sample.

An interesting work [7.13] by *Ozhogin* et al. theoretically and experimentally studied the parametric excitation of nuclear magnons in the antiferromagnetic $CsMnF_3$ under a modulating magnetic field. A good agreement of the conclusions of the theory linear in H_M with the experimental results below the supercriticality 7 dB and $H_\mathrm{M} \simeq 0.25$ Oe has been shown. In particular, the relaxation parameter of nuclear magnons obtained from the threshold of the parametric excitation by the formula $h_\mathrm{th}V = \gamma$ coincided with the quantity γ calculated from the frequency dependence using (7.3.10) to the accuracy of experimental corrections. It is significant that the last method of the determination of $\gamma(\boldsymbol{k})$ provides a higher absolute accuracy of measurements.

7.4.2 Parametric Excitation of CO of Parametric Wave System

In addition to the discussed linear interaction of the collective oscillations of parametric waves with the RF-field leading to resonance at the frequency $\Omega = \Omega_0$ there is also nonlinear interaction of the type $H_\mathrm{M}b^*b^*$ which results in the parametric resonance of the collective modes in the RF-field with the frequency $\Omega = 2\Omega_0$. As is easily seen, the instability of the initial mode d with the frequency Ω excited by the RF-field with respect to the decay into two modes b with the frequency $\Omega/2$ also refers to this effect. The action of both mechanisms when some critical amplitude of the RF-field is exceeded brings about the *double parametric resonance of parametric waves*, i.e. the simultaneous excitation of oscillations at SHF with the frequency $\omega_\mathrm{p}/2$ and at RF with the frequency $\Omega/2$.

To obtain the threshold of instability with respect to the parametric excitation of collective oscillations we seek the solution of the equation of motion (7.4.1) in the following form:

$$c(t) = c + d(\Omega)\exp(-i\Omega t) + d(-\Omega)\exp(i\Omega t) \\ + b(\Omega/2)\exp(-i\Omega t/2) + b(-\Omega/2)\exp(i\Omega t/2) \ . \quad (7.4.6)$$

By employing the relation (7.4.5) for the ground state and linearizing (7.4.1) with respect to the small amplitudes b and d we obtain

$$[\partial/\partial t + \gamma + 2i(T+S)N]b_+ + [(2iTN - \gamma)\exp(-i\Psi)]b_-^* \\ + i\omega_+ b_- + [iP_+ \exp(-i\Psi)]b_+^* = 0 \ , \qquad b_\pm = b(\pm\Omega/2, t) \ , \\ \omega_+ = U^* H_M^*/2 + 2(2T+S)(c_0 d_+ + c_0 d_-) \ , \qquad \omega_- = \omega_+^* \ , \\ P_\pm = 2(2T+S)c_0 d_\pm \exp(i\Psi) \ , \qquad d_\pm = d(\pm\Omega, t) \ . \quad (7.4.7)$$

It should be recalled that the oscillation amplitudes $d_\pm$ are specified by (7.2.7, 8). Equation (7.4.7) can be written in a more symmetrical form if we pass from the variables $b_\pm$ to the normal variables a via the following u-v-transformation

$$a_+ = ub_+ \exp(i\Psi/2) - ub_-^* \exp(-i\Psi/2) \ , \\ a_-^* = -vb_+ \exp(i\Psi/2) + u^* b_-(-i\Psi/2) \ , \\ u = \sqrt{\frac{2(T+S)N + \Omega_0}{2\Omega_0}} \ , \quad v = -\sqrt{\frac{2(T+S)N - \Omega_0}{2\Omega_0}} \ . \quad (7.4.8)$$

Then for the renormalized $a_\pm$ we have

$$(\partial/\partial t + \gamma + i\Omega_0)a_+ + iA_+ a_+^* + iB_+ a_- - \gamma a_-^* = 0 \ , \\ A_\pm = 2uv\omega_\pm + u^2 P_+ + v^2 P_-^* \ , \\ B_\pm = v^2 \omega_\pm + uv(P_+ + P_-^*) \ , \quad C_\pm = \nu + \gamma + i(\Omega_0 \pm \Omega/2) \quad (7.4.9)$$

Assuming $a_\pm(t) = a_\pm \exp(+i\Omega t/2 + \nu t)$ and writing the system of four equations for the amplitudes a_+, a_+^*, a_- and a_-^* we find the instability increment ν from the condition

$$\begin{bmatrix} C_+ & iA_+ & iB_+ & -\gamma \\ -iA_+^* & C_+^* & -\gamma & -iB_+^* \\ -iB_+ & -\gamma & C_- & iA_- \\ -\gamma & -iB_+ & -iA_-^* & C_-^* \end{bmatrix} = 0 \ . \quad (7.4.10)$$

The instability threshold ($\nu = 0$) for the given values Ω and Ω_0 was calculated on a computer using (7.4.10) (Fig 7.1). The minimum threshold has been proved to be reached at the resonance $\Omega = 2\Omega_0$ and to be only weakly dependent on Ω: $\min\{H_{M,\text{th}}\} \simeq 0.4\gamma/v \simeq \gamma/g$. The second minimum of the threshold at the frequency $2\Omega_0$ manifests itself only at a sufficient Q-factor of collective oscillations when $\gamma \ll \Omega_0$.

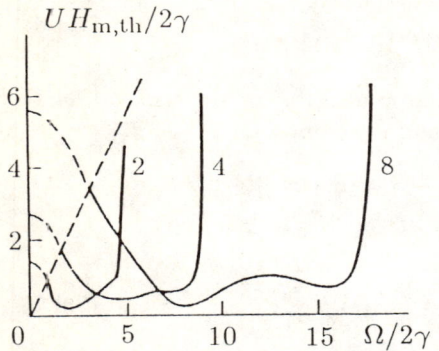

Fig. 7.1. Theoretical dependences (obtained by numerical solution of (7.4.10)) of the threshold of double parametric resonance $H_{\mathrm{m,th}}$ on the RF-frequency. The numbers of the curves (2, 4 and 8) denote the value of dimensionless eigenfrequency of collective oscillations $\Omega_0/2\gamma$. The applicability region of the theory is limited from above by a dashed line

The parametric excitation of collective oscillations in the system of parametrically excited magnons, i.e. the *double parametric resonance* was discovered and experimentally studied by *Zautkin* [7.10, 14]. These results will be described in Chap. 9.

7.5 Transient Processes when Pumping is Turned On

7.5.1 Small Supercriticality Range

This subsection deals with the transition to the S-theory steady stationary state starting from the level of the thermal noise $n(\boldsymbol{k}) = n_0$ under small supercriticality $hV - \gamma \ll \gamma$. For increasing small N up to $N \ll N_1$ (N_1 is the total number of parametric waves in the ground state) the amplitudes of the pairs will increase as predicted by the linear theory, i.e. with the increment $hV - \gamma$. At the same time a narrow packet of parametric waves with the width $\Delta\omega \simeq (hV - \gamma)$ emerges in the $\boldsymbol{k}$-space. Its subsequent behavior is described by (5.4.13) from which it is clear that the relaxation times of the amplitudes and the phases are of the order of $(hV - \gamma)^{-1}$ and γ^{-1}, respectively. As a result the phases $\Psi(\boldsymbol{k})$ can be considered to follow the the amplitudes adiabatically, i.e. $\partial\Psi/\partial t$ in (5.4.13) can be neglected. The fact that the packet is narrow $\Delta\omega \ll \gamma$ enables one to expand the trigonometrical functions in (5.4.13) into a series and to represent these equations in the following form (for more detail, see [7.15]):

$$-\frac{\partial f(x_1,\tau)}{f(x_1,\tau)\partial \tau} = x^2 + x(2r-1)\left(\int f(x_1,\tau)dx - 1\right) + r\int f(x_1,\tau)dx_1$$
$$-r + r^2\left[\int f(x_1,\tau)dx_1 - 1\right]^2 + \int x_1 f(x_1,\tau)dx_1 \,. \qquad (7.5.1)$$

Here $f = n(\boldsymbol{k}, \tau)/N_1$, $x = [\omega(\boldsymbol{k}) - \omega(\boldsymbol{k}_0)]/2SN_1$, $\tau = t(2SN_1/\gamma)$ and $r = (2T + S)/S$. Note that the dependence on supercriticality $p = h^2/h_{\text{th}}^2$ in this equation disappeared and the parameters of parametric waves system enter into this equation only via the ratio r. Equation (7.5.1) has a self-similar solution

$$f(x,\tau) = A(\tau)\sqrt{\tau/\pi}\exp\{-[(x - x_0(t))/d(\tau)]^2\}\,,\qquad (7.5.2)$$

where $A(\tau) = \int f(x,\tau)dx$ is the total number of parametric waves, $x_0(\tau) = A^{-1}\int f(x,\tau)x\,dx$ is the position of the center of gravity and $d(\tau)$ denotes the width of the packet. They satisfy the equations

$$\frac{dA}{Ad\tau} + \frac{1}{2\tau} - \frac{d(\tau x_0^2)}{d\tau} = -Ax_0^2 - rA(A-1)[r(A-1)+1]\,,$$

$$\frac{d(\tau x_0)}{d\tau} = -(A-1)\left(r - \frac{1}{2}\right),\qquad d(\tau) = \frac{1}{\sqrt{\tau}}\,. \qquad (7.5.3)$$

These equations at $\tau \to \infty$ have the following asymptotic behavior

$$x_0 = -1/2\tau\,,\quad A - 1 = -[2(4r-1)\tau^2]^{-1}\,,\qquad d(\tau) = 1/\sqrt{\tau}\,. \qquad (7.5.4)$$

Therefore not very high above the threshold the arbitrary distribution function of the pairs $n(\boldsymbol{k})$ relaxes to the stationary state (5.4.13) having the form of the δ-function by the power laws of (7.5.4)

$$N_1 - N \propto N_1/[(hV - \gamma)t]^2\,,$$
$$[\omega(\boldsymbol{k}) - \omega(\boldsymbol{k}_0)] \propto \sqrt{\gamma/(hV - \gamma)}/t\,, \qquad (7.5.5)$$
$$\omega^2(\boldsymbol{k}) - [\langle\omega(\boldsymbol{k})\rangle]^2 \propto \sqrt{\gamma[(hV - \gamma)]}/t\,.$$

To study the non-stationary behavior of the system in detail under the arbitrary hV, S and T Equations (5.4.13) were numerically solved. Fig. 7.2 shows the distribution functions $n(\boldsymbol{k}, t)$ for two successive times and values of the parameter $r = 1/6$ and $r = 1$. The homogeneous initial distribution was selected, $hV = 1.4\gamma$. First the waves increase exponentially, the maximum of $n(\boldsymbol{k}, T)$ is on the surface $\omega(\boldsymbol{k}) = \omega_\text{p}/2$, as follows from the linear theory. Here the form of the function $n(\boldsymbol{k}, t)$ is naturally independent of the parameter r. Later, when $N(t)$ is not small in comparison with N_1 the behavior of $n(\boldsymbol{k}, t)$ significantly depends on the ratio of S and T describing the interaction of waves. Thus at $T = 0$ the packet $n(\boldsymbol{k}, t)$ keeps increasing and converging, still remaining on the surface $\omega(\boldsymbol{k}) = \omega_\text{p}/2$. At $T = 0$ the packet increases and moves as a whole, deforming a little, to the surface $\omega(\boldsymbol{k}) + 2TN_1 = \omega_\text{p}/2$ (i.e. towards the larger or smaller k depending on the sign of T). When the maximum of the packet $n(\boldsymbol{k}, t)$ is near this surface the packet begins to converge asymptotically. Note that the shape of the curve $\ln n(\boldsymbol{k}, t)$ (see. Fig. 7.2) at large t is close to a parabola, which confirms that it tends to the self-similar solution (7.5.2), i.e. the Gaussian packet.

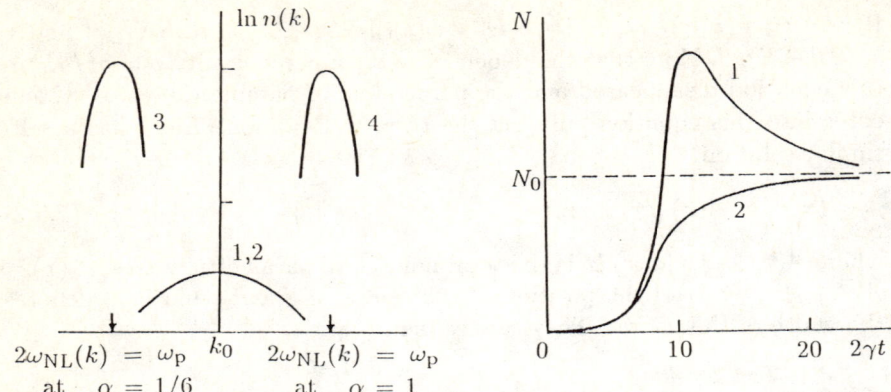

Fig. 7.2. (left) Time evolution of the parametric wave system – data of computer simulation of (5.4.13): distribution of waves $n(k)$ at $p = 2$. Linear stage: (1) at $r = 1/6$, (2) at $r = 1.0$. Asymptotic stage: (3) at $r = 1/6$, (4) at $r = 1.0$

Fig. 7.3. (right) Time dependence of the total number of parametric waves at $p = 2$. (1) at $r = 1/6$, (2) at $r = 1.0$

It can be seen from Fig. 7.3 that there are three transition stages to the steady stationary state from the thermal level when the parametric pumping is turned on. At the first – linear – stage the amplitude exponentially increases and the wave system does not depend on the nonlinear characteristics T and S. At this stage the two curves $N(t)$ (for $r = 1/6$ and r=1) coincide. At the second – nonlinear – stage the relative nonlinear frequency shift and the parametric interaction of wave pairs become significant. At the same time there is no compensation of wave damping $\gamma(\boldsymbol{k})$ by the total pumping $P(\boldsymbol{k})$ typical of the stationary state in the basic S-theory. The behavior of the system is most complex at this stage and can be simulated only on computer. At $T < 0$ the total number of waves N passes over the maximum and under $T > 0$ it monotonically increases. At the third – asymptotic – stage $|P(\boldsymbol{k})| - \gamma(\boldsymbol{k}) \ll \gamma(\boldsymbol{k})$ and the system slowly tends to the stationary state (5.5.7) described by the basic S-theory. Note that the sign of the difference $N - N_1$ in the numerical experiment (see Fig. 7.3) coincides with the sign that follows from the analytical asymptotics (7.5.4).

7.5.2 High Supercriticality Range

The numerical experiment revealed significant qualitative differences in the non-stationary behavior of the parametric wave system at small $(hV - \gamma \ll \gamma)$ and $(hV - \gamma) \gg \gamma$ supercriticalities (Fig. 7.4 and 7.5 for $hV = 4\gamma$) at the second – nonlinear – stage of the transition process. The packet $n(\boldsymbol{k}, t)$ does not behave like a single whole when the amplitude $N(t)$ is no longer small in comparison with N_1 and the second maximum of the function $n(\boldsymbol{k}, t)$

"emerges" at the point where $\omega_{\rm NL}(\boldsymbol{k},t) = \omega_{\rm p}/2$ at a given time. Then the amplitude of the second maximum increases and the amplitude of the first maximum decreases. On Fig. 7.4 at $\gamma t = 6$ the second maximum is already larger than the first. In addition, as seen from Fig. 7.5, the total amplitude oscillates as it approximates N_0. In the limiting case of the non-dissipative medium when $(\gamma/hV) \to 0$ the parametric wave system does not reach the stationary state at all and it oscillates infinitely long. This is due to the fact that at $\gamma = 0$ there exists the motion integral $\mathcal{H}_{\rm S}$ (5.4.17). At the initial time $\mathcal{H}_{\rm S} = 0$ and does not coincide with the value $\mathcal{H}_{\rm S} = -(2T + S)N_1$ in the ground state which therefore cannot be attainable at $\gamma = 0$.

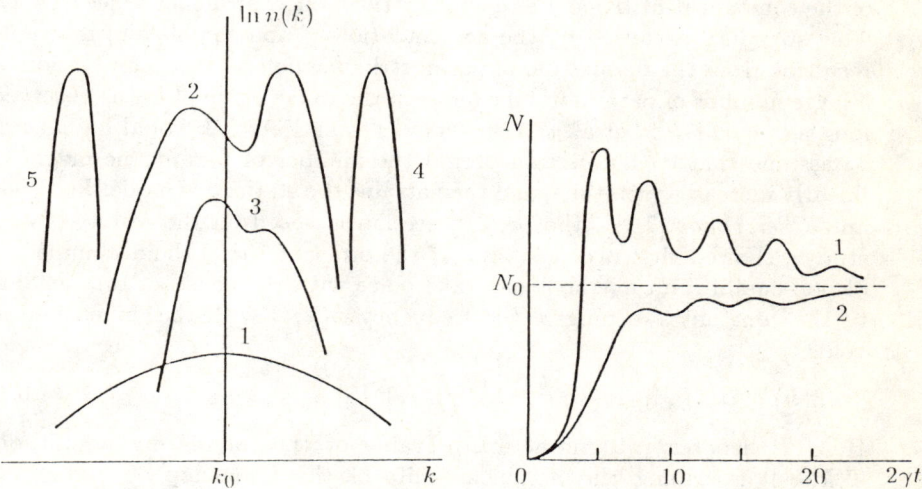

Fig. 7.4. (left) Time evolution of the parametric wave system – data of computer simulation of (5.4.13): distribution of waves $n(k)$ at p=16. Linear stage: (1) at $r = 1/6$, and $r = 1.0$. Nonlinear stage: (2) at $r = 1/6$, (3) at $r = 1.0$. Asymptotic stage: (4) at $r = 1/6$, (5) at $r = 1.0$

Fig. 7.5. (right) Time dependence of the total number of parametric waves at $p = 16$. (1) at $r = 1/6$, (2) at $r = 1.0$

At the last asymptotic stage, as can be seen from Fig. 7.4, the packet $n(\boldsymbol{k},t)$ is Gaussian. This should be expected because this stage is described by (7.5.1) whose solution (7.5.2) is self-similar and is the Gaussian packet. Indeed the validity criterion (7.5.1) is the smallness of the ratio $S(N - N_1)/\gamma \ll 1$. According to the asymptotics (7.5.4) $N(t)$ comes closer to N_0, from above or from below, as can be seen from Fig. 7.5 depending on the value of T/S. The cause of the oscillations $N(t)$ in the transition mode is obvious: it is the impact excitation of the collective oscillations. Their damping decrement is γ, and the frequency $\Omega_0 = \sqrt{2S(2T+S)}N_1 \simeq hV$ at $hV \gg \gamma$. Therefore the number of oscillations is of the order of hV/γ.

7.6 Parametric Excitation Under Sweeping of Wave Frequency

The analysis of the experimental data of *Zhitnik* and *Melkov* [7.16] show that under parametric excitation of magnons in YIG, a sweeping of the eigenfrequency of magnons $\omega(\boldsymbol{k})$ in the $\boldsymbol{k}$-space appears in some value range of the constant field H, i.e. there appears the monotonic dependence $\omega(\boldsymbol{k},t)$. This sweeping is caused by the accumulation of non-equilibrium magnons resulting from the dissipation of parametric magnons. The accumulation of a large number of parametric magnons in the range of small k was observed long ago by *Le-Hall* et al. [7.17], *Venitsky* et al. [7.18]. It would be natural to assume that over a certain period the number of parametric magnons linearly increases with time, and then attains the stationary mode. *Podivilov* and *Cherepanov* [7.19] (the present section is based on the results of this study), showed that over the time of the order of $10^2/\gamma$ the accumulation of non-equilibrium magnons can take place with the constant rate leading to the constant sweeping of the frequency $\omega(\boldsymbol{k},t)$ with the dimensionless velocity

$$\eta(\boldsymbol{k}) = \partial \omega(\boldsymbol{k},t)/2\gamma(\boldsymbol{k})\partial t \simeq 2T\sqrt{p-1}/|S|\ . \tag{7.6.1}$$

Here T denotes the characteristic value of the interacting amplitudes $T(\boldsymbol{k},\boldsymbol{k}')$ describing the nonlinear shift of the frequency of parametric magnons under the influence of non-equilibrium magnons. Note that since the values T and S, generally speaking, are of the same order, the estimation (7.6.1) reveals the possibility of considerable frequency sweeping, under which $\omega(\boldsymbol{k},t)$ changes by the value $\gamma(\boldsymbol{k})$ over the time $1/\gamma(\boldsymbol{k})$. Clearly, if this sweeping is taken into account this must lead to a significant change in the results of the basic S-theory, in particular, to the appearance of the finite width of parametric wave packet. The modification of the basic S-theory of parametric excitation under the linear dependence of their frequency on the time (linear sweeping of the frequency) will be considered in the present section.

7.6.1 Qualitative Analysis of the Problem

In this section we shall proceed from the following equations of motion

$$\begin{aligned}\frac{\partial n(\boldsymbol{k},t)}{2\partial t}&+\gamma(\boldsymbol{k})n(\boldsymbol{k},t)+\operatorname{Im}\{P^*(\boldsymbol{k},t)\sigma(\boldsymbol{k},t)\}=\gamma(\boldsymbol{k})n_0(\boldsymbol{k})\ ,\\ \frac{\partial\sigma(\boldsymbol{k},t)}{2\partial t}&+\{\gamma(\boldsymbol{k})+i[\omega_{\mathrm{NL}}(\boldsymbol{k},t)-\frac{\omega_{\mathrm{p}}}{2}]\}\sigma(\boldsymbol{k},t)=iP(\boldsymbol{k},t)n(\boldsymbol{k},t)\ .\end{aligned} \tag{7.6.2}$$

On the one hand, these equations can be considered to be the generalization of the non-stationary equations of the basic S-theory (5.4.11) (at $n(\boldsymbol{k},t) = n(-\boldsymbol{k},t)$) to the case of the non-zero temperature of the thermal bass. Unlike (5.4.11) Eqs. (7.6.2) contain the additional term $\gamma(\boldsymbol{k})n_0(\boldsymbol{k})$ describing the energy flux from the thermal bass to the system of parametric waves. In the absence of the pumping this term leads to the relaxation $n(\boldsymbol{k},t)$ not to zero but to a thermodynamic equilibrium distribution of $n_0(\boldsymbol{k})$. On the other hand, (6.7.2) are the generalization of the stationary equations (6.4.24) allowing for the interaction with the thermal bass to the non-stationary case. Unlike (6.4.24) Eqs. (7.6.2) contain additional terms with time derivatives $\partial n(\boldsymbol{k},t)/\partial t$ and $\partial \sigma(\boldsymbol{k},t)/\partial t$. In principle, using the methods described in Sect. 6.4 equations (7.6.2) can be derived beginning with (6.4.7). We shall not do that here but convince ourselves that the above comments are correct.

Let us proceed to the analysis of the basic equations (7.6.2). To this end, let us go over to the new variables $\varepsilon(\boldsymbol{k}) = [\omega_{\rm NL}(\boldsymbol{k}) - \omega_{\rm p}/2]$, i.e. the frequency shift, and $\Omega = (\Theta, \varphi)$, i.e. the polar and azimuthal angles of the wave vector $\boldsymbol{k}$. Using these variables (7.6.2) change over to

$$\frac{\partial n(\varepsilon, \Omega, t)}{2\partial t} + \gamma(\Omega)n(\varepsilon, \Omega, t) + {\rm Im}\{P^*(\Omega,t)\sigma(\varepsilon,\Omega,t)\} = \gamma(\Omega)n_0 \ ,$$

$$\frac{\partial \sigma(\varepsilon, \Omega, t)}{2\partial t} + \{\varepsilon(\Omega) + i[\varepsilon + g(t)]\}\sigma(\varepsilon,\Omega,t) \qquad (7.6.3)$$

$$+ iP(\Omega,t)n(\varepsilon,\Omega,t) = 0 \ .$$

$$g(t) = \omega_{\rm NL}(\boldsymbol{k},t) - \omega(\boldsymbol{k}) = \alpha t + c \ . \qquad (7.6.4)$$

Here $g(t)$ is the stationary sweeping (departure) of the eigenfrequency. In the absence of the sweeping (at $g(t) = 0$) (7.6.3) have the single solution

$$n(\varepsilon, \Omega) = n_0(\varepsilon)\{1 + |P(\Omega)|^2/[\varepsilon^2 + \gamma^2(\Omega) - |P(\Omega)|^2]\} \ . \qquad (7.6.5)$$

In the limit $n_0 \to 0$ this solution goes over to the solution of the basic S-theory which satisfies the condition of the external stability (5.5.4). Indeed, $n(\varepsilon, \Omega)$ is at maximum on the resonance surface $\varepsilon = 0$ and in the limit of small n_0 it can differ from zero only at the points where $|P(\Omega)| = \gamma(\Omega)$. On the remaining part of the surface the condition $|P(\Omega)| < \gamma(\Omega)$ must be satisfied, otherwise the contradiction may arise: $n(\Omega) < 0$. At $|P(\Omega)| > \gamma(\Omega)$ the solution of the system (7.6.3) increases exponentially with the increment

$$\nu(\Omega) = -\gamma(\Omega) + \sqrt{|P(\Omega)|^2 - \varepsilon^2} \ . \qquad (7.6.6)$$

The instability region with respect to the frequencies $\omega(\boldsymbol{k})$ is determined by the inequality $\nu(\Omega) > 0$. Hence we have

$$\varepsilon < \sqrt{|P(\Omega)|^2 - \gamma^2(\Omega)} \ . \qquad (7.6.7)$$

Under linear sweeping of frequency the packet of parametric waves can be such that its frequency varies with the constant velocity and the form of the packet is unchanged. In other words (7.6.3, 4) must have stationary solutions of the following form:

$$n(\varepsilon, \Omega, t) = n(\varepsilon + \alpha t + c, \Omega) . \qquad (7.6.8)$$

Let us qualitatively analyze this solution. Under sweeping at any time some waves enter the instability range near the resonance surface and others leave it. Incoming waves increase exponentially with the increment $\nu(\Omega, t)$ (7.6.6) beginning with the level of the thermal fluctuations. They pass the instability region (7.6.7) over the time

$$\tau_0 = 2\sqrt{|P(\Omega)|^2 - \gamma^2(\Omega)}/|\alpha| \simeq 2\sqrt{2\gamma(|P| - \gamma)}/|\alpha| , \qquad (7.6.9)$$

and at the same time their level increases up to

$$\begin{aligned} n_{\max} &= n(\Omega, \tau_0) = n_0 \exp\left[2\int_0^{\tau_0} \nu_0(t)dt\right] \\ &\simeq n_0 \exp[4\sqrt{2\gamma}(|P| - \gamma)^{3/2}/|\alpha|] . \end{aligned} \qquad (7.6.10)$$

Subsequently, these waves leave the instability region and damp over the time τ_1 with the average decrement $\nu(\tau_1) \simeq (\alpha\tau_1)^2/2\gamma$. Hence and from (7.6.10) we obtain the estimation for τ_1:

$$\tau_1 = \sqrt{\gamma/|\alpha|}\sqrt{|P|^2 - \gamma^2} . \qquad (7.6.11)$$

Far from the instability region the sweeping can be neglected and according to (7.6.5) $[n(\Omega) - n_0] \propto \varepsilon^{-2}$. The maximum intensity of the packet is given by (7.6.10) and is on the back boundary of the instability region (7.6.7)

$$\Delta\omega(\boldsymbol{k}) \simeq \sqrt{|P|^2 - \gamma^2}\alpha/|\alpha| . \qquad (7.6.12)$$

The packet width δ is determined by the time over which parametric waves decrease e times passing the boundary of the instability region:

$$\delta^2 \simeq |\eta|\gamma^3/\sqrt{|P|^2 - \gamma^2} , \qquad \eta = \alpha/2\gamma^2 . \qquad (7.6.13)$$

When the condition of the adiabatic slowness of the sweeping is satisfied

$$\eta < (|P|^2 - \gamma^2)^{3/2}/\gamma^3 , \qquad (7.6.14)$$

under which the above considerations hold true, the packet width (7.6.13) is much less than its shift $\Delta\omega(\boldsymbol{k})$ (7.6.12). The total number of waves $N(\Omega)$ can be readily estimated

$$N(\Omega) \simeq n_{\max}\delta \simeq n_0 \exp\left[\frac{4(|P|^2 - \gamma^2)^{3/2}}{3|\alpha|\gamma}\right] . \qquad (7.6.15)$$

7.6 Parametric Excitation Under Sweeping of Wave Frequency

The pre-exponential factor in (7.6.15) is not defined at our level of rigor. For us it is only important that in the absence of thermal noise ($n_0 = 0$) the number of parametric waves $N(\Omega) = 0$. Note also that (7.6.15) is not the complete solution of the problem, since to determine the number of parametric waves one must self-consistently obtain the renormalized pumping $P(\Omega)$.

In order to qualitatively analyze the number and the phase of parametric waves above the threshold it is convenient to make use of the condition of the equality of the pumped and dissipated energy fluxes

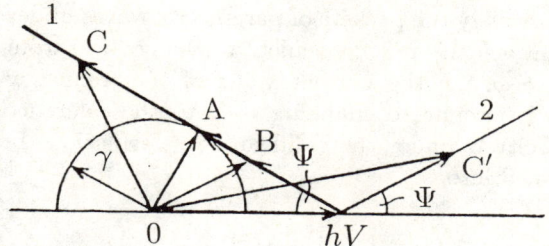

Fig. 7.6. Diagram of the pumping vectors: hV is the external pumping, $S\sigma$ is the pumping arising due to four-wave interaction, vector OA is the self-consistent total pumping without sweeping, vectors AB and AC, AC' are the same at positive and negative sweeping, γ is damping and Ψ is the pair phase. Line (1) corresponds to the continuous modification of the basic S-theory, line (2) represents the additional solution at strong negative sweeping

$$hV \sin \Psi = \gamma , \qquad (7.6.16)$$

where Ψ is the phase difference of the anomalous correlator of parametric waves and the pumping. Therefore on the vector diagram (Fig. 7.6) the vector $S\sigma = \sum_{k'} S(k, k')\sigma(k')$ is inclined at the angle $\Psi = \arcsin(\gamma/hV)$ with they pumping P (for definiteness, we assume $S > 0$). The sum $hV + S\sigma$ is the renormalized pumping. Let us find now the angle between P and $S\sigma$. From (7.6.3) it follows that in the stationary case

$$\sigma(k) = -iP(k)n(k)/\Delta(k) , \qquad \Delta(k) = \gamma(k) + i[\omega_{\mathrm{NL}}(k) - \omega_{\mathrm{p}}/2] ,$$

$$\mathrm{Re}\{P^*(k)\sigma(k)\} = -|P(k)|^2[\omega_{\mathrm{NL}}(k) - \omega_{\mathrm{p}}/2]n(k)/|\Delta(k)|^2 . \qquad (7.6.17)$$

If the frequency sweeping is not too fast the packet is narrow with respect to the frequencies, therefore in the center of the packet

$$|P(k)|^2 = \gamma^2(k) + (\Delta\omega)^2 , \qquad \mathrm{Re}\{P^*(k)\sigma(k)\} = -\Delta\omega n(k) . \qquad (7.6.18)$$

Since under positive sweeping ($\Delta\omega > 0$), the angle between $S\sigma$ and P is obtuse. Therefore the quantity is less than its values in the absence of the sweeping, i.e. $|S\sigma| < \sqrt{|hV|^2 - \gamma^2}$. In the case of negative sweeping ($\Delta\omega < 0$, the angle between $\boldsymbol{P}$ and $S\boldsymbol{\sigma}$ is acute and, consequently, $S\sigma > \sqrt{|hV|^2 - \gamma^2}$ at any sweeping velocity. Strictly speaking, under negative sweeping the condition (7.6.16) can be satisfied also when the vector $S\boldsymbol{\sigma}$ is inclined to hV at the angle $(\pi - \text{arctg}\,\gamma/hV)$ (straight line (2)). For such a solution to exit the sweeping must be sufficiently large. This additional solution is, however, unstable, as it is shown below.

7.6.2 Basic Equations of S-Theory Under Frequency Sweeping

In order to investigate quantitatively the packets of parametric waves under the sweeping with the constant velocity it is convenient to change over from the variable ε to the variable ω, i.e. the current value of the frequency detuning $\omega = \varepsilon + \alpha t$. This is similar to changing over to the reference system moving with the velocity α under the Galilean transformation. In this case (7.6.3) are transformed into

$$\left[\frac{\alpha\partial}{2\partial\omega} + \gamma(\Omega)\right] n(\omega, \Omega) + \text{Im}\{P^*(\Omega)\sigma(\omega, \Omega)\} = \gamma(\Omega)n_0 ,$$

$$\left[\frac{\alpha\partial}{2\partial\omega} + \gamma(\Omega) + i\omega\right] \sigma(\omega, \Omega) + iP(\Omega)n(\omega, \Omega) = 0 , \qquad (7.6.19)$$

$$\left[\frac{\alpha\partial}{2\partial\omega} + \gamma(\Omega) - i\omega\right] \sigma^*(\omega, \Omega) - iP^*(\Omega)n(\omega, \Omega) = 0 .$$

The stationary sweeping of the packet is described by the solutions (7.6.19) which do not explicitly depend on time. Apparently they must satisfy the boundary conditions

$$[n(\omega, \Omega) - n_0] \to 0 , \qquad \sigma(\omega, \Omega) \to 0 , \qquad \sigma^*(\omega, \Omega) \to 0 \qquad (7.6.20)$$

at $\omega \to \pm\infty$. The stationary solutions of the homogeneous system (7.6.19) at $n_0 = 0$, $|\omega| \to \infty$ have the following asymptotic behavior

$$\text{Im}\{P^*(\Omega)\sigma(\omega, \Omega)\} \propto \frac{1}{\omega^3} \exp\left[\frac{(-\gamma\omega \pm i\omega^2)}{\alpha}\right] . \qquad (7.6.21)$$

None of these solutions tends to zero at $\omega/\alpha \to \pm\infty$, therefore no linear combination of the solutions satisfies the boundary conditions (7.6.20). Physically, this means that in the absence of thermal noise the parametric excitation of waves under frequency sweeping is impossible. Therefore to determine the packet of parametric waves under sweeping we must find the regular stationary solution of the non-homogeneous system (7.6.19) satisfying the boundary conditions (7.6.20). The requirement that the solution should be regular is equivalent to the condition of the external stability in the absence of sweeping.

Under stationary frequency sweeping (7.6.19) can be transformed to the dimensionless form (7.6.24) introducing the following new variables:

$$x(\Omega) = \omega/\gamma(\Omega), \quad \eta(\Omega) = \alpha/2\gamma^2(\Omega), \quad \Pi(\Omega) = P(\Omega)/\gamma(\Omega),$$
$$U(x) = \mathrm{Re}\{\Pi^*(\Omega)\sigma(x)\}, \quad V(x) = -\mathrm{Im}\{\Pi^*(\Omega)\sigma(x)\}, \quad (7.6.22)$$

where $m(x) = n(x) - n_0$. Note that the angle Ω in (7.6.19) is a parameter and the dependence on this angle can be omitted when the eigenfrequency shape of the packet is studied at fixed Ω. It is convenient to perform also the Fourier transform which enables us to reduce the order of the differential operator and to make one more substitution of a variable

$$f(x) = \int f(k)\exp(ikx)dk, \quad z = (1+i\eta k)/i\eta. \quad (7.6.23)$$

These allow to express (7.6.19) in the form:

$$\frac{4\partial^2 V}{\partial z^2} - \left(1 - \frac{i|\Pi|^2}{\eta z}\right)V = 2i|\Pi|^2 \frac{n_0(2z+i)}{4\eta z}. \quad (7.6.24)$$

The solutions of this equation are the Whittaker functions with the index $m = 1/2$ [7.20]: $W_{y,1/2}(z)$ and $W_{-y,1/2}(z)$:

$$W_{y,1/2}(z) = \left[\exp\frac{-z/2}{\Gamma(1-y)}\right]\int [\frac{1+z}{t}]^y \exp(-t)\,dt. \quad (7.6.25)$$

where $y = -i|\Pi|^2/4$ and $\Gamma(x)$ is the gamma-function. This solution of (7.6.24) was obtained by *Podivilov* and *Cherepanov* [7.19].

7.6.3 Solution of S-Theory Equations

Note that below the threshold of the parametric instability and under a low rate of sweeping the ordinary below-threshold heating of the wave system takes place. For the adiabatic sweeping when the condition (7.6.14) is satisfied we have

$$n(\Omega) = \frac{\pi|\Pi(\Omega)|^2 n_0}{\sqrt{|\Pi(\Omega)|^2 - 1}} \exp\frac{2[|\Pi(\Omega)|^2 - 1]^{3/2}}{3\eta(\Omega)]}. \quad (7.6.26)$$

This case was quantitatively discussed in Sect. 7.6.1 and the solution (7.6.26) here qualitatively coincides with the estimation of (7.6.15) obtained earlier. Under non-adiabatic sweeping when $|\eta\Pi| > (|\Pi|^2 - 1)^{3/2}$ the number of parametric waves is limited by the level close to the thermal level

$$n(\Omega) \simeq \pi|\Pi(\Omega)|^2 n_0(1 + |\eta|^{-1/3}). \quad (7.6.27)$$

As the sweeping velocity increases the integrated value $\mathrm{Re}\{P^*\sigma\}$ decreases as $|\eta|^{-1}$. There is one more important case when under fast sweeping ($\eta \gg$

1, $|\Pi|$) the number of parametric waves is large. This is the case of high supercriticality $|\Pi|^2 \gg \eta > 1, |\Pi|$. In this case:

$$n(\Omega) = \frac{4k^2(\Omega)\gamma(\Omega)\sqrt{\eta(\Omega)}}{v(\Omega)|\Pi(\Omega)|^2} \exp \frac{|\Pi(\Omega)|^2}{2\eta(\Omega)} ,$$

$$\mu(\Omega) = 2\frac{|\Pi(\Omega)|^2}{\eta(\Omega)} \ln \frac{\eta(\Omega)}{\Pi(\Omega)} . \tag{7.6.28}$$

Now let us proceed to the problem of the shape of the parametric wave packet. In Sect. 7.6.1 it was qualitatively shown that when the condition of the adiabatic sweeping (7.6.14) satisfied the packet width of parametric waves (7.6.13) is much less than its shift from the resonance $\Delta\omega(\mathbf{k})$ (7.6.12). The exact expressions for δ and $\Delta\omega(\mathbf{k})$ coincide with these estimations with the accuracy up to the constant. When the sweeping velocity is high and the supercriticality is large we can obtain

$$\begin{aligned}\Delta\omega(\mathbf{k}) &= -|P|^2 L/2\alpha , \quad \delta^2 = |P|^2[1 + L/4] , \\ L &= \ln\{\alpha^2/\gamma^2[|P|^2 + 2\alpha\gamma\exp(-CP)]\} .\end{aligned} \tag{7.6.29}$$

Here $C = 0.577...$ is the Euler constant. Clearly, as under adiabatic sweeping, the condition $\Delta\omega|(\mathbf{k})| \gg \gamma(\mathbf{k})$ is satisfied. Strictly speaking, this is the specific part of the S-theory connected with the sweeping. Further we deal with the standard procedure of the self-consistency of the number of parametric waves with the amplitude of the renormalized pumping P.

7.6.4 Dependence of the Number of Waves on the Pumping Amplitude

The simplest case to study is the one with spherical symmetry (e.g., easy-plane antiferromagnets) when in order to obtain the total number of parametric waves N it is sufficient to multiply (7.6.26) and (7.6.28) by 4π. In the case of the axial symmetry (characteristic of the cubic ferrimagnet YIG used in the experimental study of the frequency sweeping of parametric magnons) Eqs. (7.6.26, 28) can also be rather easily integrated with respect to the angles. Under $|\eta_0| \ll (|\Pi_0|^2 - 1)$ this yields

$$\begin{aligned}N &= N_T\sqrt{\frac{\pi\eta_0}{\beta}} \frac{|\Pi_0|^{3/2}}{[|\Pi_0|^2 - 1]^{9/16}} \exp\left|\frac{3[|\Pi_0|^2 - 1]^{3/2}}{\eta_0}\right| , \\ \mu &= \sqrt{|\Pi_0|^2 - 1}\operatorname{sign} g\eta , \quad N_T = 4\pi^2 k^2 n_0 \gamma/[\partial\omega(k)/\partial k] .\end{aligned} \tag{7.6.30}$$

Here N_T is the number of thermal waves in the layer of the width $\Delta\omega(\mathbf{k}) = \pi\gamma(\mathbf{k})$ near the resonance surface. If $|\Pi_0|^2 \gg |\eta_0| \gg 1, |\Pi_0|$ then

$$N = 2N_T \frac{|\eta_0 \Pi_0|}{\sqrt{\pi\beta}} \exp\left(\pi\frac{|\Pi_0|^2}{2|\eta_0|}\right) , \quad \mu = \left(2\frac{|\Pi_0|}{\eta_0}\right)\ln\left(\frac{|\eta_0|}{|\Pi_0|}\right) . \tag{7.6.31}$$

In these expressions Π_0 and η_0 are the values of $\Pi(\Omega)$ and $\eta(\Omega)$ on the resonance surface and the coefficient $\beta \simeq 2$ characterizes the angular dependence $\Pi(\Omega)$:

$$|\Pi(\Omega)|^2 = |\Pi_0| - \beta \cos \Omega , \quad \Omega = \Theta, \Psi . \tag{7.6.32}$$

It is clear that in different cases of symmetry, spherical and axial symmetry, the respective formulae are different only in the pre-exponential factor. Further transformation will lead only to an insignificant difference in the logarithmic factor. Therefore we shall subsequently confine ourselves to the case of axial symmetry experimentally studied by *Zautkin* et al. [7.21].

The next stage of the self-consistency procedure is to make use of the relationship between P and hV. From (5.5.6) we can obtain

$$|hV|^2 = |P|^2 + 2\mu\gamma|S|N + (SN)^2(1+\mu^2)/|P|^2 , \tag{7.6.33}$$

where $\mu = \text{Re}\{P^*\Sigma/\gamma N\}$. The solution of (7.6.33) has the following form:

$$SN_\pm = \frac{|P|^2}{1+\mu^2}\left[-\mu \pm \sqrt{\mu^2 + \frac{1+\mu^2}{|P|^2}(h^2V^2 - |P|^2)}\right] . \tag{7.6.34}$$

If the condition of adiabatic sweeping (7.6.14) is satisfied, the number of parametric waves N is large in comparison with the thermal noise and $\mu = \alpha\sqrt{|P|^2 - \gamma^2}/|\alpha|\gamma$. Then it follows from (7.6.34) that

$$|S|N_\pm = -\frac{\alpha}{|\alpha|}\sqrt{|P|^2 - \gamma^2} \pm \gamma\sqrt{p-1} , \quad p = (h/h_{\text{th}})^2 . \tag{7.6.35}$$

The simplest case for the analysis is $S\alpha > 0$. Then in (7.6.35) only the solution N_+ is meaningful. At the low level of the thermal noise when

$$\varepsilon = |S|N_T/\gamma \ll 1 \tag{7.6.36}$$

it follows from (7.6.35) and (7.6.30) that

$$|S|N_+ = \gamma\sqrt{p-1} - \frac{\alpha}{4\gamma^2}(\ln[\frac{(p-1)}{p}])^{1/3} . \tag{7.6.37}$$

The first term here corresponds to the well-known result of the S-theory. As the sweeping velocity increases, the numbers of parametric waves decreases. Under a very small supercriticality when $p - 1 < \eta^2$ (7.6.37) is inapplicable because the condition (7.6.14) is violated. In this case the number of parametric waves is not big in comparison with the number of thermal waves. Under high supercriticality the applicability of condition (7.6.14) is also violated. Then (7.6.31) must be used. The qualitative behavior of the dependence of N on the sweeping rate is retained (Fig. 7.7). The negative sweeping rate results in a much more complicated situation since both solutions (7.6.35) are meaningful. At not too large levels of noise and not too

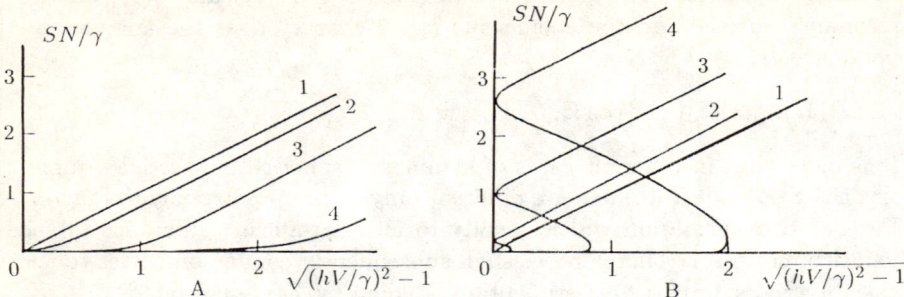

Fig. 7.7. Theoretical dependences of the total number of parametric waves N on the pumping amplitude h at different rates of sweeping: lines (1) correspond to $\eta = 0$, lines (2) to $|\eta| = 10^{-2}$, lines (3) to $|\eta| = 3 \cdot 10^{-2}$, lines (4) to $|\eta| = 0.1$. Fig. A corresponds to the positive sweeping, Fig. B to the negative sweeping

large sweeping rates there exists a single solution corresponding to the root of N_+ in (7.6.35) and described by (7.6.37). At $S\alpha \to 0$ this expression naturally coincides with the basic S-theory result. If the noise level is small and the sweeping rate is sufficiently big, there are three solutions of the wave number of N and the renormalized pumping $|\Pi|$. The smallest solution corresponds to the situation when the noise level is so low and the sweeping rate is so large that the amplitude of parametric waves over the time of passing the instability region has no time to increase significantly. The two other solutions correspond to high N, the intermediate value is unstable. Fig. 7.7 shows a set of dependences of the number of waves N on the pumping power p calculated for different values of the sweeping velocity α and the thermal noise amplitude $\varepsilon = 10^{-3}$. Clearly, there is a hysteresis of the dependence of the number of parametric waves on the pumping intensity under a fixed rate of sweeping. When the pumping intensity is less than a certain value, the critical level of parameter waves is close to the thermal level. Under $p > p_1$ where

$$p_1 \simeq 1 + [(3/4)(|\alpha|/\gamma^2)\ln(|\alpha|/4\gamma^2\varepsilon)]^{3/2} , \qquad (7.6.38)$$

two stable states of parametric waves can exist. As the pumping intensity increases under $p > p_2$ where

$$p_2 = 1 + [(3/4)(|\alpha|/\gamma^2)|\ln \varepsilon|]^{2/3} , \qquad (7.6.39)$$

the single state remains and under $\alpha \to 0$ changes over to the standard solution of the basic S-theory. The number of waves in this solution is close to the well-known expression $|S|N = \gamma\sqrt{p-1}$. It must be emphasized that the number of parametric waves for the solution (7.6.37) corresponding to the basic S-theory in the absence of sweeping indefinitely increases under increasing sweeping velocity and under fixed pumping intensity.

Concluding, the expression for the frequency of the homogeneous collective oscillations for the case of the adiabatic sweeping is presented

$$\Omega_\pm = -i\gamma \pm \sqrt{4[SN + \Delta\omega(\boldsymbol{k})](2T+S)N - \gamma^2} \,, \qquad (7.6.40)$$

where $\Delta\omega(\boldsymbol{k})$ is the shift of the packet of parametric waves specified by (7.6.12). It must be noted that expressions (7.6.40) were obtained by *Podivilov* and *Cherepanov* by awkward but rather accurate calculations [7.19]. The frequency [7.6.40] can also be obtained from the simple equations of the basic S-theory without frequency sweeping for the narrow packet of parametric waves localized on the surface $\omega_{\mathrm{NL}}(\boldsymbol{k}) = \Delta\omega(\boldsymbol{k}) + \omega_{\mathrm{p}}/2$. Incidentally, the same is true of the expression for the general number of parametric waves N and their phase Ψ. Thus in the cases when the number of parametric waves N is large in comparison with N_T the role of the frequency sweeping mainly amounts to the modification of the condition of external stability; instead of $\omega_{\mathrm{NL}}(\boldsymbol{k}) = \omega_{\mathrm{p}}/2$ we obtain $\omega_{\mathrm{NL}}(\boldsymbol{k}) = \Delta\omega(\boldsymbol{k}) + \omega_{\mathrm{p}}/2$ where the position of the packet center $\Delta\omega(\boldsymbol{k})$ is given by the expressions (7.6.12) and (7.6.29).

Later, in Chap. 9, we shall compare the theory developed in this chapter with the experimental data by *Zautkin* et al. [7.21]. A full qualitative and good quantitative agreement between theory and experiment was established.

7.7 Problems

Problem 7.1. Take into account the influence of the interaction between the resonator and the SHF-magnetic field of the pumping on the frequency of collective oscillations of parametric waves excited by parallel pumping. Investigate the instability region of the coupled oscillations.

Problem 7.2. Obtain the expressions for the frequencies of the coupled collective oscillations of parametric magnons and homogeneous precession under transverse pumping: first, for the processes of the first order far from the resonance of the homogeneous precession and then for the second-order processes at the resonance of the homogeneous precession.

Problem 7.3. By making use of the Hamiltonian of the interaction of magnons with the magnetic field (4.3.18) and (4.3.19) obtain expressions (7.2.4) for the "external force" acting on the system of parametric magnons (7.2.1) in the presence of radio-frequency and SHF magnetic fields.

Problem 7.4. Obtain the expression of the Hamiltonian of magnon–sound interaction in the easy-plane antiferromagnets. (*Lutovinov* [7.6])

Hint. Use the expression for the magnetoelastic energy and the canonical variables for the sound and magnons in easy-plane antiferromagnets.
Answer: See (7.2.6).

Problem 7.5. Obtain the condition of the external stability of the stationary state of parametric magnons (7.4.2) under radio-frequency modulation $H_M(t) = H_M \cos \Omega t$:
Answer:

$$\Delta(k) = \omega_{NL}(\boldsymbol{k}) - \omega_p/2 = \frac{|U(k)H_M|^2}{|\Delta(\Omega)|^2}\left\{T(\Omega^2 + 4S^2N^2)\right.$$
$$\left. + 4S\left[\gamma^2 + 2S^2N^2 + \frac{(2\gamma SN)^2}{\Omega^2 + 4\gamma^2}\right]\right\}. \tag{7.7.1}$$

Problem 7.6. Obtain the expression for the frequencies of the spatially non-homogeneous collective oscillations $\Omega(\boldsymbol{\kappa})$ ($\boldsymbol{\kappa}$ is the wave vector of collective oscillations) under the parametric excitation of waves in the case of spherical symmetry: $V(\boldsymbol{k}) = V$, $S(\boldsymbol{k}, \boldsymbol{k'}) = S$, $T(\boldsymbol{k}, \boldsymbol{k'}) = T$. This case is realized in easy-plane antiferromagnets, e.g., $CsMnF_3$, $MnCO_3$, etc.
Hint. For the analysis use (7.1.8)
Answer

$$\Omega(\kappa) = -i\gamma + \sqrt{(v\kappa)^2 \coth^2(v\kappa/\Delta) - \gamma^2},$$
$$\Delta = 4S(2T+S)N^2, \quad v = \partial\omega(k)/\partial k. \tag{7.7.2}$$

Problem 7.7. Obtain the expression for the frequencies of the spatially non-homogeneous collective oscillations $\Omega(\boldsymbol{\kappa})$ in the case of axial symmetry when parametric waves are excited only on the equator of the resonance surface and the vector $\boldsymbol{\kappa}$ is in the plane of the equator.

Problem 7.8. Find the instability increment (frequencies $\Omega_{\pm 1}(\kappa)$) of the spatially non-homogeneous mode of the collective oscillations with the axial number $m = \pm 1$ under the parametric excitation of magnons by parallel pumping in cubic ferromagnets when magnons are excited only on the equator of the resonance surface. Analyze (analytically and on the computer) the nonlinear stage of this instability in some limiting cases. Will this leads to a stationary but spatially homogeneous distribution of parametric magnons? If this is not the case, what type of self-oscillations will arise in the system?

Note. These questions have not yet been answered but are of great interest for the interpretation of the experimental data on the ferrimagnetic YIG when the mode instability $m = \pm 1$ can take place. It is a good challenge. In case of success, please publish the results!

Problem 7.9. Obtain the time dependence of the absorbed intensity at the linear development stage of the parametric instability when the pumping is turned on instantaneously.

Hint. Use (7.6.2) in the linear approximation

Answer:

$$W_+(t) = \pi\omega_{\rm p} \int d\Omega \gamma(\Omega) hV(\Omega) k^2(\Omega) n_0 F[\Omega, 2hV(\Omega)t]/2v(\Omega) ,$$

where $F(\Omega,\tau) = \int_0^\tau I_1(x)\exp[-\gamma(\Omega)x/hV(\Omega)]dx$ and $I_1(x)$ is the Bessel function of the imaginary argument (*Cherepanov*, unpublished).

8 Secondary Parametric Wave Turbulence

The present chapter deals with the non-stationary processes accompanying the parametric excitation of waves under stationary conditions when the pumping amplitude and other external parameters of the experiment are time independent. Under these conditions the system of interacting parametrically excited waves is in the flux equilibrium. Then the energy from the pumping W_+ is equal to the dissipation rate of the energy W_-: $W_+ = W_+ = W$. By the presence of the energy flux through the system the flux equilibrium differs from the thermodynamic equilibrium at which $W_+ = W_- = 0$.

The behavior of nonlinear systems in the state of the flux equilibrium is much more complicated than in the state of thermodynamic equilibrium. Thus at the thermodynamic equilibrium the average characteristics of the systems (e.g. occupation numbers) are time independent, i.e. stationary. In contrast, at the flux equilibrium as the energy flux through the system increases the stationary state under a certain $W = W_{\text{cr}}$ is no more stable. At $W \gg W_{\text{cr}}$ the behavior of these systems is usually chaotic and then their parameters (including $W(t)$) randomly depend on time. The classical examples of such systems are hydrodynamic turbulence arising when a liquid (or gas) flows past an obstacle with a large velocity and the turbulent thermal convection emerging in liquid or gas in the gravitational field and under the influence of high temperature gradients.

A nonlinear system of parametrically excited waves is not an exception. As the pumping amplitude increases (accompanied, accordingly, with the increase of the energy flux through the system) the ground state as a rule loses its stability. As a result the occupation numbers $n(\boldsymbol{k}, t)$ and phases of the pairs $\Psi(\boldsymbol{k}, t)$ at large W behave chaotically. This physical situation we shall call *secondary parametric turbulence*. The term *"turbulence"* emphasizes the similarity to the hydrodynamic turbulence. The attribute *"secondary"* points to the chaotic behavior of the already averaged (over the quasi-Gaussian ensemble of the canonical amplitudes of the waves $c(\boldsymbol{k}, t)$, $c^*(\boldsymbol{k}, t)$) quantities, i.e. double correlators $n(\boldsymbol{k}, t)$ and $\sigma(\boldsymbol{k}, t)$ of the parametrically excited waves. The secondary turbulence of the parametric spin waves has been studied in detail. It will be discussed in the following subsection.

8.1 Instability of Ground State and Auto-Oscillations

8.1.1 Properties and Nature of Spin Wave Oscillations

It is common knowledge that under parametric generation, the stationary mode often fails to be achieved and the magnetization is characterized by complicated variations with respect to some mean value. The basic experimental data on the auto-oscillations (AO) obtained for the high-quality YIG crystals under parallel pumping are as follows [8.1–3].

1. AO frequencies are within the range from 10^4 to about 10^6 Hz (depending on the pumping power p and the magnetic field H). Under small supercriticality the AO spectrum consists of a single line. As the intensity increases the number of lines also increases and they are shifted towards higher frequencies. In particular, the components emerge at multiples and at half of the frequency [8.1–3]. High above the threshold the spectrum has a noise character.

2. The AO threshold h_{cr} is usually very low, e.g. $0.1-1$ dB (with respect to the threshold of parametric excitations) with the exception of the range of small $k \simeq 10^3 - 10^4$ cm^{-1}, where the AO threshold notably increases. The threshold also increases when internal inhomogeneities are introduced into the crystal [8.1].

3. A giant crystallographic anisotropy of the AO is observed which significantly exceeds the anisotropy of the spin wave spectrum. Thus the AO intensity in YIG when the magnetization is oriented along the axis [111] is about 200 times as large as the intensity of these oscillations along [100].

The physical nature of AO was one of the main problems when parametric excitation of magnons is considered. Various hypotheses have been proposed. The simplest of them [8.4, 5] suggests the presence in the parametric magnon spectrum of several discrete frequencies corresponding to the eigen oscillations of the crystal. The beats between them are assumed to result in AO emergence. This hypothesis accounts well for some experimentally observed phenomena (AO dependence on the intensity and direction of the magnetic field) but completely ignores the problem of the AO frequency dependence on the pumping intensity and the origin of the various discrete frequencies. Note that the S-theory predicts the existence of only one frequency $\omega_{\mathrm{p}}/2$ in the stationary state. The other group of hypotheses is based on the assumption that parametric magnons influence the magnetization (see, for instance, study by *Green* and *Schlömann* [8.6]). If the average magnetization of the crystal follows the number of parametric magnons after a certain delay, magnetization AO can be built up in such a crystal. This viewpoint is expressed by *Monosov* in [8.1] who proceeded from the Bloch-Bloembergen phenomenological equations. Actually, however, the emergence of AO can be influenced only by the persistence of the thermal waves with the frequency of order ω_{p}. Therefore, the kinetic equation has to be used to study the influence of the thermal magnon persistence. It can be assumed that in most experiments the influence of the persistence can be neglected. Within the S-theory AO can be explained as a result of the instability of the stationary state described in Sec. 7.2. If at least for a single mode with m- number the instability conditions (7.1.7) are satisfied, then within the S-theory the system of parametric magnons have no stable states either. In this case there are two possibilities: either the system is pushed out of the S-theory applicability region (which is accompanied by the large amplitude increase of the excited waves) or the oscillations become steady in the stationary state. The development of these oscillations can be experimentally observed as the magnetization AO. The auto-oscillations can be expected to be, generally speaking, either regular or chaotic.

8.1.2 Numerical Simulation of Auto-Oscillation in the S-Theory

As is seen from (7.1.6) the instability of the stationary state is completely aperiodic (Re$\Omega_m = 0$). Because of this it is very difficult to solve analytically the problem of the nonlinear stage of its development. Thus computer simulation seems the best way out. However, to simulate the real situation, e.g. for YIG, would take too much computer time to be practicable. Therefore a computer simulation on the simplified models of the ground state seems reasonable. In [8.3, 7] the computer simulation of the AO excitation in the "two-beam" model is described where the parametric spin waves were taken to be concentrated at two fixed angles $\Theta_1 = \pi/2$ and $\Theta_2 = \pi/4$. The coefficients $S(\boldsymbol{k}, \boldsymbol{k}')$ and $T(\boldsymbol{k}, \boldsymbol{k}')$ were chosen close to those calculated for YIG as the orientation $\boldsymbol{H} \parallel [111]$ (see below), so that the conditions of the zeroth mode instability were satisfied.

Numerical experiment showed that for such a model there is AO of the amplitudes and wave phases on the beams (Fig. 8.1). The frequency dependence of these AO on the pumping level qualitatively agrees with the similar dependence ordinarily observed in the laboratory experiment. In addition, the two-beam model is used to simulate the development of the collective instability at $m \neq 0$. The number of parametric magnons at the nonlinear stage of this instability was studied. It is an interesting problem because at the linear stage no change of the sum $(N_1 + N_2)$ takes place. The beams were chosen where $\Theta_1 = \Theta_2 = \pi/2$, $\varphi_1 = \varphi_2 = \pi/2$. The experiment showed (Fig. 8.2) that the mode is established where both the difference and the sum $(N_1 \pm N_2)$ experience oscillations. The oscillations $(N_1 + N_2)$ are due to the interaction of the collective modes with different m.

The significant result of these computer simulations consisted in the proof of the fact that at $p < p_{\mathrm{cr}1} \simeq (1.0\text{--}1.5)$ dB the system of parametric waves enters a stable limiting cycle whose field of attraction is the entire phase space. At $p > p_{\mathrm{cr}1}$ the paths near this cycle become exponentially unstable, the average divergence increment per cycle increases proportionally to $p - p_{\mathrm{cr}1}$. At small values of this difference in the vicinity of the limiting cycle a narrow layer occupied by exponentially unstable paths emerges.

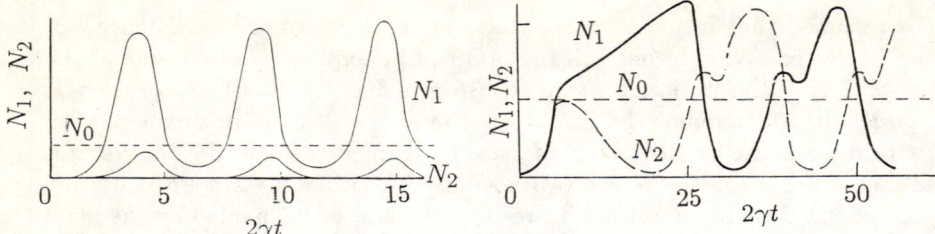

Fig. 8.1. (left) Time dependence of total numbers of pairs on beams at the instability of a zero mode. N_1 and N_2 correspond to $\Theta = \pi/2$ and $\Theta = \pi/4$; $p = 2$

Fig. 8.2. (left) Time dependence of total numbers of pairs on beams at the instability of non-zero mode. N_1 and N_2 correspond to $\phi = 0$ and $\phi = \pi/2$, $\Theta = \pi/2$, $p = 2$

The divergence of the path has been studied in a laboratory experiment by *Grankin* et al. [8.3]. They recorded the dependences $\chi''(t)$ of the different pumping pulses of one series with fixed supercriticalities and fixed other experimental conditions on one and the same screen. The only difference between successive pulses is in initial conditions at $t = 0$. At the supercriticalities $p = 2.5$ dB the successive curves $\chi''(t)$ coincide. Above this supercriticality level when the AO-Fourier spectrum undergoes a sharp broadening the curves $\chi''(t)$ for different pulses begin to diverge. At first, the divergence of nearby paths is manifested only when the time is long. Then, as the supercriticality increases, the "scattering time" becomes comparable with the AO period $\tau \simeq (3\text{--}5)/\Delta_0$ and the successive curves $\chi''(t)$ comprise a broad band.

In summary, it can be said that the computer simulation on the models reveals that within the S-theory equations the development of the internal instability of the ground state results in auto-oscillations. The properties of these AO, i.e. the dependence of the frequency and spectrum on the pumping intensity are comparable with the properties of the experimentally observed AO. In particular, under small supercriticalities both in laboratory and mathematical experiments AO prove to be periodic. An example of such a motion in the phase space is the limiting stable cycle.

In the experiments of both types the transition to the stochastic AO as the supercriticality increases takes place not via addition of new types of motion at incommensurable frequencies; it is accompanied by the broadening of the already existing spectral lines and is due to the decreasing stability of the phase paths leading to their scattering.

This implies that the secondary turbulence of spin waves arises in accordance with the concept of the strange attractor. The scenario of the route to chaos in the parametric turbulence will be treated in more detail in Sect. 8.2. For the time being discussing the rough properties of the parametric turbulence we shall only note that in computer as well as in laboratory experiments AO development does not influence significantly the

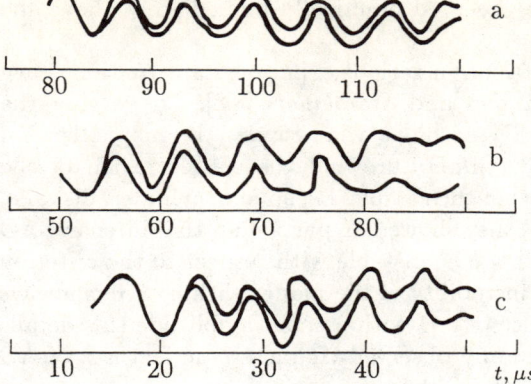

Fig. 8.3. Divergence of the trajectories $\chi''(t)$ in an experiment by *Grankin* et al. [8.3]. (a) $p = 2.6$ dB; (b) $p = 2.9$ dB; (c) $p = 3.4$ dB

average level of parametric waves. The computer simulation also shows that under the instability development the total number of parametric waves together with the observed value $\chi''(t)$ oscillate significantly at the zeroth mode and weakly when the instability is developed at higher modes. Interestingly enough, AO excitation in the computer simulation was as a rule accompanied by a decreasing average value of χ'. The same phenomenon was observed in the laboratory experiment [8.1].

8.1.3 Conditions for Excitation of Auto-Oscillations

The instability criterion (7.1.7) enables us to predict the conditions under which AO are to be observed in laboratory experiments. Let us consider the interaction amplitudes S and T for YIG calculated by *Musher, Starobinets* and *L'vov* by (7.1.5) for the typical experimental situation, i.e. $N = 1/3$ (sphere) $\omega_p/2\pi = 9.4$ gHz, $k = 0$ ($H = H_c$), $\omega_m = 4.9$ gHz, $\omega_a = 0.23$ gHz (room temperature). See Table 8.1

Table 8.1. Interaction amplitudes S_m and T_m for YIG (in units of $2\pi g^2$)

Orientation	T_0	S_0	$T_2 = T_{-2}$	S_2	S_{-2}
[100]	0.28	0.52	0.11	0.01	-0.36
[111]	-0.75	0.30	0.05	0.01	-1.27

As can be seen, not only the values, but also the signs of the amplitudes depend on the magnetization orientation. Substitution of the table values into the instability criterion (7.1.7) shows that in the "easy direction" [111] there is instability with respect to the zero mode, whereas in the "difficult directions" [100] all the modes in this situation are stable. Experimentally, at $H = H_c$ there are no AO in the difficult direction up to the excesses equal to (6–7) dB corresponding to the second threshold. At the same time in the easy direction intense AO are observed practically immediately above the threshold [8.5].

The conditions of AO excitation in various experimental situations have been studied in detail by *Zautkin* and *Starobinets* [8.8]. By varying the Hamiltonian coefficients S_m, T_m within a wide range (the magnetization being varied by changing the temperature and the concentration of the admixtures as well as the shape of the sample, i.e. sphere, cylinder, disk, i.e. the direction of magnetization) they showed, in particular, that intensive AO appear only when the zeroth mode is unstable with respect to the criterion (7.1.7). At the same time the instability of the modes with $m = 0$ is always accompanied by the appearance of AO with a small amplitude (for details see Sec. 9.7). Other experimentally observed characteristic features of AO can be easily accounted for by the S-theory.

In conclusion, it must be noted that the simple AO theory described in the present section predicts that the threshold of their excitation h_{cr} coincides with the threshold of the parametric excitation h_{cr} and the AO frequency at $h = h_{\text{cr}}$ is zero. Experimentally, we observe at the same time the finite threshold of AO and the non-zero initial frequency in the YIG crystals the threshold of auto-oscillations usually equals $(0.1–1)$ dB. The initial frequency does not correlate with the value of the threshold. Depending on the constant magnetic field it varies within the range $(10^4 – 10^5)$ Hz [8.5]. These facts can be explained by the influence of weak nonlinear damping, which does not significantly change the values χ' and χ'', by the random inhomogeneities in crystals, by the absence of exact axial symmetry, by the feedback effect on the resonator, etc. To obtain the relative contribution of those mechanisms the auto-oscillations must be further studied both theoretically and experimentally.

8.2 Route to Chaos in Dynamic Systems

We have already noted that the nonlinear system of parametric waves is an example of nonlinear dissipative systems which in the state of the flux equilibrium lose their stability as the energy flux W through the system increases. Under big W these systems are characterized by a most complicated chaotic behavior as a rule. In addition to the hydrodynamic systems, such as the atmosphere and the oceans of our planet, the chaotic behavior is characteristic of various systems in chemistry (e.g. Belousov-Zhabotinsky reactions), biology, electronics, solid-state physics, etc.

The theory of dynamic chaos as a field of theoretical mechanics and mathematics which could be applied to the above named fields has intensively developed in the last decade. Several books have been published on this topic (see, for instance, [8.9]), and the elements of this theory have already become a part of courses on theoretical physics [8.10].

A lot of interesting results were obtained when the route to chaos was studied under the parametric excitation of the spin waves [8.11–19]. This is the reason for including this section.

8.2.1 Introduction

The transition to chaotic behavior under increasing W is known to be often rather abrupt. By analogy with the radio technical generator, such excitation will be called *hard*. As a rule, the turbulence properties under hard excitation vary greatly in different systems and must be studied specifically by concrete sciences, i.e. hydrodynamic turbulence, laser physics, etc.

In other cases the route to chaos is smooth and takes place through the succession of bifurcations successively making the system's behavior more and more complex. Under small supercriticalities when the turbulence excitation is soft the number of unstable modes participating in the motion can be small, e.g. two or three. "This enables us to assume that the types of stability losses of ... the continuum could be obtained essentially in the same way as the stability analysis of the periodical movement of the dissipative discrete mechanical system described by the finite number of the equations" [8.10]. Mathematically, it means that the dimensionality of the phase space N defined as the number of the ordinary differential equations of the first order with respect to time

$$dX_j/dt = f_j(X_1, X_2, ..., X_n, t) , \quad j = 1, 2, 3, ..., N \tag{8.2.1}$$

describing the system is finite. Moreover, in some cases the important features of the chaotic behavior can be effectively described in the phase space of the low dimensionalities at $N \simeq 3\text{--}5$. In the limit of the small N different physical systems become more and more similar in many respects; the number of common features increases as the dimensionality of the phase space N decreases.

Let us consider as a limiting case the two-dimensional phase space concentrated on a plane. Let the phase trajectories be attracted inside a certain area and never come outside it. Since the phase trajectories do not intersect topologically they can have only three types of behavior. These trajectories are either attracted to the stationary limiting point (pole) (see Fig. 8.4A) or they are wound around it (focus) (see Fig 8.4B) or they are wound on a limiting cycle (Fig. 8.4C). In the first case the system asymptotically passes to the stationary state, in the second case the transition to the stationary state takes place through the damping oscillations, in the third case asymptotical behavior of the system is the non-damping periodic auto-oscillations. There are no other variants of behavior of the considered system.

It must be emphasized that our conclusion about the time behavior of the nonlinear system is unrelated to its physical nature. This conclusion is based only on the fact that its phase space is two-dimensional and on the

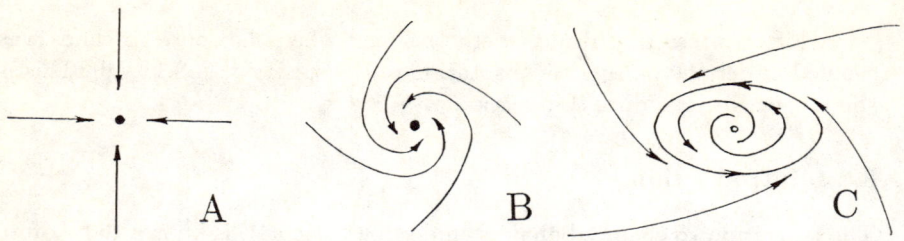

Fig. 8.4. Two-dimensional attractors: limiting point, A; stable focus, B; stable cycle, C

assumption of the presence in this space of the asymptotically attracting area, i.e. an attractor. Accordingly, experimental manifestations of a nonlinear dissipative system can help us to study the topology of its attractor on a plane, but will give us no evidence of its physical nature. Similarly, an experimental study of integers, consisting of counting apples or feeding bits of information into computers will give us no idea of the taste of apples or the physical structure of the computer memory. The different properties of integers, apples and computers are the objects of different sciences. Perhaps it is useful to know mathematics when studying apples. But will it make sense to study mathematics using only apples?

This almost absurd example can help us to understand the relation of notions in the nonlinear physics of magnets and the modern theory of the route to chaos in simple dynamic systems. Indeed, what can we learn about an antiferromagnetic $CuCl_2$ or ferromagnet $Y_3F_5O_{12}$ when we observe in them the bifurcations of doubling of the magnetization auto-oscillation periods [8.11,12]. What can be learn about the antiferromagnets $CsMnF_3$ or $(CH_3NH_3)CuCl_4$ if we observe in them other scenarios of the route to chaos [8.15]? Nothing, I think except the almost trivial fact that these objects are nonlinear systems of general position and thus manifest the general laws of the route to chaos.

Does this imply that the experimental study of chaos in magnets is not interesting? By no means. The mathematical theory of the route to chaos in simple dynamic systems requires not only computer studies but also physical experiments. Not, however, in order to check results like "two times two equals four" within the system of Peano's axioms, but, for example, in order to find new laws and scenarios. The universality of the laws of transition to chaos makes it possible to carry out these experiments on water and alcohol (in hydrodynamics) as well as on magnets. As in our previous example with the theory of integers and apples, the choice of the subject of inquiry is mostly a matter of personal likes and dislikes, availability and potential efficiency.

Sections 8.2.3 and 8.3.2 present an example of using the magnets as a subject of chaotization process studies. These are results of *Smirnov's* investigations of the route to chaos under the parametric turbulence of magnons in an antiferromagnet $CsMnF_3$. This study effectively employs some notions of the dynamic chaos theory, which will be briefly described in Sec. 8.2.2.

Those readers who, first, had the patience to read my book up to this place, and, second, are not acquainted with the theory of dynamic chaos will be rewarded by the harmonic postulates of this fashionable science. In this case my task will be fulfilled. At any rate, the presented material is sufficient to form a balanced attitude to the dynamic chaos and its relative importance in the nonlinear wave theory.

8.2.2 Elementary Concepts of Theory of Dynamic Chaos

1 Landau-Hopf scenario. In order to analyze the ways of turbulence generation, let us consider phase spaces with dimensionalities larger than two which admit asymptotic motion different from a stationary point or a limit cycle. For a long time the possible attractors for a dissipative system were considered a stationary point, a limit cycle, a two- or three-dimensional torus, etc. At the n-torus the system participates with n different periods T. From this viewpoint, stationary point and the limit cycle are 0- and 1-torus. If some of the periods prove to be commensurable, the dimensionality of the torus decreases and the torus will be called a resonance torus. Thus, for instance, the trajectory on the surface of a two-dimensional resonance torus will be closed and topologically equivalent to a circle, i.e. 1-torus. In these terms, the well-known Landau-Hopf scenario of the route to chaos consists of the successive increase of the torus' dimensionality, the torus representing the system attractor. At large N the correlation functions of the system will be damped over the time of the order of T_j. However, over a very long period T_R, the so-called Poincaré recovery time, the system trajectory will pass arbitrarily close to the initial point and the correlation functions will abruptly and drastically increase up to the order of unity. The order of magnitude is

$$T_R \simeq T \exp(an),$$

where a is of the order unity.

The behavior of the nonlinear system according to the Landau-Hopf scenario seems atypical, i.e. in the space of the system parameters it can be realized over the set of zero dimension. To put it simpler, the probability of such a scenario is vanishingly small and for it to be realized special efforts are necessary, for instance $N \to \infty$ of non-interacting generators of non-commensurable frequencies should be built which will work for the common load. For more details about this scenario see Sect. 30 in [8.10].

2 Strange attractor. In 1963 *Lorenz* [8.20] considered a limitingly truncated system of equations of thermal convection in a plane layer heated from below. In a dimensionless form this system of equations is

$$dx/dt = \sigma(y - x) ,$$
$$dx_!/dt = -y + rx - zx , \qquad (8.2.2)$$
$$dx/dt = -bz + xy ,$$

where b, r and σ are the dimensionless parameters. The phase volume of such a system decreases with time:

$$\frac{d}{dt}\mathrm{div}\, \boldsymbol{R} = -(\sigma + 1 + r) < 0 , \qquad (8.2.3)$$

where $\boldsymbol{R}$ is the vector x, y, z. From the Lorenz equations it follows also that

$$\frac{d}{2dt}[x^2 + y^2 + (z - r - \sigma)^2]$$
$$= d(r+\sigma)^2/4 - \{\sigma x^2 + y^2 + b[z-(r+\sigma)/2]^2\} . \qquad (8.2.4)$$

At large R the right-hand side is negative. This means that the distance between the points $\boldsymbol{R}$ and $\boldsymbol{R}_0 = (0, 0, \sigma + r)$ decreases with time, i.e. all the trajectories enter into some limited volume surrounding $\boldsymbol{R}_0$. Inside this volume there are no (in a certain area of changing parameters r, b and σ) stable stationary points and no stable limit cycles. As a result, phase trajectories falling into a limited volume and unable reach this stable manifold have to move in a very complicated and bizarre way. Some general considerations for the "phase portrait" for the dissipative system where the trajectories are complicated and tangled (e.g. as in [8.10]) can be given. As the system is dissipative the trajectories only enter a certain limited volume. To tangle the trajectories it is necessary that two arbitrarily close points denoting the initial conditions should move apart with time at a distance comparable with the dimensions of the attractor, i.e. *all the trajectories should be unstable*. The attracting sets of limited and unstable trajectories (the possibility of their existence has been predicted by *Lorenz* [8.20] for the case of the system (8.2.1)) are now called *strange attractors*.

Let us consider the geometrical structure in an n-dimensional space by studying the behavior of the bundle of trajectories on their way to the attractor [8.10]. In the cross-section of the bundle the trajectories fill a certain volume V in $(n-1)$-dimensional transversal subspace. Under the motion along the trajectory the volume $V(t)$ decreases because the system is dissipative (in the conservative case V is time independent by the Liouville theorem). However, in a certain amount $m < (n-1)$ of directions the bundle section broadens due to the exponential divergence of the unstable trajectories.

In the remaining $(n - m - 1)$ directions the bundle is compressed. The general compression must be stronger than the extension because the volume decreases. Along the trajectory the direction of compression and extension must change, or else the trajectories can fall outside the attractor. Therefore, the solid cross-section bends and at the same time flattens, bends again and so on. This process is sometimes called *Baker's transformation* because it resembles rolling the pastry, folding it up, rolling it again, etc. The cross-section of the bundle proves to be extended and folded many times. This happens not only to the selected bundle but to any of its parts. As a result, the attractor is a system of an infinite number of flat and infinitely thin layers connected through their sides and spaced infinitely close to each other. The general volume of the attractor in its n-dimensional space equals zero. Such sets are called the Cantor sets. They are characterized by the fractional dimensionality according to *Hausdorf* $n_H < n$:

$$n_H = \lim_{\varepsilon \to 0}[\ln N(\varepsilon)/\ln(1/\varepsilon)] . \tag{8.2.5}$$

where $n(\varepsilon)$ denotes the minimum number of the n-dimensional cubes with the edge ε required for the coverage of our set. Clearly, for the part of the plane with the area S $N(\varepsilon) = S/\varepsilon^2$, therefore the limit of (8.2.5) is equal to two. For the line $n_H = 1$, for the set of separate points $n_H = 0$. For the *Lorenz* attractor $n_H \simeq 2.07$ (for $\sigma = 10$, $b = 8/3$, $r = 28$) [8.20]

3 Feigenbaum scenario When analyzing possible scenarios of turbulence generation under the destruction of the limit cycle it proves possible to avoid analyzing the specific type of the dynamic systems, and its general behavior can be predicted by means of the Poincaré representation. This mapping is a set of intersection points of the phase trajectory with the surface in the phase surface transverse to it. The initial dynamic system makes it possible to determine basically the intersection coordinate $\boldsymbol{R}_{n+1}$ on the $(n+1)$th round as a function of coordinate $\boldsymbol{R}_n$ on the previous round. In such a function of succession arises:

$$\boldsymbol{R}_{n+1} = \boldsymbol{F}(\boldsymbol{R}_n) . \tag{8.2.6}$$

In many cases the trajectories in the transverse subspace come very close to some line. It significantly simplifies the analysis of the trajectory behavior, because the succession function turns out to be one-dimensional

$$X_{n+1} = F(X_n) . \tag{8.2.7}$$

As can be easily shown (see, for instance, [8.10]), for the motion to be stochastic $F(x)$ must be a non-monotonic function. *Feigenbaum* obtained a fundamentally important result. He showed that for all the systems with one-dimensional non-monotonic mapping (8.2.7) depending on some parameter W (e.g. on the Reynolds number, pumping amplitude or on the energy flux through the system) with the extremum of the quadratic form there exists a qualitatively and quantitatively universal scenario of transition from periodic to chaotic motion, i.e. *Feigenbaum scenario* [8.21, 22].

The most important features of this scenario are shown by the mapping

$$X_{n+1} = 2CX_n + X_n^2 . \tag{8.2.8}$$

thoroughly studied, e.g. in [8.9]. The periodic motion (cycle) corresponds to the roots of the equation(8.2.8) at $X_{n+1} = X_n$. There are two such solutions

$$X_{10} = 0, \quad X_{11} = 1/2 - C. \tag{8.2.9}$$

As can be readily seen, within the range $1/2 < C < 3/2$ the cycle X_{11} is stable, and in the range $-1/2 < C < 1/2$ the other cycle X_{10} is stable. Therefore, as C decreases from $3/2$ to $1/2$, the coordinate X of the stable cycle increases from -1 to 0 and then in the range C from $1/2$ to $-1/2$ it remains equal to zero.

What happens in the range $C = -1/2 - \varepsilon$, $\varepsilon \ll 1$? In order to answer this question, periodical points of the period 2 must be considered. It can be readily obtained from the evident condition

$$X = F_2(X), \quad F_2(X) = F(F(X)), \tag{8.2.10}$$

Here $F_2(X)$ is the square of the mapping $F(X)$, describing the mapping over two windings of the cycle

$$X_{n+1} = F_2(X_n). \tag{8.2.11}$$

At $C > -1/2$ (8.2.10) has the solution (8.2.9). At $C = -1/2$ the two additional solutions appears (namely $x_{2\pm} = \pm 2$), these solutions are stable at small ε. The generation of a pair of stable points at $C < C_1 = -1/2$ describing the cycle of double period is an example of *bifurcated doubling of cycle*. As C further decreases at some $C_2 < C_1$ the double cycle is no longer stable and a new cycle with the period 4 is generated. The infinite succession of the bifurcation of the period doubling converges to the value $C_\infty = -0.78497$. At $C < C_\infty$ the motion is chaotic. *Feigenbaum* showed that at large n the succession of bifurcation values of C_H for arbitrary mapping with a single maximum has a universal behavior

$$[C_{n+1} - C_n]/[C_n - C_{n-1}] = 1/\delta \tag{8.2.12}$$

with a universal value $\delta = 4.6692$ [8.22].

4 Reciprocal bifurcations of chaotic motion. In the previous subsection we gave a brief description of the succession of period doubling bifurcations for the mapping (8.2.8) with C changing from $C_1 = -1/2$ to $C_\infty = -0.78497$. What is the nature of motion for $C < C_\infty$?

This range has been studied by *Lorenz* [8.23], *Collet* and *Eckmann* [8.24] and *Helleman* [8.25]. In computer experiments they observed that as C decreases the bands of chaotic motion merge and experience inverse bifurcations at some points $C = C_n$. It was also found that bifurcations of the limit cycles with the period $n = 6$, 5 and 3 cutting the chaotic area. The reciprocal bifurcations of the chaotic bands obey the similarity law (8.2.12) with the same constant [8.26].

5 Pomeau-Manneville scenario The quadratic mapping of the form

$$X_{n+1} = X_n^2 + X_n + \varepsilon \tag{8.2.13}$$

demonstrate one more kind of bifurcation, i.e. the so-called reciprocal tangential bifurcation [8.27]. As values change from negative to positive the stable and anstable points of this mapping merge and vanish [8.28]. This bifurcation is characterized by another type of route to chaos-via intermittence. It is usually called the *scenario of Pomeau-Manneville*. In this case the ranges of almost chaotic motion are chaotically overlapped by the successive ranges of irregular motion. The average length of coherent beams decreases as $1/\sqrt{\varepsilon}$. Interestingly, the route to chaos in the system (8.2.8) from the integer cycles corresponding to the value $C < C_\infty$ when the C parameters increases (and not decreases) takes place in accordance with the scenario of Pomeau-Manneville without period doubling.

In conclusion it must be emphasized again that the survey of the scenario of the route to chaos in this section is by no means exhaustive or complete. My intention was to give the reader the most important mapping used in the interpretation of experiments on chaos in magnets [8.11–19], and I mostly followed the textbook [8.10] and *Smirnov*'s doctorate thesis [8.28].

8.2.3 Chaos of Parametric Magnons in CsMnF$_3$

In recent studies of parametric secondary turbulence [8.11–19] this problem was connected with the modern ideas of chaotic dynamics of nonlinear dissipative systems. The first observation of a complete doubling route to chaos was made by *Gibson* and *Jeffries* [8.12] with the second-order Suhl process in a Ga-YIG. *Waldener, Barberis* and *Yamazaki* [8.13] observed a route to chaos by irregular periods, and *Yamazaki* [8.11] observed one period doubling but no chaos in AFM CuCl$_2$. The first observation of subharmonic routes to chaos was made by *Resende* et al. in 1986 in pure YIG spheres. A detailed review of experimental studies of chaos origin in magnetic systems is given in [8.17, 19]. In this section we shall consider only *Smirnov*'s results concerning the antiferromagnet CsMnF$_3$ as an example of an advanced experimental study of this problem.

1 Main results. In [8.16] the time dependence of the power $W(t)$ absorbed by the sample of CsMnF$_3$ was studied under different amplitudes of pumping h and varying values of the external magnetic field H. Depending on the values of $p = h^2/h_{th}^2$ different modes of $W(t)$ behavior were observed (see Fig. 8.5 [8.16]). These include stationary state $W(t)$ (mode 0), periodic autooscillations of the relaxation character with the period $t \simeq 0.1$ ms mode 1. The bifurcations of doubling and quadrupling of the period of the basic motion, modes 2 and 4, were also observed as well as the cycles with periods $3T$, $5T$ and $7T$, modes 3, 5, 7. The mode of chaotic motion was observed

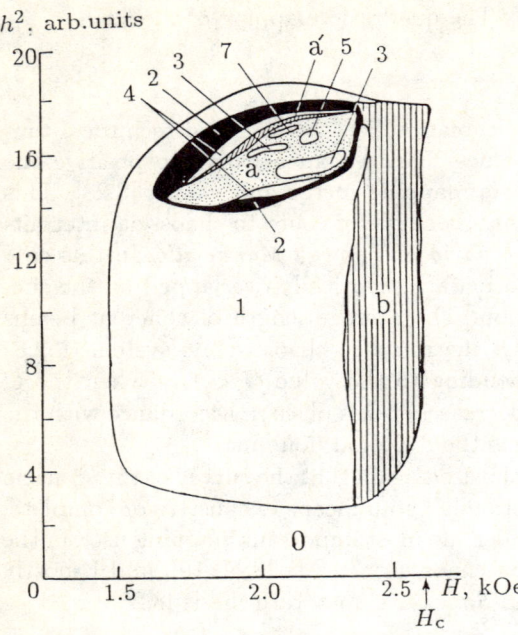

Fig. 8.5. Diagram of stationary (0), periodic (1) – (7) and chaotic (a,a' and b) regimes. Periodic situations of the corresponding multiplicity (from 1 to 7); a-chaos 1 (motion in one and two zones); a'-chaos 1 (motion in four zones); b-chaos 2.(after *Smirnov* [8.16]

whose spectrum contains merged lines at the frequencies $f = 1/T$, $1/2T$ and $1/4T$ which was called Chaos 1, and another type of chaotic motion with a wide spectrum called Chaos 2. The diagrams in Fig. 8.5 shows the general succession of bifurcations leading to changing of the modes. Fig. 8.6 shows the cross-section of this diagram (with higher resolution of p) in the field $H = 2.0$ kOe. Following [8.16], let us treat this cross-section in more detail.

2 Fine structure of the mode change of AO depending on pumping intensity.
As can be seen from Fig. 8.5, the transition from the periodic mode of auto-oscillations of the absorbed power to "chaos 1" takes place in compliance with the Feigenbaum scenario, i.e. through period doubling. As described in Sec. 8.2.2, the analysis of the quadratic mapping (8.2.8) made it possible to clarify many other details of mode change under changing C parameter which, as it turned out, appear in experiment [8.16]. Fig. 8.6 shows the temporal sequence of the different types of microwave absorption observed in the field of 2.0 kOe when the pumping power was varied. Most consistently the Feigenbaum scenario and other details of the mode change typical of the mapping (8.2.8) are manifested as the intensity changes from high to low values. The process of the inverse bifurcation of doubling - the merge of external zones of the phase-plane diagram is clearly seen. The spectra of chaotic motion at the parameter values corresponding to these transitions must have a form universal for many systems [8.29] (see [8.9], Fig. 7.23). The spectrum obtained for the value $C = C_2$ corresponding to the merging

of the four zones into two has a sharp peak at the frequencies $f/4$ and $f/2$ and broad peaks of lower amplitude at the same frequencies. The wide and sharp peaks at the frequency $f/2$ are characteristic of $C = C_1$ (two zones merge into one). These spectra qualitatively agree with the spectra experimentally observed in [8.16].

The laws of transition from the loop modes with periods 7 ,5, 3 to the chaotic modes are of considerable interest. In the experiment [8.16] the transient (from cyclic to developed chaotic period) was observed only for the cycle with the period 3. As the power decreases the period is doubled, i.e. a cycle with period 6 is generated. The exit from the cycle 3 in the opposite direction (as the power increases) takes place in a different way. The spectral lines corresponding to the cycle 3 rapidly broaden and the real signal reveals the intermittence of wave packets from the cycles 3 and chaotic intervals. Unfortunately, in experiment [8.16] the intermittence was observed in a very narrow intensity range, so that it is not possible to obtain the power dependence of average duration of coherent wave packets. However, it can be positively assumed that in this experiment intermittence is due to the merge and vanishing of stable and unstable cycles, since the place of this mode change is determined by all the remain bifurcations.

Following *Lorenz* [8.9] a dependence $y_{n+1}(y_n)$ for the chaotic mode (position 8, Fig. 8.6) can be plotted where y_n is the absolute value of the n-th minimum on the $W(t)$ dependence. Such a plot determines the Poincaré mapping for the surface parameterized by the condition of the maximum of this coordinate in the phase space. The plot of this dependence is shown in Fig 8.7A and has the form of a curve with the minimum. According to Feigenbaum's theory, the system with such Poincaré mapping during the transition to chaos must experience a cascade of duplication bifurcations. This implies [8.9] that close phase trajectories diverge exponentially with time, and that the attractor of our system is strange one. The construction of mapping $y_{n+1}(y_n)$ for the chaotic mode near the intermittence of the cycle 3 and also for "chaos 2" brings about no unambiguous dependence (Fig. 8.7B). The same situation with mapping is evidently also observed in the numerical experiment on the mapping (8.2.8), and the C value close to the value under which cycle 3 arises at the approach from the condensation point of duplication bifurcations (see [8.9] and Fig. 15 in [8.9]).

Note also that the transition from the mode of chaotic motion corresponding to the position 12 on Fig. 8.6 to the mode of the position 13 is accompanied by a small hysteresis and is usually associated with the two chaotic attractors of the system in this range of the parameter values.

A close observation of the transition from the cyclical mode of the fundamental cycle to "chaos 2" revealed that this transition also takes place through intermittence. Experiment shows that the mean value $\langle 1/\tau \rangle$ (τ is the duration of coherent wave packets) depends on $\varepsilon = (H - H_{cr})$ as ε^2 (see [8.16], Fig. 11). The "chaos 1" is characterized by the order close in time,

228   8 Secondary Parametric Wave Turbulence

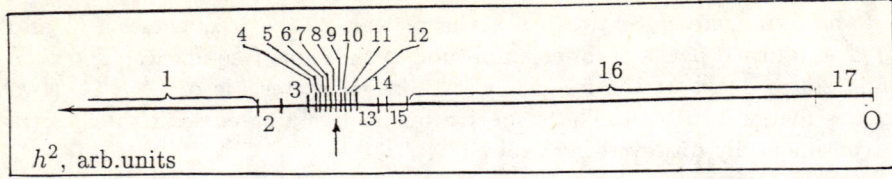

Fig. 8.6. Sequence of transformation of periodic and chaotic regimes in a field of 2.0 kOe as the power is varied. (1)-steady state; (2)-cycle of fundamental period; (3)-cycle 2; (4)-cycle 4; (5)-chaos of four zones; (6)-chaos of two zones; (7)-cycle; (8)-chaos of one zone; (9)-cycle 3; (10)-cycle $6 = 3 \times 2$; (11)-chaos of one zone; (12)-zone contraction; (13)-chaos of one narrow zone; (14)-cycle 4; (15)-cycle 2; (16)-cycle of the fundamental period; (17)-steady state. The arrow points to the value of h^2 at which the intermittency of cycle 3 is observed

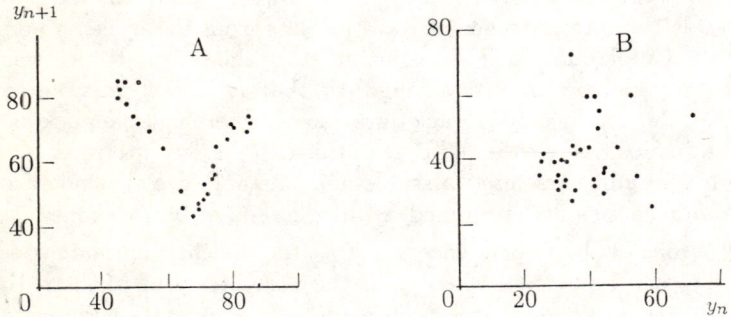

Fig. 8.7. (A) Poincaré mapping for chaotic regime 8 in Fig. 8.6, (B) for chaos (after [8.16])

which is confirmed by the observed dependence $y_{n+1}(y_n)$ (Fig. 8.7A) and also a fixed hierarchy of the mode transformations under intensity changes (Fig. 8.6).

In conclusion, we can say that in the discussed example of the secondary turbulence in the antiferromagnet $CsMnF_3$ the route to chaos follows the Feigenbaum scenario (through period duplication) if the pumping intensity is increased under an increasing magnetic field. The Pomeau-Manneville scenario is set (transition through the intermittence of coherent wave packets). Many details of the evolution of the time, spectral and amplitude motions of the system of parametric magnons are observed, theoretically predicted on the basis of the quadratic mapping (8.2.8), i.e., periodic motion with the period multiple of 3,5,7, bifurcation of the cycle 3 intermittence, merging of chaotic zones.

8.3 Geometry of Attractors of Secondary Parametric Turbulence of Magnons

8.3.1 Effective Phase Space and Dimensionality of Inclusion

As has already been noted, the modern approach to the problem of turbulence generation is based on the assumption of the finite dimensionality of the phenomena which determine the development of instabilities. Although the rigorous mathematical formulation of this statement, i.e. the theorem of central variety [8.30] has been proved only for the bifurcation of stability loss of the stationary point, there are intuitive arguments [8.10] in favor of the existence of a finite set of essential modes or degrees of freedom which determine the system dynamics over a long time and for more complicated modes of motion.

In this respect it is interesting to obtain experimentally the number of independent variables unambiguously describing the potentially infinite dimensional motion of the dissipative continuum system when the number of degrees of freedom really involved in motion is not known beforehand. To define the necessary number of such variables will mean to construct one-to-one mapping of the phase space of the asymptotic motion mode onto the Euclidean space of these variables; therefore this number can naturally be called the dimensionality of inclusion n_e [8.31-33].

The first attempts to construct an infinite dimensional phase space on the basis of measurement results were associated with the simultaneous measurements of a number of independent quantities in hydrodynamics [8.33, 34]. The next important step was to prove the fact that the phase coordinates $X_j(t)$ for time t may can be taken in the form $X_j(t) = a[t+(i-1)T]$, where $a(t)$ is the only measurable quantity and T is the shift in time [8.31, 32]. For the position P of the mapping points on the trajectory in the equivalent phase space to be in one-to-one correspondence with its position in the real phase space of the system the number of variables X_j must be not less than $(2n_H + 1)$ where n_H is the dimensionality of the attractor according to Hausdorf in the real phase space [8.32]. In cases when the attractor's geometry is simple and its mappings onto the space of the equivalent variables have no self-cross sections, the number n_e can be decreased down to n_H. In [8.32] – a fundamental research for this analysis – this procedure was shown to obtain the correct topology of the attractor as well as to enable one to accurately determine the values of all the Lyapunov exponents. The physical meaning of this procedure consists in the fact that the values of one variable (called T) are determined by the interaction with all the essential variables of the problem. Therefore the evolution history of this variable $a(t)$ contains (in indirect form) the information about the values of other variables in previous times.

In order to determine the n coordinates $X_j(t)$ of the point in the equivalent phase space measurements of the value of one coordinate $a(t)$ over n

times are used, i.e. the data are not taken from nowhere as it may seem. The experimenters must find very attractive such a construction algorithm of phase space of any dimensionality by means of measuring the time dependence of only one value, because they always have time but not always the chance to put any amount of sensors on the object of their observation.

If the number of important variables n_e is not known beforehand (as is usual in the case of distributed systems) it must also be found from the experiment how many coordinates describe the evolution of the system. We could employ here a convenient geometric criterion formulated in [8.33], it is equivalent to the distribution criterion of the joint probability [8.31]. If $n > n_e$, then any measurable quantity $y(X)$ must be a function of the constructed n-phase variables $y(X) = f(X_1, X_2, ..X_n)$. If $n < n_e$, then in the general case y will not be a function of $X_1...X_n$, than y should be added as a $n+1$ coordinate. Therefore, beginning with coordinate $n=1$ it is necessary to add new coordinates one by one and to check for the functional dependence on the previous coordinates. Under the moving point $\boldsymbol{X}(t)$ the repetition of the vector $\boldsymbol{X}$ results in the repetition of $\boldsymbol{y}$ if there is a functional dependence and brings about no such repetition if there is no such dependence. This circumstance is conveniently checked graphically. To this end we plot as abscissa the distance $r_n^2 = \sum_{j=1}^{n}(X_j(t) - X_j^0)^2$ from the current position of the point on the attractor to some fixed point $\boldsymbol{x}^0$ on this attractor. On the y-axis we plot the value $d = |\boldsymbol{X}_{n+1}(t) - X_{n+1}^0|$. If the above described functional dependence exists, then $d \to 0$ at $r_n \to 0$. Let us call this check of the functional dependence the *criterion* 1. To realize this criterion in an n-dimensional sphere with a small radius there must be only a few points on the attractor unlike the methods which determine Hausdorf's dimensionality of the attractor. The authors of [8.33] formulated another less exact criterion (let us call it *criterion* 2) which excludes all the small scales. According to the criterion 2 the dimensionality n_e is achieved if the envelope of the trajectory on the plane r,d is completely below the straight line $d = Kr$ where $K \simeq 1$. The criterion 2 can be varied usefully for large dimensionalities of inclusion when the number of the experimental points on the attractor in the volume limited by small linear dimensions is small. The dimensionality n_e obtained by the criteria 1 and 2 in some specific cases can be decreased, for instance, by means of additional cross-sections, as will be described below, or by using the rotated coordinate system.

8.3.2 Experimental Study of Attractor Structure in CsMnF$_3$

In the present section we shall dwell further on the interesting results obtained by *Smirnov* [8.6] in his research on the secondary parametric turbulence of magnons in CsMnF$_3$. Fig. 8.8 shows projections of attractors onto the plane $W(t)$, $W(t+T)$ taken from this work. Modes (1), (2), (3) on this figure correspond to the "chaos 1" on the diagram on Fig. 8.5 and to the modes (5), (8), (11) on Fig. 8.6. The modes (4) and (5) on Fig. 8.8 correspond to "chaos 2" on the mode diagram. The mode (4) is transient from the cycle of the basic period to "chaos 2".

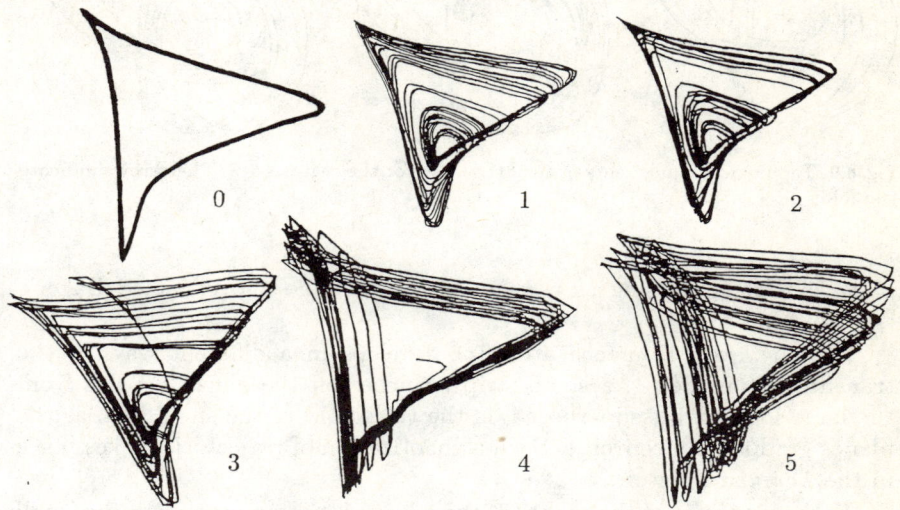

Fig. 8.8. Phase portraits for the periodic (0) and chaotic regimes (1-5) (following [8.17])

Constructing the trajectory on the plane r_n, d we obtain by the criteria the following dimensions of inclusion: for modes (1), (2), (3) $n_e = 3$, in the mode (4) $n_e = 3$, for mode (5) - $n_e = 5$ by criterion 1 and by criterion 2.

The shape of the attractors included into the 3-dimensional space can be studied in detail by means of plane cross-sections. The points where the phase trajectory crosses the intersecting plane are located on line segments, i.e. the attractor is formed from a two-dimensional band which at the intersection with cross-secting planes form the line segments. These plane bands form foldings as well as the branches in the band planes, and afterwards the branch is imposed on the main part of the band.

By constructing a number of cross-sections we can obtain an unambiguously topological structure of the attractor for each of the modes (1), (2), (3), Fig. 8.9 shows the topological equivalents of these attractors.

For the mode (1) the attractor is topologically equivalent to the Rössler attractors [8.9] for those parameter values under which the motion takes place in two zones, i.e. it is a two-loop spiral from a flat band with folds (see Fig. 55 [8.9]) In the Rössler attractor under its evolution towards the increase of the area of chaotic change of variables the foldings are embedded into each other, which results in the loop been formed from a plane strip of the flat band with a fold. The chaotization of the motion in this case occurs at the expense of the divergence of the trajectories in the band

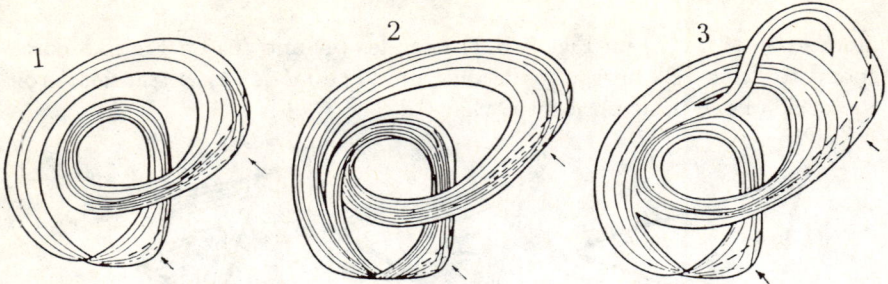

Fig. 8.9. Topological equivalents of the attractors for the regimes 1-3. The arrows indicate the folds

plane and their missing as a result of folding (the above named Baker's transformation).

A spin-wave turbulence attractor develops in a different way. In the transition to mode (2) a small strip branches off the plane, departs from the large loop und is superposed on the trajectory of the small loop as the planes graduallally converge. Such a chaotization of trajectories takes place in the *Lorenz* attractor.

In the attractor of the mode (3) the edges of the band merge at the small and large loops and the branching-off belt of the Lorenz attractor type passes over from the internal orbits to the external orbits making there one more folding.

Therefore, the attractors for the modes corresponding to the "chaos 1" can be constructed from the elements of the Rössler and Lorenz attractors. They correspond to the motion with the least state of chaos in the sense that the phase trajectories in them diverge in one direction only (they have only one positive Lyapunov exponent).

The described cross-sections also reveal the traces of the fractal structure of the attractor, i.e. in the layer which can be called two-dimensional with some finite accuracy there turns out to be one more layer than the resolution scale, which in turn, must also consist of layers, etc. Thus it can be seen on the cross section 5 that the attractor band giving a branch of the cross-section inclined to the vertical axis of the figure of about 45° is layered.

For modes (4) and (5), the cross-section (Fig. 4 in [8.16b]) show that the projections of the corresponding attractors in the three-dimensional space are solid: the points where the trajectories pierce the cross-section planes fill the two-dimensional sections. Therefore, the trajectories in mode (4) mix in the solid tube of the trajectories (convergence takes place in two directions, and the dimensionality of the flux is not less than 4) and for mode (5) the trajectories miss in the space of still larger dimensionality, although its dimensionality is limited and does not exceed 5.

In conclusion I should like to emphasize that the information on the dimensionality of the attractors of different dynamic system is undoubtedly necessary for the study of their statistical behavior. This, however, is only a small part of the information required. Again I should like to bring an analogy. Both dogs and chairs have four legs unlike the three-legged pianos and two-legged ostriches. Does this imply that the statistical behavior of dogs and chairs is similar and ostriches resemble pianos more than dogs? By no means. But, on the other hand, the information about the number of legs is not useless. Thus, a three-legged chair will fall more often and a dog devoid of one leg will also behave strangely. Similarly, the information about the attractor dimensionality is interesting not only from the geometrical, but also from general point of view. It is necessary, for instance, when seeking equations simpler than the input equation that determine the behavior of the dynamic systems in the range of supercriticalities where the dimensionality of the attractor is significantly less than the dimensionality of the phase space.

8.4 Secondary Turbulence and Collapses in Narrow Parametric Wave Packets

Throughout this book the threshold of the parametric instability has been assumed to be minimum for the wave pairs with vectors $\pm k$ filling line or surface in k-space. The phase sum in the pair is in this case a dynamic variable and individual wave phases are chaotic with a good accuracy. Now we shall describe the situation when the excitation threshold is minimum for the only pair $\pm k_0$, e.g. under parallel pumping in uniaxial ferromagnets with the anisotropy of the "easy-plane" type. The basic peculiarity of the problem is the narrowness (in each direction) of the wave packets excited in the k-space. On the one hand, in this case we cannot employ the statistical description as in the S-theory, on the other hand, this enables us to reduce the interaction Hamiltonian using other parameters, i.e. the narrowness of the packet. To this end, the problem must be formulated in terms of complex amplitudes of the envelopes of the waves.

8.4.1 Equations for Envelopes

By using the canonical equations of motion with the exact Hamiltonian $\mathcal{H}_{\text{int}} = \mathcal{H}_{\text{p}} + \mathcal{H}_4$ given by (5.2.1) and (1.1.32) and assuming that then in the k-space narrow wave packets are excited

$$a(\boldsymbol{k}) = [A(\boldsymbol{k} - \boldsymbol{k}_0) + B(\boldsymbol{k} + \boldsymbol{k}_0)]$$

we also obtain, as in Sec.4.1., the equations for the complex amplitudes of the envelopes $A(\boldsymbol{r},t), B(\boldsymbol{r},t)$. These are the Fourier components of $A(\boldsymbol{\kappa},t)$ and $B(\boldsymbol{\kappa},t)$. The resulting equation has the form

$$\left[i\left(\frac{\partial}{\partial t} + \boldsymbol{v}\cdot\nabla\right) + \frac{1}{2}\hat{L}\right]A(\boldsymbol{r},t) = -i\gamma A(\boldsymbol{r},t) + hVB^*(\boldsymbol{r},t) +$$
$$+ \left[\omega(\boldsymbol{k}_0) - \frac{\omega_p}{2} + T|A(\boldsymbol{r},t)|^2 + 2S|B(\boldsymbol{r},t)|^2\right]A(\boldsymbol{r},t) ,$$
$$\left[i\left(\frac{\partial}{\partial t} + \boldsymbol{v}\nabla\right) + \frac{1}{2}\hat{L}\right]B(\boldsymbol{r},t) = -i\gamma A(\boldsymbol{r},t) + hVA^*(\boldsymbol{r},t) +$$
$$+ [\omega(\boldsymbol{k}_0) - \omega_p/2 + T|B(\boldsymbol{r},t)|^2 + 2S|A(\boldsymbol{r},t)|^2]B(\boldsymbol{r},t) ,$$
(8.4.1)

where

$$\boldsymbol{v} = \frac{\partial \omega}{\partial \boldsymbol{k}} , \quad \hat{L} = \sum_{i,j}\frac{\partial^2 \omega}{\partial k_i \partial k_j}\frac{\partial^2}{\partial x_i \partial x_j} ,$$

$$T = T(\boldsymbol{k}_0,\boldsymbol{k}_0;\boldsymbol{k}_0,\boldsymbol{k}_0)/2 , \quad S = T(\boldsymbol{k}_0,-\boldsymbol{k}_0;\boldsymbol{k}_0,-\boldsymbol{k}_0)/2$$

(the amplitudes of wave interaction). Equations (8.4.1) have a trivial solution

$$A(\boldsymbol{r}) = |A_0|\exp(-i\Psi_1) , \quad B(\boldsymbol{r})|A_0|\exp(-i\Psi_2) ,$$
$$2|S|A_0^2 = \sqrt{(hV)^2 - \gamma^2} , \quad \sin(\Psi_1 + \Psi_2) = \gamma/hV ,$$
$$\Psi_1 = \Psi_2 = \Phi/2 ,$$
(8.4.2)

corresponding to the excitation of a standing wave $\pm \boldsymbol{k}_0$. I showed [8.35] that the solution (8.4.2) is practically always unstable with respect to increasing valuation of amplitudes and phases of envelope waves

$$\delta A(\boldsymbol{r},t) \propto \exp(i\boldsymbol{\kappa}\boldsymbol{r} + \nu(\boldsymbol{\kappa})t) , \quad \delta B(\boldsymbol{r},t) \propto \exp(i\boldsymbol{\kappa}\boldsymbol{r} + \nu(\boldsymbol{\kappa})t) .$$

The development of the instability significantly depends on the amplitudes of the interaction Hamiltonian. With the exception of the case $T > 0$, $S > 0$, which will not be discussed the instability increment is maximum on the surface $\boldsymbol{\kappa} \perp \boldsymbol{k}$ and has the form [8.35]:

$$[\gamma + \nu(\kappa)]^2 - \gamma^2 = \begin{cases} -\frac{1}{4}\hat{L}\kappa^2(\hat{L}\kappa^2 + 4TA_0^2) \\ -8S(2S+T)A_0^4 - \frac{1}{4}\hat{L}\kappa^2(\hat{L}\kappa^2 + 4(4S+T)A_0^2). \end{cases} \quad (8.4.3)$$

At the distance from the surface it decreases rapidly. Because of this the main properties of the nonlinear stage of the instability can be described by means of the two-dimensional equations (8.4.1) where A and B depend only on the coordinates x and y orthogonal to $\boldsymbol{v}$. The upper and lower lines in the expression (8.4.3) correspond to perturbations of the type $\delta A(\boldsymbol{r},T) = \pm \delta B(\boldsymbol{r},t)$. When $T > 0$, $S > 0$ the perturbation of the type $\delta A = -\delta B$ is seen to have a reserve of stability. As shown in [8.35], the relation $A(\boldsymbol{r},t) = B(\boldsymbol{r},t)$ is also valid on the nonlinear stage of motion. Limiting ourselves for simplicity by the case $\omega'' > 0$ we can reduce (8.4.1) by changing the scale

$$i\left(\frac{\partial}{\partial t} + \gamma\right)A - hVA^* + \frac{1}{2}\Delta A = [(2S+T)|A|^2 - TA_0^2]A . \quad (8.4.4)$$

Here we determined $\omega(\boldsymbol{k}_0)$ by the condition of external stability. Thus the most stable standing wave is selected for which, as shown in [8.35], the area of the positive increment in the $\boldsymbol{k}$-space is limited: $\nu(\kappa) > 0$ at $\kappa < \kappa_c$ where $\kappa_c \ll k_0$. The nonlinear instability stage of the plane standing wave for the further development will be studied within the scope of (8.4.4) formulated here.

8.4.2 Stationary Solitons

The simplest variant of nonlinear behavior of the system is its transition to the stationary state different from the plane standing wave $A = \text{const}$. Therefore in this section we shall consider the stationary solutions of the basic equations (8.4.4). As a rule, these solutions are deep non-sinusoidal variations of the amplitude $A(\boldsymbol{r})$ and phase $\Phi(\boldsymbol{r})$ and they can naturally be called *solitons*.

The solitons with a constant phase $\Phi(\boldsymbol{r}) = \Phi_0$ can be studied most conveniently. It follows from (8.4.4) that

$$hV \sin \Phi_0 = \gamma ,$$

and for $C = |A(\boldsymbol{r})|$ we have the equation

$$\Delta C = C[hV \cos \Phi_0 - TC_0^2/2 + FC^2] . \qquad (8.4.5)$$

where C_0 must be understood as the initial amplitude of the pair. Taking into account that $hV \cos \Phi_0 + SC_0^2 = 0$ let us write (8.4.5) as

$$\Delta C = -dU/dC . \qquad (8.4.6)$$

For the plane solitons $C(x)$ this equation describes the motion of the nonlinear oscillator with the coordinate C in the potential field $U(C) = FC^2(2C_0^2 - C^2)/8$ where x acts as time. The behavior of the solution is characterized by the "energy" of the oscillator. The case $F > 0$ is the most interesting because in this case there exist solutions with a little difference from the plane wave. They are realized at E close to $E_{\min} = FC_0^2/8$ and are weakly modulated plane wave $C(x) = C_0 - C_1 \cos \kappa_0 x$, where $\kappa_0^2 = -FC_0^2$. Such solutions ($C_\perp \ll C_0$) we shall call *small solitons*. As the energy increases the amplitude and the oscillation period increase. At $E = 0$

$$C(x) = 2\sqrt{C_0}/\cosh(x\kappa_0/\sqrt{2}) \qquad (8.4.7)$$

and passes to the solitary wave, i.e. the *single soliton* which is the analogue of the plane self-focusing beam in the nonlinear medium. At $E > 0$ the solution resumes its periodic structure, but now it changes within the limits symmetrical about zero. Small solitons with the variable phase

$$\Phi(\boldsymbol{r}) = \Phi_0 + \Phi_1 \cos \boldsymbol{\kappa r} , \quad C(\boldsymbol{r}) = C_0 + C_1 \cos \boldsymbol{\kappa r} \qquad (8.4.8)$$

have the size

$$|\kappa|^2 = |\kappa_1|^2 = -2SC_0^2 \;, \quad \gamma \Phi_1 = TC_0 C_1 \;. \tag{8.4.9}$$

We draw attention once more to the evident fact that the characteristic dimensions κ_0 and κ_1 of the small solitons correspond to those κ for which the increment of the plane wave described in [8.35] becomes zero:

$$(\nu + \gamma)^2 - \gamma^2 = \frac{1}{2}(\kappa^2 - \kappa_0^2)(\kappa^2 - \kappa_1^2) \;, \quad \kappa_0^2 = -(2S + T)C_0^2 \;.$$

When studying very deep solitons with a variable phase when in (8.4.4) the damping and pumping can be neglected we again come to (8.4.6) with the potential

$$U = -FA^4/8 + \alpha^2/2A^2 \;, \tag{8.4.10}$$

where α is an arbitrary constant. There are also deep solitons moving with a constant velocity. In this case the term $v^2 C^2/8$ is added to the potential of the deep solitons.

The analysis performed by *L'vov* and *Rubenchik* [8.36] enables us to assume that all the stationary solitons in (8.4.4) are unstable. In the same work the initial nonlinear stage of this instability was treated and special attention was paid to the case when the increment was abnormally small.

Interestingly, the nonlinear interaction becomes significant at very small amplitudes. Nevertheless, it does not lead to the limitation of the amplitudes. There is only a decrease in the rate at which the initial perturbation grows thus slowing it down significantly. In this case the amplitude increases as $\sqrt{t}$. A narrow in the $\boldsymbol{k}$-space wave packet is generated ($\kappa \sim \kappa_0 \ll k_0$, $\Delta \kappa \sim \kappa_0$). Such a state is highly turbulent and will be studied in the next section.

8.4.3 Average Characteristics of Secondary Turbulence

First, let us estimate the width of the area excited in the $\boldsymbol{k}$-space at arbitrary supercriticality. It follows from the basic S-theory that in the case under discussion when $V(\boldsymbol{k})$ has its maximum at the point $\boldsymbol{k}_0$ the packet of parametrically excited waves relaxes to the standing monochromatic wave with $\boldsymbol{k} = \boldsymbol{k}_0$ if the individual wave phases can be considered random. It is sufficient for the phase randomness that the phases of two waves in a packet have to diverge at a value of about unity over the time less than the characteristic time of the nonlinear interaction. This takes place in a packet with $(\Delta k)^2 \gg \kappa_0^2 = SA_0^2/\omega''(\boldsymbol{k})$. Therefore the packet with $\Delta k \gg \kappa_0$ narrows up to the size $\sim \kappa_0$ and its mean amplitude relaxes to the value $\sim A_0$. But if $\Delta k \ll \kappa_0$, then such a packet is unstable with the increment (8.4.3) with respect to the perturbations with $\kappa \sim \kappa_0$ and, consequently, will broaden up to $\Delta k \sim \kappa_0 \sim \sqrt{SA_0^2/\omega''}$. Note that in the whole area of the turbulent motion the instability increment (8.4.3) is positive in the narrow layer $\kappa v \leq SA_0^2$ close to the plane $\boldsymbol{\kappa} \perp \boldsymbol{v}$. So the turbulence considered is almost two-dimensional, namely:

$$(\kappa_\parallel/\kappa_\perp)^2 \sim SA_0^2/\omega(k) \ll 1 \ . \tag{8.4.11}$$

The mean level of turbulence A^2 cannot differ greatly from A_0^2 given by the (8.4.2). Indeed, as has already been noted, it was shown in [8.35] that the monochromatic plane wave is stable with respect to the short wave perturbations with $\kappa \simeq k_0$ only if the amplitude equals to A_0. Evidently, such instability is retained also for the pair modulated with $\kappa \ll k_0$. So if A_0^2 differs appreciably from A^2, short wave modulations are excited, which contradicts the above-found narrowness of the packet. Therefore, the instability development of the plane wave leads to a strong quasi-two-dimensional wave turbulence of the modulation $A(\boldsymbol{r}, t)$ with a mean level

$$\langle A^2 \rangle \simeq \langle A_0^2 \rangle = \sqrt{h^2 V^2 - \gamma^2}/2S \ , \tag{8.4.12}$$

and with the modulation depth of the order unity, characteristic frequency of motion $(hV - \gamma)$ and with the characteristic scale in the coordinate space $r_\perp \sim \kappa_0^{-1} > k_0^{-1}\sqrt{\omega(k)/2A_0^2}$. We can say that there arises a dynamic soliton structure with the coherence length of the order of the soliton size, i.e. $\sqrt{\omega''/S}/A_0$ which changes significantly in space over a time $1/(hV - \gamma)$.

8.4.4 Destruction of Parametric Solitons with Large Amplitude

In the areas where the soliton amplitude during the turbulent motion proved to be anomalously large $A \gg A_0$ damping and pumping in the equations of motion can be neglected, because over the characteristic time of the problem $1/SA^2$ the system will not have enough time for significant energy exchange with the thermal bath and pumping. In this approximation (8.4.1) describe the non-stationary behavior of wave pair in a conservative medium. The behavior of one almost monochromatic wave in the nonlinear medium has been experimentally and theoretically studied using computers in association with the problems of nonlinear optics [8.37], plasma and hydrodynamics [8.38, 39]. The phenomenon of light self-focusing was discovered [8.40]. Later it was shown that the self-focused light beam is unstable, in some cases this instability leads to the short-time collapse of the beam [8.41, 42].

It will be shown below that similar phenomena take place also in our case of the wave pair. The direct calculation can show that (8.4.1) have the following integral of motion:

$$\begin{aligned}I = & \frac{\omega''}{2} \int (|\nabla B|^2)\, d\boldsymbol{r} + \\ & + \frac{T}{2} \int (|A|^4 + |B|^4)\, d\boldsymbol{r} + 2S \int |A|^2 |B|^2\, d\boldsymbol{r} \ .\end{aligned} \tag{8.4.13}$$

Let us show that its sign significantly determines the system evolution. Let us consider the second derivative of the clearly positive R^2

$$R^2 = \frac{\omega''}{2} \int r^2(|A|^2 + |B|^2)\,d\mathbf{r} > 0 \,. \tag{8.4.14}$$

Direct calculation with the help of (8.4.1) where (with respect to the parameter $\sqrt{SA_0^2/\omega} \ll 1$) the second derivatives with respect to z can be neglected shows (see Sec. 1.5.4) that $d^2R^2/dt^2 = 2I$. Thus,

$$R(t) = It^2 + 2\alpha t + \beta \,, \tag{8.4.15}$$

where α and β are the constant integrations. Clearly, over a finite time $R^2(t)$ becomes negative when $I < 0$; this fact contradicts (8.4.14). This means that the solution of (8.4.1) "breaks" over the finite time.

Let us study this phenomenon in more detail. To this end, within the frame of the two-dimensional (8.4.4) (where $A(x,y) = B(x,y)$) let us compute the time evolution of the axially symmetrical initial distribution

$$A(\mathbf{r},0) = A_0[1 - K(a-r)\exp(-2r^2/\sqrt{\pi}a^2)] \,, \tag{8.4.16}$$

simulating the local increase of the amplitude spontaneously arising during the turbulent motion. It would be natural to select a quarter of the length of the envelope corresponding to the maximum increment as a characteristic length a where $A(r,0) - A_0$ decreases to zero. It gives the estimate

$$a^2 = \pi^2 S^2 \omega'' / \sqrt{(4S+T)^2(h^2V^2 - \gamma^2)} \,. \tag{8.4.17}$$

Figures 8.10, 11 show the results of the computer simulation [8.43]: the evolution of the amplitude $|A(0,t)|$ and of the phase $\Psi(0,t)$ in the center of the packet (Fig 8.10), amplitude distribution $|A(\mathbf{r},t)|$ for some characteristic times (Fig. 8.11). Clearly, there is some "critical" modulation depth K_c, such that at $K > K_c$ the packet collapses over the finite time whereas the amplitude in the center increases indefinitely. The critical values of K_c are given in Table 8.2.

Table 8.2. Dependence of critical initial amplitude K_c on the pumping amplitude: $K_1 < K_c < K_2$ (for $K = K_1$ the packet still spreads, whereas for $K = K_2$ it collapses)

$\frac{hV}{\gamma} - 1$	K_1	K_2	$\frac{hV}{\gamma} - 1$	K_1	K_2
10^{-3}	23.0	24.0	1	3.0	3.1
10^{-2}	10.5	11.0	2	2.6	2.7
10^{-1}	5.4	5.5	3	–	0.5
0.4	3.4	3.5			

It can be seen that at $hV - \gamma \ll \gamma$ the critical amplitude $K_c \gg 1$. To understand this fact, note that the phase $\Psi(\mathbf{r},t)$ near the focus $r = 0$ of

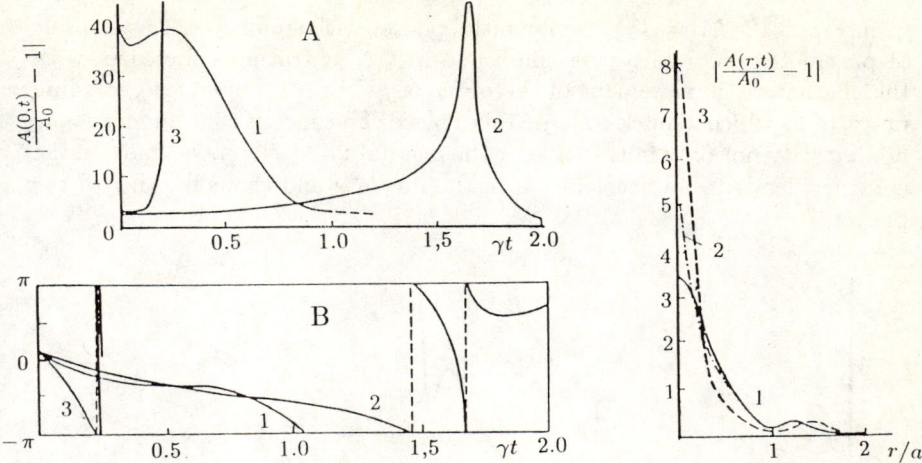

Fig. 8.10. (left) Evolution of the amplitude $A(0,t)$ (A) and phase $\Psi(0,t)$ (B) at the center of a packet for $hV = 2\gamma$, $T = -S > 0$. Curves 1, 2 and 3 correspond to $k = 2.3$ and 3.5, respectively. For the sake of clarity the vertical scale for curve 3A is increased by a factor of 20

Fig. 8.11. (right) Distribution $A(r,t)$, for $hV = 2\gamma$, $T = -S > 0$, $k = 3.5$. Curves 1, 2 and 3 correspond to $\gamma t = 0.05$, 0.125 and 0.2, respectively; $a = 0.8\sqrt{\omega''/\gamma}$

the collapsing packet increases monotonically. This was analytically shown in [8.44] and is clear from Fig. 8.11 at $t > 0.2$. The rotation of the packet phase with respect to the pumping phase leads to the fact that the energy flux to the vicinity of the focus stops; therefore for the collapse the value SA^2 must be at least of the order γ, which significantly exceeds the mean level of turbulence $SA_0^2 \simeq \sqrt{h^2V^2 - \gamma^2}$ in the range $hV - \gamma \ll \gamma$. This result (($K_c \gg 1$ at $hV - \gamma \ll \gamma$) is weakly dependent on the choice of the initial distribution phase.

8.4.5 Soliton Mechanism of Amplitude Limitation

As already noted the depth of the amplitude modulation in the turbulent motion is of the order unity, and therefore the probability of formation of solitons with the amplitude significantly exceeding A_0, is exponentially small. This implies (see Table 8.2) that slightly above the threshold there ($hV - \gamma \leq \gamma$) is almost no soliton collapse. It is clear from Fig. 8.10 that the solitons with the amplitude $1 < K < K_c$ are dispersed. With increasing h the critical value K_c decreases and there is a characteristic amplitude h_c at which $K_c \simeq 1$. Note that the kind of initial distribution (8.4.16) is to a large extent arbitrary, and afterwards under amplitude h_c we shall understand such value of the pumping amplitude above which practically any area with the characteristic diameter $1/\kappa_0$ is covered during the collapse. As is seen from Fig. 8.12 ($hV = 4\gamma$) for the initial packet with small $K < 1$ as a result of parametric instability the amplitude in the central area increases with the characteristic increment of the order of hV up to the values exceeding unity, after which a quick collapse takes place. This means that under $h > h_c$ nonlinearity not only fails to hinder the instability development, but, on the contrary, leads to an acceleration of the increase and the collapsing of the packet.

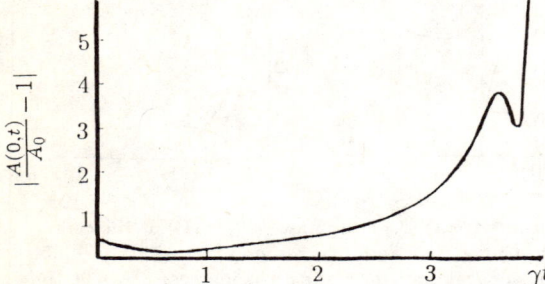

Fig. 8.12. Evolution of $A(0,t)$ at the center of a packet for $hV = 4\gamma$, $T = -S$

Let us consider the phenomena occurring at $h > h_c$. Evidently on studying the evolution of the collapsing soliton the influence of the pumping and damping can be neglected. Equation (8.4.4) in this case is transformed to the nonlinear parabolic equation. The properties of such equations have been studied in detail in connection with the problem of the light self-focusing. As shown in [8.40], the amplitude in the soliton center rapidly increases in time: $A(0,t) \sim (t-t_0)^{-2/3}$. At the same time its radius decreases so that the process of collapse entraps the strictly definite amount of energy

$$I_c = \omega \int |A|^2 \, d\mathbf{r} = 1.86 \, \omega'' \omega / |2S + T| \, . \tag{8.4.18}$$

When the wave amplitude in the collapsing soliton is large enough the nonlinear damping which leads to the fast damping of the soliton energy becomes essential. The effective nonlinear damping can be estimated as $\gamma_{\rm NL} \sim I_c \kappa_0^2 / \tau \omega A_0^2 \sim 1/\tau$, where τ is the time between two successive collapses in the area of the size κ_0^2. According to the dimensional estimation $\tau \sim 1/hV$ at $h > h_c$. Taking into account this mechanism of energy dissipation leads to the fact that the mean amplitude of the turbulent pulsations $\langle A \rangle$ becomes less than A_0, and the susceptibility χ'' with the increasing amplitude h does not decrease but reaches the plateau whose order of magnitude coincides with the maximum χ''.

A promising method of experimental research of the strong soliton turbulence of parametrically excited spin waves is the measurement of the spectral density of the electro-magnetic irradiation of the ferromagnet at the frequencies close to the pumping frequency $\omega_{\rm p}$. At $h < h_c$ when the collapses are rare the spectral density of the noise $(h^2)_\omega$ is close to the Gaussian with the width $|\omega - \omega_p| \sim \gamma\sqrt{p-1}$. At $h > h_c$ the significant contribution to irradiation is made by the collapsing solitons where the phase of the pair $\Psi(r,t)$ breaks away from the pumping phase and begins to rotate quickly by the law $\Psi(t) \sim (t-t_c)^{-1/3}$ (see [8.41]). The time of breaking away of the phase and several first turns are distinct in Fig. 8.11.

The fast rotation of the phase leads to the significant broadening of the irradiation spectrum $(h^2)_\omega$ which can be used in recording the collapses. By making use of [8.44] we can show that $(h^2)_\omega \sim (\omega - \omega_p)^{-7/4}$. Non-linear damping limiting the amplitude in the collapse $A < A_{\max}$ must cut the irradiation at the frequency $SA_{\max}^2 \sim |\omega_{\max} - \omega_p|$.

9 Experimental Investigations of Parametrically Excited Magnons

9.1 Experimental Investigations of Parametric Instability of Magnons

9.1.1 Methods and Materials Investigated

As already mentioned in Chap. 5, the first experiments on the excitation of magnons by parallel pumping were carried out quite a while ago (in 1960) [9.1–3] on Yittrium Iron Garnet (YIG) which still remains one of the most popular objects used in this type of research. Today, parametric instability has been observed in a number of magnets. In addition to ferrites, these also included low-dimensional ferromagnets and antiferromagnets as well as three-dimensional antiferromagnets with various magnetic symmetries.

Research is carried out in the range of the microwave frequencies. Usually simple microwave spectrometers with direct amplification are used. Figure 9.1 (left) outlines such a setup [9.4]. The sample, oriented in a certain direction, is placed into the resonator at the antinode of the microwave magnetic field $h \parallel H$. The operating conditions of the generator are selected in accordance with the specific aim of the investigations. Parametric excitation of magnons was registered when the additional absorption appeared at sufficiently high microwave power. Under pulsed operation of the generator the development of the instability in time can be traced. The amplitude of the microwave magnetic field h in the sample is calculated from the magnitude of the power in the resonator and the parameters of the resonator. The threshold field h_{th} above which the parametric excitation appears is definitively related to the damping of the excited magnons $\gamma(\boldsymbol{k})$ by (4.3.21).

The calculation of $\gamma(\boldsymbol{k})$ from h_{th} is a traditional procedure in the investigations of damping. Most of the research conducted on parallel pumping is based on such traditional studies and looks for the specific mechanisms of magnon interaction with other excitations and with the inhomogeneities of the sample. The nature of the stationary state above the threshold of the parametric excitation was studied only on YIG samples and easy-plane antiferromagnets, such as $MnCO_3$, $CsMnF_3$ and $FeBO_3$. In the present chapter, we shall be concerned primarily with the experimental study of these substances. Let us first consider the measurement of constants in the spectrum of magnons $\omega(\boldsymbol{k})$ because this is of special importance for what follows.

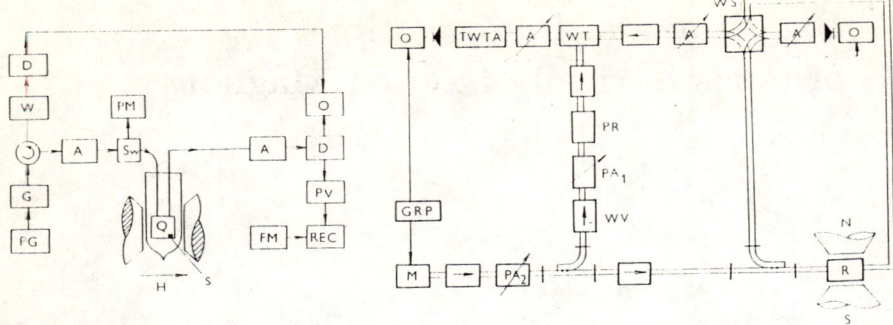

Fig. 9.1. On the left: Block diagram of the spectrometer for observing parametric excitation of magnons: G, microwave generator; A, attenuator; D, crystal detector; Sw, switch; W, wavemeter ; PM, peak-value voltmeter; FM, field meter; REC, recorder; PG, pulse generator; O, oscilloscope; S, sample (*Kotyuzhansky* and *Prozorova* [9.4]). On the right: Block diagram of the experimental setup for measuring the susceptibility χ (*Melkov* and *Krutsenko* [9.24]: M, magnetron; GRP, generator of rectangular synchronization pulses; WV, waveguide valve; WT, waveguide tee; PA, precision attenuator; PR, phase rotator; TWTA, travelling wave tube amplifier; O, oscillograph; A, attenuator; WS, waveguide switch; R, resonator

9.1.2 Measurements of Constants in Spin Wave Spectra

There exist numerous experimental methods to determine the constants of the spin wave spectrum. In order to obtain "the spectral gap", i.e. the frequency of the uniform oscillations, the electron magnetic resonance (either ferromagnetic or antiferromagnetic) is usually employed to determine the frequency. This method provides maximum accuracy. The frequencies of the inhomogeneous exchange ω_{ex} can be obtained from the temperature dependences of the specific heat and magnetization, data on inelastic neutron scattering, two-magnon light scattering, etc. We shall dwell on two kinds of experiments where these constants are obtained by using the parametric excitation of spin waves by parallel pumping.

1 Size effect This effect was discovered by *Jants* and *Schneider* and *Andlauer* in experiments on mono crystalline spheres of YIG [9.5] and afterwards it was observed by *Kotyuzhansky* and *Prozorova* on FeBO$_3$ plates [9.6]. The effect is as follows: At low damping of spin waves when their free path is longer than the sample dimensions the power passing through the resonator is not a monotonic function of H, but has dips of a resonance type under certain values of H_n. The observed phenomenon can naturally be attributed to the effect of the boundary conditions on the sample surface leading to the space resonance of the spin wave at

$$n\lambda = 2d \, . \tag{9.1.1}$$

Here d is the dimension of the sample, $\lambda = 2\pi/k$ is the spin wave length and n is an integer. (As a rule, in these experiments $d \simeq 0.1$ cm, $k = 10^4-10^5$ and, consequently, $n \simeq 10^3$). This explanation was confirmed by the experiments on samples of different sizes. The results of one such experiment are shown in Fig. 9.2. The distance between the dips ΔH_n can be calculated from condition (9.1.1) and the formulae describing the spin wave spectra. From the comparison of the experimental and theoretical values of ΔH_n the frequency ω_{ex} can be calculated.

2 Threshold anomaly in the intersection region of spin waves and phonon spectra. The magnon and phonon branches of the energy spectra can have points of intersection. If the crystal symmetry allows the interaction of these branches, in the vicinity of these intersection points the spectra will be distorted. There are "mixed" magnetoelastic oscillations in these regions whose damping differs from the damping of magnons. Therefore h_{th} can change significantly near such intersection points. This phenomenon was first observed in monocrystals. Later, this effect was studied on easy-plane antiferromagnets (in particular, on $CsMnF_3$ and $MnCO_3$ [9.8–11]) whose spectra were intersection points at nonzero fields.

Figure 9.3.a shows such anomalies which are peak-shaped dependence of the power absorbed by the sample on magnetic field H changed. The experiments were performed by *Kotyuzhansky* and *Prozorova* [9.11] in $CsMnF_3$ at various pumping frequencies ω_p. The wave numbers at which anomalies were observed versus the spin wave frequencies are plotted in Fig. 9.3b. This dependence agrees with the dispersion law of the elementary excitation with which spin waves interact. The direct proportionality of ω and k observed in this case confirms that spin waves interact with sound waves. Using the value of the sound velocity k at the intersection point one can obtain the value of the frequency ω_{ex} in the spin wave spectrum.

9.1.3 Spin Wave Damping

The rate of damping $\gamma(\boldsymbol{k})$ with $da/dt = -\gamma a$ where a is the amplitude of the wave, is a very important characteristic of large practical significance. A detailed discussion of experimental research into the nature of relaxation is beyond the scope of our book. We shall therefore restrict ourselves to a summary of the principal relaxation processes in the magnets under study. At k not too high, far from the Brillouin zone edge, magnon damping is determined by three kinds of processes:

1 Intrinsic processes including three- and four-magnon relaxation as well as the relaxation of magnons caused by their interaction with phonons and nuclei;

2 Two-magnon processes of inelastic scattering by inhomogeneities (pores, polycrystallinity, deviations from stoichiometry, etc.);

3 "Slow" and "fast" relaxation, i.e. processes involving impurity ions with strong spin-orbit coupling.

Quite naturally we shall be interested mostly in the "eigen" relaxation because the other contributions to the damping can be substantially reduced by the selection of more perfect samples. In addition, for the investigation of the above-threshold state it is of great importance to know the dependence of the magnon damping on their number.

The simplest and most universal method for determining $\gamma(\boldsymbol{k})$ of the magnons is via calculation on the basis of the experimental values of the threshold of parametric excitation of magnons h_{th}. The absolute accuracy of this method is not high (the error is usually 20–30%) because when obtaining h_{th}, several quantities are employed which were measured with an accuracy of several per cent. These include the power input of the resonator, the coupling coefficient with the resonator, the Q factor of the resonator. In addition, the distortion of the field distribution in the resonator due to the magnetic sample cannot accurately be allowed for. However, the relative error of this method for finding the damping is only about 2%, which makes it very convenient for studying the relaxation dependence on various external parameters. Calculations are performed according to (4.3.21): $h_{\text{th}} V(\boldsymbol{k}) = \gamma(\boldsymbol{k})$. The coefficient $V(\boldsymbol{k})$ in this formula depends on the kind of magnetic system being investigated and on the method of parametric excitation of magnons. The coefficients $V(\boldsymbol{k})$ for the cubic ferrimagnet and for the easy-plane antiferromagnet were given in Sect. 4.3. All the relaxation process that can occur in real experiments on these magnets have been theoretically calculated. By comparing the experimentally obtained functional dependences $\gamma(\boldsymbol{k}, H, T)$ with the theoretical prediction, the particular contributions of the specific relaxation process to the total value γ can be found. It must be noted that for YIG, $MnCO_3$, $CSMnF_3$ and $FeBO_3$ theoretical and experimental data are in fairly good agreement.

A. Yittrium Iron Garnet The most comprehensive research on relaxation in YIG was performed experimentally by *Anisimov* and *Gurevich* [9.12, 13] and theoretically (allowing for the complex magnetic structure of YIG) by *Kolokolov, L'vov* and *Cherepanov* [9.14, 15]. Figure 9.4 shows in a plot of temperature T versus wave vector k the regions where from the theoretical view-point of [9.15] the contribution of one or the other process predominates. For long spin waves $k < k_e$ ($ak_e \simeq 2\omega_{\text{ex}}/\omega_i$) the three-magnon processes in the quasi-ferromagnetic branch of the spectrum are forbidden by the conservation laws, and the amplitude of the exchange four-magnon processes is very small. The damping $\gamma(\boldsymbol{k})$ under $k \to 0$ was first experimentally studied by *Kasuya* and *Le Craw* [9.16] who assumed that the damp-

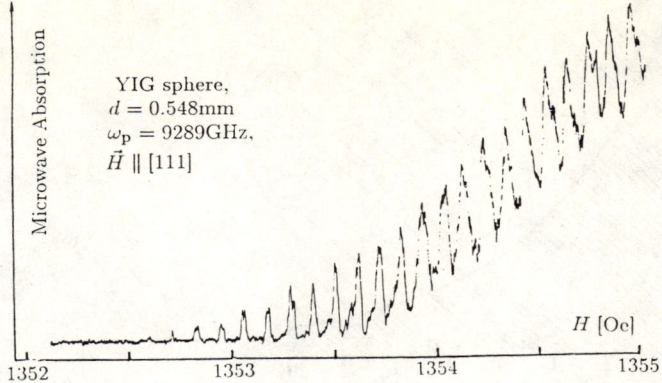

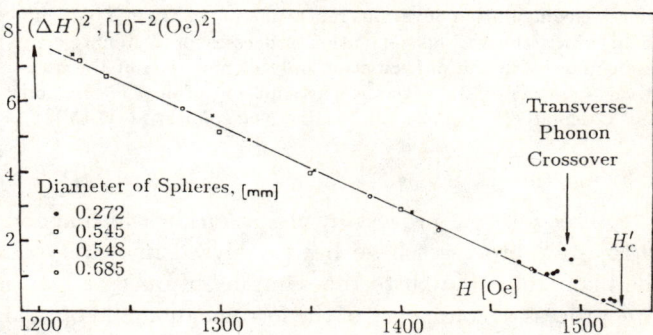

Fig. 9.2. (a) Absorbed microwave power versus magnetic field upon excitation of spin waves by parallel pumping in a YIG sphere (*Jantz* et al. [9.5]). (b) Intervals ΔH between absorption peaks versus magnetic field. Results obtained for four spheres of different diameters d, normalized to $d = 0.3$ mm

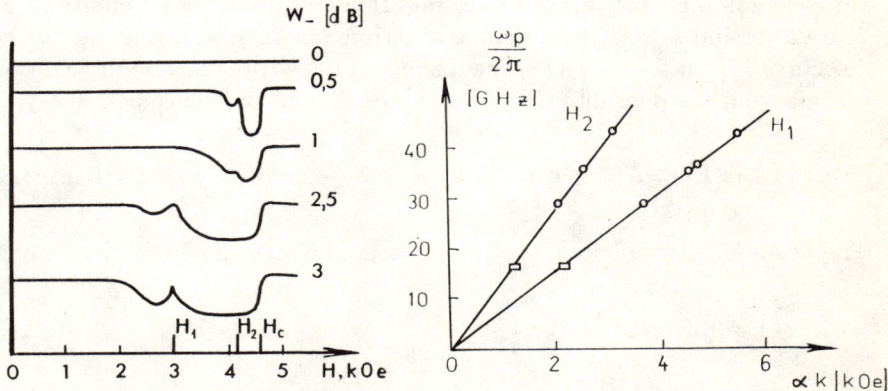

Fig. 9.3. (a) Absorbed microwave power versus magnetic field upon excitation of spin waves by parallel pumping in a YIG sphere. (b) Intervals ΔH between absorption peaks versus magnetic field. Results obtained for four spheres of different diameters d, normalized to $d = 0.3$ mm (*Jantz* et al. [9.5])

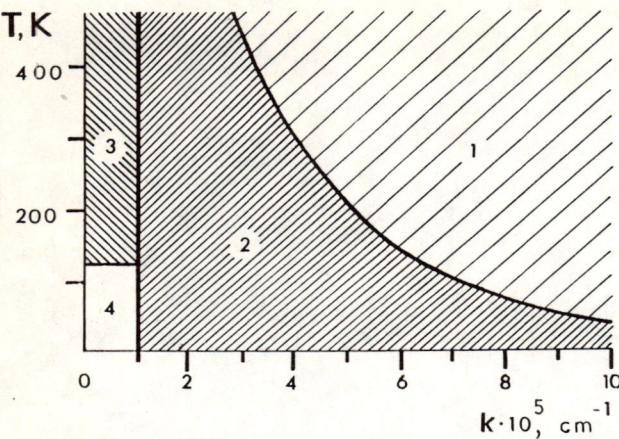

Fig. 9.4. Diagram of the relative contribution of various relaxation processes in YIG. Numbers indicate the region in which the various relaxation processes predominate: (1), four-magnon; (2), three-magnon with the participation only of magnons of the quasi-ferromagnetic branch of the spectrum; (3), with the participation of optical magnons; (4), scattering by defects. Calculated for $\omega_k/2\pi = 4.7$ GHz (*Kolokolov* et al. [9.15])

ing was caused by the three-particle process involving an optical magnon and a phonon. In [9.15] such processes have been analyzed in detail and it was shown that the main contribution to the damping of magnons with $k \to 0$ was made by the process of confluence of the ferromagnon and optical magnons. The amplitude of this process is proportional to the energy of a uniaxial crystallographic anisotropy of ions Fe^{3+} in the octahedral sites. At low temperatures $T < 120$ K the damping $\gamma(0)$ exponentially decreases. In this temperature range (range 4) the main eigen process of relaxation is the four-magnon magneto-dipole scattering. However, under real conditions at $T < 120$ K and $k < k_e$ the damping of ferromagnons is as a rule due to the defects [9.17]. In the range (2) the magneto-dipole three-magnon relaxation prevails which is determined by two processes, i.e. confluence and decay

$$\omega(\mathbf{k}) + \omega(\mathbf{k'}) = \omega(\mathbf{k} + \mathbf{k'}), \quad \omega(\mathbf{k}) = \omega(\mathbf{k'}) + \omega(\mathbf{k} - \mathbf{k'}) . \qquad (9.1.2, 3)$$

The decay processes are allowed at $k > k_S$. The quantity k_S is given by

$$\omega(k_S) = 2\omega(k_S/2) . \qquad (9.1.4)$$

Both in confluence and decay processes $\gamma(\mathbf{k}) \propto T$, but the dependence γ of k differs for these processes. Experimentally, at $k = k_S$ a characteristic point is observed in the experimental γ–versus–k curve. In the range (1) the four-magnon exchange scattering is the main process and

$$\gamma(\boldsymbol{k}) = \text{const} \cdot \omega(\boldsymbol{k})(ak)^2 \left(\frac{T}{\omega_{\text{ex}}}\right)^2 [\ln^2\left(\frac{T}{\omega(\boldsymbol{k})}\right)$$
$$- 3.3\ln\left(\frac{T}{\omega(\boldsymbol{k})}\right) - 0.3] \text{ at } T < 200 \text{ K}, \qquad (9.1.5)$$
$$\gamma(\boldsymbol{k}) = \text{const } T\omega(k)(ask)^2(T/\omega_{\text{ex}})^4 \text{ at } 350 \text{ K} > T > 200 \text{ K}$$

At room temperature in YIG $\gamma(0) = 2.4 \cdot 10^6$ s^{-1}, $\gamma(k)$ is the k-dependent part of the damping γ being of the same order of magnitude at $k \simeq 10^5$ cm.

B Antiferromagnets with anisotropy of the "easy-plane" type: MnCO$_3$ and CsMnF$_3$.

Relaxation of magnons in these antiferromagnets was investigated within the temperature range 1.2–4.2 K and in the frequency range $\omega(\boldsymbol{k})$ from 10 to 20 Hz [9.8–12]. It was shown that under these conditions $\gamma(\boldsymbol{k})$ varies from 0.1 to 10 MHz. For the relaxation of magnon at liquid helium temperatures the following facts are essential: (i) The presence of a low-frequency quasi-ferromagnet branch of the spectrum ($\omega_2(0)/2\pi \simeq 10^{11}$ Hz); (ii) The velocities of magnons and phonons are close; (iii) There is strong hyperfine interaction in antiferromagnets with Mn^{2+} ions. According to theory the following processes of relaxation are important:

1. confluence of two magnons of the quasiferromagnetic branch into a magnon of the quasiferromagnetic branch γ_{3m},

2. magnon-phonon interaction $\gamma_{m,ph}$,

3. magnon scattering on the paramagnetic subsystem of ^{55}Mn nuclei- γ_{mn}.

The relaxation parameters γ_{3m}, $\gamma_{m,ph}$ and γ_{mn}, have different functional dependences, and therefore the contributions of each mechanism can be separated. The first process is basic at comparatively high temperatures and strong magnetic fields $\gamma_{3m} \propto H^3 \exp T$. The quality γ_{3m} coincides with the theoretical value γ_{3m} calculated by *Sobolev* and given in [9.19]. For the second process *Lutovinov* obtained $\gamma_{m,ph} \sim T\Theta^2$ [9.19] (where Θ is the constant of the magnetoelastic interaction). As the temperature decreases the process of magnon scattering on magnetization fluctuations of the nuclear subsystems becomes predominant. As follows from theoretical research by *Woolsey* and *White* $\gamma_{mn} \sim AI(I+1)k/\omega_p$ [9.21], (where A denotes the constant of the hyperfine interaction, I is the spin of the nucleus), in all the investigated substances the experimental value of γ_{mn} is in good agreement with the theoretical result [9.21]. Therefore all these processes in combination enable us to describe the relaxation at helium temperatures in the antiferromagnets MnCO$_3$ and CsMnF$_3$.

FeBO$_3$: The magnon damping was studied in a wider temperature range, i.e. from 1.2 to 150 K [9.18] (The measurements were taken at the pumping frequency $\omega_p/2\pi = 36$ GHz, γ was changed within the range 1 –50 MHz). Unfortunately it was not possible in this case to avoid the "slow relaxation" caused by the presence of Fe^{2+} ions which results in maximum at $T \simeq 18$ K. At higher temperatures, the damping is determined by three-magnon processes, at lower temperatures it is mainly influenced by the magnon-phonon processes. The energy spectrum qualitatively differs from the $MnCO_3$ and $CsMnF_3$ spectra – the magnon and phonon branches do not intersect, because the limiting "magnon velocity" $a\omega_{ex} > u$ (u is the sound velocity). In this connection, in addition to the process of confluence involving the phonons in the region of fields where the group velocity of magnons is lower than the sound velocity, the process of decay into a magnon and phonon is also allowed.

9.2 Nonlinear Behavior of Parametric Magnons – General Information

9.2.1 Measuring Technique for Susceptibilities χ' and χ''

A traditional method for investigating the above-threshold state of magnons is to measure the high-frequency susceptibility $\chi = \chi' + i\chi''$ above the parametric excitation of magnons. These values are connected by (5.5.31) with the distribution function $\sigma(\boldsymbol{k})$. It is possible to say that the value χ is determined by the total number of parametric waves N and by its phase Ψ. The shape of χ' and χ'' dependences on the pumping intensity significantly depends on the mechanism of amplitude limitation (see (5.5.34, 35)). In traditional experimental measurements of the magnetic susceptibility the reaction of a high-quality resonator to the changed state of the sample inside it is usually employed. The resulting quality change of the resonator determines the imaginary part of the susceptibility χ'', and the change in the eigenfrequency determines the real part of χ'. The measuring technique of the nonlinear susceptibilities under parallel pumping differs from the standard measuring techniques in some specific features, i.e. pulsed operation of the microwave generator, wide field range where the energy absorption is observed, χ' and χ'' dependence on the pumping intensity. The latter leads to the fact that the power of the generator is not a good measure of the pumping field amplitude. The field h in the resonance is more conveniently determined from the output power of the resonator. This enables one to automatically allow for the feedback effect of spin waves on the pumping which is substantial near the threshold even if the filling factor of the resonator is small.

The typical experimental setup for the study of parallel pumping is shown in Fig. 9.1 (on the left) [9.22]. The principle of its operation is as follows. The excitation of spin waves deteriorates the loading quality of the resonator Q and its eigenfrequency ω_0 is changed, which brings about the echo signal informing us about the values of χ' and χ''. The quantity χ''

is connected with the reflection coefficient Γ in the exact resonance (at $\omega = \omega_0$) by the following relation [9.23]

$$\chi'' = AQ_H\Gamma/(1-\Gamma) \,, \tag{9.2.1}$$

where $A = 2\pi \int_{VS} h^2 d\boldsymbol{r} / \int_{VC} h^2 d\boldsymbol{r}$ is the filling factor of the resonator equal to the ratio of the integrals over the sample volume VS and resonator volume VC. The real part of the susceptibility χ' is directly determined by the frequency shift $\Delta\omega = \omega_{\text{res}} - \omega_{\text{res},0}$ where $\omega_{\text{res},0}$ is the eigenfrequency of the empty resonator and ω_{res} is the eigenfrequency of the resonator-sample system. These frequencies correspond to the minimum of the reflection coefficients (for the empty resonator and for resonator, wich contains a sample). For several reasons, mainly because of the inaccurately determined coefficient A, the measuring error of absolute values χ' and χ'' is usually of the order of 20–40 %. All this has result in a wide spread of experimental data of different researchers and no unambiguous interpretation of the nature of the above-threshold state has been put forward.

Melkov and *Krutsenko* in their research of the above-threshold state in YIG [9.24] significantly improved the measuring technique and the measuring error was reduced to several per cent. The block diagram of the experimental setup is shown in Fig. 9.1 (on the right). The amplitude and phase of oscillations in the resonator are measured by comparing the signal that had passed through it with the reference signal. When the pumping amplitude is below the threshold value ($h < h_{\text{th}}$) the signal is compensated to zero. When $h > h_{\text{th}}$ it results in the magnetization in the sample $m_z(\omega_p) = \chi h(\omega_p)$ changing with the pumping frequency and directed along $\boldsymbol{H}$. The magnetization m_z brings about the additional microwave magnetic field in resonator $h_p(m_z)$. It causes the changes in the amplitude and phase of the total self-consistent magnetic field of the resonator with respect to the amplitude and the phase of the pumping. These changes can be measured by a phase rotator PR and amplitude attenuator PA placed in the reference signal channel by the compensation of the mismatch signal due to the parametric excitation of magnons. Evidently, if the change in the amplitude $a = h/h_{\text{th}}$ and in the field phase ψ in the resonator are known, information about the microwave magnetic field of reaction $h_p(m_z)$ and, consequently, about the nonlinear susceptibility can be obtained

$$\chi' = -4\pi\omega_p^2 aA \sin\psi \,, \quad \chi'' = 4\pi\omega_p^2 aA \sin\psi[\text{ctg}\psi - (\sin\psi)/a] \,. \tag{9.2.2}$$

Clearly, from χ the phase of the pair Ψ can be calculated since $\text{tg}\Psi = \chi''/\chi'$. Measurements were usually carried out at $\omega_p/2\pi = 9.37$ GHz during the pulsed operation (pulse duration was 200 ms, the pulse repetition frequency was equal to 50 Hz). Spherical YIG samples with a diameter from 1 to 4 mm were mostly employed. The characteristic orientations of the magnetic field $\boldsymbol{H}$ are [100], [111] and [110].

9.2.2 Comparison of the S-Theory and Experiment for Susceptibilities

With such a comparison for cubic ferromagnets in view, let us represent (5.5.13) for χ in the following form (allowing for the axial symmetry by means of the invariant phase (5.5.8)):

$$\chi' = \frac{2}{h}\int_{-1}^{1} V(x)n(x)\cos\Psi_{\text{inv}}(x)dx$$

$$= \frac{2}{h^2}\int_{-1}^{1} S_{\text{inv}}(x,x_1)n(x)n(x_1)\cos[\Psi_{\text{inv}}(x)-\Psi_{\text{inv}}(x_1)]dxdx_1, \quad (9.2.3)$$

$$\chi'' = \frac{2}{h}\int_{-1}^{1} V(x)n(x)\sin\Psi_{\text{inv}}(x)dx = \frac{2}{h^2}\int_{-1}^{1}\gamma(x)n(x)dx.$$

Clearly, the imaginary part of the susceptibility χ'' characterizes only the total number of parametric waves, at the same time the real part of χ' significantly depends on the phase relations between the pairs. Therefore, the quantity χ' is a finer characteristic of the system sensible to the details of the pair distribution in space, to auto-oscillations, inhomogeneities, etc. This can account for the considerable divergences in experimental values of χ' in different publications. Thus, for instance, in one of the earlier works by *Hartwick* et al. [9.25] it was reported that in YIG the relation $\chi'/\chi'' = 0.1$ and that depends only weakly on the magnetic field. Because of the smallness of this relation, different kinds of nonlinear damping leading to $\chi' = 0$ were suggested as the limiting mechanism (*Schlömann* [9.27]). The description of the behavior of spin waves above the threshold of parametric excitation given by *Monosov* in [9.28] and in particular the mechanism of the amplitude "self-suppression" was also essentially based on the assumption that $\chi' \ll \chi''$. However, subsequent careful investigation of the behavior of χ' and χ'' in different experimental situation showed that in perfect monocrystals in the absence of auto-oscillations the value χ' is not small and can even exceed the value of χ''.

Figure 9.5 shows the characteristic dependences of $\chi'(h^2)$ and $\chi''(h^2)$ for a YIG sphere in three basic crystallographic directions, i.e. in the direction [100] when there are no magnetization oscillations and in the directions [111] and [110] when intense auto-oscillations are observed. One can see, first, that the auto-oscillations of the magnetization does not significantly affect the value of χ'' and, second, that AOs reduce χ' to a fraction of its value. This can be explained by the fact that high above the instability threshold

$$\sin\Psi = h_{\text{th}}/h \ll 1, \quad |\cos\Psi| \simeq 1-(1/2)(h_{\text{th}}/h)^2, \quad (9.2.4)$$

i.e. the cosine of the phase shift of pairs with respect to the pumping which determines the value χ' is close to its extremum. Thus the auto-oscillations

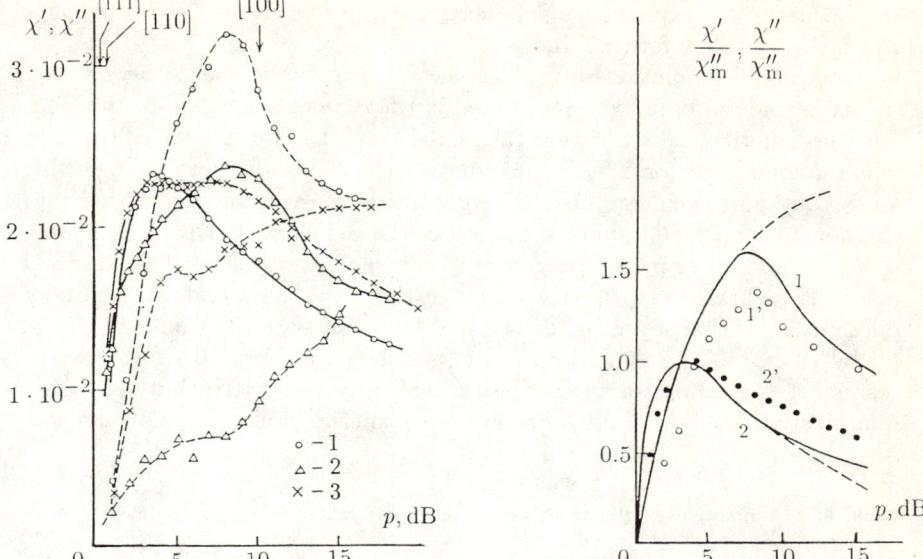

Fig. 9.5. (left) Experimental dependences of the real part of nonlinear susceptibility χ' (dashed lines) and the imaginary part of nonlinear susceptibility χ'' (solid lines) on the pumping power for a YIG sphere in three basic crystallographic directions: (1) – direction [100]; (2) – [111]; (3) – [110]. Arrows denote the thresholds of auto-oscillation excitation (*Zautkin* et al. [9.22])

Fig. 9.6. (right) Real part of nonlinear susceptibility χ' (1, 1') and imaginary part χ'' (2, 2') versus pumping power for a YIG sphere at $H = H_c - 100$ Oe, $M \parallel [100]$. The calculation results are shown by solid lines; dashed lines correspond to the calculations in the model of one group of pairs; dots denote experimental results (*Zautkin* et al.[9.22])

(which lead to the periodical changes of the angle Ψ) do not change the mean value of $\sin\Psi$, and reduce the mean value of the $\cos\Psi$ together with the susceptibility χ'.

The fact that $\chi' \simeq \chi''$ is an unambiguous evidence of the significant phase mismatch of the pumping and magnon pairs which was predicted by the S-theory. The comparison of its prediction with the experimental dependences $\chi'(h)$ and $\chi''(h)$ is significantly facilitated under small supercriticality ($h < h_2$) when only the pairs on the equator with $\Theta = \pi/2$ are excited. In this case it follows from (9.2.3) and (5.5.12) that

$$\chi'' = \frac{2V_1^2}{|S_{11}|}\frac{h_{\mathrm{th}}\sqrt{h^2 - h_{\mathrm{th}}^2}}{h^2}, \quad \chi'' = \frac{2V_1^2}{S_{11}}\frac{(h^2 - h_{\mathrm{th}}^2)}{h^2}. \quad (9.2.5)$$

Fig. 9.6. shows the qualitative agreement between the theoretical and experimental curves (see Fig. 9.5 for $M \parallel [100]$). For example, according to the theory the curve $\chi'(h)$ intersects the curve $\chi''(h)$ at the maximum. The discrepancy between the theoretical and experimental curves $\chi'(h)$ at $h > 8$ dB can be naturally explained by the excitation of the second group of pairs which do not allow for (9.2.5).

Computer calculations aimed at the comparison of the S-theory with the experimental data were performed by *Zautkin* et al. [9.22]. To this end, first the functions $S(x, x')$ were calculated by (3.1.21) under specific experimental conditions for YIG, i.e. magnetization, field of the crystallographic anisotropy and exchange field. The obtained values were substituted into the nonstationary equations of motion of the S-theory (5.4.13) which were solved on computer by time iteration with respect to the level of the thermal noise. The obtained stationary values of the amplitudes and phases under different supercriticalities made it possible to calculate the values of χ' and χ'' by (9.2.3). The calculation results are shown in Fig. 9.6 as well as the results of the laboratory experiment. Not only qualitative but also good quantitative agreement of theoretical and experimental results is evident.

Table 9.1. Comparison of the theoretical and experimental values of nonlinear susceptibility $\chi''_m = \max_h \chi''(h)$ for spherical samples of different crystals: (1) $Y_3Fe_5O_{12}$; (2) $Y_3Fe_{4.35}O_{12}$; (3) $Bi_{0.2}Ca_{2.8}Fe_{3.6}V_{1.4}O_{12}$; (4) $Li_{0.5}Fe_{2.5}O_4$; (5) $NiFe_2O_4$. (After *Zautkin* et al. [9.22])

Crystal	$4\pi M$ [Oe]	H_a [Oe]	$2\Delta H(k)$ [Oe]	χ''_m Exper.	χ''_m Theory	Direction
1.	1750	84	0.12	24	21	[100]
2.	1500	8	0.36	23	22	[100]
3.	650	58	0.45	5.0	7.5	[100]
4.	3700	580	0.80	80	70	[111]
5.	3200	490	1.40	25	19	[100]
5.	—	—	—	55	84	[111]

Table 9.1 shows the comparison of the theoretical and experimental values of the maximum susceptibility $\chi'' = \max \chi''(h)$ for various spherically shaped cubic ferromagnets. Absolute susceptibility measurements were performed for YIG by using the standard technique described in the previous section. The values χ'' for other crystals were measured on the same setup comparing it to YIG. The theoretical values were calculated by (9.2.5)

$$\chi''_m = V_1^2/|S_{11}|, \qquad (9.2.6)$$

which for cubic ferromagnets in the case of axial symmetry $M \parallel [111], [100])$ after the expressions for V and S are substituted assumes the following form (*Zautkin* et al. [9.22]):

$$8\pi\chi''_m = [N_z - 1 + \delta(\omega_a/\omega_m) + \sqrt{1 + (\omega_p/\omega_m)^2}]^{-1}, \qquad (9.2.7)$$

where δ is given by Table 3.1.

As can be seen from Table 9.1 the simple formula (9.2.7) gives a good description of the absolute values of the above-threshold susceptibility for a wide class of cubic ferromagnets. A certain discrepancy between theory and experiment for $NiFe_2O_4$ (orientation [111]) is evidently due to the fact that in this case the susceptibility maximum is above the generation threshold of the second group of pairs when (9.2.7) cannot be applied.

Green and *Healy* [9.29] measured the susceptibility χ''_m for the uniaxial ferromagnet $Ba_2Zn_2Fe_{12}O_{14}$ with easy-plane anisotropy and obtained an abnormally high value of $\chi''_m \simeq 0.2$. The theoretical estimation of χ''_m from (9.2.6) with the coefficients V_1 and S_{11} and calculated not allowing for the dipole-dipole interaction (at $\omega_m < \omega_a$, $\omega_p \leq \omega_a$) is

$$2\pi\chi''_m = \omega_m \omega_a / \omega_p^2 \; .$$

Taking from [9.29] the parameter values for $Ba_2Zn_2Fe_{12}O_{14}$, $4\pi M = 2850$ Oe, $\omega_a/g = 9900$ Oe and the pumping frequency $\omega_p/g = 6300$ Oe we obtain $\chi''_m \simeq 0.1$. There is a qualitative agreement between the theoretical and experimental values of χ'' also under the parametric excitation of magnons in antiferromagnets (*L'vov* and *Shirokov* [9.30]; *Prozorova* and *Smirnov* [9.31]).

It is rather interesting to compare theory and experiment for the signs of the real susceptibility χ'. According to (5.5.35) the signs of the of χ' and S coincide for cubic ferromagnets. In accordance with the theoretical predictions $\chi' > 0$ for the weakly anisotropic crystal (enumerated in Table 9.1. as 1, 2, 3) and $\chi' > 0$ for crystals with high anisotropy (number 4, 5) for orientation [100] and $\chi' < 0$ for orientation [111].

9.2.3 Measurements of Interaction (Frequency Shift) Amplitude

If $n(\boldsymbol{k},t)$ rapidly changes with the development of the parametric instability, then the eigenfrequency of magnons and the pumping frequency become mismatched. This is revealed also when the pulse after passing through the resonator is somewhat distorted. However, frequency renormalization in such an experiment cannot be studied with sufficient accuracy since as the instability develops, the microwave field amplitude and magnon damping change. *Prozorova* and *Smirnov* [9.31] employed the following method for the measurements of the nonlinear magnon frequency shift $\omega_{NL}(\boldsymbol{k}_1,t)$ under a drastic change in the number of other magnons with the frequency $\omega_{NL}(\boldsymbol{k}_2,t)$. Two pumping signals with the frequencies $\omega_{p1} = 2\omega(\boldsymbol{k}_1)$ and $\omega_{p2} = \omega(\boldsymbol{k}_2)$ (parametrically exciting magnons with different frequencies) were applied to the crystal under investigation. The sample was placed into a rectangular resonator containing a copper strip. The dimensions of the resonator were chosen so that the frequency TE_{021} of the volume mode of the resonator should correspond to ω_{p2} and the frequency of the eigen mode of the band resonator should correspond to a lower frequency ω_{p1}.

The crystal was placed in the antinode of both high frequency magnetic fields and was set up to satisfy parallel pumping conditions. The use of the microwave filters facilitated the separate reception of signals. The measurements were performed on the $CsMnF_3$ monocrystals at $T = 1.62$ K at the frequencies $\omega_{p1}/2\pi = 21.36$ GHz and $\omega_{p2}/2\pi = 35.1$ GHz in the magnetic fields allowing parametric excitation of magnons at both frequencies.

Figure 9.7 shows the oscillograms of the microwave pulses that had passed through the resonator. The upper beam corresponds to the signal with the frequency ω_{p1}, and the lower beam to the signal with the frequencies ω_{p2}. One can see in the upper pulse two transient processes. The first one is usual transient processes when the microwave power with frequency is ω_{p1} switched on. The second one corresponds to the time of an avalanche-type increase of the amplitude of magnons with the frequency $\omega(\boldsymbol{k}_2)$. This can be interpreted as the transition of the system of PM1 excited by the pumping with the frequency ω_{p1} into a new stationary state associated with the renormalization of the frequency $\omega_{NL}(\boldsymbol{k}_1)$ because of the amplitude increase of PM2. In the stationary state the following conditions

$$\partial \Psi(k_1,t)/\partial t = 0 , \quad \partial n(k_1,t)/\partial t = 0$$

and relation (5.5.12) must be satisfied. If (5.5.12) is violated, the phase $\Psi(\boldsymbol{k}_1)$ deviates from its stationary value and, according to (5.3.4) the energy flux W into the sample changes, which shows in the pulse passed through the resonator.

The nonlinear frequency shift was measured using the fact that the similar transient process must be observed at the pulse with the frequency ω_{p1} when the pumping frequency discontinuously changes by amount

$$\delta\omega_{p1} = -2\Delta , \tag{9.2.8}$$

because in this case the relation (5.5.12) is also violated. In order to change ω_{p1} an additional rectangular pulse with variable slope of its front edges was fed to the repeller plate of the corresponding klystron. Setting the time of the pumping frequency change equal to the time of the avalanche development and the magnon lifetime, respectively, and selecting $\delta\omega_{p1}$ such that the transient processes should be completely identical, we can obtain the frequency shift Δ from (9.2.8). The quantity Δ was found to be proportional to the number of PM2. After that, the coefficient $T(\boldsymbol{k}_1,\boldsymbol{k}_2,H)$ was calculated (for the $CsMnF_3$ sample with the volume of 1 cm in the field of 1.1 kOe, $T(\boldsymbol{k}_1,\boldsymbol{k}_2,H)/2\pi \simeq 3 \cdot 10^{-12}$ Hz). The quantity $n(\boldsymbol{k}_1)$ was used in calculations, obtained from (5.5.12). Theoretically, the dependence $T(\boldsymbol{k}_1,\boldsymbol{k}_2,H)$ for the antiferromagnets of the "easy-plane" type was calculated by *L'vov* and *Shirokov* [9.30]. Within the experimental accuracy there is an agreement of the theoretical and experimental results.

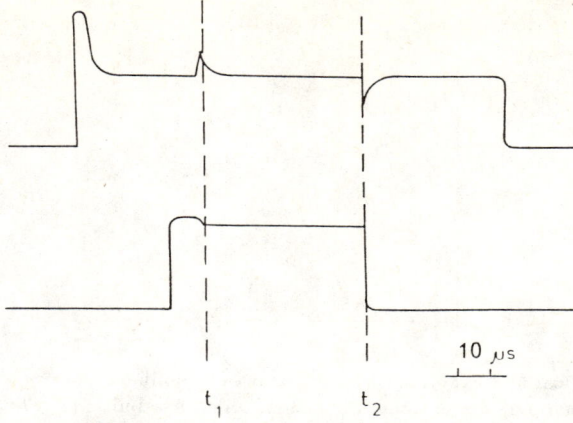

Fig. 9.7. Oscillogram of pulses after passing through the resonator. The lower beam ($\omega_p/2\pi = 35.1$ GHz) corresponds to the excitation of PM1, which causes a shift of the magnon's spectrum; the upper beam ($\omega_p/2\pi = 21.36$ GHz) corresponds to the excitation of PM2, which detects this shift (*Prozorova* and *Smirnov* [9.31])

9.2.4 Nonlinear Ferromagnetic Resonance

In this section the nonlinear theory of the ferromagnetic resonance by *L'vov* and *Starobinets* [9.32] is compared with experiments [9.33]. Unlike the early theory of *Suhl* [9.34] taking into account the interaction of only one uniform precession with parametric magnons, the nonlinear theory [9.32] allows also for the interaction of parametric magnons with each other within the scope of the basic S-theory (see Sec. 6.3). In the experiments of *Gurevich* and *Starobinets* [9.33] under the second-order instability the amplitude in the uniform precession above the threshold drastically increases which can be interpreted as the qualitative confirmation of the mechanism of magnon–magnon interaction. The shape of the resonance curve under large p is usually much more complex than shown in Fig. 6.6. There are dips on the curve and at some detunings some auto-oscillations are observed. In our opinion, they are due to the loss of stability. This is confirmed by the numeric experiment on nonlinear equations (6.3.3) describing the nonlinear ferromagnetic resonance performed by *Zakaidaikov* and *Musher*. The auto-oscillations have been treated in detail in Chap. 7 for a simpler situation, i.e. parallel pumping of magnons in cubic ferromagnets.

Let us compare the obtained results with the measurements by *Gurevich* and *Starobinets* [9.32], who measured the nonlinear susceptibilities χ' and χ'' depending on the amplitude of the pumping p under the first-order instability when the uniform precession is not at resonance $\delta \gg 1$ (see Fig. 9.8). This figure shows the theoretical dependences calculated by (6.3.21, 23), and the parameter d entering into the equations of the theory was obtained from the conditions of equality of the experimental and theoretical

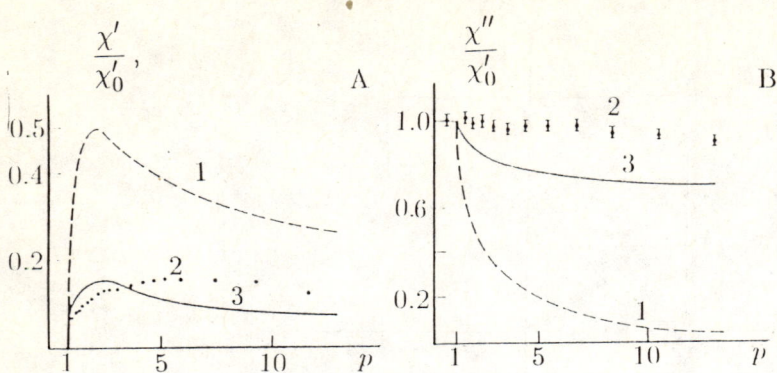

Fig. 9.8. Comparison of theoretical and experimental dependences of nonlinear susceptibilities χ'' (A) and χ' (B) on pumping power under Shul's first order instability: (1) *Shul* theory [9.34], (2) experiment [9.32], (3) *L'vov – Starobinets* theory at $d = 3 \cdot 10^{-4}$

values of $\chi''_{\max}$. It turned out that $d \simeq 3 \cdot 10^{-4}$. This order of magnitude is in good agreement with the theoretical results (see (6.3.22)). In spite of the smallness of the value, to take it into account (i.e. allowing for magnon–magnon interaction) significantly influences the form of the dependences $\chi'(p)$, $\chi''(p)$. Fig. 9.8 for comparison shows the theoretical dependences for the case $d = 0$, corresponding to *Suhl* [9.34].

9.3 Investigations of Stationary State With One Group of Pairs

From the viewpoint of the detailed comparison of the basic S-theory with the experiment the state with one group of similar pairs is undoubtedly of greater interest. First, because this mode serves as an illustrative example of a high-degree self-ordering of the flux equilibrium in the system with a large number of degrees of freedom. Second, because the distribution function of pairs is so simple that it is easily to find simple analytical expressions for the values which can be experimentally measured. The possibility of unambiguous interpretation provides the basis for consistent and detailed experimental study of the behavior of the nonlinear system of interacting parametric magnons. Such research has been carried out mostly in the former USSR by *Starobinets, Melkov, Prozorova, Ozhogin, Smirnov, Zautkin* and their colleagues. Some of these experiments are described in this chapter. We shall begin with a more detailed discussion of the nonlinear susceptibility of χ in the state with one group of pairs.

9.3.1 Nonlinear Susceptibility in the One-Group State

In Sec. 9.2.2 it was shown that the numerical values of susceptibilities agree with the formulae of the basic S-theory and that there is a qualitative agreement of theory and experiment for χ' and χ'' dependence on the pumping intensity. For the quantitative comparison of the χ'' dependence on $p = h^2/h_{\mathrm{th}}^2$ it is convenient to represent the experimental data as the dependence of $f = \chi'' p$ on $\chi = (p-1)$ because in the S-theory this dependence (9.2.5) is the power function

$$f = \chi'' p = \chi''_{\mathrm{m}} \sqrt{p-1} \ . \tag{9.3.1}$$

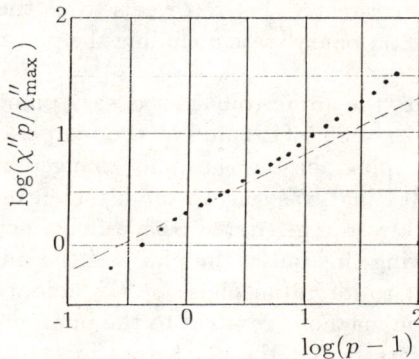

Fig. 9.9. Comparison of theory (dashed line) and experiment (dots) for absorbed power $W \propto p\chi''/\chi''_{\mathrm{max}}$ versus supercriticality $(p-1)$ (in the log-log plot), YIG sphere, $H_{\mathrm{c}} - H = 100$ Oe, $M \parallel [100]$. After Zautkin et al. [9.35]

For a detailed comparison of the theory with the experiment Zautkin et al. [9.35] represented (see Fig. 9.9) the experimental data for YIG (at $M \parallel [100]$ when intense auto-oscillations are absent. The value of the magnetic field H was selected equal to $H_{\mathrm{c}} - 100$ Oe, so that the nonlinear damping in the "rectifying" logarithmical coordinates is insignificant. The linear dependence obtained by them implies that

$$f_{\exp} = p\chi'' = C(p-1)^z \ . \tag{9.3.2}$$

At the same time the constant C characterizing the numerical value of the susceptibility, as it was shown in Sec. 9.2.2, practically coincides with its theoretical value. Therefore we can say that there is complete agreement of the experiment with the theory (9.3.1) if we neglect a certain difference in the exponent of z (tangents of the straight lines slope in Fig. 9.9). A small divergence $z_{\exp} = 0.7$ and $z_{\mathrm{theor}} = 0.5$ is not too surprising if we take into consideration that the basic S-theory assumes that the medium is ideally homogeneous and real crystals contain some inhomogeneities, impurities, etc. Two magnon elastic scattering (on inhomogeneities, etc.) weaken the rigid phase correlations in the pairs. This results in some weakening of the phase mechanism of their amplitude limitation and, consequently, faster increase ($z > 0.5$) of the absorbed power as the power of the pumping level increases.

9.3.2 Direct Measurement of Pair Phase

Prozorova and *Smirnov* [9.36, 37] developed the "method of the transient processes" for the investigation of the stationary state. This method is based on the following. In the stationary state the system of parametric waves is at equilibrium with the pumping field. The change in the pumping parameters (phase, frequency, amplitude), which is fast in comparison with the lifetime of waves, leads to the violation of the equilibrium and to the change in the absorbed power. Then a certain transient process is observed and over the time $\tau = 1/2\gamma$ the system reaches the stationary state again. The reaction of the parametric waves system to the parameter change reveals to us the information about the properties of the stationary state including the phase of parametric waves.

Experiments were performed on the antiferromagnetic easy-plane $MnCO_3$ at the pumping frequency $\omega_p/2\pi = 35.5$ GHz and at the temperature of the liquid helium [9.36]. The pumping phase was rapidly changed in the following way a short pulse ($\tau_p \simeq 0.1$ ms) was sent to the klystron repeller plate. During the pulse time the klystron generated with a frequency differing from ω_p by $\Delta\omega_p$. Thus, following this pulse, the phase of the microwave signal was shifted with respect to its initial phase by the amount $\beta \simeq \Delta\omega_p \tau_p$. It appeared that parametric magnons respond to the pumping change phase depending on β. Immediately after the phase-rotating pulse the absorbed power can either be more or less than in the stationary state. Moreover, at definite values β the spin system radiates, delivering the energy to the resonator. Then, during a time of approximately 1 ms, the system returns to the stationary state. By the reaction to the pumping phase change, the stationary phase of the parametric spin waves Ψ was found. Fig. 9.10 shows the results of the performed experiment: the change in the absorbed power in the sample versus the pumping phase change β. It follows from the analysis of these results that Ψ differs from the optimum one (corresponding to the maximum power selection) the larger the ratio h_{th}/h. This weakens the connection with the pumping and, consequently limits the amplitude of parametric waves. The obtained parametric dependence $\Psi(p)$ is shown in Fig. 9.11. Similar experiments on YIG were carried out by *Melkov* and *Krutsenko* [9.24] and also by *Prozorova* and *Smirnov* [9.36]. Their results are shown in the same figure.

It is clear that within the experimental error the points are located at the bisector of the coordinate angle as follows from (5.5.7) for antiferromagnets and (5.5.12) for ferromagnets in the state with a single group of pairs.

It confirms the validity of the S-theory for the correct description of the significant aspects of the above-threshold behavior of parametric waves. The deviation of the results for YIG at the supercriticality over 10 dB ($h > 3h_{th}$)

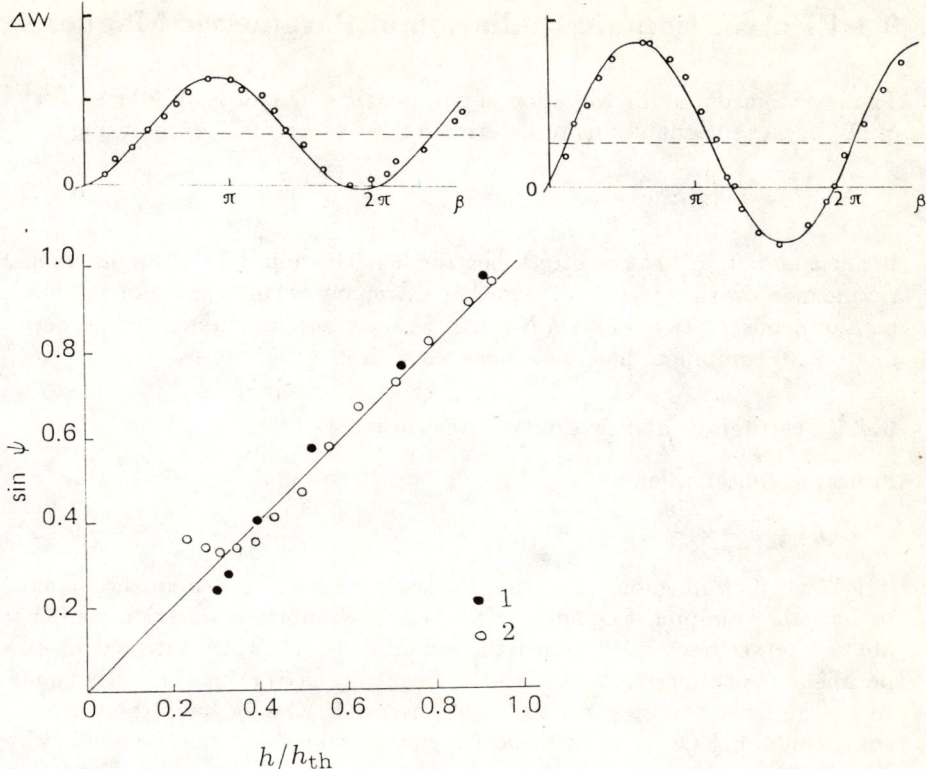

Fig. 9.10. (above) Variation in the power ΔW absorbed by CsMnF$_3$ versus variation of the pump phase. (A) $h/h_{\text{th}} = 1.08$; (B) $h/h_{\text{th}} = 2.25$, $T = 1.62$ K, $H = 2.25$ kOe (*Prozorova, Smirnov* [9.36])

Fig. 9.11. (below) Stationary phase Φ of a pair versus supercriticality. (1) results of an experiment conducted on CsMnF$_3$ monocrystals at $T = 1.62$ K and $H = 2.55$ kOe (*Prozorova* and *Smirnov* [9.36]). (2) Results for YIG sphere, 2 mm in diameter, $\boldsymbol{H} \parallel [100]$, $H = H_c - 500$ Oe (*Melkov* and *Krutsenko* [9.24])

from the simple formula (5.5.12) is associated with the excitation of the second group of pairs with $\Theta = \pi/2$. This phenomenon was described in Sec. 5.5.3 of the basic S-theory. Its experimental confirmation will be given in Sec. 9.6.

9.4 Electromagnetic Radiation of Parametric Magnons

The investigation of the radiation of parametric magnons is of interest, first of all, because it enables us to determine their spectrum, i.e. the value

$$N(\omega) = \int n_p(\boldsymbol{k},\omega)\, d\boldsymbol{k} \ . \tag{9.4.1}$$

It must be noted that the direct electromagnetic radiation of the parametric magnons with $k \simeq 10^4$–10^6 cm^{-1} is extremely weak because of the momentum conservation $k_p \ll k$. Nevertheless, recently manufactured modern microwave equipment has made possible such investigations.

9.4.1 Frequency of Parametric Magnons

From the conservation laws

$$\omega(\boldsymbol{k}) + \omega(-\boldsymbol{k}) = \omega_p \tag{9.4.2}$$

it follows that magnons are parametrically excited at a frequency equal to half the pumping frequency. However, an important question remains about the accuracy with which the conditions (9.4.2) are satisfied in experiment. Are only magnons with the frequency $\omega(\boldsymbol{k})$ excited exactly equal to $\omega_p/2$? Does the spectral distribution $n(\omega(\boldsymbol{k}))$ really have the form of the δ-function? Or does excitation occur in some frequency band $\delta\omega(\boldsymbol{k})$? The fact that the excited magnons have the frequency $\omega(k) = \omega_p/2$ can be confirmed with a certain accuracy by the region of existence of additional absorption (if it exists) or from the experimentally observed peculiarity of parametric magnon relaxation at the intersection point of the magnon and (known from other experiments) phonon spectra [9.7–9]. The frequency of parametric magnons with $k \simeq 10^4$–10^6 cm^{-1} can be obtained with more accuracy in the experiments on Mandelstam-Brillouin scattering of light on these magnons [9.38, 39]. However, the error in determining the frequency of magnons in these experiments (1 GHz) is still too large for measuring $\delta\omega(k)$.

The most direct and accurate method of obtaining the spectral distribution of parametric magnons is to study the spectrum of the resulting radiation $I(\omega)$. We readily have:

$$I(\omega) = 8\omega^4(k)M|\Delta M|N(\omega)/3c^3 k^2 V^{2/3} \ , \tag{9.4.3}$$

where ΔM is the magnetization change when one magnon is excited. Interestingly, in antiferromagnets with weak ferromagnetism ($H_D \neq 0$) this value can significantly exceed the Bohr's magneton μ_B. Indeed,

$$\Delta M = -\hbar \frac{\partial \omega(k)}{\partial H} = -\frac{g(H+H_D)}{2\omega(k)}\mu_B \ . \tag{9.4.4}$$

Therefore, in order to observe radiation, it is convenient to increase the pumping frequency and use substances with a large value of H_D.

The study of parametric magnon radiation was carried out by *Kotuzhansky* et al. [9.41, 42]. The monocrystals of $FeBO_3$ were used (cylindrical samples with a diameter 2 mm and a height of 3 mm). The axis of the cylinder coincided with the principal axis of the crystal. The microwave pumping was produced by the continuous-action magnetron oscillator with the frequency $\omega_p/2\pi = 35.6$ GHz and 10 W power. The cylindrical resonator of the transmission-type microwave spectrometer was tuned so that the frequency of its mode TE_{012} coincided with ω_p. The samples are placed in the exit coupling window of the resonator symmetrically about its wall with the thickness of 0.5 mm. Therefore, one part of the sample protruded into the standard waveguide at 1.5 cm of the microwave range. The fields h and H were in the basal plane of the crystal and were parallel to each other. The signals with the frequencies ω_p and $\omega_p/2$ were separated by the microwave filters. The decoupling between the channels was 40 dB. The signal at the frequency $\omega_p/2$ was received by a superheterodyne receiver with the sensitivity 10^{-14} W and was analyzed by the spectrum analyzer.

It was experimentally found that above the threshold of the parametric excitation as the microwave pumping power is absorbed at the frequency ω_p simultaneously appears the radiation from the sample at the frequency $\omega_p/2$. The radiation intensity $I(\omega)$ depended on time in a random manner. The oscillogram of the output signal from the receiver had the form of splashes with duration $\tau \simeq 10^{-5}$–10^{-6} s. The principally important result of the experiment was the fact that the center of the radiation band coincided with $\omega_p/2$ with the accuracy $\pm 2\pi \cdot 20$ kHz, i.e. with the relative accuracy of the order of 10^{-6}. This was proved as follows: The radiation signal and the signal from the additional microwave klystron oscillator, which was tuned such that its second harmonics coincided with the pumping frequency, were simultaneously applied to the output of the receiver. The klystron was tuned to the zero beats of voltage from the microwave detector where the signals from the klystron and magnetron became detuned. Within the error of the frequency measurement due to the parasitic deviation of the klystron and magnetron frequencies, the klystron frequency coincided with the radiation frequency.

9.4.2 Frequency Width of Parametrically Excited Magnons

1 Limitation of the number of magnons by nonlinear damping in the easy-plane antiferromagnet $FeBO_3$ The previously cited paper (*Kotyuzhansky* et al. [9.41]) also discusses the shape of the wave packet $N(\omega)$. It was found that the width of the wave packet $\Delta\omega$ increases linearly with h/h_{th} (see Fig. 9.12). The broadening of the radiation line, $N(\omega)$, in $FeBO_3$ was investigated theoretically by *Mikhailov* and *Chubukov* [9.42]. The theoretical estimate of the linewidth $\Delta\omega$ obtained in this paper is by a factor of 25 less than the experimentally observed value. The form of the h-dependence of $\Delta\omega$ obtained by *Mikhailov* and *Chubukov* [9.42] coincides with the experimental data only for $p > 16$. This disagreement is evidently due to certain simplifications in the theory [9.42]. In particular, the assumption of the equilibrium of the photons. As shown by *L'vov* [9.43], the shape of the wave packet $N(\omega)$ is not determined by any specific mechanism of nonlinear damping and is obtained in accordance with (6.4.16) by the general structure of the Green's function (6.4.15). *Mikhailov* and *Chubukov* [9.42] showed (see Fig. 9.13) that the experimentally observed line shape $N(\omega)$ coincides with the theoretical data (See (6.5.26), *L'vov* [9.43]).

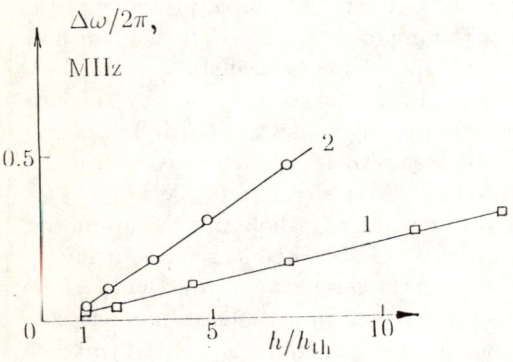

Fig. 9.12. Magnon radiation linewidth in $FeBO_3$ versus pumping amplitudes for $H = 380$ Oe. (Width was measured at the level $0.1\, I_{\max}$). (1) $T=1.2$ K; (2) $T=4.2$ K (*Kotyuzhancky* et al. [9.41])

2 Limitation of the number of magnons by phase mechanism in the ferromagnet YIG. *Krutsenko, L'vov* and *Melkov* [9.44] applied another procedure in order to investigate the shape of the wave packet $N(\omega)$ in YIG. To this end, electromagnetic radiation from the ferrite was recorded at the frequencies close to half pumping frequency. This phenomenon is due to the two-magnon scattering of parametric magnons by static inhomogeneities. In this case for the complex envelope of the electromagnetic wave $U(t)$ which is radiated by the crystal one has

$$U(t) = \sum_{n} \int g(\boldsymbol{k}) \exp(i\boldsymbol{k}\boldsymbol{r}_n) a(\boldsymbol{k},t) d\boldsymbol{k} \;, \qquad (9.4.5)$$

where $g(\boldsymbol{k})$ is the transformation amplitude of the magnon with $\boldsymbol{k} = 0$ into a homogeneous oscillations mode (magnons with $\boldsymbol{k} = 0$) on a single scattering center, $\boldsymbol{r}_n$ denote the coordinates of these centers. By employing (9.4.5) we obtain the expression for the correlation function of the radiation $K(\tau)$

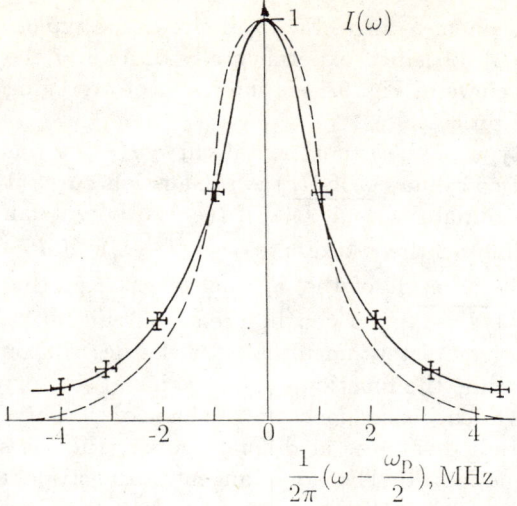

Fig. 9.13. Magnon radiation lineshape in FeBO$_3$. The solid line is the theoretical relation (calculated from equations given by the author [9.43] and later by *Mikhailov* and *Chubukov* [9.42]), in which the half-width is taken equal to the experimentally observed value (*Kotyuzhansky* et al. [9.41]); the dashed line is a Lorentzian of the same width

$$K(\tau) = \langle U(t)U^*(t+\tau)\rangle = N_{\text{def}} \int |g(\boldsymbol{k})|^2 n(\boldsymbol{k},\tau)\, d\boldsymbol{k} \ . \tag{9.4.6}$$

Here N_{def} is the number of scattering centers, $n(\boldsymbol{k},\tau)$ designate the different-time correlation function of magnons; $n(\boldsymbol{k}) = n(\boldsymbol{k},\tau)$ at $\tau = 0$. From (9.4.6) it can be seen that with the accuracy of the time-independent scale factor, the function $K(\tau)$ is equal to the correlation function of magnons integrated over k: $N(\tau) = \int n(\boldsymbol{k},\tau)\, d\boldsymbol{k}$, and therefore the spectral density $K(\omega)$ differs from the spectral density of the magnons $N(\omega)$ specified by relation (9.4.1) only in the constant factor. This conclusion is essentially due to the fact that the two-magnon scattering by the static inhomogeneities is not accompanied by the change in oscillation frequency. Thus, the static properties of parametric magnons can be investigated by treating the electromagnetic radiation from the ferrite.

The samples used in experiments [9.44] were ferrite spheres made of YIG monocrystals, 2.5 to 2.9 mm in diameter. They were studied at room temperature. The pumping frequency $\omega_{\text{p}}/2\pi = 9.37$ GHz. The spheres were oriented along the hard direction of magnetization [100] $\parallel \boldsymbol{H}$ to exclude the effect of the low-frequency self-oscillations of the magnetization. The electromagnetic radiation from the ferrite is maximum when the frequency of this radiation is equal to the frequency of one of the magnetostatic modes. The registering circuit consisted of a 6 cm-range amplifier, protected against the pumping frequency by a low frequency filter, a square-law detector and a spectrum analyzer of single-frequency range. Fig. 9.14 shows the typical dependence $I(\epsilon)$, $\epsilon = (\omega - \omega_p/2)$, obtained experimentally on one of the studied YIG spheres. The solid curve in Fig. 9.14 is the result of averaging over 10 realization of stochastic process.

In order to determine the type of the experimental curve $I(\epsilon)$, it was plotted on the various rectified coordinates. If $I(\epsilon)$ is a Gaussian curve it will be a straight line on the coordinates $\sqrt{\ln(1/I)}, \epsilon$; if $I(\epsilon)$ is a Lorentzian curve this yields the straight line on the coordinates $\sqrt{1/I - 1}, \epsilon$; if $I \sim \sqrt{\epsilon/\mathrm{sh}\epsilon}$ the straight line will be a result of the dependence $I(\epsilon)$ plotted on the coordinates $\ln[1/I + \sqrt{1/I^2 - 1}]$. As can be seen from Fig. 9.14, the first standard functions are completely unsuitable for the description of the experiment, at the same time the function $\sqrt{\epsilon/\mathrm{sh}\epsilon}$ is in satisfactory agreement with the experiment. Such a coincidence of the shape of the curve $I(\epsilon)$ was observed for all the studied samples at different supercriticalities and values of the constant magnetic field. The slope tangent of the straight line 3 in Fig. 9.14 determines the width of the curve $\Delta\epsilon$ which because of the peculiarities of the quadratic detector is $\sqrt{6}$ times as large as the real width of the electromagnetic radiation curve $\Delta\omega$ coinciding with the width of the spectral density of parametric magnons. Fig. 9.15 shows the distribution width of parametric magnons $\Delta\omega$ versus the supercriticality $p = (h/h_{\mathrm{th}})^2$. The results obtained are in good agreement with the S, T^2-theory which will be considered in Sec. 10.2.

9.5 Collective Resonance of Parametic Magnons

This section deals with homogeneous collective oscillations, i.e. the simplest type of the oscillations corresponding to the gap in their spectrum. The resonance in the system of parametric magnons resulting from the excitation of homogeneous collective oscillations is very intense and can therefore be observed more readily. *Zautkin* and *Starobinets* [9.46] observed experimentally the reaction of the sample to the additional alternating field with the frequency $\omega_p \pm \Omega$ under parallel pumping of magnons in YIG monocrystals. The mixing of this field with the pumping field at the frequency ω_p in the nonlinear system brings about a signal of low combinatorial frequency Ω which results in the collective resonance of parametric magnons. The resonance was detected by the peaks which appeared in the absorption spectrum of the weak signal and from oscillation excitation of the longitudinal magnetization of the sample at a low frequency.

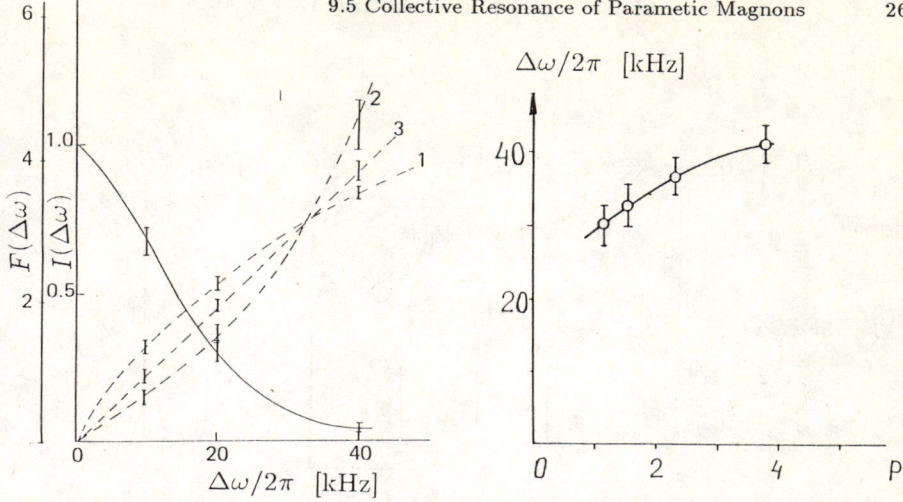

Fig. 9.14. (left) Solid line: spectral power $I(\epsilon)$ of the parametric magnon radiation (YIG sphere, $\Delta H(k) = 0.15$ Oe, $p = 6$ dB). Dashed curve: build-up of $I(\epsilon)$ in the coordinates: (1) $F(\epsilon) = 2\sqrt{-\ln I(\epsilon)}$; (2) $F(\epsilon) = \sqrt{I^{-1}(\epsilon) - 1}$; (3) $F(\epsilon) = \ln[I^{-1}(\epsilon) + \sqrt{I^{-2}(\epsilon) - 1}$ (After *Krutsenko* et al. [9.44])

Fig. 9.15. (right) Width of the parametric magnon radiation $\Delta\omega/2\pi$ versus pumping power p for YIG sphere (After *Krutsenko* et al. [9.44])

9.5.1 Experimental Technique

Figure 9.16 shows the experimental setup for observing the resonance of collective oscillations of parametric magnons. The rectangular resonator was connected by two holes with two waveguide channels over which two microwave signals were transferred to the sample. One of them, namely the powerful magnetron pulse at the frequency $\omega_p = 2\pi \cdot 9.37 \cdot 10^9$ s^{-1}, was employed as usual for parallel pumping of magnons and the second one was an additional weak signal whose magnetic field was polarized to the constant field and the pumping field. As a source of the weak signal a klystron with 3 cm range of wave length was used. The pulses of the sawtooth voltage of the oscillograph sweeping (which is turned on simultaneously with the pulse of the intense pumping sent to the sample) were sent to the klystron repeller plate. As a result, over the time of the pumping influence the frequency of the weak signal changed within a small range from $\omega_p - \Omega$ to $\omega_p + \Omega$, whose limits were regulated by the sweep generator from zero to several MHz. The central frequency at the same time corresponded to the peak of the klystron generation zone and was selected equal to the frequency of pumping and the eigenfrequency of the resonator in its operation mode. The sweeping duration of the weak signal with respect to frequency corresponded to the duration of the pumping pulse (10^2–10^4 ms at the repetition frequency 25 Hz). After passing through the resonator with the sample the weak signal came to the directional coupler, connected with the waveguide channel of

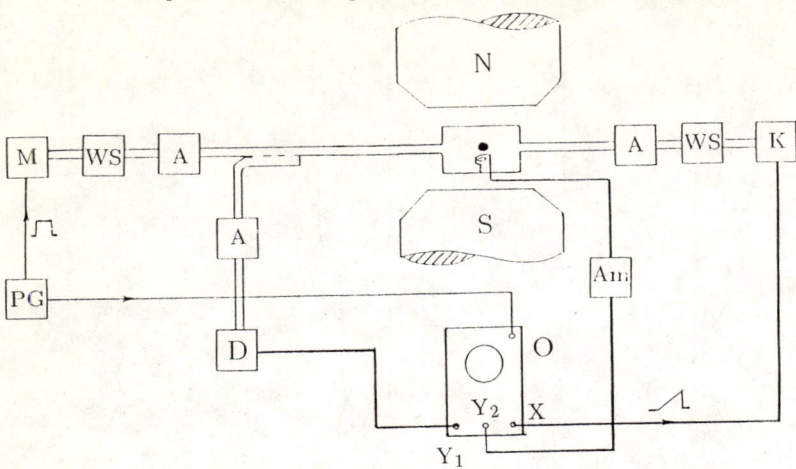

Fig. 9.16. Block diagram of experimental setup for investigations of collective resonance of parametric magnons (*Zautkin* et al. [9.46]). M, magnetron; K, klystron; PG, pulse generator; WS, waveguide switch; A, attenuator; R, cavity resonator; D, cristal detector; Am, amplifier; O, oscillograph.

the magnetron; then it was registered and observed on the screen of the oscillograph as a curve proportional to the plot of the amplitude of passage through the resonator versus frequency $D^2(\omega + \Omega)$. Below the threshold of the parametric excitation of the spin wave the oscillogram have a shape of an ordinary resonance curve of the resonator on the peak of wich there are zero beats indicating that at the moment (at the given point of the oscillogram) the frequencies of the magnetron and klystron are equal ($\Omega = 0$). Above the threshold the resonance curve is overlapped by the absorption spectrum of the weak signal due to its interaction with the system of parametric magnons. In order to facilitate the observation of the passing signal the repelled signal was successfully decreased to a level not exceeding 10% of the useful signal reaching the detector by means of matching the resonator with the waveguide.

For the registration of the oscillation variation at low frequency a detector coil was placed into the resonator near the sample, the axis of the pumping was parallel to the constant magnetic field. At the oscillating longitudinal magnetization in the signal is induced in the coil. After the amplification by the low-noise amplifier this signal was transmitted to the second channel of the oscillograph and simultaneously registered with the absorption spectrum of the weak channel. The observation and measurements were performed on several YIG samples shaped as discs and spheres at different orientations of the external magnetic field with respect to its crystallographic axes. Most interesting are the measurements at the orientation along the axis [100] when above the threshold up to the supercriticalities

9 dB in the system of parametric magnons there are no auto-oscillations of the magnetization.

Figure 9.17 shows the oscillograms of the frequency dependences of the coefficient of the weak signal passage through the resonator $D^2(\omega + \Omega)$ and of the oscillations of the longitudinal magnetization $m_z(\Omega)$. The oscillograms were obtained at different fixed supercriticalities of pumping. It can be seen that immediately above the threshold a comparatively narrow absorption peak appears on the curve $D^2(\omega + \Omega)$. This peak increases as the pumping power increases and is shifted towards higher frequencies. Simultaneously with the absorption peak the resonance increase of the low-frequency oscillations amplitude is observed on the oscillograms $m_z(\Omega)$, the maximum of the amplitude showing the behavior similar to the absorption spectrum. Such a simple picture is observed up to the supercriticalities of the order of 9 dB. At large pumping power the spectrum is significantly distorted: the intense peak is broadened, new peaks appear in the vicinity of the center frequency, on the oscillogram $m_z(\omega)$ there appear respective additional symmetrical maxima. This mode is accompanied by the emergence of chaotic auto-oscillations modulating the reflected pulse of the magnetron and hindering the isolation of peaks in the spectrum $D^2(\omega + \Omega)$. It must be noted that the complex absorption spectrum is observed also in the case $M \parallel [111]$, when immediately above the threshold intense auto-oscillations are excited.

9.5.2 Frequency of Collective Resonance

Let us obtain the resonance frequency of the collective oscillations Ω_{res}. It can be found through direct measurement of the absorption peak shift from the center frequency on the oscillogram $D^2(\omega + \Omega)$. Fig. 9.18A plots the measured resonance frequency versus the pumping intensity for two values of the constant field. The dependences prove to be linear within a wide range of supercriticalities corresponding to one peak in the absorption spectrum of the weak signal. The slope of the plots in Fig. 9.18A increases as the constant field decreases.

The experimental dependence of the resonance frequency on the constant field is shown in Fig. 9.18B for the supercriticality $p = 4$ dB. In order to verify the assumption of the space homogeneity of the excited oscillations based on the high intensity of the observed resonance, the dependences of the resonance frequency on the constant field were measured on the samples shaped as spheres of different diameter. Fig. 9.18B confirms this assumption because the collective oscillations are practically not influenced by the size of the crystal, they do not "feel" the boundaries of the sample.

Figure 9.18C shows the experimental data which enable one to form an opinion about the influence of the sample shape on the resonance frequency of collective oscillations. The curves were obtained on the YIG samples shaped as a sphere and disc, the direction of the sample magnetization was

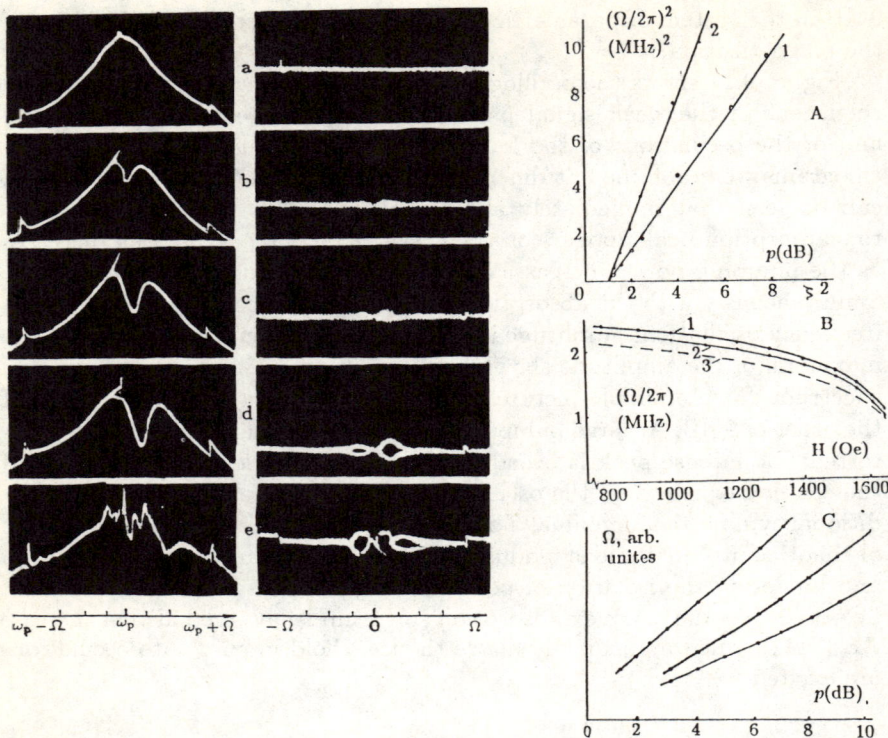

Fig. 9.17. (left) Oscillograms of the frequency dependences: of the coefficient of weak signal passage through the resonator (on the left), of the amplitude of longitudinal magnetization (on the right) under different pumping power; (a) 0 dB, (b) 2 dB, (c) 4 dB, (d) 6 dB, (e) 13 dB. The speed of the oscilloscope unfold was 1.5 MHz /point (*Zautkin* et al. [9.46])

Fig. 9.18. (right) [A] Resonant frequency of collective oscillation versus supercriticality for YIG sphere at $M \parallel [100]$ and different value of magnetic field: (1) $H_c - H = 100$ Oe, (2) $H_c - H = 300$ Oe.
[B] Resonant frequency of collective oscillation versus magnetic field for YIG spheres at $p = 4$ dB, $M \parallel [100]$ and diameters $d = 1.002$ mm (plot 1), $d = 1.56$ mm (plot 2); plot 3 is the result of calculation according to (7.4.1b) and (7.1.5) at $m = 0$.
[C] Theory (lines) and experiment (dots) for resonant frequency of collective oscillation versus supercriticality for YIG at $M \parallel [100]$: plot 1 at $N_z = 0$ (disc), plot 2 at $N_z = 1/3$ (sphere), plot 3 at $N_z = 1$ (disc). After *Zautkin* et al. [9.46]

selected everywhere along the axis [100] in order to avoid the influence of its crystal anisotropy on the comparison of results. All three curves were measured at the same values $(H_c - H)$, which ensured the equality of the wave numbers of spin waves. The very fact of the sample shape influence on Ω_{res} is not surprising because this influence is experienced, as we have already seen, also by the amplitudes of the interaction Hamiltonian of parametric magnons S_0 and T_0 which determine the eigenfrequency of collective oscillations Ω_0 in the system.

To explain these experiments let us employ the theory of collective oscillations developed in Sect. 7.1.2. As can be seen from (7.2.10) for the nonlinear susceptibility to the weak signal $\chi''(\omega_p + \Omega)$ the resonance frequency Ω_{res} differs from the eigenfrequency Ω_0 of the collective oscillations by the value of the order γ^2/Ω_0 which under the considered experimental conditions is 0.1 MHz at $p = 1.5$ and rapidly decreases as the supercriticality increases ($\gamma^2/\Omega_0^2 \simeq 1/4p$). Therefore we can compare the measurement results with the simple formula (7.1.4), which, if we allow for the well-known dependence $N(p)$ (5.5.7) assumes a convenient form

$$\Omega_0^2 = 4\gamma^2[1 + 2T_0/S_0](p-1) \ . \tag{9.5.1}$$

The linear dependence of the squared eigenfrequency of the collective oscillations on the supercriticality is confirmed by the straight-line experimental plots in Fig. 9.18A. The slope of the plot, say, for $H_c - H = 100$ Oe, equal to $(1.5 + 0.2)$ MHz is in good agreement with the theoretical value

$$\partial(\Omega_0)^2/\partial p = 4\gamma^2[1 + 2T_0/S_0] = 1.33 \text{ MHz} \ .$$

In the last estimation the value $T_0/S_0 = 0.54$ was employed founded from (7.1.5) and the value $\gamma = 0.4$ MHz calculated from the absolute measurements of the threshold of parametric magnons.

Formula 9.5.1 naturally explains the field dependence of the resonance frequency of collective oscillations of parametric magnons by the dependence of the damping parameter of spin waves $\gamma(k)$ on their wave number k determined by the value of the constant field. In order to compare the experimental and theoretical data, the dependence $\Omega_0(H)$ is plotted in Fig. 9.18B, calculated from (9.5.1) allowing for the real values of $\gamma(k)$ obtained from the threshold measurements $h(H)$. With an accuracy of 10% the calculation gives correct values of the resonance frequency over the entire studied field range.

By calculating T_0/S_0 from (7.1.5) one can obtain from (9.5.1) the resonance frequencies of collective oscillations of parametric waves for the disc magnetized parallel to the plane of the sphere and the disc magnetized perpendicular to the plane. Calculation shows that these frequencies relate as $3.4 : 2.3 : 1$ respectively. The same type of influence of the sample shape on was also observed experimentally (see Fig. 9.18C).

Formula (9.5.1) thoroughly verified and experimentally validated enables us to suggest the resonance method of obtaining the relaxation parameter of spin waves $\gamma(k)$ at known value of T_0/S_0. And on the contrary, if $\gamma(k)$ is known, relation (9.5.1) allows to obtain experimentally the important parameter T_0/S_0 characterizing the interaction of pairs of spin waves above the threshold of the parametric excitation.

9.5.3 Susceptibility to Field of Weak Microwave Signal

Measurements show that the intensity of the absorption on the curve $D^2(\omega_p + \Omega)$ is strictly proportional to the intensity of the weak signal. This circumstance enables us to introduce the susceptibility of the parametric magnon system at the frequency of the weak signal $\chi''(\omega_p + \Omega)$ as the ratio of the absorbed intensity of this signal to the intensity incident on the sample from the klystron. This susceptibility can be experimentally obtained from the form of the frequency dependence of the coefficient of passage through the resonator [9.23]. Fig. 9.19 shows measurement results for the susceptibility to the weak signal field as a function of the pumping supercriticality. The susceptibility $\chi''(\omega_p + \Omega)$ is represented in terms of the susceptibility to the main pumping χ'' because this susceptibility provides a natural scale for their comparison. The dependence proves to be linear, and the numeric values are rather large (at $p = 8$ exceeding the value of χ'' more than by an order of magnitude, which is in agreement with the assumption of intense homogeneous collective oscillations).

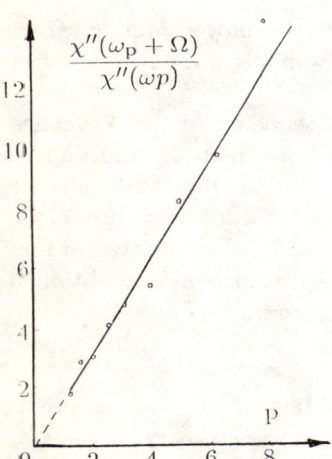

Fig. 9.19. Susceptibility to weak RF signal $\chi(\omega_p + \Omega)$ over susceptibility to pumping $\chi(\omega_p)$ versus supercriticality p for YIG sphere at $M \parallel [100]$, $H_c - H = 100$ Oe. After *Zautkin* et al. [9.46]

According to the theory developed in Sec. 7.2, the susceptibility to the weak signal must have a resonance character (7.2.10) with the maxima near $\omega_p \pm \Omega$ and the halfwidth γ. At the resonance points $\chi''(\omega_p \pm \Omega)$ is given by (7.2.11). Thus, the resonance susceptibility $\chi''(\omega_p + \Omega)$ is p times as large as the susceptibility to the pumping with respect p to the order of magnitude. This fact is also manifested experimentally. From Fig. 9.19 it can be seen that the ratio $\chi''(\omega_p + \Omega)/\chi''$ near the threshold is equal to 2 and at $p=8$ is 14.

According to the theory (7.2.11) the experimental dependence $\chi''(\omega + \Omega)/\chi''$ on the supercriticality of the magnon system p is linear. The theoretical value of the slope equals to 2 at $T_0/S_0 = 0.54$ (7.1.5) is in agreement with the experimental value 1.65 for the plot in Fig. 9.19.

Asymmetry of absorption spectrum of weak signal (see Fig. 9.17) is the peculiarity of the collective resonance of spin waves. This fact can be easily accounted for theoretically. As follows from (7.2.11), the signs of susceptibilities of $\chi''(\omega_p + \Omega)$ and $\chi''(\omega_p - \Omega)$ are different. This means that at the frequency $(\omega_p + \Omega)$ the absorption takes place, and at the frequency $(\omega_p - \Omega)$ the weak signal is amplified (at positive T and S). The magnitudes of these susceptibilities in accordance with (7.2.11) also differ greatly. At $T_0/S_0 = 0.54$ (YIG sphere) the amplification turns out to be much less than the absorption $\chi''(\omega_p - \Omega)/\chi''(\omega_p + \Omega) = -0.03$ and therefore cannot be registered on the background of the resonance curve of the coefficient of the passage through the resonator.

9.5.4 Linewidth of Collective Resonance

Measurements by *Zautkin* et al. [9.46] show that within the experimental accuracy the resonance width of collective oscillations is independent of the pumping intensity up to the supercriticality $p \simeq 9$ dB. Fig. 9.20 shows the dependence of the frequency width of resonance (measured on the oscillograms $D^2(\omega + \Omega)$) on the constant field. According to theory (see (7.2.10) the resonance halfwidth which determines the damping of collective oscillations proved to be equal to the parametric spin damping of waves making up the collective modes of oscillations. This fact was experimentally confirmed in Fig. 9.21 where the dependences of the magnon relaxation frequencies on the external magnetic field are plotted based on the experimental measurements of their threshold (curve 1) and of the collective resonance width (curve 2). These curves coincide within the limits of experimental error. Therefore the measurement of the width of the absorption peak $D^2(\omega + \Omega)$ under the immediate observation of its shape on the screen of an oscillograph provides an illustrative method for obtaining the relaxation time of spin waves. It is important that this method involves no absolute measurements of the microwave field inside the sample, which is a difficult task in the general case.

Since the damping γ, as is known [9.13, 15, 47], increases linearly as the number of magnons increases, according to the above stated correspondence a similar linearity with respect to k should be expected for the linewidth of the collective resonance. The experimental plot of the field dependence of the linewidth (Fig. 9.20) also points to its dependence on k, at the same time the value γ plotted in Fig. 9.21 as a function of $\sqrt{H_c - H} \sim k$ really proves to be linear.

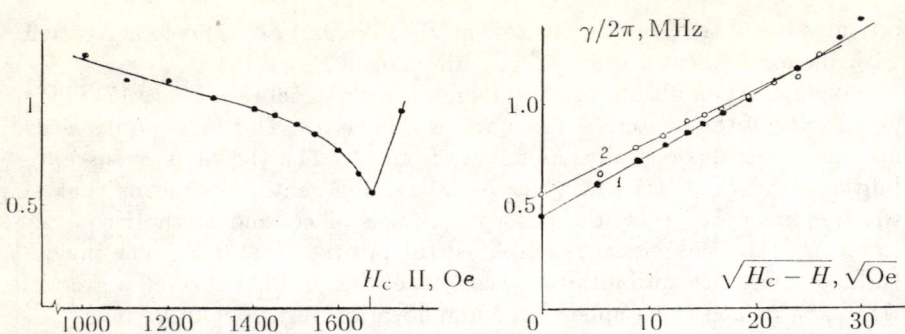

Fig. 9.20. (left) Frequency width of collective resonance of parametric magnons in YIG sphere versus magnetic field at $M \parallel [100]$. After *Zautkin* et al. [9.46]

Fig. 9.21. (right) Frequency of magnon damping in YIG sphere versus magnetic field at $M \parallel [100]$, obtained from parametric threshold data (curve 1) and from data for width of collective resonance (curve 2). After *Zautkin* et al. [9.46]

9.5.5 Oscillations of Longitudinal Magnetization

The dependence of the oscillation amplitude of the longitudinal magnetization registered by e.m.f. in the induction coil on the frequency of the weak signal and supercriticality is qualitatively characterized by a set of oscillograms in Fig. 9.17. Quite surprising is the drastic difference in the behavior of $m_z(\pm\Omega)$ and $D^2(\omega_p \pm \Omega)$ which consists, as becomes clear from Fig. 9.17, in the fact that once more a lower maximum of $m_z(-\Omega_{\text{res}})$ corresponds to the resonance $m_z(+\Omega_{\text{res}})$, at the same time the absorption peak $D^2(\omega_p + \Omega_{\text{res}})$ has no symmetrical satellite at the frequency $\omega_p - \Omega_{\text{res}}$.

Fig. 9.22 plots the intensity of low-frequency oscillations versus the intensity of the weak signal. The plot shows that within the measurement range the oscillations are linear with respect to the field of the signal exciting these oscillations. This fact enables us to simplify the task of the development of collective resonance theory and to confine ourselves to the linear approximation for the oscillation amplitude of the wave amplitude with respect to the stationary state.

In order to find $m_z(\Omega)$, in (3.4.8) for the longitudinal magnetization, value $a(\boldsymbol{r},t)$ must be expressed in terms of $a(\boldsymbol{k},t)$. Then the circular canonical variables $a(\boldsymbol{r},t)$ must be expressed in terms of to the normal canonical variables of the quadratic Hamiltonian. Subsequently we must substitute into the obtained expression the expansion (7.2.7) and retain only the relevant for us resonant terms proportional to $\exp[-i\Omega t]$ to oscillate with the low frequency Ω. As a result, we obtain in linear approximation with respect to the amplitude of the weak signal:

$$m_z(\Omega) = U_0[c_0^* d(\Omega) + c_0 d^*(-\Omega)] \,. \tag{9.5.2}$$

Here $U_0 \simeq g$ is given by (7.2.4), $|c_0|^2 = N$ denotes the total number of parametric magnons, $d(\Omega)$ is specified by (7.2.3, 7). By employing all these formulae we readily obtain from (7.2.8)

$$m_z(\Omega) = U_0 N \frac{h_s}{h_{th}} \frac{\gamma[\Omega + 2S_0 N - 2i\gamma]}{\Delta_0^2 - \Omega^2 - 2i\gamma\Omega}, \tag{9.5.3}$$

Here h_s is the amplitude of the weak signal with the frequency $(\omega_p + \Omega)$, $\Delta_0 = 2N\sqrt{S_0(2T_0 + S_0)}$ designates the frequency of the collective resonance. As it should be expected the amplitude $m_z(\Omega)$ is at maximum if $\Omega = \pm\Delta_0$. From (9.5.3) an expression for the ratio of the resonance maxima at the image frequencies $\omega_p \pm \Omega$ can be obtained:

$$\frac{m_z(\Delta_0)}{m_z(-\Delta_0)} = \sqrt{\frac{1 + 2(p-1)[1 + t + \sqrt{1+2t}]}{1 + 2(p-1)[1 + t - \sqrt{1+2t}]}}. \tag{9.5.4}$$

Here $t = T_0/S_0 \simeq 0.54$ under experimental conditions (YIG sphere, $M \parallel$ [100]). At $p = 2$, for example, ratio (9.5.4) equals 2.4 which is in agreement with the experimental value of the ratio of the maxima, equal to 2.

The consistent explanation (from the same position) of the weak asymmetry $m(\pm|\Omega|)$ and the strong asymmetry $\chi''(\omega_p \pm |\Omega|)$ is a significant merit of the statement about collective resonance of parametric magnons developed in Chap. 7.

The observed linearity of the signal with respect to the weak signal field (Fig. 9.22) is in agreement with (9.5.3) and does not require separate discussion. Note that the proportionality m_z and h_s can be employed in order to measure the amplitudes of the weak microwave signals without registering, in the linear mode. Note also that according to (9.5.3) the measurements of the low-frequency oscillations amplitude m_z is a very sensitive method for obtaining stationary magnetization variation Δm_z, which is proportional to the number of parametrically excited magnons.

9.5.6 Other Methods for Excitation of Collective Oscillations

In Sec. 7.2.3 the theory of the direct excitation of collective oscillations by the longitudinal radio-frequency field at the frequency was worked out. *Orel* and *Starobinets* [9.48] experimentally observed this phenomenon. The detailed comparison of the theoretical and experimental data has been performed in Ref. [9.48] and we shall not dwell on it here. Note only that their experimental data on the dependence of χ'' on supercriticality at resonance are completely described by (7.2.16). The observed frequency of collective oscillations also qualitatively coincides with the frequency calculated from (9.5.1) as well as the experimental data of [9.46] obtained by the above-described method of excitation by a weak microwave signal. Therefore, the S-theory adequately describes the microwave and radio frequency resonance

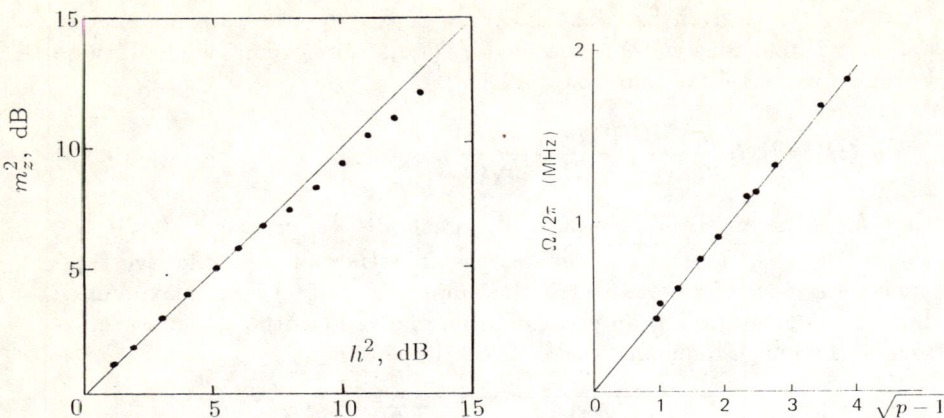

Fig. 9.22. (left) Power of low-frequency oscillations versus power of weak signal for YIG sphere at $M \parallel [100]$, $H_c - H = 100$ Oe. After *Zautkin* et al. [9.46]

Fig. 9.23. (right) Resonant frequency of collective oscillation versus supercriticality for antiferromagnets $MnCO_3$ (*Prozorova* and *Smirnov* [9.46])

of collective oscillations, which agrees with the experimental data and requires no fitting parameters. This confirms the correctness of our physical concept of the nature of collective oscillations.

Closely related to the resonance methods of exciting collective oscillations is the method of excitation of collective oscillations by means of a drastic change in the pumping phase. This method was realized by *Prozorova* and *Smirnov* [9.36]. Their data on the natural frequency of collective oscillations in the antiferromagnet $MnCO_3$ are also in complete agreement with the theoretical dependence (9.5.1) (see Fig. 9.23).

9.6 Stepwise Excitation in YIG

As has been shown in Chap. 5, one of the main conclusions of the basic S-theory is the stage-by-stage excitation of wave packets that are singular in k-space. This means that when the microwave power is increased smoothly, first a narrow packet is excited, for which h_{th} is minimum. When the power is increased further this packet is not broadened. Only when the microwave field reaches a strength h_2 exceeding h_{th} by 8 to 10 dB, is another group of waves created discontinuously. The observation of the influence of the size effect and of magnon-phonon peaks in the relation $h_{\text{th}}(\boldsymbol{H})$ shows that this conclusion is correct.

The simple distribution form of magnons at $h < h_2$ makes it possible to obtain analytical expressions for the most important characteristics of the system of parametric magnons within the wide range of the pumping intensity. The good agreement between calculated and measured results for the susceptibility and collective resonance of parametric magnons (see Sec. 9.2, 3 and 5) indirectly confirm this conclusion. Special experiments are required for the direct proof of the excitation of waves with $\Theta = \pi/2$. They will be described below.

9.6.1 Re-Radiation into the Transverse Channel

Zautkin et al. [9.49] found a direct experimental proof of the stage-by-stage excitation. A YIG sample was put into a cylindrical resonator with two degenerate orthogonal modes (TE$_{112}$), whose magnetic fields $\boldsymbol{h}_\parallel$ and $\boldsymbol{h}_\perp$ were parallel and perpendicular, respectively, to the static field $\boldsymbol{H}$. The required polarization of the modes and their decoupling was achieved by the definite orientation of the waveguides connecting the resonator with the generator and input device.

The parallel channel ($\boldsymbol{h}_\parallel$) was used for the parametric excitation of magnon pairs with the frequency $\omega(\boldsymbol{k})/2\pi = 4.7$ GHz; the perpendicular channel ($\boldsymbol{h}_\perp$) registered the radiation of the sample at the pumping frequency $\omega_\mathrm{p} = 2\omega(\boldsymbol{k})$. Up to the magnon excitation threshold, the decoupling between the channels was 55 dB. A drastic increase in radiation power was observed (see Fig. 9.24) in the perpendicular channel at a supercriticality $p_\mathrm{cr} \simeq 8$–12 dB. (Different values of p_cr are due to the variations in the constant field, and the orientation and shape of the samples.) The radiation in the perpendicular channel is the result of the interaction of a magnon pair $a(\boldsymbol{k}), a(-\boldsymbol{k})$ with the uniform precession of the magnetization a_0 a which is described by the Hamiltonian

$$\mathcal{H}_\perp = \frac{1}{2} \sum_{\boldsymbol{k}} [u^*(\boldsymbol{k}) a_0^* a(\boldsymbol{k}) a(-\boldsymbol{k}) + \text{c.c.}] \,, \qquad (9.6.1)$$

where $u(\boldsymbol{k}) = u \sin(2\Theta) \exp(i\Phi)$. The radiation power is determined by the expression

$$P_\perp = h_\perp^2 \simeq |\sum_{\boldsymbol{k}} u^*(\boldsymbol{k}) a(\boldsymbol{k}) a(-\boldsymbol{k})|^2 \,, \qquad (9.6.2)$$

which vanishes for $\Theta = 0$ and $\Theta = \pi/2$. The dependence $P_\perp(h_\parallel)$, plotted in Fig. 9.24 is a conclusive evidence of the fact that in the p range from 0 to 8 dB, only magnons pairs propagating at an angle $\Theta = \pi/2$ are excited. The radiation appearing at higher supercriticalities is naturally associated with the excitation of the second group of pairs with $\Theta \neq \pi/2$. The presence of the second threshold can also be observed by the characteristic distortion of the pumping pulse. A series of consecutive thresholds, first observed in this manner by *Petrakovsky* and *Berzhansky* [9.50] in YIG, is evidently associated with the stage-by-stage excitation of parametric magnons.

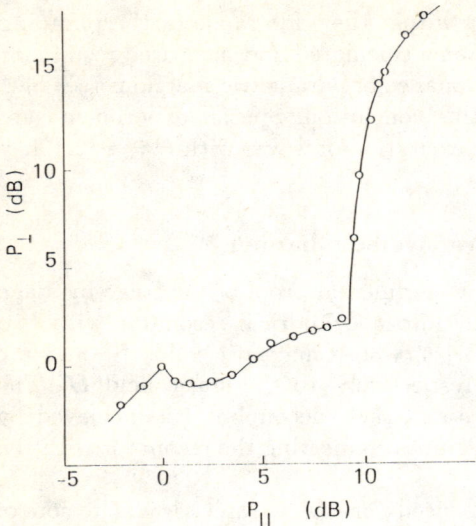

Fig. 9.24. Radiation power in transverse channel versus pumping power for YIG sphere at $M \parallel [100]$, $H_c - H = 100$ Oe, $\omega_p/2\pi = 9.4$ GHz. After *Zautkin* et al. [9.49]

9.6.2 Interaction of Second–Group Magnons and Transverse Signal

Methodologically, this experiment can be considered an "active" version of the above experiment. The main part of the setup is, as before, the bimodal cylindrical resonator TE$_{112}$. The mode $h_\parallel \equiv h_s$ is intended for parallel pumping of magnons, and the mode excited by the klystron is used for generation of an additional weak signal probing the system of parametric magnons. The interaction with the transversal weak signal is described (as in the previous case) by the Hamiltonian $\mathcal{H}_\perp$ (9.6.1) with the amplitude $u(\mathbf{k}) \simeq \sin 2\Theta$ and is allowed, therefore, only for the spin waves with $\Theta \neq \pi/2$. This interaction leads either to absorption or to amplification of the weak signal caused by the excitation of collective oscillations in the system of parametric magnons. In order to detect them we employ the sweeping of the klystron frequency by changing the voltage on its repeller plate with respect to the pumping frequency (coinciding with the frequency of the resonator). The signal with low combination frequency periodically changing within the range of 1 MHz excites the collective oscillations of magnons with $\Theta \neq \pi/2$. At the same time an additional peak of the weak signal absorption appears on the curve of the signal reflected from the resonator and having the shape of the resonance curve of the resonator.

The resonance peak of collective oscillations of the new group of parametric spin waves ($\Theta \neq \pi/2$) appears at a threshold h_{cr} as a weak distortion of the resonance curve in its center. As h increases and becomes higher and higher above the threshold of the interaction with the field of the transverse signal h_{cr} the peak becomes greater and is shifted from the central frequency.

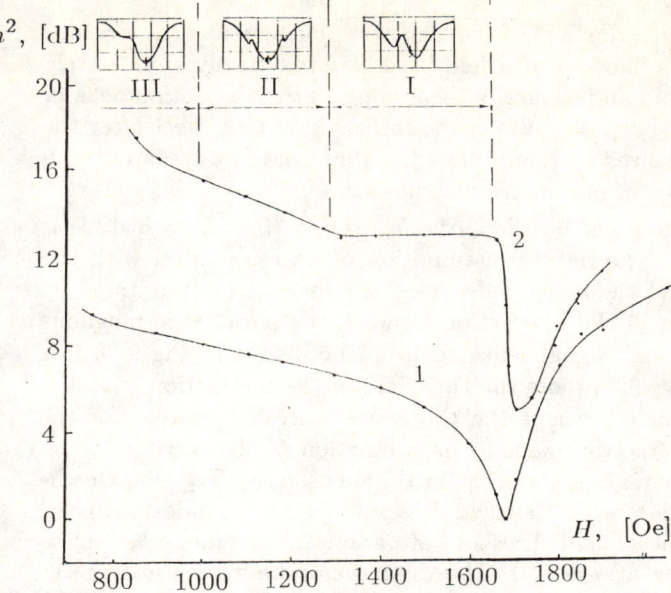

Fig. 9.25. Threshold of parametric excitation of magnons h_{th}^2 (curve 1) and threshold of their interaction with transverse weak signal h_{cr}^2 (curve 2) versus magnetic field for YIG sphere at $M \parallel [100]$. After *Zautkin* et al. [9.51]

Figure 9.25 plots the experimental power dependences of the thresholds h_{th} and h_{cr} on the constant magnetizing field obtained by *Zautkin* [9.45] for the YIG sphere with the orientation $M \parallel [100]$. The first threshold h_{th} was obtained with an accuracy of 0.1 dB from the distortion of the magnetron pulse reflected from the resonator. The threshold h_{cr} was registered with an accuracy of 0.5 dB from the distortion of the resonance curve on the oscillogram of the signal reflected from the resonator from the klystron side, i.e. using the above-described procedure.

The absence of the weak signal absorption at $h_{\text{th}} < h < h_{\text{cr}}$ over the entire range of the constant fields $H < H_{\text{c}}$ is the direct proof of the fact that below the threshold h_{cr} only the pairs with $\Theta = \pi/2$ are excited in the system (recall that H_{c} is the magnetic field under which $\omega_{\text{p}} = 2w(\boldsymbol{k})$ at $k = 0$ and $\Theta = \pi/2$). The values $h_{\text{cr}}/h_{\text{th}} \simeq 7\text{--}12$ dB (for different magnitudes of the field H) agree with the experimental data for re-irradiation into the transverse channel.

The phenomena observed in the range $H > H_{\text{c}}$ (Fig. 9.25) show that only the waves with $\Theta \neq \pi/2$ actually interact with the transverse weak signal. Indeed, at $H > H_{\text{c}}$ the bottom of the spin wave spectrum (the value $\omega(k,0)$ at $k = 0$, $\Theta = \pi/2$) exceed $\omega_{\text{p}}/2$. It means that only the long wave part of the spectrum with some $\Theta < \Theta_* < \pi/2$ appears to be at the parametric resonance with the pumping. This removes the exclusion of the magnon interaction with the field of the transverse channel and, as a result, collective oscillations induced by the field h_s emerge practically immediately above the first threshold and naturally there appears the absorption peak of the weak signal. The sharp decrease of the threshold of this effect after the passage through H_c leaves no doubt that the threshold is associated with the distribution change of parametric magnons in Θ.

For this reason the point on the curve $h_{\mathrm{cr}}(H)$ at $H_c - H = 350$ Oe is of particular interest. The natural assumption of its association with the changed distribution of the registered parametric magnons is confirmed by the changed character of the interaction between the parametric magnons and the field of the weak signal reflected from the inserts in Fig 9.25 (for details see [9.51]). Fig. 9.26 plots the threshold of the interaction with the transverse weak signal h_{cr} versus the theoretical threshold values for the higher groups of parametric magnons as a function of the constant magnetic field. The comparison shows that in the fields $H_c - H \geq 300$ Oe the threshold of interaction with the field h_{cr} practically coincides with the calculated value of the second threshold of parametric magnons h_2, calculated in Sect. 6.1 (see also [9.51]). The discrepancy between them in the fields $H_c - H < 300$ Oe can be explained by the fact that the susceptibility of the second packet of parametric magnons to the transverse signal $\chi''(\omega_p + \Omega)$ is proportional to $u^2(\boldsymbol{k}) \sim \cos^2\Theta \sin^2\Theta$ (see (9.6.1)) and decreases as the magnetic field increases (see the points for the $\cos\Theta_2$ in Fig 9.26B). The calculation of the second packet susceptibility shows that for the fields $H_c > H > H_c - 300$ Oe and the pumping power $h^2 < 10 h_{\mathrm{th}}^2$ the magnitude of $\chi''(\omega_p + \Omega)$ is too small: $\chi'' \simeq 3 \cdot 10^{-3}$. Consequently, within this range of magnetic field values the distribution change in parametric magnons above the second threshold is not registered experimentally.

In this case the generation of the third group of parametric magnons at $h = h_3 > h_2$ is experimentally perceived as the "second threshold". In order to be able to interpret the experimental results, the threshold of generation of the third group of parametric magnons h_3 must be known, because h_{cr} must coincide with h_3.

This calculation of the threshold has been carried out by *Podivilov* [9.52] who widely used the theory of change in the distribution function of magnons above the intermediate threshold described in Sec. 6.1. His calculations of the thresholds and angles of generation of the third group of parametric magnons are shown in Fig. 9.26. In this figure the point of the third threshold calculated in [9.51] at $H_c - H = 100$ Oe when the second group emerges on the equator of the resonance surface leading only to the broadening of the parametric magnon packet with respect to the angle Θ is also shown.

The complete plot of the dependence $h_{\mathrm{cr}}(H)$ in Fig 9.26 shows that in the fields $H < H_c - 300$ Oe the value h_{cr} corresponds to the excitation threshold of the second group of parametric magnons h_2 and for $H > H_c - 300$ Oe it

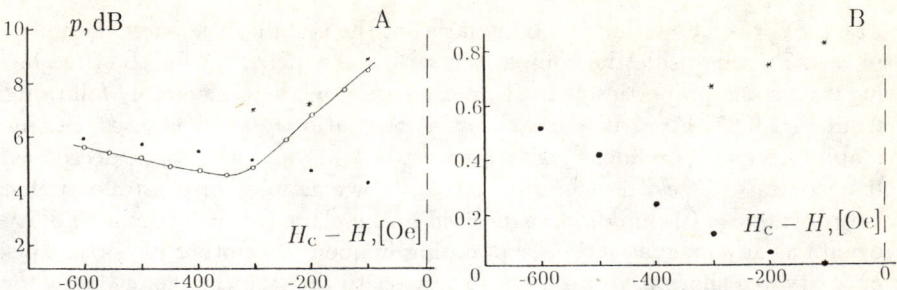

Fig. 9.26. Magnetic field dependence of the threshold of parametric excitation (Fig. A) and of the cosine of polar angles (Fig. B) in YIG sphere at $M \parallel [100]$. Dots 1 are theory for a second group of pairs, dots 2 are theory for a third group p_3 and dots 3 are experimental data for the threshold of the parametric magnon interaction with weak signal. After Zautkin et al. [9.51, 52]

corresponds to the generation threshold of the third group h_3. In this case the salient point of the experimental curve discussed above can naturally be accounted for by the different field dependence of the threshold of the second and third groups of spin waves.

9.7 Conditions of Excitation of Auto-Oscillations of Magnons

As shown in Sec 8.1.3, the auto-oscillations under parametric excitation of waves are due to the instability of the ground state with respect to the variation of amplitudes and phases of the parametric waves, i.e, to the instability of the collective oscillations. When there is axial symmetry in the problem (e.g. under parallel pumping of magnons in cubic ferromagnets at $M \parallel [100]$ or $[111]$) the criterion of generation of auto-oscillations according to this theory has the following form (7.1.1):

$$(S_m + S_{-m})(S_m + S_{-m} + 4T_m) < 0 . \qquad (9.7.1)$$

Here m is the number of the axial mode, S_m and T_m denote the axial Fourier harmonics of the interaction amplitudes of the waves. Values S_m and T_m in the cubic ferromagnets differ from zero at $|m| \leq 2$. (7.1.5) give the explicit expressions for these amplitudes through such parameters of the problem as the magnetization, pumping frequency, demagnetization factors (sample shape), etc. From these formulae and inequality (9.7.1) it is possible to predict the emergence of auto-oscillations in different experimental situations. Fig 9.27 shows the instability region of the mode $m = 0$ as a phase diagram on the plane $(\omega_p/\omega_m , N_{z0,\text{eff}})$ [9.52]. The "instability phase" over which auto-oscillations take place is between the lines $S_0 = 0$ and

$2T_0 + S_0 = 0$, i.e. within the boundaries of the stability region of the mode $m = 0$. Putting aside the comparison with the experiment for the time being note some properties of the zero oscillation mode immediately following from Fig. 9.27. First, it is clear that as the ratio ω_p/ω_m increases, the instability region "broadens" with respect to $N_{z0,\text{ef}}$ and as $|N_{z0,\text{ef}}|$ decreases, it becomes wider with respect to ω_p/ω_m. For example, for crystallographic isotropic ($\omega_a = 0$) thin disc magnetized in the plane ($N_z = 0$) the instability occurs at any magnetization or pumping frequency. Another important peculiarity is a sharp asymmetry with respect to the sign change of the factor $N_{z0,\text{ef}}$: the negative values of $N_{z0,\text{ef}}$ correspond to the larger part (area) of the instability region. This accounts for the experimentally observed [9.28, 51] abnormally high crystallographic anisotropy of auto-oscillations in YIG. However, the well-known rule that auto-oscillations exist at the orientation $\boldsymbol{M} \parallel [111]$ and practically vanish at $\boldsymbol{M} \parallel [100]$ is not always observed, as can be seen from Fig. 9.27, it is also confirmed experimentally (see Table 9.2 below). This fact can now be easily explained: the parameter of the crystallographic anisotropy enters into the instability criterion (9.7.1) in a complex combination with other parameters (7.1.5) and the result is determined only by the relation of their numerical values.

Making use of the explicit form of the functions $S_{\pm 2}$ and $T_{\pm 2}$ (7.1.5) and of the criterion (9.7.1) we can also obtain the instability region of the modes $m = \pm 2$ (see Fig. 9.28). In comparison with the zero mode whose instability condition is independent of the magnitude of the wave vector k and parametric spin waves, modes $m = \pm 2$ are unstable only within some interval Δk corresponding to a certain region of positive values of $N_{z2,\text{ef}}$. Therefore, for $\boldsymbol{M} \parallel [100]$ (magnetization in the "difficult" direction, where the zero mode is often stable) the auto-oscillations of modes $m = \pm 2$ are usually arise.

The considered mode $m = 0$ and even modes $m = \pm 2$ (inhomogeneous in the azimuthal angle) are homogeneous in the sample volume, i.e. have a zero wave vector. Naturally, along with these modes, there can exist spatially inhomogeneous modes of auto-oscillations caused by the instability of some part of the spectrum of the collective oscillations propagating with the wave vector $\boldsymbol{\kappa} \neq 0$. Indeed, if the space dispersion is taken into account, to each mode of collective oscillations there corresponds a whole branch of the spectrum $\Omega_m(\kappa)$, whose gap $\Omega_m = \Omega_m(0)$ is given by (7.6.1). The spectrum of collective waves is described in Sec. 7.1.3. In the simplest case when $\boldsymbol{\kappa} \parallel \boldsymbol{M}$, $m = 0$ it is specified by (7.1.13):

$$\Omega_0(\kappa) = -i\gamma \pm \sqrt{[2(T_0 + S_0)N + \omega'' \kappa^2/2]^2 - 4T_0^2 N^2 - \gamma^2}, \qquad (9.7.2)$$

where $\omega'' = \partial^2 \omega(k)/\partial k_z^2$. For the transverse perturbations ($\boldsymbol{\kappa} \perp \boldsymbol{M}$) the branches of oscillations with different m prove to be connected with each other and the normal modes appear to be their linear combinations. In this case a simple analytical expression cannot be written.

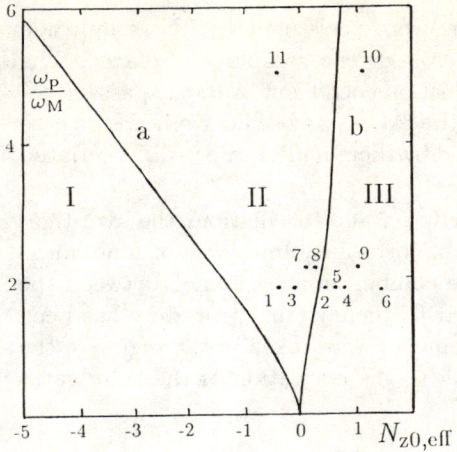
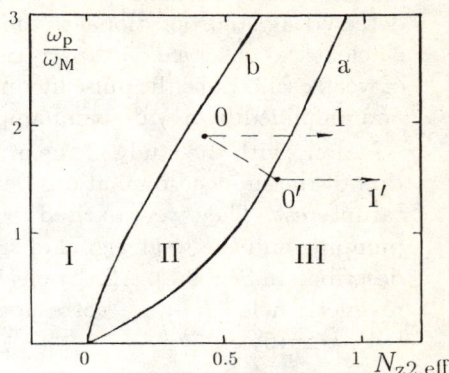

Fig. 9.27. (left) Theoretical stability phase diagram for collective mode $m=0$ in YIG. I and III are the regions of stability, II is the unstable region (intense auto-oscillations are predicted and observed). After *Zautkin* et al. [9.52]

Fig. 9.28. (right) Theoretical stability phase diagram for collective mode $m = \pm 2$ in YIG. I and III are the regions of stability, II is the unstable region (weak auto-oscillations are predicted and observed). After *Zautkin* et al. [9.52]

Figure 9.29 shows the possible variants of behavior $\Omega^2(\kappa)$ at $\gamma = 0$. The region of the negative values of Ω^2 corresponds to the instability of the ground state, which leads to the self-excitation of collective waves. Note that the ground state of the parametric wave system is always stable with respect to perturbations with large κ.

9.7.1 Experimental Setup

Auto-oscillations were investigated [9.52] for the system of parametric magnons excited by parallel pumping (see, for instance, setup in Fig. 9.1). The electromagnetic field with the frequency $\omega_p/2\pi = 9.37$ GHz was excited in the resonator by a magnetron with pulsed operation (pulse duration was 0.1–1 ms) or in a continuous mode. To avoid excessive sample heating in the continuous mode of operation the measurements were taken with an intensity not high above the threshold of parametric excitations. In addition the sample was cooled by compressed air. Samples of cubic ferromagnets with garnet structures were investigated with different magnetizations and anisotropic fields (see Table 9.1). The samples had a spherical shape 0.8–2.2 mm in diameter or were shaped as discs with the diameter of 4 mm and thickness of 0.1–0.2 mm. Magnetization orientation was selected along the most symmetrical directions, i.e. along the "easy" [111] and "difficult" [100] axes of the crystals.

Auto-oscillations of magnetization were registered by the amplitude modulation of the microwave field in which the sample was placed. The detected signal proportional to the reflection coefficient or to the passage coefficient was observed on the screen of the two-beam oscillograph. In the case of weak oscillations the pulse modulated by them had been pre-differentiated and amplified by a wide-band amplifier.

Along with the study of the properties of auto-oscillations the investigation of the induced oscillations near the instability limit was of fundamental interest. They were excited by the combination resonance between the pumping and the weak signal of similar frequency (this procedure has been described in Sec. 9.5). All the measurements were taken in the region of the magnetic fields $H < H_c$ corresponding to the excitation of the spin waves with $k \simeq 10^5$ cm^{-1}.

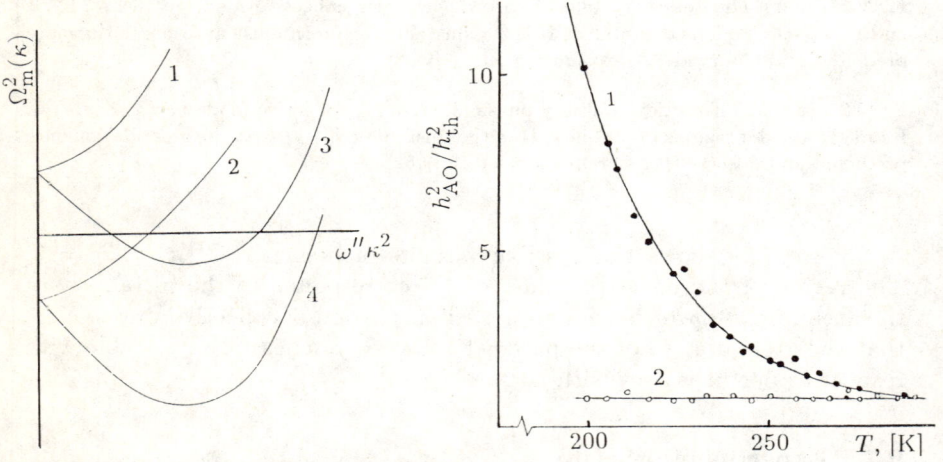

Fig. 9.29. (left) Examples for spectra of collective oscillations (9.7.2) (curves 1 and 2 correspond to condition $(T + S)\omega'' > 0$, curves 3 and 4 to $(T + S)\omega'' < 0$

Fig. 9.30. (right) Temperature dependences of auto-oscillation threshold in YIG sphere at $M \parallel [100]$ (curve 1) and $M \parallel [111]$ (curve 2). After *Zautkin* et al. [9.52]

9.7.2 Intensive Auto-Oscillations of Mode $m = 0$

In the early experiments by *Zautkin* et al. [9.52] auto-oscillations of comparatively large amplitude were investigated which could be observed directly as a low-frequency envelope of the microwave pumping pulse. As a whole, the picture of instability development observed in these experiments was qualitatively similar to those already known and described, e.g., in [9.28,

51, 52]. The attention was mainly paid to the validation of the instability criterion (9.71).

Table 9.2. Strong auto–oscillations (AO) in different cubic ferromagnets: YIG – $Y_3Fe_5O_{12}$ ($4\pi M = 1750$ Oe, $H_a = 84$ Oe); YIScG – $Y_3Fe_{4.35}Sc_{0.65}O_{12}$ ($4\pi M = 1500$ Oe, $H_a = 8$ Oe); ViCaIVG – $Vi_{0.2}Ca_{2.8}Fe_{3.6}V_{1.4}O_{12}$ ($4\pi M = 650$ Oe, $H_a = 58$ Oe) (After *Zautkin* and *Starobinets* [9.52], $\omega_p/2\pi = 9.8$ GHz, $T = 300$ K)

Crystal	Number of expt.	N_z	Direction of magnet.	Sign of $S_0(2T_0 + S_0)$	Strong AO
YIG	1	0	[111]	−	present
	2	0	[100]	+	absent
	3	1/3	[111]	−	present
	4	1/3	[100]	+	absent
	5	1	[111]	+	absent
	6	1	[100]	+	absent
YIScG	7	0	[100]	−	present
	8	1/3	[100]	−	present
	9	1	[100]	+	absent
ViCaIVG	10	1/3	[100]	+	absent
	11	1/3	[111]	−	present

To this end, *Zautkin* and *Starobinets* [9.52] realized 11 different experimental situations and calculated from (7.1.5) the quantity $S_0(2T_0 + S_0)$ for each of them. The results tabulated in Table 9.1 show that in all cases there is a complete agreement between the stability of the observed state of the parametric magnon system and the sign of the product $S_0(2T_0 + S_0)$. The experimental situations in the table are in correspondence with the points on the state diagram of the homogeneous collective oscillations ($m = 0$) plotted in Fig. 9.27. Clearly, the intense oscillations are excited only when the corresponding point on the Fig. 9.27 is in the instability region of the zero mode. Therefore, the shape of the sample, its magnetization and its crystallographic anisotropy which determine the position of the point on the diagram can have a strong impact on the excitation of auto-oscillations of the system of parametric magnons. The diagram in Fig. 9.27 with experimental points is an attempt to systematize the data on auto-oscillations which proved to be successful owing to the introduction of the criterion (9.7.1). It must also be noted that if intense auto-oscillations exist, they are observed over the entire range of the magnetic field according to the fact that the instability condition of the zero mode is independent of the value of the wave vector of spin waves.

9.7.3 Crossing the Instability Boundary and Spatially Inhomogeneous Auto-Oscillations

Let us compare theory with experiment on a more profound level. The points 2 and 8 on the phase diagram are of a special interest (Fig. 9.27). These points are close to the stability boundary ($2T_0 + S_0 = 0$) and at a slight change of temperature (increase for point 2 and decrease for point 8) they cross this boundary. Such transitions involving the state change have been experimentally observed. Fig. 9.30 shows the behavior of the excitation threshold of the intense auto-oscillations at the transition of point 8 (YIG with the addition of scandium at $M \parallel [100]$) to the stability region accompanied by a temperature decrease. It turns out in this case that auto-oscillations do not disappear abruptly, their amplitude decreases smoothly as the transition to the stability region of the mode $m = 0$ proceeds.

In the second experiment (point 2 – disc-shaped YIG sample) auto-oscillations in the sample emerged at heating up to 360 K. The calculated value of the transition temperature obtained from the condition $2T_0 + S_0 = 0$ allowing for (7.1.5) is 330 K which is in satisfactory agreement with the experiment.

Such temperature transitions over the instability boundary $2T_0 + S_0 = 0$ enable us to verify the following statement. If the intense auto-oscillations are really the result of the instability of the zero mode of collective oscillations then at transition from the stable state into the unstable state the eigenfrequency $\Omega_0 = 2\sqrt{S_0(2T_0 + S_0)N}$ of the homogeneous mode must tend to zero. The check experiment [9.52] was based on the collective resonance of parametric magnons described in Sec. 9.5. It turned out that as the temperature increases the resonance feature of the absorption peak really decreases which is very important for the interpretation of the experiment, but does not tend to zero. In addition, near the critical temperature in the spectrum of the weak signal passage at the frequency $\omega_p + \Omega$ alongside with the dip there appears a peak of negative absorption at the image frequency $\omega_p - \Omega$ corresponding to the signal amplification.

The amplification of the weak signal is basically important from the viewpoint of the concept of collective resonance developed in Sec. 7.2. Let us turn to (7.2.11) for the resonance susceptibility of the system of parametric spin waves in the field of the weak signal. The value of the radicand in it for the YIG is of the order of unity and therefore the absolute value of the negative susceptibility $\chi''(\omega_p - \Omega)$ is much less than the positive susceptibility $\chi''(\omega_p + \Omega)$. This can account for the absence of the amplification effect in experiments described in Sec. 9.5.3. However, in the cases when $S_0 \simeq 0$ or $2T_0 + S_0 \simeq 0$, the negative susceptibility according to (7.2.11) becomes in principle arbitrarily great which explains the possibility of the electromagnetic radiation being amplified by the collective oscillations of parametric magnons close to the boundary of their instability. The opposite statement also holds true: the emergence of amplification indicates that the value $S_0(2T_0 + S_0)$ approximates zero. The fact that we do not experimentally observe a total vanishing of the resonance absorption frequency of the weak signal can be explained by Fig. 9.29. As can be seen from this figure, the continuous spectrum of collective oscillations always include such regions of the wave vectors κ for which instability appears earlier than for the homogeneous mode that can interact with the weak signal. For example, if for the branch 3 in Fig 9.29 the size of the gap decreases when the experimental conditions (temperature) are changed, then the inequality $\Omega_0^2 < 0$ is satisfied earlier for some $\kappa \neq 0$. The resulting auto-oscillations will be significantly spatially inhomogeneous; their amplitude obtained from the modulation depth of the pumping field depends on the "admixture" of the homogeneous mode which increases as it goes deeper into the instability region.

The spatial inhomogeneity of the auto-oscillations must also be manifested in the dependence of their frequency on the sample size d. Experimentally, it is observed as the increased frequency of the auto-oscillations when the sample size is decreased, but the frequency behavior cannot be described by a simple dependence proportional to $1/d$ and is significantly dependent on the length of the wave vector of spin waves.

Within the suggested model the frequencies of the collective oscillations (9.7.2) increase as the general number of magnons N becomes larger. In Fig. 9.29 it corresponds to the increased gap and the upward shift of the entire considered spectrum branch of the collective oscillations. Therefore, not far from the instability boundary the absence of the auto-oscillations can be reached by increasing the pumping intensity. This effect is really observed in the longitudinally magnetized YIG disc (point 2 in Fig. 9.27).

An interesting situation occurs when crossing the boundary $S_0 = 0$ (until now all the transitions referred to crossing the boundary $2T_0 + S_0 = 0$ on the phase diagram. At $S_0 \to 0$ the second threshold corresponding to the excitation of a new group of magnons with $\Theta \neq \pi/2$ tends to the first threshold (the self-consistent interaction of magnons limiting the energy flux from the pumping into the pairs with $\Theta \neq \pi/2$ disappears). The emergence of a new group of parametric magnons ordinarily is accompanied by the excitation of auto-oscillations. Therefore the auto-oscillations exist usually on the both sides of the boundary $S_0 = 0$. It makes impossible to observe experimentally crossing this boundary.

9.7.4 Instability of Higher Collective Modes

Now let us consider the auto-oscillations due to the instability of higher modes of collective oscillations with which $m \neq 0$ disturb the isotropic distribution of pairs in the azimuthal angle and therefore weakly interact with the homogeneous pumping field. Experiments carried out by *Zautkin* and *Starobinets* [9.52] on the setup with the improved sensitivity revealed the presence of weak auto-oscillations in some regions of magnetic fields. Earlier such auto-oscillations had been described in [9.28]. The authors of [9.52] choose a situation when the theory predicts stability for the mode $m = 0$ and instability for the modes $m = \pm 2$. This is the case of the YIG sphere at the orientation $\boldsymbol{M} \parallel [100]$. Fig. 9.31 shows the observed dependence of the threshold of weak auto-oscillations on the magnetic field in this case. It can be seen here that there are auto-oscillations in two regions of the fields. In the intermediate region auto-oscillations arise at the supercriticality 8 dB, but it is naturally associated with the generation of a new group of magnons above the second threshold and, consequently, has nothing in common with the modes $m = \pm 2$. Therefore these points are not shown on the plot. Auto-oscillations in the region I in Fig. 9.31 can be easily explained by the phase diagram depicting the instability region of the modes $m = \pm 2$ in Fig. 9.28. This diagram shows the trajectory of the point corresponding to the YIG sphere under the changed magnetic field. The initial point 0 corresponding to $k = 0$ (at $H = H_c$) is in the instability region. As the magnetic field H decreases, the wave number k, and, consequently, the coefficient $N_{z2,\text{ef}}$ increase (see (7.15)) and the point moving along the trajectory 01 comes out of the instability region. The theoretical width of the instability region found from condition (9.7) allowing for (7.1.5) for YIG at room temperature is equal to 550 Oe. The experimental value of the region 1 width in Fig 9.31 depending on the shape and size of the sample changes within the range 150–250 Oe.

As the temperature drops this region narrows, and region I vanishes completely at temperatures below 275 K. This is due to the increased anisotropy and magnetization leading to the shift of the initial point on the phase diagram along the line 00' (see Fig. 9.28). Now it is clear that the trajectory 0'1' corresponding to the change of field H at low temperatures does not cross the instability region of the modes $m = \pm 2$ at all.

The position of region II in Fig. 9.31 coincides with the field range where a hard excitation of parametric spin waves caused by the negative nonlinear damping [9.53] is observed. The auto-oscillations in this region are of a complex nature due to the interaction of the parametric and thermal spin waves with delayed reaction of the waves to the change in the state of the parametric subsystem. The generation of such auto-oscillations has been analyzed by *L'vov* in [9.54]. The peculiarity of these oscillations as distinct from the oscillations in region I is a continuous noise spectrum immediately

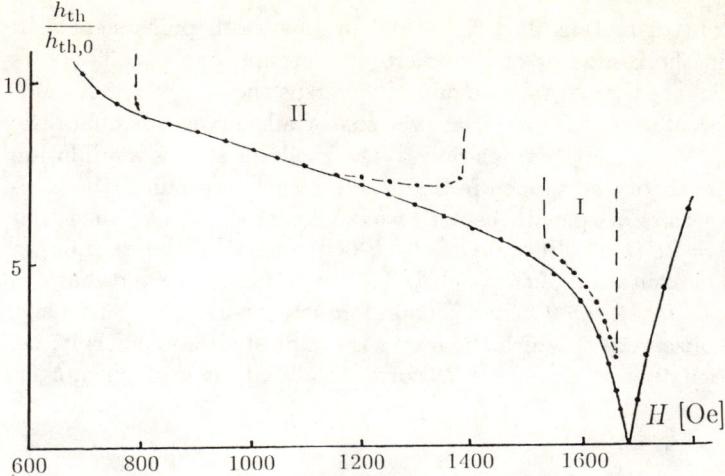

Fig. 9.31. Threshold of parametric excitation of magnons h_{th}^2 (solid curve) and threshold of weak auto-oscillations (dashed curve) versus magnetic field for YIG sphere at $M \parallel [100]$. After *Zautkin* et al. [9.52]

above the threshold of the oscillations generation in a wide range of fields H coinciding with the threshold of parametric excitation.

In conclusion, we can say that the comparison of the theory of auto-oscillation generation under the parametric excitation of waves with a whole series of experiments on cubic ferromagnets described in Sec. 9.7 shows that the S-theory naturally gives an adequate quantitative description of various fine and often unexpected effects and *without making use of fitting parameters*. Among those effects are the influence of crystallographic anisotropy, temperature and shape of the sample on the properties of the auto-oscillations.

9.8 Effect of Radio-Frequency Field Modulation on Parametric Resonance

9.8.1 Suppression of Parametric Instability by Modulation

In Sec. 7.3 the influence of the periodic modulation of external magnetic field $\boldsymbol{H}(t)$ on the threshold of parametric excitation of magnons was studied theoretically. In order to verify the conclusions of this investigation, *Zautkin* and *Orel* [9.55] performed experiments on different types of modulation under parallel pumping of magnons in YIG at the frequency of 9370 MHz. The measurement procedure was standard, employing the pulsed operation with a pulse duration of 0.5 ms. The modulating field was produced by generator current of rectangular, sinusoidal or saw-tooth pulse signals in the coil built in the resonator of the microwave pumping. Fig. 9.32 shows the threshold field of the parametric pumping versus the amplitude of modulation of different types. The initial sections of all curves are quadratic in H_m. At $H_m \gg 0.3$ Oe the transition to the mode of strong modulation takes place. The theory developed in Sec. 7.3 in this case predicts the excitation of two packets of spin waves with wave vectors $\boldsymbol{k}_1$ and $\boldsymbol{k}_2$ such that $\omega(\boldsymbol{k}_1) - \omega(\boldsymbol{k}_2) = 2U(k)H$. Because of the $\boldsymbol{k}$-dependence of the damping γ the thresholds of excitation of the packets $\boldsymbol{k}_1$ and $\boldsymbol{k}_2$ must be somewhat different. Indeed, at $H_m \geq 0.3$ Oe in a certain frequency region two successive thresholds are observed, of which the larger is manifested more sharply, i.e. the time of the instability development corresponding to it is much shorter.

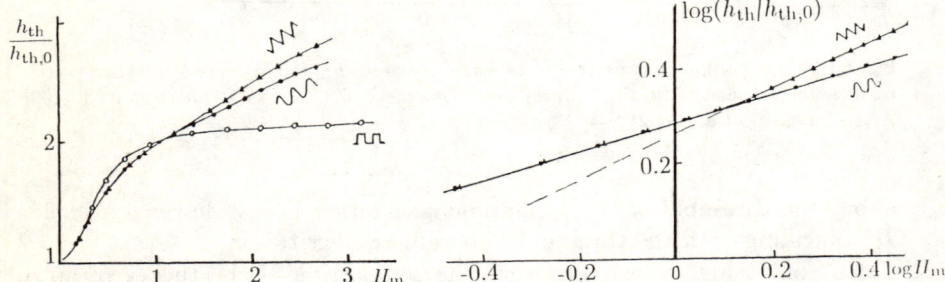

Fig. 9.32. (left) Threshold of parametric excitation of magnons h_{th}^2 versus amplitude of magnetic field modulation of different forms. Experiment in YIG sphere at $M \parallel [100]$. $H_c - H = 100$ Oe, Modulation frequency $\Omega/2\pi = 0.3$ MHz. After *Zautkin* et al. [9.55]

Fig. 9.33. (right) Log-log plot for threshold of parametric excitation of magnons h_{th}^2 versus modulation frequency (saw-tooth and sinusoidal signals) Experiment in YIG sphere at $M \parallel [100]$. $H_c - H = 100$ Oe, After *Zautkin* et al. [9.55]

In the mode of strong modulation the dependence $h_{th}(H_m)$ for the RF signal of rectangular shape in full agreement with the theory reaches the plateau when $h_{th}/h_{th,0} = 2$ is attained ($h_{th,0}$ – the threshold amplitude at $H_m = 0$). Two other modulation types lead to the monotonic increase of the threshold; at the same time the strongest influence at large H_m is exercised by the saw-tooth modulation. For the qualitative comparison of the obtained dependences with the theory it is convenient to approximate them by the functions $h_{th} \sim H_m^r$ for which the calculated values h_{th} (7.3.6, 7) yield the values $r = 2$ for the small amplitude of any shape, $r = 1/3$ for the sinusoid of large amplitude and $r = 1/2$ for the saw-tooth pulses. Fig. 9.33 shows the experimental curves in rectifying the logarithmic coordinates. Clearly, the dependences $h_{th}(H_m)$ are really adequately described by the power functions, the slopes of straight line segments correspond to the exponents of power $r_{\exp} = 2$ for small modulation amplitudes of any shape, $r_{\exp} = 0.34$ for a strong sinusoidal modulation and $r_{\exp} = 0.45$ for a strong saw-tooth modulation. Therefore, the ideas developed in Sec. 7.3 are experimentally verified. It must be noted, however, that as the modulation frequency changes, the numerical values of the parameters also change somewhat. In particular, the saturation of the threshold in the case of rectangular modulation takes place at $h_{\text{th}}/h_{\text{th},0} = 1.8$–$2.4$ with the frequency variation being within the range $\Omega/2\pi = 0.2$–0.5 MHz. Evidently, this is due to the limited frequency region to which the approach of Sec. 7.3 can be applied assuming that the excited wave packet has no time for the shift in the k-space when the field $H_{\text{m}}(t)$ is changed.

9.8.2 Stationary State of Parametric Magnons Under Modulation of Their Frequency

This state has been treated theoretically in Sec. 7.4.1. From the obtained equations (7.4.3-5) we can find the nonlinear susceptibilities χ' and χ'' of the system of parametric magnons under the modulation of their frequency. This yields awkward expressions with the complex coefficients Γ and S (7.4.3,4). It is much more convenient to represent the p-dependences χ' and χ'' graphically for the given values of the amplitude and modulation frequency calculated on a computer. Figures 9.34A, 35A show the theoretical dependences of the susceptibility on the supercriticality of the pumping for $H_{\text{m}} = 0.15$ Oe, $\Omega/2\pi = 1.6$ MHz and $\Omega_{\text{m}}/2\pi = 0.8$ MHz as well as the curves $\chi(p)$ obtained from the formulae of the basic S-theory (5.5.35) for the parallel pumping of magnons in YIG without the field modulation plotted here for the sake of comparison.

Figures 9.34B, 35B show the experimental dependences χ' and χ'' on the supercriticality for $H_{\text{m}} = 0$, $H_{\text{m}} = 0.15$ Oe and some values of Ω [9.56]. The experimental data were obtained like in Sec. 9.11 with the constant field being oriented along the axis [001] of the YIG crystal.

All the curves in Figs. 9.34, 35, theoretical as well as experimental, show that the reaction of parametric magnons to the modulating radio-frequency field has a resonance character. As can be expected the maximum deflections of the susceptibilities from their stationary values are near the resonance of collective oscillations when the supercriticality attains the value under which the eigenfrequency of these oscillations $\Omega_0(p)$ determined by it becomes equal to the modulation frequency Ω. Therefore as the frequency of the RF modulation increases, the maximum of the resonance change in susceptibilities shifts towards higher supercriticalities. Comparing the calculated dependences in Fig. 9.34 and 9.35 with the experimental values of the corresponding parameters of the RF signal (curve 3 in Fig 9.34B and curve 2 in Fig. 9.35B) we can see the same order of magnitude of the effect, i.e. RF addition to the susceptibility, effect sign at the resonance point and the characteristic change of sign at a distance from the resonance due to

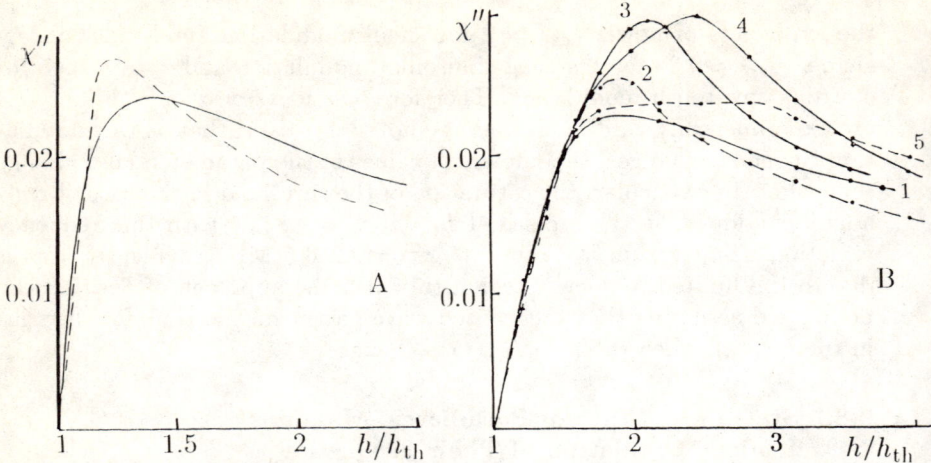

Fig. 9.34. Theory (A) and experiment (B) for imaginary part of nonlinear susceptibility χ'' versus pumping power for a YIG sphere ($H = H_c - 100$Oe, $M \parallel [100]$) under modulation of the magnon frequency. Solid curve in (A) corresponds to $H_m = 0$, dashed line in (A) to $H_m = 0.14$ Oe. Curves 1, 2, 3, 4 and 5 in (B) correspond to $\Omega/2\pi$ 0.0, 0.8, 1.2, 1.6, and 3.2 MHz. ($H_m = 0.14$ Oe). After *Zautkin* et al. [9.56]

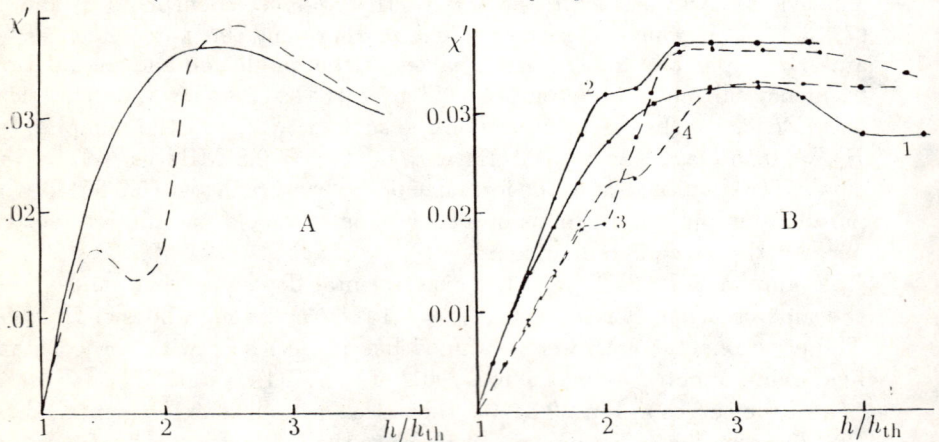

Fig. 9.35. Theory (A) and experiment (B) for real part of nonlinear susceptibility χ' versus pumping power for a YIG sphere ($H = H_c - 100$Oe, $M \parallel [100]$) under modulation of the magnon frequency. Solid curve in (A) corresponds to $H_m = 0$, dashed line in (A) to $H_m = 0.14$ Oe. Curves 1, 2, 3 and 4 in (B) correspond to $\Omega/2\pi$ 0.0, 0.8, 1.6 and 2.4 MHz. ($H_m = 0.14$ Oe). After *Zautkin* et al. [9.56]

the supercriticality change. Therefore, we can conclude that the agreement of theory and experiment is satisfactory.

9.9 Double Parametric Resonance and Inhomogeneous Collective Oscillations of Magnons

Parametric excitation of collective oscillations in the system of parametrically excited waves – *double parametric resonance* was observed and experimentally studied by *Zautkin* et al. [9.56]. In their investigation three parallel magnetic fields were applied to a spherical sample of YIG, i.e. H – a constant magnetizing field; $h \exp(-i\omega_p t)$ – the microwave pumping; $H_m \exp(-i\Omega t)$ – RF pumping. The RF field was produced by a miniature coil located in the center of the microwave resonance. The measurements were taken in the RF range from 0.1 to 10 MHz. Parametric oscillations at the frequency $\Omega/2$ appeared at some critical value of the RF field. The signal at the frequency $\Omega/2$ induced in the coil was amplified by the selective receiver and registered on the oscillograph. The emergence of the signal had a threshold character. As H_m increased the signal almost disappeared in the threshold manner, i.e. its amplitude became 10 times smaller.

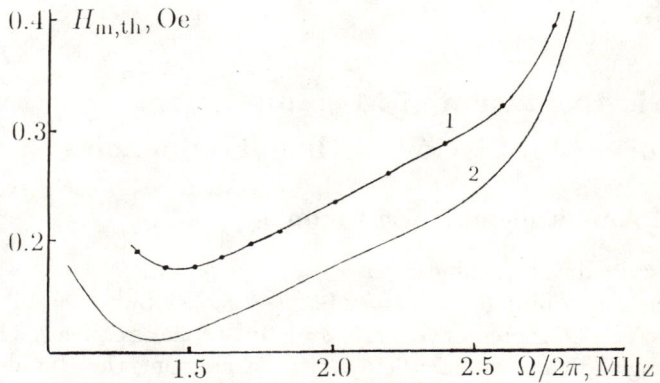

Fig. 9.36. Threshold of double parametric resonance $H_{m,th}$ versus modulation frequency $\Omega/2\pi$ in YIG sphere at $M \parallel [100]$. $H_c - H = 100$ Oe. Curve 1 is experiment at $p = 2.5$ (4 dB), curve 2 – theory for $\Omega_0/\gamma = 3.4$ ($\gamma/2\pi = 0.37$ MHz). After *Zautkin* et al. [9.46]

Figure 9.36 shows one of the experimental dependences $H_{m,th}(\Omega)$ (at $H = H_c - 100$ Oe, supercriticality equal to 4 dB) and the respective theoretical curve obtained from the condition of the zero determinant (7.4.10) (see also Fig. 7.1). The two curves are in good agreement with the exception, perhaps, of the low frequency region where the signal is very weak and no reliable measurements of the threshold can be performed. Note that the theory developed in the approximation quadratic in H_m predicts the existence of only one threshold $H_{m,th}$ under the fixed frequency Ω and supercriticality p. At the same time, in the experiment the second threshold is observed, i.e. the threshold of disappearance of collective oscillations as H

increases. Physically, this threshold is connected with the H_m dependence of the frequency of collective oscillations leading to the violation of the condition of parametric resonance. However, its actual calculation is difficult because higher-order terms in H_m must be allowed for $H_{m,th}(\varphi)$ and can be entirely impossible within the perturbation theory with respect to UH_m/γ.

In conclusion, note that we considered the spatially homogeneous collective oscillations with the wave vector $\kappa = 0$. Alongside in the system of parametric magnons there exists a wide spectrum of collective oscillations (see Fig. 9.29). Some intervals of this spectrum in principle can be parametrically excited by a homogeneous RF field at the frequency satisfying the condition $\Omega = \Omega(\kappa) + \Omega(-\kappa)$. These oscillations with $\kappa = 0$ have no stable connection with the resonator and consequently the radiation at the half-frequency will be significantly weakened. *Zautkin* and *Orel* [9.57] managed not only to register the fact of the existence of inhomogeneous collective oscillations at the double parametric resonance but also to investigate the spectrum. They employed a series of indirect data and analyzed experimentally obtained dependences of the frequency and amplitude of collective oscillations on H, H_m and h.

9.10 Parametric Excitation of Magnons Under Noise Modulation of their Frequencies

9.10.1 Threshold Amplitude of Noise Pumping

In Sec. 6.7 we described the nonlinear theory of the parametric excitation of waves by the noise pumping $h(t) = h(t)\exp(-i\omega_p t)$. Experimentally (see [9.58]) it proved to be more convenient to modulate the frequency of magnons by the longitudinal magnetic field $H_m(t)$. Employing the Hamiltonian (4.3.19) we can easily see that

$$\omega(\boldsymbol{k},t) = \omega(\boldsymbol{k}) + H_m(t)U(\boldsymbol{k}) , \qquad (9.10.1)$$

Here $U(\boldsymbol{k})$ is the interaction amplitude of the longitudinal field $H_m(t)$ with magnons. Using the canonical equations of motion (5.2.2) with the frequency (9.10.1), and after the substitution of variables

$$b(\boldsymbol{k},t) = c(\boldsymbol{k},t)\exp[-i\varphi(t)], \quad \varphi(t) = U(\boldsymbol{k})\int_{-\infty}^{t} H_m(\tau)d\tau , \qquad (9.10.2)$$

one has the equations with constant frequency and the pumping

$$h(t) = h\exp[-2i\varphi(t)] . \qquad (9.10.3)$$

Therefore, the modulation of the magnon frequency (9.10.1) really proves to be equivalent to the modulation of the pumping phase (9.10.3).

The experiment by *Zautkin* et al. [9.58] was performed on the YIG crystals under parallel pumping at the frequency of 9.37 GHz. The experimental setup includes a rectangular resonator and ordinary waveguide elements connected in a circuit "for reflection". The pulsed operation of the magnetron with a pulse frequency of 25 Hz and a duration of 0.6 ms was employed. The threshold of parametric excitation of spin waves and their susceptibility were determined by the procedure described in Sec. 9.1.1, i.e. by the change of signal reflected from the resonator. For the noise modulation of the field through the coil built in the resonator a current from the noise generator was transmitted through a set of filters of lower frequencies and a wide-band amplifier. The generator produces a noise signal with a constant spectral density in the range $\Delta f = 0$–7 MHz. Such a noise can be considered "white" since its band significantly exceeds the relaxation frequency $\gamma/2\pi \simeq 0.4$ MHz and other characteristic frequencies, as will be shown below. Such conditions are most interesting from the viewpoint of the concept developed in Sec. 6.7. The filters reducing the frequency band of the noise signal provide for the determination of the influence of the problem parameters on the process of parametric excitations.

Figure 9.37 shows the experimental dependences of the threshold amplitude (on the coordinates $h_{th}^2/h_{th,0}^2$) on the spectral density of the noise field $H_m^2(0)$. For the rectangular shape of the spectrum $H_m^2(\Omega)$ with the width Δ (in the linear frequency) $H_m^2(0) = \langle H_m^2 \rangle/2\Delta$; $\langle H_m^2 \rangle$ is the mean square value of $H_m(t)$. First it must be noted that as Δ increases, the modulation influence on the threshold field becomes stronger (see curves 1,2 and 3 for $\Delta = 0.8, 2.0$ and 4.0 MHz). Then this dependence is saturated and curves 3 and 4 for $\Delta = 4.0$ and 8.0 MHz coincide. Apparently, the noise modulation at $\Delta < 3$ MHz is not "sufficiently white".

Compare curves 3 and 4 with the theoretical dependence (6.7.1) of the threshold field

$$h_{th}^2 V^2 = \gamma(\gamma + \Delta) , \qquad (9.10.4)$$

which is accurate if the phase fluctuations of the pumping field $h(t)$ (i.e. the process $H(t)$ are the white Gaussian noise. In (9.10.4) Δ is the correlator of phase fluctuations, i.e.

$$\langle \varphi(t)\varphi(t') \rangle = \Delta \delta(t - t') . \qquad (9.10.5)$$

Allowing for (9.10.2) the threshold formula (9.10.4) can be rewritten as

$$[h^2/h_{th,0}^2 - 1] = U^2 H_m^2/\gamma . \qquad (9.10.6)$$

Here $h_{th,0}^2 = \gamma^2/V^2$ is the threshold power at $H_m = 0$. Clearly, in full agreement with the theoretical dependence (9.10.6) lines 3 and 4 in Fig. 9.37 are straight. According to (9.10.6) the tangent of the slope $k = U^2/\gamma$. For the yttrium garnet under experimental conditions $U = 2 \cdot 10^7$ sec$^{-1}\cdot$Oe^{-1}

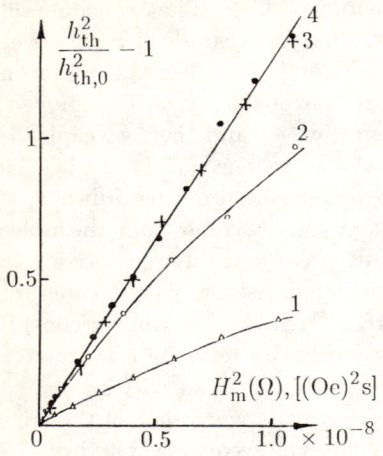

Fig. 9.37. Experimental dependences of the threshold of parametric excitation of magnons h_{th}^2 on the spectral power of noise $H_m^2(0)$ for YIG sphere at $M \parallel [100]$, $H_c - H = 100$ Oe. Curves 1, 2, 3 and 4 correspond to spectral widths Δ equal to 0.8, 2.0, 4.0 and 8.0 MHz. After *Zautkin* et al. [9.58]

and for the slope of the straight line at the value $2\gamma = 2\pi \cdot 0.75 \cdot 10^6$ sec^{-1} we obtain the value $K = 1.7 \cdot 10^8$ sec$^{-1} \cdot$ Oe^{-2}. The experimental data (see Fig. 9.37) are $K_{\text{exp}} = (1.3 \pm 0.3) \cdot 10^8$ sec$^{-1} \cdot$ Oe^{-2}. Despite a large error in obtaining K_{exp} due to the large error of the modulation coil calibration we must emphasize the qualitative agreement of the theoretical value K_{theor} and experimental results.

9.10.2 Efficiency of Phase Mechanism Under Noise Pumping

The most important characteristics of the above-threshold state of the system of parametric magnons from the viewpoint of the S-theory are the total number N and the phase shift Ψ of the pair with respect to the pumping phase. Therefore, most significant in the investigation of the nonlinear behavior of magnons above the threshold is the measurement of the complex susceptibility whose imaginary part χ'' characterizes its total number N and the real part χ' essentially depends on the phase relations in the system.

Zautkin et al. [9.58] measured χ'' and χ' by a standard method (see Sec. 9.1.1), their experimental setup is shown in Figs. 9.1. Figures 9.38, 39 show their experimental dependences of the susceptibilities χ' and χ'' on the supercriticality at different spectral densities of the noise field H_m for the largest band of the noise frequencies $\Delta f = 7$ MHz. For the set of the given dependences it is characteristic that not high above the threshold the noise significantly reduces the susceptibility χ'', and at greater p it somewhat increases it. As a result, with increasing H the whole curve χ'' is shifted towards large p, its shape remaining similar to the respective curves for the coherent pumping (curve 1 in Fig 9.38). The behavior of the real part of the susceptibility χ' is similar: noise influence is strongest near the threshold, at the same time χ' is significantly reduced, but as the pumping intensity increases, its difference from the susceptibility in the coherent mode decreases. At the maximum point the difference is less than 30%. Note that even at the highest value of the spectral density of the noise the maximum value of χ' (Fig. 9.39) exceeds χ'' for the same value of H_m.

Even without a quantitative comparison of the advanced S-theory conclusions with the noise pumping considered in Sec. 6.7 (it will be performed later) we can assert that the above results point to the presence of phase correlations in the system of parametric spin waves and in the case of the noise pumping. The main evidence of the effect of the phase mechanism of the amplitude limitation is that $\chi' \neq 0$. Other mechanisms, as already mentioned, are purely dissipative and result in a zero real susceptibility. In the experiment [9.58] χ' is not only non-zero, but is essentially not small, it exceeds the imaginary part of χ'' similarly to the case of coherent pumping. In addition, the significant similarity of the behavior of $\chi'(p)$ and $\chi''(p)$ with the respective dependences obtained in the absence of noise reveals that the mechanism limiting the parametric magnon increase in this case is the same.

In order to compare theory with experiment in more detail, the nonlinear susceptibilities χ', χ'' must be calculated under noise pumping. Employing the ideas developed in Sec. 5.5.5, 6.7, *Cherepanov* [9.58] got

$$\chi'' = \chi_m \begin{cases} (p-1)^{3/2}/3\sqrt{3}p & \text{at } (p-1) \ll 1 \\ 3\sqrt{3}\sqrt{p-1}/16p & \text{at } p \gg 1 \end{cases} \tag{9.10.7}$$

$$\chi' = \chi_m \begin{cases} 2(p-1) & \text{at } p-1 \ll 1 \\ 1 & \text{at } p \gg 1 \end{cases} \tag{9.10.8}$$

Here $\chi_m = 2V^2/|S|$. For comparison, we represent the nonlinear susceptibilities under coherent pumping (5.5.35) in the same notation

$$\chi'' = \chi_m \sqrt{p-1}/p, \quad \chi' = \chi_m(p-1)/p. \tag{9.10.9}$$

For the sake of convenience of the comparison of theory with experiment, Fig. 9.40 shows also the experimental values of the quantity $\chi''p$ proportional to the absorbed power as a function of $p-1$. The graph is plotted in the logarithmic coordinates where (9.10.7, 8) specify the linear relation of these values. The singularities of the noise pumping expressed in the finite formulae of the theory (9.10.7, 8) are clearly seen on the resulting diagram. The diagram has two rectangular sections, the tangent of its slope at small $p-1$ being equal to 1.4 which agrees with the exponent of power equal to 1.5 in (9.10.7). At large p the slope coincides with the slope of the respective straight line for the coherent pumping depicted for better comparison with the same coordinates in Fig. 9.40. Equation (9.10.7) for $\chi''(p)$ also coincides (up to the constant factor) with $\chi''(p)$ for coherent pumping (9.10.9).

Now let us compare the numerical values of the measured susceptibilities with their calculated values. Equation (9.10.7) for the imaginary part of the

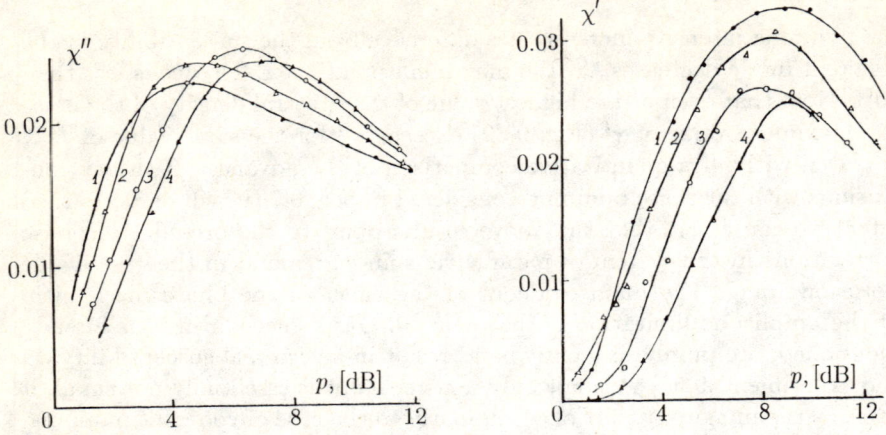

Fig. 9.38. (left) Experimental dependences of the imaginary part of the nonlinear susceptibility χ'' on the pumping power p under noise modulation in YIG sphere at $M \parallel [100]$, $H_c - H = 100$ Oe. Curves 1, 2, 3 and 4 correspond to $H_m^2(0)$ equal to 0.0, $1.7 \cdot 10^{-9}$, $3.4 \cdot 10^{-9}$ and $5.1 \cdot 10^{-9}$ (Oe)$^2 \cdot$ s. After *Zautkin* et al. [9.58]

Fig. 9.39. (right) Experimental dependences of the real part of the nonlinear susceptibility χ' on pumping power p under the same experimental conditions as at Fig. 9.38

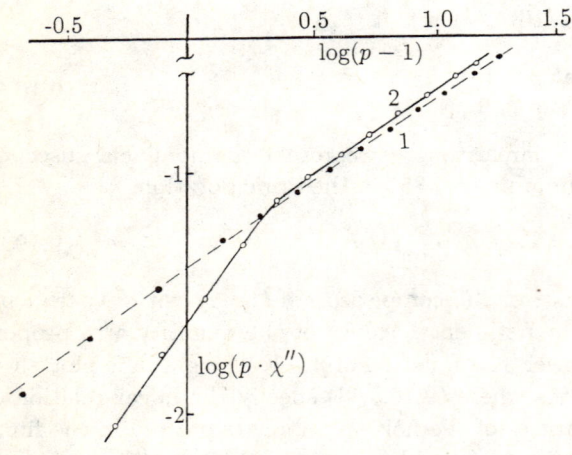

Fig. 9.40. Experimental dependences of the imaginary part of the absorption power $\propto \chi'' \cdot p$ on the pumping power $p - 1$ (in the log–log plot) under noisy modulation in YIG sphere at $M \parallel [100]$, $H_c - H = 100$ Oe. Curves 1 and 2 correspond to $H_m^2(0)$ equal to 0.0 and $5.1 \cdot 10^{-9}$, (Oe)$^2 \cdot$ s. After *Zautkin* et al. [9.58]

susceptibility is in good agreement with the experimental data for the region of small supercriticalities. For instance, for $p = 2$ dB it follows from this formulae that $(\chi''/\chi''_m)_{\text{theor}} = (4/3\sqrt{3})(p-1)^{3/2} = 0.4446$, at the same time the experimentally obtained relation is $(\chi''/\chi''_m)_{\text{exp}} = 0.44(\pm 7\%)$. In the region of large supercriticalities there is a considerable discrepancy with (9.10.8), which yields $(\chi''/\chi''_m)_{\text{exp}} = 0.324$, at the same time $(\chi''/\chi''_m)_{\text{theor}} = 0.8$. As for the real part of the susceptibility, under high pumping intensity $(\chi')/\chi'_m)_{\text{theor}} = 1$ and $(\chi')/\chi'_m)_{\text{exp}} = 0.7$; such an agreement can be considered satisfactory. In the region of small supercriticalities there is a significant discrepancy at $p = 2$ dB we have $(\chi')/\chi'_m)_{\text{theor}} = (2/3)(p-1) = 0.48$, although $(\chi')/\chi'_m)_{\text{exp}} = 0.1$. These discrepancies are apparently caused by the fact that in the experiment the value $\Delta_{\text{ef}}/\gamma \simeq |h_{\text{th}} V^2|/\gamma^2$, was not sufficiently large (< 3), and at the same time expressions (9.10, 7, 8) had been obtained on the assumption that $\Delta_{\text{ef}}/\gamma \gg 1$,. Nevertheless, the agreement of the basic functional dependences (9.10.7,8) with the experimental results shows that the excitation level of parametric magnons under noise excitation is limited, as in the case of coherent excitation, by the mechanism of the phase detuning from the pumping, and our theory correctly describes their general behavior above the threshold.

10 Nonlinear Kinetics of Parametrically Excited Waves

10.1 General Equations

In the previous chapters of this book the nonlinear theory of the parametric excitation of waves was given in the mean–field approximation. It was called the S-theory, after the amplitudes of the interaction of wave pairs, $S(\bm{k},\bm{k}')$, which plays a decisive role in it. As has been shown in Chap. 9, the S-theory is in good qualitative and quantitative agreement with the whole set of experiments on parametric excitation of magnons in ferromagnets and antiferromagnets. At the same time there are some other experiments and it is possible to realize special experiments which show a necessity of overcoming the framework of the S-theory. Indeed, the S-theory describes only the total characteristics of the parametrically excited waves and does not allow for the width of its distribution in ω, $\bm{k}$-space. In the approximation of the S-theory (the first order of perturbation theory with respect to $\mathcal{H}_{\text{int}}$) the stationary state of parametric waves is singular:

$$n(\bm{k},\omega), \quad \sigma(\bm{k},\omega) \propto \delta(\omega - \omega_{\text{p}}/2)\delta[\omega_{\text{NL}}(\bm{k}) - \omega_{\text{p}}/2], \qquad (10.1.1)$$

where $n(\bm{k},\omega)$ and $\sigma(\bm{k},\omega)$ are the Fourier transform of the non-simultaneous normal and anomalous correlators (see Sect. 6.4.1, eqs. (6.4.1–4)). Actually, singularities of the parametric wave distributions (10.1.1) never occur. To confirm this point it is sufficient to attempt to estimate the effect of the next order in $\mathcal{H}_{\text{int}}$ with the help of the kinetic equation [10.1]. Its collision term becomes divergent for the solution (10.1.1). This means that the mutual scattering of the parametric waves should broaden their distribution function (10.1.1).

There have been many attempts to improve the approximation of the S-theory, allowing for the second-order terms in $\mathcal{H}_{\text{int}}$ like derivation of the kinetic equation. But this approach is not correct because the kinetic equation can be applied only if the wave distribution has sufficiently large width in frequencies to guarantee the stochastization of wave phases; at the same time the distribution (10.1.1) has only one frequency $\omega=\omega_{\text{p}}/2$. As a rule, in attempts to overcome the framework of second order approximation in $\mathcal{H}_{\text{int}}$ it is necessary to use a diagrammatic technique, which is the regular method to formulate a perturbation theory with respect to $\mathcal{H}_{\text{int}}$.

There are several types of diagrammatic techniques applied to describe non-equilibrium systems (see, e.g., [10.2, 3]). For classical problems it is more natural to use the diagrammatic technique (DT) suggested by *Wyld* in 1961 [10.2] which has become a regular procedure for investigation of developed hydrodynamic turbulence. The Wyld DT is very similar to the well known Feynman DT for quantum electrodynamics and other field theories: the rules of diagram reading are the same in both DTs, the Dyson equation for the Green's function is also the same. The principal feature of the Wyld DT (as well as that of any technique for strongly non-equilibrium systems [10.3]) consists in constructing two diagram series for the Green's functions (6.4.1) and for the pair correlators (6.4.4). In the thermodynamic equilibrium the Green's function $G(\boldsymbol{k},\omega)$ and double correlator $n(\boldsymbol{k},\omega)$ related by the universal relationship (by the fluctuation-dissipation theorem [10.3]) and two types of function reduce to one type. Under parametric excitation of waves there is no such relation.

Our goal in this chapter is to describe the consistent nonlinear theory of parametric excitation of waves which takes into account not only the mean–field S-interaction of pairs, but also the T^2-scattering of parametric waves from each other and their interaction with the thermal bath of the thermal waves. The latter interaction leads to the damping of the parametric waves. This theory was named S,T^2-theory [10.4–6]. The formalism of the S,T^2-theory is essentially more complicated than that of the S-theory. To keep the review of the S,T^2-theory within reasonable limits, the discussion in this chapter presupposes a higher standard of knowledge than in the previous part of this book. In particular, the reader is assumed to be familiar with the ideas of the Feynman diagrammatic technique [10.3]. A systematic derivation of the main equation of the S,T^2-theory, making use of the Wyld DT for non-equilibrium processes was carried out by *L'vov* [10.6]. To describe this procedure let us consider the motion equation (6.4.8) for $b_j \equiv b(\boldsymbol{k}_j,\omega_j)$ (Fourier transform of the canonical variables $b(\boldsymbol{k}_j,t)$):

$$b(\boldsymbol{k},\omega) = G_0(\boldsymbol{k},\omega)[-hV(\boldsymbol{k})b^*(-\boldsymbol{k},\omega_\mathrm{p}-\omega)$$
$$-\frac{1}{2} \sum_{\boldsymbol{k}+\boldsymbol{k}_1=\boldsymbol{k}_2+\boldsymbol{k}_3} \int T(\boldsymbol{k},\boldsymbol{k}_1;\boldsymbol{k}_2,\boldsymbol{k}_3) b^*(\boldsymbol{k}_1)b(\boldsymbol{k}_2)b(\boldsymbol{k}_3) \quad (10.1.2)$$
$$\times \delta(\omega+\omega_1-\omega_2-\omega_3)\, d\omega_1 d\omega_2 d\omega_3 + f(\boldsymbol{k},\omega)]\ .$$

Here $G_0(\boldsymbol{k},\omega) = [\omega - \omega(\boldsymbol{k}) + i\gamma(\boldsymbol{k})]^{-1}$ is the zeroth Green's function of (6.4.8), which describes the response of the field $b(\boldsymbol{k},\omega)$ to the external force $f(\boldsymbol{k},\omega)$ at $T(\boldsymbol{k},\boldsymbol{k}_1;\boldsymbol{k}_2,\boldsymbol{k}_3)=0$ and $hV(\boldsymbol{k})=0$. There is a Langevin random force $f(\boldsymbol{k},\omega)$, which simulates the interaction of a wave system with the thermal bass, in the right-hand part of this equation. To develop the diagrammatic technique one can obtain the formal solution of these equations in the form of a series in degrees of $f(\boldsymbol{k},\omega)$:

$$b(\boldsymbol{k},\omega) = b_1(\boldsymbol{k},\omega) + b_3(\boldsymbol{k},\omega) + b_5(\boldsymbol{k},\omega) + \ldots,$$
$$b_1(\boldsymbol{k},\omega) = G_0(\boldsymbol{k},\omega)f(\boldsymbol{k},\omega) + hV(\boldsymbol{k})G_0^*(-\boldsymbol{k},\omega_{\rm p}-\omega)f^*(-\boldsymbol{k},\omega_{\rm p}-\omega),$$

$$\begin{aligned}
b_3(\boldsymbol{k},\omega) = & G_0(\boldsymbol{k},\omega) \int T(\boldsymbol{k},\boldsymbol{k}_1;\boldsymbol{k}_2,\boldsymbol{k}_3)G^*(\boldsymbol{k}_1,\omega_1) \\
& \times G(\boldsymbol{k}_2,\omega_2)G(\boldsymbol{k}_3,\omega_3)f^*(\boldsymbol{k}_1,\omega_1)f(\boldsymbol{k}_2,\omega_2)f(\boldsymbol{k}_3,\omega_3) \\
& \times d\boldsymbol{k}_1 d\boldsymbol{k}_2 d\boldsymbol{k}_3 d\omega_1 d\omega_2 d\omega_3/(2\pi)^3 + \ldots \\
\Rightarrow & G_0[TG^*G^2 f^* f^2] + hVG_0^*[T^*G(G^*)^2 f(f^*)^2], \\
b_5(\boldsymbol{k},\omega) \Rightarrow & G_0\{TG^*Gf^*f[G_0(TG^*G^2 f^* f^2)]\} + \ldots
\end{aligned} \qquad (10.1.3)$$

Then one can build a series for $b(\boldsymbol{k},\omega)f^*(\boldsymbol{k}_1,\omega_1)$, $b(\boldsymbol{k},\omega)f(\boldsymbol{k}_1,\omega_1)$, $b(\boldsymbol{k},\omega)b^*(\boldsymbol{k}_1,\omega_1)$, and $b(\boldsymbol{k},\omega)b(\boldsymbol{k}_1,\omega_1)$ then can average over the Gaussian ensemble of the random force f. Using definitions (6.4.1) for the normal and anomalous Green's functions $G(\boldsymbol{k},\omega)$ and $L(\boldsymbol{k},\omega)$, definitions (6.4.4) for normal and anomalous double correlators $n(\boldsymbol{k},\omega)$ and $\sigma(\boldsymbol{k},\omega)$, one can derive a series for these functions. The next step in the derivation of the Wyld DT is the Dyson summation of a weakly linked (reducible) diagram which results in the following system of the Dyson equations for the Green's functions:

$$\begin{aligned}
G(\boldsymbol{k},\omega) = & G_0(\boldsymbol{k},\omega)[1 + \Sigma_{\rm c}(\boldsymbol{k},\omega)G(\boldsymbol{k},\omega) + \Pi_{\rm c}(\boldsymbol{k},\omega)L^*(\boldsymbol{k},\omega)], \\
L(\boldsymbol{k},\omega) = & G_0^*(-\boldsymbol{k},\omega_{\rm p}-\omega)[\Pi_{\rm c}^*(-\boldsymbol{k},\omega_{\rm p}-\omega)G(\boldsymbol{k},\omega) \\
& + \Sigma_{\rm c}^*(-\boldsymbol{k},\omega_{\rm p}-\omega)L^*(\boldsymbol{k},\omega)],
\end{aligned} \qquad (10.1.4)$$

and the Wyld equations for double correlators:

$$\begin{aligned}
n(\boldsymbol{k},\omega) = & [|G(\boldsymbol{k},\omega)|^2 + |L(\boldsymbol{k},\omega)|^2][\Sigma_{\rm d}(\boldsymbol{k},\omega) + f^2(\boldsymbol{k},\omega)] \\
& + G(\boldsymbol{k},\omega)\Pi_{\rm d}(\boldsymbol{k},\omega)L^*(-\boldsymbol{k},\omega_{\rm p}-\omega) \\
& + L(\boldsymbol{k},\omega)\Pi_{\rm d}^*(\boldsymbol{k},\omega)G^*(\boldsymbol{k},\omega), \\
\sigma(\boldsymbol{k},\omega) = & G(\boldsymbol{k},\omega)L(\boldsymbol{k},\omega)[\Sigma_{\rm d}(\boldsymbol{k},\omega) + f^2(\boldsymbol{k},\omega)] \\
& + L(-\boldsymbol{k},\omega_{\rm p}-\omega)G(-\boldsymbol{k},\omega_{\rm p}-\omega) \\
& \times [\Sigma_{\rm d}(-\boldsymbol{k},\omega_{\rm p}-\omega) + f^2(-\boldsymbol{k},\omega_{\rm p}-\omega)] \\
& + G(\boldsymbol{k},\omega)\Sigma_{\rm d}(\boldsymbol{k},\omega)G(-\boldsymbol{k},\omega_{\rm p}-\omega) + L^2(\boldsymbol{k},\omega)\Sigma_{\rm d}^*(\boldsymbol{k},\omega).
\end{aligned} \qquad (10.1.5)$$

Here $f^2(\boldsymbol{k},\omega)$ is the random force correlator (6.4.12); the following notations are introduced for mass operators (MO) of the compact diagram sums:

$\Sigma_{\rm c}(\boldsymbol{k},\omega)$: normal causal mass operator,
$\Pi_{\rm c}(\boldsymbol{k},\omega)$: anomalous causal mass operator,
$\Sigma_{\rm d}(\boldsymbol{k},\omega)$: normal distributive mass operator,
$\Pi_{\rm d}(\boldsymbol{k},\omega)$: anomalous distributive mass operator.

Zakharov and *L'vov* [10.7] described the derivation procedure of these equations in detail. To close (10.1.4, 5), the mass operators must be expressed in terms of the Green's functions and correlators. These expressions are partially summarized but, in fact, they are infinite series of the perturbation theory. However, it is sufficient to retain first-order diagrams for small amplitudes. Diagrams of second order in the interaction of waves are retained in the S, T^2-theory. In this approximation the simplest distributive mass operators are those which determine the distribution function of parametric waves in accordance with (10.1.5):

$$\Sigma_{\mathrm{d}}(\boldsymbol{k},\omega) = 2\int \big[|T(\boldsymbol{k},\boldsymbol{k}_1;\boldsymbol{k}_2,\boldsymbol{k}_3)|^2 n_1 n_2 n_3 + 2T(\boldsymbol{k},\boldsymbol{k}_1;\boldsymbol{k}_2,\boldsymbol{k}_3)$$
$$\times T^*(\boldsymbol{k},-\boldsymbol{k}_2;-\boldsymbol{k}_1,\boldsymbol{k}_3)\sigma_1^*\sigma_2 n_3\big]\delta(\boldsymbol{k}+\boldsymbol{k}_1-\boldsymbol{k}_2-\boldsymbol{k}_3)$$
$$\times \delta(\omega+\omega_1-\omega_2-\omega_3)d\boldsymbol{k}_1 d\boldsymbol{k}_2 d\boldsymbol{k}_3 d\omega_1 d\omega_2 d\omega_3/(2\pi)^3 \,,$$
$$\Pi_{\mathrm{d}}(\boldsymbol{k},\omega) = 2\int \big[T^2(\boldsymbol{k}_1,\boldsymbol{k}_2;\boldsymbol{k}_3\boldsymbol{k}_4)\sigma_1^*\sigma_2\sigma_3 + 2T(\boldsymbol{k}_1,\boldsymbol{k}_2;\boldsymbol{k}_3,\boldsymbol{k}_3)$$
$$\times T(-\boldsymbol{k},\boldsymbol{k}_2,\boldsymbol{k}_1,-\boldsymbol{k}_3)n_1 n_2 \sigma_3\big]\delta(\boldsymbol{k}+\boldsymbol{k}_1-\boldsymbol{k}_2-\boldsymbol{k}_3)$$
$$\times \delta(\omega+\omega_1-\omega_2-\omega_3)d\boldsymbol{k}_1 d\boldsymbol{k}_2 d\boldsymbol{k}_3 d\omega_1 d\omega_2 d\omega_3/(2\pi)^3 \,. \tag{10.1.6}$$

Here $n_j = n(\boldsymbol{k}_j,\omega_j), \sigma_j = \sigma(\boldsymbol{k}_j,\omega_j)$. The expressions for the causal mass operator, which defines the causal Green's function, are, in accordance with (10.1.4), somewhat more complicated:

$$\Sigma_{\mathrm{c}}(\boldsymbol{k},\omega) = 2\int T(\boldsymbol{k},\boldsymbol{k}')n(\boldsymbol{k}')d\boldsymbol{k}' d\omega'/2\pi$$
$$+ 2\int \big\{[|T(\boldsymbol{k},\boldsymbol{k}_1;\boldsymbol{k}_2,\boldsymbol{k}_3)|^2]G_1^* n_2 n_3$$
$$+ n_1(G_2 n_3 + n_2 G_3)\big] + 2T(\boldsymbol{k}_1,\boldsymbol{k}_2;\boldsymbol{k}_3,\boldsymbol{k}_4)T^*(\boldsymbol{k},\boldsymbol{k}_2;\boldsymbol{k}_1,\boldsymbol{k}_3) \tag{10.1.7a}$$
$$\times \big[\sigma_1^*\sigma_2 G_3 + (L_1^*\sigma_2 + \sigma_1^* L_{\bar 2})n_3\big]\big\}\delta(\boldsymbol{k}+\boldsymbol{k}_1-\boldsymbol{k}_2-\boldsymbol{k}_3)$$
$$\times \delta(\omega+\omega_1-\omega_2-\omega_3)d\boldsymbol{k}_1 d\boldsymbol{k}_2 d\boldsymbol{k}_3 d\omega_1 d\omega_2 d\omega_3/(2\pi)^3 \,,$$

$$\Pi_{\mathrm{c}}(\boldsymbol{k},\omega) = hV(\boldsymbol{k}) + \int S(\boldsymbol{k},\boldsymbol{k}')\sigma(\boldsymbol{k}')d\boldsymbol{k}' d\omega'/2\pi$$
$$+ 2\int \big\{T^2(\boldsymbol{k}_1,\boldsymbol{k}_2;\boldsymbol{k}_3,\boldsymbol{k}_4)\big\}\big[L_1^*\sigma_2\sigma_3 + \sigma_1^*(L_2\sigma_3+\sigma_2 L_3)\big]$$
$$+ 2T(\boldsymbol{k}_1,\boldsymbol{k}_2;\boldsymbol{k}_3,\boldsymbol{k}_4)T(\boldsymbol{k},-\boldsymbol{k}_2;\boldsymbol{k}_1,\boldsymbol{k}_3)\big[n_1 n_2 L_{\bar 3} \tag{10.1.7b}$$
$$+ (G_1^* n_2 + n_2 G_2)\sigma_3\big]\big\}\delta(\boldsymbol{k}+\boldsymbol{k}_1-\boldsymbol{k}_2-\boldsymbol{k}_3)$$
$$\times \delta(\omega+\omega_1-\omega_2-\omega_3)d\boldsymbol{k}_1 d\boldsymbol{k}_2 d\boldsymbol{k}_3 d\omega_1 d\omega_2 d\omega_3/(2\pi)^3 \,,$$

Here $\bar{j} \equiv -\boldsymbol{k}_j, \omega_{\mathrm{p}} - \omega_j$. If there were random inhomogeneities, i.e. defects, pores, deformation fields, in a medium, then the elastic scattering from them leads to an additive contribution to the Hamiltonian,

$$\mathcal{H}_{\mathrm{el}} = \int g(\boldsymbol{k},\boldsymbol{k}')\eta(\boldsymbol{k}-\boldsymbol{k}')b^*(\boldsymbol{k},t)b(\boldsymbol{k}',t)d\boldsymbol{k} d\boldsymbol{k}' \,, \tag{10.1.8}$$

where $g(\boldsymbol{k},\boldsymbol{k}')$ is the interacting amplitude which characterizes the scattering of waves from the inhomogeneities, and $\eta(\boldsymbol{k})$ is their amplitude. For point defects we have the following expression: $(2\pi)^3 \eta(\boldsymbol{k}) = v_0 \sum_{r_n} \exp(i\boldsymbol{k}\cdot\boldsymbol{r}_n)$, where $\boldsymbol{r}_n$ are the coordinates of defects, and v_0 is the volume of the elementary cell. The detailed analysis [10.6, 7] shows that in interesting cases it is enough, as a rule, to retain the lowest diagrams in $g(\boldsymbol{k},\boldsymbol{k}')$ which give an additive contribution into mass operators (c is the concentration of defects):

$$\Sigma_{c,el}(\boldsymbol{k},\omega) = c\int |g(\boldsymbol{k},\boldsymbol{k}')|^2 G(\boldsymbol{k}',\omega)d\boldsymbol{k}',$$
$$\Pi_{c,el}(\boldsymbol{k},\omega) = c\int [g(\boldsymbol{k},\boldsymbol{k}')]^2 L(\boldsymbol{k}',\omega)d\boldsymbol{k},$$
$$\Sigma_{d,el}(\boldsymbol{k},\omega) = c\int |g(\boldsymbol{k},\boldsymbol{k}')|^2 n(\boldsymbol{k}',\omega)d\boldsymbol{k}',$$
$$\Pi_{d,el}(\boldsymbol{k},\omega) = c\int [g(\boldsymbol{k},\boldsymbol{k}')]^2 \sigma(\boldsymbol{k}',\omega)d\boldsymbol{k}'.$$
(10.1.9)

10.2 Limit of the S-Theory

10.2.1 Form of the Green's Function

One can represent the solution of the Dyson equations (10.1.3,4) in the form, similar to (6.4.5)

$$G(\boldsymbol{k},\omega) = [\omega_p - \omega - \omega_{NL}(\boldsymbol{k}) - i\Gamma(\boldsymbol{k})]/\Delta(\boldsymbol{k},\omega),$$
$$L^*(\boldsymbol{k},\omega) = \Pi_c^*(-\boldsymbol{k},\omega_p-\omega)/\Delta(\boldsymbol{k},\omega),$$
(10.2.1a)

$$\Delta(\boldsymbol{k},\omega) = [\omega - \omega_{NL}(\boldsymbol{k}) + i\Gamma(\boldsymbol{k})][\omega_p - \omega - \omega_{NL}(\boldsymbol{k}) - i\Gamma(\boldsymbol{k})] - |\Pi_c(\boldsymbol{k},\omega)|^2,$$
$$\omega_{NL}(\boldsymbol{k}) = \omega(\boldsymbol{k}) + \mathrm{Re}\{\Sigma_c[\boldsymbol{k},\omega(\boldsymbol{k})]\},$$
$$\Gamma(\boldsymbol{k}) = \gamma(\boldsymbol{k}) - \mathrm{Im}\{\Sigma_c[\boldsymbol{k},\omega(\boldsymbol{k})]\}.$$
(10.2.1b)

It is obvious that $\mathrm{Im}G(\boldsymbol{k},\omega)$ and $L(\boldsymbol{k},\omega)$ in the middle of the frequency packet, i.e., when $\omega=\omega_p/2$ have the Lorentzian form with the width $\Delta k = \nu/v$ (v is the group velocity and $\nu^2 = \Gamma^2 - |\Pi|^2$). This width is essentially less than Γ/v because of the compensation of damping by the pump according to (10.10a). The packet width of the frequency $\omega_{NL}(\boldsymbol{k})$ in the center of the packet, i.e. on the resonance surface, is even smaller: $\Delta\omega \approx \nu^2/\Gamma$. The normal Green's function $G(\boldsymbol{k},\omega)$ looks like the free Green's function far from the resonance surface, but the function must have the renormalized frequency and damping:

$$G(\boldsymbol{k},\omega)) \to 1/[\omega - \omega_{NL}(\boldsymbol{k}) + i\Gamma(\boldsymbol{k})]$$
(10.2.2)

and the anomalous Green's function $L(\boldsymbol{k},\omega)$ is small: $|L(\boldsymbol{k},\omega)|^2 \ll |G(\boldsymbol{k},\omega)|$.

Concluding this section it is necessary to note that the Green's functions (10.2.1a) coincide with (6.4.15), if one puts

$$\omega_{\mathrm{NL}} = \omega(\boldsymbol{k}) + 2\int T(\boldsymbol{k},\boldsymbol{k}')n(\boldsymbol{k}')d\boldsymbol{k}', \quad \Gamma(\boldsymbol{k}) = \gamma(\boldsymbol{k}),$$
$$\Pi_{\mathrm{c}}(\boldsymbol{k},\omega) = P(\boldsymbol{k}) = hV(\boldsymbol{k}) + \int S(\boldsymbol{k},\boldsymbol{k}')\sigma(\boldsymbol{k}')d\boldsymbol{k}', \qquad (10.2.3)$$

i.e. neglects terms proportional to T^2 in (10.1.7). This is an approximation of the S-theory for the Green's functions!

10.2.2 Separation of the Waves into Parametric and Thermal

This problem has been discussed in Sect. 6.4.3 in the mean–field approximation of the basic S-theory. By analogy with (6.4.19) let us put $n_{\mathrm{p}}(\boldsymbol{k},\omega) = n(\boldsymbol{k},\omega) - n_{\mathrm{T}}(\boldsymbol{k},\omega)$, where

$$n_{\mathrm{T}}(\boldsymbol{k},\omega) = \frac{2\Gamma(\boldsymbol{k})n_0(\boldsymbol{k})}{[\omega_{\mathrm{NL}}(\boldsymbol{k}) - \omega]^2 + \Gamma^2(\boldsymbol{k})}. \qquad (10.2.4)$$

Here n_{p} and n_{T} are the distribution functions of parametric and thermal waves. It is necessary to note that functions $\Gamma(\boldsymbol{k})$ and $\omega_{\mathrm{NL}}(\boldsymbol{k})$ in (10.2.4) are not to be calculated in the thermodynamic equilibrium but using the real spectrum $n_{\mathrm{T}}(\boldsymbol{k},\omega)$, $n_{\mathrm{p}}(\boldsymbol{k},\omega)$ and $\sigma(\boldsymbol{k},\omega)$ which was calculated in the presence of pumping. In this definition n_{T} is everywhere a smooth function of $\boldsymbol{k}$ but it moves to the equilibrium spectrum asymptotically beyond the resonance surface for $[\omega_{\mathrm{NL}}(\boldsymbol{k}) - \omega_{\mathrm{p}}/2] \gg \Gamma$. As for the quantities

$$n_{\mathrm{p}}(\boldsymbol{k}) = \int n_{\mathrm{p}}(\boldsymbol{k},\omega)d\omega/2d\pi, \quad \sigma(\boldsymbol{k}) = \int \sigma(\boldsymbol{k},\omega)d\omega/2\pi, \qquad (10.2.5)$$

they rapidly decrease beyond the resonance surface. It may be shown that $n_{\mathrm{p}}(\boldsymbol{k})$ and $\sigma(\boldsymbol{k}) \propto 1/[\omega_{\mathrm{NL}}(\boldsymbol{k}) - \omega_{\mathrm{p}}/2]^2$. It must be noted that in the above arguments we have nowhere used the actual form of the interaction of waves. The definitions of parametric and thermal waves in (10.2.4) are, therefore, valid for every interaction, particularly in cases where three-wave processes and interactions with phonons, etc. are essential. Now let us use the method developed for studying the packet of parametric waves at relatively small amplitudes of pumping, where the scattering of the thermal waves by each other forms the main contribution to the mass operator. It is necessary to substitute $n(\boldsymbol{k},\omega) = n_{\mathrm{T}}(\boldsymbol{k},\omega) + n_{\mathrm{p}}(\boldsymbol{k},\omega), \sigma(\boldsymbol{k},\omega)$ into the expressions for Σ_{c}, Π_{c}, Σ_{d} and Π_{d} and to study the obtained expressions. The result obtained in the zero approximation in n_{p}, σ is known from (10.2.4a). Formula

$$\omega_{\mathrm{T}}(\boldsymbol{k}) = \omega(\boldsymbol{k}) + 2\int T(\boldsymbol{k},\boldsymbol{k}')n_{\mathrm{T}}(\boldsymbol{k}',\omega')d\boldsymbol{k}'d\omega'/2\pi$$

describes the frequency dependence of the waves on the medium temperature in the first order in $\mathcal{H}_{\text{int}}$. Let us assume that this dependence has already been included in the definition of $\omega(\boldsymbol{k})$, so that

$$\omega_{\text{NL}}(\boldsymbol{k}) = \omega(\boldsymbol{k}) + 2\int T(\boldsymbol{k},\boldsymbol{k}')n_{\text{p}}(\boldsymbol{k}')d\boldsymbol{k}' \ . \tag{10.2.6}$$

Expressions (10.1.6) for Σ_{d} and Π_{d} are quadratic in the amplitudes of the interaction Hamiltonian $T(\boldsymbol{k},\boldsymbol{k}_1;\boldsymbol{k}_2,\boldsymbol{k}_3)$. At the first stage of the investigation this allows one to calculate them in the zero approximation in the amplitude of the parametric turbulence. The applicability framework of this approximation and the effects appearing for large amplitudes will be considered below. In this approximation $\Pi_{\text{d}}=0$ and

$$\begin{aligned}\Sigma_{\text{d}} =& 2\int |T(\boldsymbol{k},\boldsymbol{k}_1;\boldsymbol{k}_2,\boldsymbol{k}_3)|^2 n_{\text{T}}(\boldsymbol{k}_1)n_{\text{T}}(\boldsymbol{k}_2)n_{\text{T}}(\boldsymbol{k}_3)\delta(\boldsymbol{k}+\boldsymbol{k}_1-\boldsymbol{k}_3-\boldsymbol{k}_4)\\ & \times \delta\bigl[\omega(\boldsymbol{k})+\omega(\boldsymbol{k}_1)-\omega(\boldsymbol{k}_2)-\omega(\boldsymbol{k}_3)\bigr] d\boldsymbol{k}_1 d\boldsymbol{k}_2 d\boldsymbol{k}_3 \ . \end{aligned} \tag{10.2.7}$$

This expression is the contribution to the correlator of random force $f^2(\boldsymbol{k})$, arising due to four-wave scattering. Let us assume that the contribution (10.2.7) has already been included in $f^2(\boldsymbol{k})$. Then in equations (10.1.5) it is necessary to put not only $\Pi_{\text{d}}=0$ but also $\Sigma_{\text{d}}=0$. Substituting the expressions (10.2.1a, 3) for the Green's functions into (10.1.5) for double correlators (at $\Pi_{\text{d}}=\Sigma_{\text{d}}=0$) it is easy to see that the resulting equations coincide with (6.4.16), which has been obtained in the temperature S-theory in Sect. 6.4. That means the approximation of the basic S-theory may be obtained from the S,T^2-theory if we neglect the influence of parametric waves in the expressions for mass operators of second order in the vertex $T(\boldsymbol{k},\boldsymbol{k}_1;\boldsymbol{k}_2,\boldsymbol{k}_3)$. The parameter of this approximation will be given later.

10.3 Nonlinear Theory for Parametric Excitation of Waves in Random Media

Except for some rare cases, real media possess various inhomogeneities which destroy their ideal translation symmetry. The nature of the inhomogeneities in ferromagnetic crystals and their influence on the ferromagnetic resonance and spin waves were studied in a large number of works: see, e.g., *Spark's* monograph [10.8] and *Schlöman's* paper [10.9]. A more detailed information can be found in *Gurevich* [10.10]. Many experimental works (see [10.10, 12 – 14]) are devoted to the question of the influence of inhomogeneities on parallel pumping of magnons. This section will discuss the theory of the phenomenon and make a comparison with experiment.

10.3.1 General Equations in the S, g^2-Approximation

Let us formulate the so-called S, g^2-*approximation* in the nonlinear theory for the parametric excitation of waves in random media [10.15]. We start with the diagrammatic equations (10.1.5–9) and (10.2.1) in which we will take into account:

(1) interaction of the parametric and thermal waves which leads to damping of parametric waves and to a dependence of the spectrum on the temperature,

(2) mean–field S-interaction between the parametric wave pairs, resulting in a renormalization of the pumping,

(3) elastic scattering of waves from inhomogeneities in Born's approximation proportional to $|g(\boldsymbol{k}, \boldsymbol{k}')|^2$.

In such a S, g^2-theory we obtain the following equations:

$$\begin{aligned} \Gamma(\boldsymbol{k}, \omega) &= \gamma(\boldsymbol{k}) - \mathrm{Im} \Sigma_{\mathrm{c,el}}(\boldsymbol{k}, \omega), & \Pi_{\mathrm{c}}(\boldsymbol{k}, \omega) &= P(\boldsymbol{k}) + \Pi_{\mathrm{c,el}}(\boldsymbol{k}, \omega), \\ \Sigma_{\mathrm{d}}(\boldsymbol{k}, \omega) &= \Sigma_{\mathrm{d,el}}(\boldsymbol{k}, \omega), & \Pi_{\mathrm{d}}(\boldsymbol{k}, \omega) &= \Pi_{\mathrm{d,el}}(\boldsymbol{k}, \omega), \end{aligned} \quad (10.3.1)$$

where $\gamma(\boldsymbol{k})$ and $P(\boldsymbol{k})$ are damping and pumping of parametric waves in the basic S-theory and where the MOs Σ, Π are given by (10.1.9). It is easy to see that if $n(\boldsymbol{k}, \omega)$ and $\sigma(\boldsymbol{k}, \omega) \propto \delta(\omega - \omega_{\mathrm{p}}/2)$, Σ_{d}, and Π_{d} are also $\propto \delta(\omega - \omega_{\mathrm{p}}/2)$. This means that elastic scattering of parametric waves on inhomogeneities does not disturb the uniform character of the parametric turbulence of waves inherent in the S-theory. With the one-frequency approximation the Wyld equations (10.1.5) may be integrated over the modulus k. As a result we have:

$$n(\Omega) = \frac{\pi k^2(\Omega) \Gamma(\Omega)}{v(\Omega)\nu^3(\Omega)} \left\{ \Gamma(\Omega) \Sigma_{\mathrm{d}}(\Omega) + \mathrm{Im}[\Pi_{\mathrm{c}}^*(\Omega) \Pi_{\mathrm{d}}(\Omega)] \right\}, \quad (10.3.2a)$$

$$\Gamma(\Omega)\sigma(\Omega) + i\Pi_{\mathrm{c}}(\Omega)n)\Omega) = 0, \quad \nu^2(\Omega) = \Gamma^2(\Omega) - |\Pi_{\mathrm{c}}(\Omega)|^2;$$

$$\begin{aligned} \Gamma(\Omega) &= \gamma(\Omega) + \pi c \int |g(\Omega, \Omega_1)|^2 \frac{\Gamma(\Omega_1) k^2(\Omega_1)}{\nu(\Omega_1) v(\Omega_1)} d\Omega_1, \\ \Pi_{\mathrm{c}}(\Omega) &= P(\Omega) + \pi c \int g(\Omega, \Omega_1) g(\bar{\Omega}, \bar{\Omega}_1) \frac{\Pi_{\mathrm{c}}(\Omega_1) k^2(\Omega_1)}{\nu(\Omega_1) v(\Omega_1)} d\Omega_1, \end{aligned} \quad (10.3.2b)$$

$$\begin{aligned} \Sigma_{\mathrm{d}}(\Omega) &= \Sigma_{\mathrm{d,el}}(\Omega) = c \int |g(\Omega, \Omega_1)|^2 n(\Omega_1) d\Omega_1, \\ \Pi_{\mathrm{d}}(\Omega) &= \Pi_{\mathrm{d,el}}(\Omega) = c \int g(\Omega, \Omega_1) g(\bar{\Omega}, \bar{\Omega}_1) \sigma(\Omega_1) d\Omega_1. \end{aligned} \quad (10.3.2c)$$

Here $n(\Omega)$ and $\sigma(\Omega)$ are the distribution functions $n_{\mathrm{p}}(\boldsymbol{k}, \omega)$ and $\sigma(\boldsymbol{k}, \omega)$ (6.4.21) integrated in ω and module k and depending only on the angular coordinates $\Omega = \Theta, \varphi$, on the resonant surface; $\bar{\Omega} = \pi - \Theta, \varphi + \pi$; $k(\Omega)$ and $v(\Omega)$ are the wave vector and group velocity of the waves at the point on the resonant surface with the angular coordinate Ω.

Equations (10.3.2) represent the S, g^2-theory, a closed system of integral equations which enables one to describe the system of interacting parametric waves in a medium with random inhomogeneities. They were obtained in [10.15] for the first time.

Elastic scattering of waves may be characterized by the decrement of damping. From the conventional perturbation theory it follows:

$$\gamma_{\text{el}}(\boldsymbol{k}) = \pi c \int \left|g(\boldsymbol{k}, \boldsymbol{k}_1)\right|^2 \delta\left[\omega(\boldsymbol{k}) - \omega(\boldsymbol{k}_1)\right] d\boldsymbol{k}_1 \simeq \pi c \Delta^2 k^2 g^2 / v, \qquad (10.3.3)$$

where $g \approx g(\boldsymbol{k}, \boldsymbol{k}_1)$, and $\Delta^2 \leq 4\pi$ is a characteristic scattering space angle (Δ is the scattering angle). Thus, the S, g^2-theory includes three dimensionless parameters: the degree of homogeneity of the medium $\gamma_{\text{el}}/\gamma$, the scattering angle Δ, and the supercriticality p.

10.3.2 Distribution Function of Parametric Waves

From (10.3.2) there follows the integral relation

$$\int \gamma(\Omega) n(\Omega) d(\Omega) + \text{Im}\left\{\int hV^*(\Omega)\sigma(\Omega)d\Omega\right\} = 0, \qquad (10.3.4)$$

which describes the energy balance in the system of parametric waves: the total dissipation of energy due to external relaxation mechanisms is equal to the total energy flow into all the pairs. Elastic scattering does not enter into this relation because it occurs with frequency conservation and, hence, does not expel energy out of a system of parametric waves. Such a scattering results in two effects: isotropization of the distribution functions and destruction of the phase correlations in the pairs, which leads to a decrease of the ratio $\sigma(\Omega)/n(\Omega)$.

Let us briefly discuss now the simplest and interesting limiting case where $\gamma_{\text{el}} \gg \gamma$. Then it follows from (10.3.2)

$$\sigma(\Omega) = -in(\Omega)\left[\Pi_c(\Omega)/\Gamma(\Omega)\right], \quad n(\Omega) = N/4\pi; \qquad (10.3.5)$$

$$\Gamma(\Omega) = 4\pi^2 c \left\langle \left|g(\Omega, \Omega_1)\right|^2 \frac{k^2(\Omega_1)}{v(\Omega_1)} \right\rangle_{\Omega_1}, \quad \langle f(\Omega)\rangle_\Omega \equiv \int f(\Omega) \frac{d\Omega}{4\pi}; (10.3.6)$$

$$\Pi_c(\Omega) = P(\Omega) + \langle K(\Omega, \Omega_1)\Pi_c(\Omega_1)\rangle_{\Omega_1},$$

$$K(\Omega, \Omega_1) \equiv \frac{g(\Omega, \Omega_1)g(\bar{\Omega}, \bar{\Omega}_1)k^2(\Omega_1)}{\langle|g(\Omega_1, \Omega_2)|^2 k^2(\Omega_2)/v(\Omega_2)\rangle_{\Omega_2} v(\Omega_1)}. \qquad (10.3.7)$$

With the exception of the so-called *degenerate cases*, in which the operator $K(\Omega, \Omega_1)$ has the eigenvalue 1 (for instance, at $K = 1$), it follows from (10.3.7) that $\Pi_c \approx \Pi \approx P$. Allowing for (10.2.5, 6) this yields

$$|P|^2 \simeq \gamma_{\text{el}}\gamma, \quad |\sigma(\Omega)| \simeq n(\Omega)\sqrt{\gamma/\gamma_{\text{el}}}, \qquad (10.3.8)$$

i.e. the destruction of phase correlations and, accordingly, the increase of the excitation threshold of waves. Instead of the estimate $\langle h_{th} V \rangle = \langle \gamma \rangle$, which would have held for $|\sigma|=n$, it follows from (10.3.4) that

$$h_{th}^2 \langle |V(\Omega)|^2 \rangle \simeq \langle \gamma(\Omega) \rangle \gamma_{el} \,. \tag{10.3.9}$$

This formula has been qualitatively confirmed in a direct experiment by *Smirnov* and *Petrov* [10.4] who independently measured the values h_{th}, γ_{el} and $\langle \gamma(\Omega) \rangle$ in the antiferromagnet $CsMnF_3$. A quantitative analysis of the behavior of parametric waves requires knowledge of the function $g(\boldsymbol{k}, \boldsymbol{k}_1)$ which is determined by the destructive character of the inhomogeneity of the medium. After that there are no fundamental difficulties to carry out this analysis.

10.3.3 Behavior of Parametrically Excited Waves Beyond the Threshold

In the study of the behavior of parametrically excited waves the most interesting case is that of large scattering intensities, where a stronger influence of two-magnon scattering may be expected. In the simple model, with $S(\Theta, \Theta') \simeq V(\Theta)V(\Theta')$, $g(\boldsymbol{k})=g$, and $\gamma(\boldsymbol{k})=\gamma$, it follows from (10.3.2)

$$SN = (15 \gamma_{el}/8) \sqrt{p-1}\,, \qquad h_{th} V = \sqrt{15 \gamma \gamma_{el}/8}\,. \tag{10.3.10}$$

In the case of scattering from point defects, when $g(\boldsymbol{k}, \boldsymbol{k}_1)$ is proportional to $(\boldsymbol{k}\boldsymbol{k}_1)$, one can obtain for the fluctuation of the exchange constant an expression close to (10.3.10)

$$SN = (9\gamma_{el}/8)\sqrt{p-1}\,. \tag{10.3.11}$$

For intense small-angle scattering, when $\gamma \Delta^2 > \gamma$, one can obtain [10.3]:

$$SN = \frac{45}{16} \gamma_{el} \Delta^2 \sqrt{p-1}\,. \tag{10.3.12}$$

In all the described cases the excitation level of parametric waves for strong scattering from inhomogeneities proves to be a factor of $\gamma_{el} \Delta^2/\gamma$ greater (at the same supercriticality) than in uniform medium. This is caused by partial destruction of the phase correlations, which leads to a weakening of the phase mechanism of the amplitude limiting. The specific form of the dependence N on supercriticality in (10.3.10–13) is not universal, but originates from the specific form of the function $S(\Theta, \Theta_1)$. In other cases (see, e.g., [10.16]) the dependence of N on h is more complicated and it reproduces these dependences only qualitatively.

In all the described cases one can obtain from (10.3.2) for the nonlinear susceptibility the following expression

$$\chi = \left[2V^2(h^2 - h_{\text{th}}^2) + ih_{\text{th}}\sqrt{h^2 - h_{\text{th}}^2}\right]/h^2 \,. \tag{10.3.13}$$

This result coincides with the known expressions (5.5.35) for χ, which were obtained within the basic S-theory for an uniform medium. It means that the dependences of the nonlinear susceptibilities χ' and χ'' on the value (h/h_{th}) are not significantly changed by the scattering of waves from inhomogeneities. Of course, the value of the threshold field in a nonuniform medium itself is greater than that in an uniform medium. The literal coincidence of the formulae for χ should not be overestimated. In more complicated situations (see, e.g., [10.16]) the dependence $\chi(h/h_{\text{th}})$ resembles (10.3.13) only qualitatively.

10.4 Consistent Nonlinear Theory for Parametric Excitation of Waves

We know that the S-theory takes into account only the mean–field interaction of $\pm \boldsymbol{k}$–pairs of parametrically excited waves. In order to describe phenomena beyond the framework of such approximation it is necessary to include into the formalism of the theory the scattering of individual parametric waves and their interaction with the thermal bath. This was possible with the help of the Wyld diagrammatic technique [10.2]. An account of such a consistent theory (S, T^2–theory) based on the Wyld DT is given in my book [10.3] in Russian. Here I represent only a short review of its results.

10.4.1 Spectral Density of Parametrically Excited Waves

As has already been pointed out, the elastic scattering does not change the number and frequency of parametric waves but only destroys their coupling to the pump, the process leading to the dephasing of wave pairs, and to the isotropization of the parametric wave distribution. If the frequency of the elastic scattering distributed is greater than all other relaxation frequencies, the distribution of the parametric waves is isotropic, and the influence of the parametric pumping and of the parametric wave scattering from each other can be taken into account as small perturbations. A simple equation appears in this approximation for the distribution function of parametric waves in frequencies. One can solve it analytically and isolate the only stable solution from the stationary solutions. We discuss here the results of the S, T^2-theory for this case, which looks at first sight complicated. In the mass operator we will take into consideration the contributions of elastic scattering Σ_{el}, Π_{el}, the contribution of the interaction of parametric waves with thermal waves $\gamma(\boldsymbol{k})$ and $f^2(\boldsymbol{k},\omega)$ and the contribution of the interaction of parametric waves among themselves, (10.1.6). In the limit $\gamma_{\text{el}} \gg \gamma$ the equations (10.1.4, 5) have the isotropic solution $n(\boldsymbol{k},\omega) - n(k,\omega)$. It allows one to integrate them in the general form not only in k but also in the solid angle Ω. Ultimately, after some modifications, we obtain the equation for the spectral density of parametric waves $n(\omega) = \int n(\boldsymbol{k},\omega)d\boldsymbol{k}$,

$$n(\epsilon) \equiv n(\omega - \omega_p/2) = \frac{\Gamma_1}{\epsilon^2 + \eta^2} \left\{ \frac{4\pi^2 k^2 n_0}{v} \right.$$
$$\left. + \frac{T^2}{kv} \int n(\epsilon_1) n(\epsilon_2) n(\epsilon_3) \delta(\epsilon + \epsilon_1 - \epsilon_2 - \epsilon_3) d\epsilon_1 d\epsilon_2 d\epsilon_3 \right\}. \quad (10.4.1)$$

Here T^2 is the mean value of the square $T(\boldsymbol{k},\boldsymbol{k}_1;\boldsymbol{k}_2,\boldsymbol{k}_3)$. Γ_1 has the order of magnitude $(\gamma_{el})^2/\gamma$. The second term in (10.4.1) may be neglected at small supercriticality. In this case the distribution $n(\epsilon)$ is a Lorentzian with width $\eta = \eta_T$, which may be determined by integrating (10.4.1) in ϵ. Taking (10.3.12) into account we have

$$\eta_T = \frac{\Gamma_1 \gamma}{kv} \frac{4\pi^2 k^3 n_0}{N} \simeq \frac{r}{\sqrt{p-1}}. \quad (10.4.2)$$

Here r is the small parameter (6.4.28c), characterizing the influence of the thermal bass: $r = SN_T/kv$, $N_T = 4\pi^2 n_0 k^3$. In the opposite case of large supercriticality, the thermal term in (10.4.1) can be neglected. *Cherepanov* and *L'vov* [10.17] showed that this equation has a one-parametric set of solutions. However, only one of them is stable. It is a *spectral soliton*:

$$Tn(\epsilon) = \frac{kv}{2\Gamma_1 \cosh(\pi\epsilon/2\eta)}, \quad \eta \simeq \frac{(\gamma_{el})^2}{\gamma} \sqrt{\frac{\gamma}{kv}(p-1)}. \quad (10.4.3)$$

In almost uniform crystals, when $\gamma_{el} \ll \gamma$, the study of spectral solitons becomes very complicated because of the anisotropy of $n(\boldsymbol{k},\omega)$. Therefore *Krutensko* et al. [10.18] limited themselves to an axially symmetric situation which is realized in isotropic and cubic ferromagnets. In the region of supercriticalities $p_1 < p < p_2$ (here $p_1 - 1 \simeq 2p_s(\gamma_{el}/\gamma)^{3/4}$, $p_2 - 1 \simeq p_s(\gamma_{el}/\gamma)^{3/4}$, $p_s \simeq kv/\gamma$) the broadening in ω is determined by the T^2-scattering of parametric waves and the line shape of $n(\omega)$ is close to (10.4.3) with the effective width, η_{int}:

$$\eta_{int} = \gamma \left(\frac{\gamma_{el}}{\gamma}\right)^{1/4} \sqrt{\frac{p-1}{p_s}}. \quad (10.4.4)$$

Broadening in angles is determined by elastic two-wave scattering. Both thermal fluctuations and two-wave scattering may be neglected at greater supercriticalities $p > p_2$. Then

$$\Delta\Theta(\boldsymbol{k}) \approx \sqrt{\frac{\eta_{int}}{\gamma}}, \quad \Delta\omega \approx \eta_{int} \approx \gamma \left[\frac{p-1}{p_s}\right]^{2/3}. \quad (10.4.5)$$

The line shape of $n(\omega)$ remains similar to that in (10.4.3). In the applicability framework of the theory $(p \simeq p_s)$ $\Delta\Theta(\boldsymbol{k}) \simeq \pi$. However, such supercriticalities are of interest only from an academic point of view since auto-oscillations arise first, leading to sharp broadening of the spectrum in $\Delta\Theta(\boldsymbol{k})$ and $\Delta\omega(\boldsymbol{k})$.

In conclusion we point out that the spectral solitons (10.4.3) at the parametric excitation of the magnons in YIG have been experimentally investigated in detail by *Krutensko* et al. [10.18]. Good qualitative and quantitative agreement with the above described theory has been observed (see Sect. 9.4.2 and Figs. 9.12, 13). One can see that the data concerning $n(\omega)$, which are given in Fig. 9.13 in "straightening" coordinates (chosen in such a way that the Gaussian in coordinates 1, the Lorentzian in coordinates 2, and function (10.4.3) in coordinates 3 will be straight lines) lie on a straight line only in coordinates 3. This and other facts give reason to believe that the theory elaborated on describes the reality well.

10.4.2 Structure of the Distribution Function in $\boldsymbol{k}$-Space

It has already been pointed out that the T^2-scattering of parametric waves leads to the finite width of the distribution function $n(\boldsymbol{k},\omega)$ of parametric waves, not only in ω but also in k. There $\Delta\omega \simeq \nu^2/2\gamma$, which is much less than $\Delta\omega(\boldsymbol{k}) \simeq \nu$, when the supercriticality is not large. It gives reason to think that the study of the distribution function structure $n(\boldsymbol{k}) = \int n(\boldsymbol{k},\omega)d\boldsymbol{k}$ may be restricted to the so called *one-frequency turbulence* approximation. This is the assumption that

$$n(\boldsymbol{k},\omega) = n(\boldsymbol{k})\delta(\omega - \omega_{\rm p}/2), \quad \sigma(\boldsymbol{k},\omega) = \sigma(\boldsymbol{k})\delta(\omega - \omega_{\rm p}/2). \tag{10.4.6}$$

Such an approximation allows one to analyze equations of the S, T^2-theory effectively in practically all interesting cases, as will be shown in this section. In the one-frequency approximation the Wyld equations (10.1.5) have the simple form:

$$\begin{aligned}
n_{\rm p}(\boldsymbol{k}) =& \frac{2}{\Delta^2(\boldsymbol{k})} \Big\{ \varGamma^2(\boldsymbol{k})\varSigma_{\rm d}(\boldsymbol{k}) \\
&+ \Big[\omega_{\rm NL}(\boldsymbol{k}) - \frac{\omega_{\rm p}}{2}\Big] {\rm Re}\Big\{ \varPi_{\rm c}^*(\boldsymbol{k})\varPi_{\rm d}(\boldsymbol{k}) \Big\} + \varGamma(\boldsymbol{k}) {\rm Im}\Big\{ \varPi_{\rm c}^*(\boldsymbol{k})\varPi_{\rm d}(\boldsymbol{k}) \Big\} \Big\}, \\
\sigma(\boldsymbol{k}) =& \frac{1}{\Delta^2(\boldsymbol{k})} \Big\{ 2\varPi_{\rm c}(\boldsymbol{k})\varSigma_{\rm d}(\boldsymbol{k})\Big[\omega_{\rm NL}(\boldsymbol{k}) - \frac{\omega_{\rm p}}{2} - i\varGamma(\boldsymbol{k})\Big] \\
&+ \varPi_{\rm c}^2(\boldsymbol{k})\varPi_{\rm d}^*(\boldsymbol{k}) + \Big[\omega_{\rm NL}(\boldsymbol{k}) - \frac{\omega_{\rm p}}{2} - i\varGamma(\boldsymbol{k})\Big]^2 \varPi_{\rm d}(\boldsymbol{k}) \Big\}, \quad (10.4.7) \\
\Delta(\boldsymbol{k}) =& \Big[\omega_{\rm NL}(\boldsymbol{k}) - \frac{\omega_{\rm p}}{2}\Big]^2 + \nu^2(\boldsymbol{k}), \quad \nu^2(\boldsymbol{k}) = \varGamma^2(\boldsymbol{k}) - |\varPi_{\rm c}(\boldsymbol{k})|^2.
\end{aligned}$$

Here the mass operators $\varSigma_{\rm d}(\boldsymbol{k})$ and $\varPi_{\rm d}(\boldsymbol{k})$ are determined by the equations:

$$\Sigma_{\mathrm{d}}(\boldsymbol{k},\omega) = \Sigma_{\mathrm{d}}(\boldsymbol{k})\delta(\omega - \omega_{\mathrm{p}}/2), \quad \Pi_{\mathrm{d}}(\boldsymbol{k},\omega) = \Pi_{\mathrm{d}}(\boldsymbol{k})\delta(\omega - \omega_{\mathrm{p}}/2). \quad (10.4.8)$$

The values $\Sigma_{\mathrm{d}}(\boldsymbol{k})$ and $\Pi_{\mathrm{d}}(\boldsymbol{k})$ can be taken on the resonant surface: $\Sigma_{\mathrm{d}}(\boldsymbol{k}) \Rightarrow \Sigma_{\mathrm{d}}(\Omega)$, $\Pi_{\mathrm{d}}(\boldsymbol{k}) \Rightarrow \Pi_{\mathrm{d}}(\Omega)$. Contributions to the mass operators Σ_{d} and Π_{d} arise due to elastic two-wave scattering (see expressions (10.3.2) for $\Sigma_{\mathrm{d,el}}$ and $\Pi_{\mathrm{d,el}}$) and due to four-wave scattering of parametric waves:

$$\Sigma_{\mathrm{d}}(\Omega) = \Sigma_{\mathrm{d,el}}(\Omega) + \Sigma_{\mathrm{d,int}}(\Omega), \Pi_{\mathrm{d}}(\Omega) = \Pi_{\mathrm{d,el}}(\Omega) + \Sigma_{\mathrm{d,el}}(\Omega). \quad (10.4.9)$$

Expressions for $\Sigma_{\mathrm{d,int}}(\Omega)$ and $\Pi_{\mathrm{d,int}}(\Omega)$ follow from (10.1.6):

$$\begin{aligned}\Sigma_{\mathrm{d,int}}(\Omega) =& 2\int \Big[|T(\Omega, \Omega_1; \Omega_2, \Omega_3)|^2 n(\Omega_1) n(\Omega_2) n(\Omega_3) \\ &+ 2\Gamma(\Omega, \Omega_1; \Omega_2, \Omega_3) \Gamma^*(\Omega, \bar{\Omega}_2; \bar{\Omega}_1, \Omega_3) \sigma^*(\Omega_1) \\ &\times \sigma(\Omega_2)\sigma(\Omega_3) \delta(\boldsymbol{n} + \boldsymbol{n}_1 - \boldsymbol{n}_2 - \boldsymbol{n}_3) d\Omega_1 d\Omega_2 d\Omega_3, \\ \Pi_{\mathrm{d,int}}(\Omega) =& 2\int T(\Omega, \Omega_1; \Omega_2, \Omega_3) \big[T(\Omega, \Omega_1; \Omega_2, \Omega_3) \sigma^*(\Omega_1) \\ &\times \sigma(\Omega_2)\sigma(\Omega_3) + 2T(\Omega, \bar{\Omega}_2; \bar{\Omega}_1, \Omega_3) n(\Omega_1) n(\Omega_2) \\ &\times \sigma(\Omega_3) \big] \delta(\boldsymbol{n} + \boldsymbol{n}_1 - \boldsymbol{n}_2 - \boldsymbol{n}_3) \, d\Omega_1 d\Omega_2 d\Omega_3 \, . \end{aligned} \quad (10.4.10)$$

Here $\boldsymbol{n} = \boldsymbol{k}/k$, $\boldsymbol{n}_j = \boldsymbol{k}_j/k$. The values $\Sigma_{\mathrm{c,int}}$ and $\Pi_{\mathrm{c,int}}$, which are the renormalization of the pumping and damping (on the parametric wave scattering) are small (in comparison with the $\gamma(\boldsymbol{k})$ and $P(\boldsymbol{k})$) and will not be taken into consideration. It is very simple to analyze solutions of one-frequency equations of the S, T^2-theory in a rough form. First of all, assuming that $\Sigma_{\mathrm{d,int}}=0$, $\Pi_{\mathrm{d,int}}=0$, we can verify that $\nu(\Omega)=0$ for those directions where $n(\Omega) \neq 0$. This means that the distribution of parametric waves in the $\boldsymbol{k}$-space is singular: $n(\boldsymbol{k}) \neq 0$ only on the resonance surface which satisfies the condition of external stability of the basic S-theory. The distribution $n(\Omega)$ on this surface and the integral quantity N are defined by (10.4.7 – 10), which reduces to equations of the basic S-theory in the considered approximation. Next, integrating the first of Eqs. (10.4.7), one gets an estimate for the quantity ν/γ which characterizes the relative part of damping not compensated by the pumping:

$$(\nu/\gamma)^3 \simeq (TN)^2/(\gamma k v) \simeq (\gamma/k v)(p - 1). \quad (10.4.11a)$$

It is necessary to remember that the spectral width in ω–space $\Delta\omega \simeq \nu^2/2\gamma$; so

$$\Delta\omega \simeq \gamma[(\gamma/kv)(p - 1)]^{2/3} \, . \quad (10.4.11b)$$

This result is in a good agreement with the experimental data for parametric magnons in YIG [10.18] shown in Fig. 9.13.

From the one-frequency equations (10.4.7 – 10) it follows that the distribution $n(\boldsymbol{k})$ in the modulus k close to the resonant surface is the squared

10.4 Consistent Nonlinear Theory for Parametric Excitation of Waves

Lorentzian with width $\Delta\omega(\boldsymbol{k}) \approx \nu$ in $\omega(\boldsymbol{k})$ and a width of the order of ν/γ in $\Theta(\boldsymbol{k})$ (under conditions of axial symmetry). It may be seen from (10.4.7–10) that the relative difference of their coefficients from those of the basic S-theory for the parameter $(\nu/\gamma)^2$ is small and, hence, for total values (like the total number of parametric waves N, etc.) the difference in the approximate results of the basic S-theory and the accurate results of the S, T^2-theory is also small for the same parameter. In particular, from (10.4.7) follows that $1 - |\sigma|/n \approx \nu^2/2\gamma$, i.e. at $\nu \ll \gamma$ the phase correlations in pairs are retained almost completely. An analysis of the diagrams which were neglected when solving (10.4.6) and which are proportional to T^3, T^4, etc. shows [10.1] that they are arranged in a series with the parameter $\lambda = (\Gamma/\nu)(TN/kv)$ and, consequently, for $\nu \leq \gamma \cdot \lambda \leq \sqrt{\gamma/kv} \ll 1$. This means that the equations of the S, T^2-theory are correct and that the integral quantities N, χ' and χ'' are well described by the corresponding formulae of the S-theory right up to the amplitudes $h \approx h_s$ which is determined by the condition

$$h_s = h_{th}\sqrt{kv/\gamma}. \tag{10.4.12}$$

As a specific example of a solution of the one-frequency equations of the S, T^2-theory, parallel pumping of the magnons in a cubic ferromagnet for $\boldsymbol{M}\|[100]$ and $[111]$ was considered by *Cherepanov* and *L'vov* [10.19]. Since the cubic anisotropy is very small, the magnon distribution on the resonant surface comprises a set of long stripes with $\Delta\varphi \gg \Delta\Theta$ which are stretched along the equator. Analyzing the distribution in Θ one can therefore consider the distribution in φ to be isotropic. From this assumption it follows that

$$\Delta\Theta \approx \frac{\nu_0}{\gamma} \approx \left[\frac{\gamma}{kv}(p-1)\right]^{1/3} \tag{10.4.13}$$

and the distribution in φ for $\boldsymbol{M}\|[100]$ has the form of a smeared cross with

$$\Delta\varphi = \sqrt{2}\nu_0/\nu_1, \quad \nu^2(\varphi) = \nu_0^2 + \nu_1^2\left[\sum_j \sin^2(\varphi - \varphi_j) - 2\right]. \tag{10.4.14}$$

However, for $\boldsymbol{M}\|[111]$, when the distribution $n(\varphi)$ has a shape of a smeared star of six vertices, the estimate for ν_0 and consequently for all the quantities connected with it, is quite different [10.19]:

$$\nu_0 \approx \gamma[\mu^2(p-1)/\gamma kv]^{1/5}, \quad \mu = \partial^2\gamma(\varphi)/\partial\varphi^2. \tag{10.4.15}$$

In some cases (e.g. for real $T(\boldsymbol{k},\boldsymbol{k}_1;\boldsymbol{k}_2,\boldsymbol{k}_3)$ and for spherical symmetry of the problem) the contribution to the parametric magnon scattering proportional to T^2 is almost completely canceled. Then the T^3 scattering should be taken into account and it is possible to prove that $\nu \approx \gamma\sqrt{\gamma/kv}(p-1)^{3/8}$. The latter two examples show that the general estimates (10.4.11) for $\Delta\omega(\boldsymbol{k}) = \nu$ and for $\Delta\omega$ (10.4.13) for $\Delta\Theta$ may prove to be incorrect in some specific situations because of unexpected cancellations. Therefore, in spite of the basic understanding of the main statements of the S, T^2-theory, the investigation of the parametric excitation of waves in other media may still lead to the discovery of new effects.

References

Chapter 1

1.1 V.S. L'vov: *Nonlinear Spin Waves* (Nauka, Moscow 1987) [in Russian]
1.2 L.D. Landau and E.M. Lifshitz: *Course of Theoretical Physics. Vol. 1: Mechanics* (Pergamon, Oxford 1966)
1.3 V.E. Zakharov: Izv. Vuzov, Radiofizika **17**, No 4, 431–453 (1974)
1.4 V.E. Zakharov and E.A.Kuznetsov: Hamiltonian Formalism for System of Hydrodynamic Type, in *Soviet Scientific Reviews.– Section C.– Mathematical Physics Reviews*, ed. by S.P. Novikov (OPA, Amsterdam 1984)
1.5 V.S. L'vov: *Lectures on Nonlinear Physical Phenomena* (University of Novosibirsk Press, Novosibirsk 1977)
1.6 V.P. Krasitsky: Zh. Eksp. Teor. Fiz. **97** in press (1991)
1.7 G.Lamb: *Hydrodynamics* (Dover, N.-Y 1930)
1.8 V.E. Zakharov and N.N. Filonenko: Dokl. Akad. Nauk SSSR **170**, 1292 (1966)
1.9 H.W. Wyld: Ann of Phis. **14**, 143, (1961)
1.10 N. Bloembergen: *Nonlinear Optics*, (W.A. Benjamin, Inc., New York, 1965)
1.11 V.E. Zakharov: "The Inverse Scattering Method" in *Solitons*, ed. by R.K. Bullough and Caudrey, *Topics in Current Physics Vol. 17* (Springer, Berlin, Heidelberg 1980) pp.243–285
1.12 S.P. Novikov, S.V. Manakov, L.P. Pitaevskii and V.E. Zakharov: *Theory of Solitons*, (Plenum, New York 1984)
1.13 V.E. Zakharov and S.V. Manakov: Zh. Eksp. Teor. Fiz. **69**, 1654 (1975)
1.14 V.E. Zakharov: Prikl. Meck.Techn. Fiz. **2**, 86 (1968)
1.15 L.D. Landau and E.M. Lifshitz: *Course of Theoretical Physics, Vol.5, Fluid Mechanics* (Pergamon, Oxford 1986)
1.16 V.E. Zakharov and A.B. Shabat: Zh. Eksp. Teor. Fiz. **61**, 118 (1971)
1.17 V.I. Vlasov, V.I. Talanov, V.A. Petrishev: Izv. Vuzov, Radiofizika **14**, 1353 (1971)
1.18 S.A. Ahmanov, A.P. Suhorukov and F.V. Hohlov: Usp. Fiz. Nauk **93**, 19–70 (1967)
1.19 V.E. Zakharov and V.S. Synakh: Zh. Eksp. Teor. Fiz. **68**, 940–948 (1975)

Chapter 2

2.1 J.H. Van Vleek: *The Theory of Electromagnetic Susceptibilities*, (Oxford University Press, Oxford 1932)
2.2 A.A. Abragam: *The Principles of Nuclear Magnetism*, (Oxford University Press, Oxford 1961)
2.3 M. Sparks: *Ferromagnetic Relaxation Theory*, (Mc-Graw-Hill, New York 1964)
2.4 C.D. Mattis: *The Theory of Magnetism*, (Mc-Graw-Hill, New York 1965)
2.5 J.S. Smart:*Effective Field Theories of Magnetism* (Saunders, Philadelphia 1966)
2.6 B. Lax and K.J. Button: *Microwave Ferrites and Ferrimagnetics*, (Mc-Graw-Hill, New York, London 1962)

2.7 S.V. Tyablicov: *Methods for Quantum Theory of Magnetism*, (Nauka, Moscow 1965), [in Russian]
2.8 A.I. Akhiezer, V.G. Bar'yakhtar and S. V.Peletmisky: *Spin Waves*, (Nauka, Moscow 1967), [in Russian], [English transl.: North–Holland, Amsterdam; Wiley, New York, 1968]
2.9 S. V.Vonsorsky: *Magnetism*. (Nauka, Moscow 1971), [in Russian]
2.10 *Handbuch der Physik*. Bd XVIII/I:*Magnetism* (Springer Verlag, Berlin, Heidelberg 1968)
2.11 *Magnetism. A Treatise on Modern Theory and Materials*, ed. by G.T. Rado and H. Suhl (Acad. Press, New York 1963–1966), Vols. I-IV
2.12 *Physics of Magnetic Dielectrics*, ed. by G.A. Smolensky (Nauka, Moscow 1974), [in Russian]
2.13 V.G. Bar'yakhtar, K.N. Krivorychko and D.A. Yablonsky:*Green Functions in the Theory of Magnetism*, (Naukova Dumka, Kiev 1984), [in Russian]
2.14 K. Binger and A.P. Young: *Spin Glasses: Experimental Facts , Theoretical Concepts and Open Questions*, Rev. Mod. Phys. **58**,No 4, 801–976 (1986)
2.15 L.D. Landau and E.M. Lifshitz: Phys. Z6. Sowjet **8**, 153 (1935), see also: L.D. Landau: *Collected Works*, Vol. 1,(Nauka, Moscow 1969) pp.127–143
2.16 I.E. Dzyaloshinsky: Zh. Eksp. Teor. Fiz. **32**, No 6, 1547–1563(1957)
2.17 P.W. Anderson: *Solid State Physics* , vol. 14 (Acad. Press, New York 1963)
2.18 G.M. Nedlin: Fiz. Tverd. Tela **66**, 1822 (1974).
2.19 J.C. Slater: *Quantum Theory of Molecules and Solids*, (McGraw-Hill, N.Y. 1963)
2.20 E.A. Turov: *Physical Properties of Magnets*, (Nauka, Moscow 1963)
2.21 I.V. Kolokolov, V.S. L'vov, V.B. Cherepanov: Sov. Phys.– JETP **57**, 605 (1983)
2.22 I.V. Kolokolov, V.S. L'vov, V.B. Cherepanov: Sov. Phys.– JETP **59**, 1131 (1984)

Chapter 3

3.1 A.I. Akhiezer, V.G. Bar'yakhtar and S.V. Peletmisky: *Spin Waves* (North–Holland, Amsterdam, 1968)
3.2 L.D. Landau and E.M. Lifshitz: *Course of Theoretical Physics, Vol.V, Part 1 Statistical Physics, 3rd edition* (Pergamon, Oxford 1980)
3.3 P.G. De Gennes, P.A. Pincus, F. Hartmann- Bountron and J.M. Winter: Phys. Rev. **129** , 1105–1115 (1963)
3.4 V.A. Tulin: *Nuclear Spin Waves in Magnets*, Fis. Niskikh Temp. **5**, 965–993 (1979)
3.5 E. Schlömann, J.H. Saunder, M.H. Sirvetz: Trans. IRE, MTT–8 No 1, 96 (1960)
3.6 V.G. Bar'yakhtar and D.A. Yablonsky: Teor. Mat Fiz. **25**, 250 (1975)
3.7 I.V. Kolokolov, V.S. L'vov, V.B. Cherepanov: Sov. Phys.–JETP **57**, 605 (1983)
3.8 V.S. Lutovinov and V.L. Safonov: Fiz. Tverd. Tela **22**. 2640–2650 (1980)

Chapter 4

4.1 V.E. Zakharov, V.S. L'vov and S.S. Starobinets: *Instability of the Monochromatic Spin Waves*, Fiz. Tverd. Tela 11, No 10, 2924-2930 (1969)
4.2 A.G. Gurevitch: Fiz. Tverd. Tela **6**, No 8, 2388 (1964)
4.3 E. Schlömann, R.J. Joseph: J. Appl. Phys. **30** Suppl. 177–178 (1959)
4.4 S.K. Turitsin and G.E. Falkovitch: Zh. Eksp. Teor. Fiz. **87**, No 7, 1061–1065 (1985)
4.5 V.I. Ozhogin and A.Yu. Yakubovsky: Zh. Eksp. Teor. Fiz. **67**, No 1 (7), 287–308(1974)
4.6 C.S. Gardner, I.M. Green, M.D. Kruskal and R.M. Miura: Phys. Rev. Lett. **19**, 1095 (1967)

4.7 S.P. Novikov, S.V. Manakov, L.P. Pitaevskii and V.E. Zakharov: *Theory of Solitons*, (Plenum Publ., New York 1984)
4.8 V.E. Zakharov: "The Inverse Scattering Method" in *Solitons*, ed. by R.K. Bullough and Caudrey, Topics in Current Physics, Vol. 17 (Springer,Berlin, Heidelberg 1980)
4.9 B.B. Kadomsev and V.I. Petviashvili: Dokl. Akad. Nauk SSSR **192**, 753 (1970)
4.10 V.E. Zakharov: Sov.Phys.– JETP **35**, 908 (1972)
4.11 H. Suhl: Phys. Chem. Sol **1**, 209–227 (1957)
4.12 F.R. Morgenthaler: J. Appl. Phys. **31S**, 95S–97S (1960)
4.13 E. Schlömann, J.H. Saunder, M.H. Sirvetz: Trans. IRE, MTT–8 No 1, 96 (1960)
4.14 V.I. Ozhogin: Zh. Eksp. Teor. Fiz. **58**, No 6, 2079–2089 (1970)

Chapter 5

5.1 H. Suhl: *Note on the Ferromagnets*, J. Appl. Phys. **30**, No 12, 1961–1964 (1959)
5.2 E. Schlömann, J.H. Saunder, and M.H. Sirvetz: *L–Band Ferromagnetic Resonance at High Peak Power Level*, Trans. IRE,MTT **8**, No 8, 96–100 (1960)
5.3 V.E. Zakharov, V.S. L'vov and S.S. Starobinets: Preprint No 227 (Institute for Nuclear Physics, Novosibirsk 1968)
5.4 V.E. Zakharov, V.S. L'vov, S.S. Starobinets: Fiz. Tverd. Tela **11**, 2047 (1969)
5.5 V.E. Zakharov, V.S. L'vov and S.S. Starobinets: *Instability of the Monochromatic Spin Waves*, Fiz. Tverd. Tela **11**, No 10, 2924–2930 (1969)
5.6 V.E. Zakharov, V.S. L'vov and S.S. Starobinets: *Stationary Nonlinear Theory of Parametric Excitation of Waves*, Sov. Phys.–JETP **32**, 656 (1971)
5.7 V.V. Zautkin, V.S. L'vov, S.L. Musher, S.S. Starobinets: *Proof of Stage-by-Stage Excitation of Parametric Spin Waves*, Sov. Phys.–JETP, Lett. **14**, 206 (1971)
5.8 V.V. Zautkin, V.S. L'vov and S.S. Starobinets: *Resonance Phenomena in Parametric Spin Wave System*, Sov. Phys.–JETP **36**, No 1, 96–99 (1973)
5.9 V.V. Zautkin, V.E. Zakharov, V.S. L'vov, S.L. Musher and S.S. Starobinets: Sov. Phys.–JETP **35**, 926 (1972)
5.10 V.V. Zautkin and S.S. Starobinets: Fiz. Tverd. Tela **16**, No 3, 678–686 (1974)
5.11 V.E. Zakharov and V.S. L'vov: Sov. Phys.–JETP **33**, 1113–1119 (1971)
5.12 V.E. Zakharov, V.S. L'vov and S.L. Musher: Sov. Phys.–Solid State **14**, 710 (1973)
5.13 V.E. Zakharov and V.S. L'vov: Sov. Phys.– Solid State **14**, 2513 (1973)
5.14 V.S. L'vov : Sov. Phys.–Solid State **13**, 2949 (1971)
5.15 V.S. L'vov and S.S. Starobinets: Sov. Phys.– Solid State **13**, 418–425 (1971)
5.16 V.S. L'vov: Preprint No 68–72 (Institute for Nuclear Physics, Novosibirsk 1972)
5.17 V.S. L'vov and A.M. Rubenchik: Preprint No 1–72 (Institute for Nuclear Physics, Novosibirsk 1972)
5.18 V.S. L'vov, S.L. Musher and S.S. Starobinets: Sov. Phys.– JETP **37**, 546 (1973)
5.19 V.S. L'vov, A.M. Rubenchik, V.S. Sobolev and V.S. Synakh: Fiz. Tverd. Tela **15**, No 3, 793–800 (1973)
5.20 V.S. L'vov and A.M. Rubenchik: *Nonlinear Theory of the Parametric Instability of Waves in a Plasma*, Sov. Phys.–JETP **37**, No 2, 263–268 (1973)
5.21 V.S. L'vov and M.I. Shirokov: *Nonlinear Theory of Parametric Excitation of Spin Waves in Antiferromagnets*, Sov. Phys.– JETP **40**, 960 (1975)
5.22 T.S. Harvick, E.R. Peressini and M.T. Weiss: *Subsidiary Resonance in YIG* . J. Appl. Phys. **32**, No 3, 223–224 (1961)
5.23 V.E. Zakharov, V.S. L'vov and S.S. Starobinets: *Spin–Wave Turbulence Beyond the Parametric Excitation Threshold*, Sov. Phys.–USP **17**, 896 (1975)
5.24 V.V. Zautkin, V.S. L'vov, B.I. Orel and S.S. Starobinets: Sov. Phys.–JETP **72**, 272–284 (1977)
5.25 V.S. L'vov and A.M. Rubenchik: Sov. Phys.–JETP **45**, 67–74 (1977)

5.26 V.S. L'vov, V.B. Cherepanov: *Highly Anisotropic Distributions of Parametrically Excited Waves in Near-Isotropic Media* Sov. Phys.–JETP **49**, 1145 (1979)
5.27 V.S. L'vov and G.E. Falkovich: *On the Interaction of Parametrically Excited Spin Waves with Thermal Spin Waves*, Sov. Phys.– JETP **55**, No 5, 904–912 (1982)
5.28 V.V. Zautkin, V.S. L'vov, E.V. Podivilov: Sov. Phys. - JETP **96**, 177 (1989)
5.29 V.G. Morozov and A.I. Muhay: Teor. Mat. Fiz. **51**, 234 (1982)
5.30 A.S. Michailov: Zh. Eksp. Teor. Fiz. **69**, No 2(8), 523–524 (1975)
5.31 V.M. Tsukenic and P.M. Yankelevich: **68**, No 6, 2116–2124 (1975)
5.32 I.A. Vinikovetsky, A.M. Frishman and V.M. Tsukernic: Zh. Eksp. Teor. Fiz. **76**, 2110–2125 (1979)
5.33 A.S. Bakay: Zh. Eksp. Teor. Fiz. **74**, No 3, 933–1004 (1978)
5.34 A.S. Michailov: Fiz. Tverd. Tela **18**, No 2, 494–502 (1976)
5.35 B.Ya. Kotuyzhansky and L.A. Prozorova: Zh. Eksp. Teor. Fiz.,Lett. **30**, No 8, 430–432 (1971)
5.36 B.Ya. Kotuyzhansky, L.A. Prozorova: Zh. Eksp. Teor. Fiz. **62**, No 6, 1199 (1972)
5.37 B.Ya. Kotuyzhansky and L.A. Prozorova: Zh. Eksp. Teor. Fiz. **65**, No(6) 12, 2470–2478 (1973)
5.38 B.Ya. Kotuyzhansky, L.A. Prozorova: Zh. Eksp. Teor. Fiz. Lett.**32**, 254 (1980)
5.39 B.Ya. Kotuyzhansky and L.A. Prozorova: Zh. Eksp. Teor. Fiz. **81**, No 5, 1931 (1981)
5.40 B.Ya. Kotuyzhansky, L.A. Prozorova and L.E. Svistov: Zh. Eksp. Teor. Fiz., Lett. **37**, No 12, 585–588 (1983)
5.41 B.Ya. Kotuyzhansky and L.A. Prozorova: Zh. Eksp. Teor. Fiz. **84**, 658–664 (1984)
5.42 B.Ya. Kotuyzhansky, L.A. Prozorova and L.E. Svistov: Zh. Exsp. Teor. Fiz. **86**, 1101–1116 (1984)
5.43 I.V. Krutsenko, V.S. L'vov, G.A. Melkov: Sov. Phys.– JETP **48**, 561 (1978)
5.44 I.V. Krutsenko and G.A. Melkov: Fiz. Tverd. Tela **21**, No 1, 271–274 (1979)
5.45 N.G. Kutuzoy and G.A. Melkov: Fiz. Tverd. Tela **17**, No 3, 958–960 (1975)
5.46 A.V. Lavrinenko, V.S. L'vov, G.A. Melkov and V.B. Cherepanov: *"Kinetic Instability" of a Strongly Nonequilibrium System of Spin Waves and Tunable Radiation of a Ferrite*. Sov. Phys.– JETP **54**, No 3, 542–549 (1981)
5.47 G.A. Melkov: Sov. Phys.– JETP **34**, 198 (1972)
5.48 G.A. Melkov: Fiz. Tverd. Tela **17**, No 6, 1728–1732 (1975)
5.49 V.L. Grankin, G.A. Melkov and V.A. Ruban: Fiz. Tverd. Tela **15**, 632–634 (1973)
5.50 G.A. Melkov and V.L. Grankin: Sov. Phys.– JETP **42**, 721 (1975)
5.51 G.A. Melkov: Zh. Eksp. Teor. Fiz. **70**, 1324–1329 (1976)
5.52 G.A. Melkov and I.V. Krutsenko: Sov. Phys.– JETP **45**, 295 (1977)
5.53 V.I. Ozhogin: Zh. Eksp. Teor. Fiz. **58**, No 6, 2079–2089 (1970)
5.54 V.I. Ozhogin and A.Yu. Yakubovsky: Zh. Eksp. Teor. Fiz. **63**, 2205 (1972)
5.55 V.I. Ozhogin and A.Yu. Yakubovsky: Zh. Eksp. Teor. Fiz. **67**, No 1 (7), 287–308 (1974)
5.56 V.I. Ozhogin, S.M. Suleymanov and A.Yu. Yakovbovsky: Zh. Eksp. Teor. Fiz., Lett **32**, 308–311 (1980)
5.57 V.I. Ozhogin , S.M. Suleymanov and A.Yu. Yakubovsky: Zh. Eksp. Teor. Fiz.,Lett **34**, No 11, 606–608 (1981)
5.58 B.I. Orel and S.S. Starobinets: Sov. Phys.– JETP **41**, 154 (1975)
5.59 S.V. Petrov and A.I. Smirnov: Zh. Eksp. Teor. Fiz. **80**, No 4, 1628–1638 (1981)
5.60 L.A. Prozorova, A.S. Borovik–Romanov: Zh. Eksp. Teor. Fiz., Lett. **10**, 316 (1969)
5.61 L.A. Prozorova and A.I. Smirnov: Zh. Eksp. Teor. Fiz. **69**, No 2(8), 758–763 (1975)
5.62 L.A. Prozorova and A.I. Smirnov: Zh. Eksp. Teor. Fiz. **74**, 1554–1561 (1978)
5.63 L.E. Svistov and A.I. Smirnov: Zh. Eksp. Teor. Fiz. **74**, 1554–1561 (1978)
5.64 A.I. Smirnov: Zh. Eksp. Teor. Fiz. **73**, 2255–2263 (1977)
5.65 A.I. Smirnov: Zh. Eksp. Teor. Fiz.,Lett. **27**, 177–181 (1977)
5.66 A.I. Smirnov: Zh. Eksp. Teor. Fiz. **84**, 2290–2305 (1983)

5.67 A.I. Smirnov: Zh. Eksp. Teor. Fiz. **88**, 1369–1381 (1985)
5.68 A.I. Smirnov: Zh. Eksp. Teor. Fiz. **90**, 385–397 (1986); **94**, No 5, 185–193 (1988)
5.69 V.S. L'vov: "Solitons and Nonlinear Phenomena in Parametrically Excited Spin Waves", in *Solitons*, ed. by S.E. Trullinger, V.E. Zakharov and V.L. Pokrovsky, (Elsevier, Amsterdam 1986) Ch.5 pp 241–300
5.70 V.S. L'vov and L.A. Prozorova: "Spin Waves Above the Threshhold of Paramagnetic Excitation", in *Spin Waves and Magnetic Excitation 1*, ed. by A.S. Borovik-Romanov and S.K. Sinha, (Elsevier, Amsterdam 1988) Ch.4 pp. 233–285
5.71 V.I. Belinicher and V.S. L'vov: *Spin–diagram technique for non-equilibrium processes in the theory of magnetism*, Sov. Phys.–JETP **59** 564–571 (1984)
5.72 E. Schlömann: J.Appl.Phys. **33**, No 2, 527-534 (1962)
5.73 P. Gottlib and H. Suhl: J.Appl.Phys. **33**, No 4, 1508 (1962)
5.74 I.M. Halatnicov: *Theory of Superfluidity*, (Nauka, Moscow 1971) [in Russian]
5.75 V.S. L'vov: *Lectures on Nonlinear Physics Phenomena* (University of Novosibirsk Press., Novosibirsk 1977)
5.76 V.V. Zautkin, V.S. L'vov and E.V. Podivilov, Sov. Phys.–JETF **69**, No 1, 177–185 (1989)
5.77 L.D. Landau and E.M. Lifshitz: *Course of Theoretical Physics, Vol.5, Statistical Physics, Part 1 3rd Ed.* (Pergamon, Oxford 1980)

Chapter 6

6.1 V.E. Zakharov, V.S. L'vov and S.S. Starobinets: *Spin–Wave Turbulence beyond the Parametric Excitation Threshold*, Sov. Phys.–USP **17**, 896 (1975)
6.2 V.V. Zautkin, V.S. L'vov, E.V. Podivilov: Sov. Phys.– JETP **96**, 177 (1989)
6.3 V.S. L'vov and G.E. Falkovich: *On the Interaction of Parametrically Excited Spin Waves with Thermal Spin Waves*, Sov. Phys.– JETP **55**, No 5, 904–912 (1982);
6.4 H.Le Gall, B. Lemair and D. Sere: Solid State Com. **5**, No 12, 919 (1967)
6.5 V.S. L'vov: *On the Interaction of Parametrically Excited Spin Waves with Thermal Spin Waves*, Preprint No 69–72 (Inst. for Nuclear Physics, Novosibirsk 1972)
6.6 V.S. L'vov, G.E. Falkovich: *Parametric-Wave Distribution Under Nonlinear Damping*, Prep. No 220 (Inst. of Automation and Electrometry, Novosibirsk 1986)
6.7 V.S. L'vov and S.S. Starobinets: Fiz. Tverd. Tela **13**, 523–533 (1971)
6.8 H. Suhl: Phys. Chem. Sol. **1**, 209–227 (1957)
6.9 E. Schlömann: Phys. Rev. **16**, No 4, 828–837 (1959)
6.10 E. Schlömann, J. Green and V. Milano: *Recent Developments in Ferromagnetic Resonance At High Power Levels*, J. Appl. Phys. **31**, Supple., 386S–395S (1960)
6.11 R.W. Damon: in *Magnetism* ed. by G.T.Rado and H. Suhl Vol.1, Ch.11 (1963)
6.12 A.G. Gurevich and S.S. Starobinets: Fiz. Tverd. Tela **3**, No 7, 1 995–1988 (1961)
6.13 B. Lax and K.J. Button: *Microwave Ferrites and Ferrimagnets*, (Mc-Graw-Hill, New York 1962)
6.14 V.E. Zakharov and V.S. L'vov: Sov. Phys.–JETP **33**, 1113–1119 (1971)
6.15 A.D. Piliya: *Proc. 10th Int. Conf. on Ionised Gases*, (Oxford 1971) p. 320
6.16 A.A. Galeev and R.Z. Sagdeev: Nuclear Synthesis, **13**, 603 (1973)
6.17 V. Aiksman and G.F. Shaydurov: Dokl. Akad. Nauk SSSR **180**, 1315 (1968
6.18 L.M. Gorbunov: Zh. Eksp. Teor. Fiz. **67**, 1386 (1974)
6.19 V.S. L'vov: "Nonlinear Theory of Parametrically Excited Waves"; D. Sc. Thesis, Institute for Nuclear Physics Ac.Sci. SSSR, Novosibirsk (1973), Ch.9
6.20 V.S. L'vov and A.M. Rubenchik: *Nonlinear Theory of the Parametric Instability of Waves in a Plasma* Sov. Phys.– JETP **37**, No 2, 263–268 (1973)
6.21 V.B. Cherepanov: Fiz. Tverd. Tela **21**, No 3, 641–647 (1979)

6.22 V.V. Zautkin, B.I. Orel, V.B. Cherepanov: Zh. Eksp. Teor. Fiz. **85**, 708 (1983)
6.23 I.B. Levinson: Zh. Eksp. Teor. Fiz. **65**, No 1, 331–342 (1973)

Chapter 7

7.1 B.Ya. Kotuyzhansky, L.A. Prozorova: Zh. Eksp. Teor. Fiz. **62**, No 6, 1199 (1972)
7.2 L.A. Prozorova and A.I. Smirnov: Zh. Eksp. Teor. Fiz. **69**, No 2(8), 758–763 (1975)
7.3 V.S. L'vov and A.M. Rubenchik: Sov. Phys.–JETP **45**, 67–74 (1977)
7.3 V.V. Zautkin, V.S. L'vov and S.S. Starobinets: *Resonance Phenomena in Parametric Spin Wave System*, Sov. Phys.–JETP **36**, No 1, 96–99 (1973)
7.5 B.I. Orel and S.S. Starobinets: Sov. Phys. - JETP **41**, 154 (1975)
7.6 V.S. Lutovinov: Fiz. Tverd. Tela **20**, 1807–1815 (1978)
7.6 V.B. Cherepanov: Zh. Eksp. Teor. Fiz. **90**, 153–157 (1985)
7.7 A.I. Smirnov: Zh. Eksp. Teor. Fiz. **84**, 2290–2305 (1983)
7.8 H. Suhl: Phys. Rev. Lett. **6**, 174–176 (1961)
7.9 T.S. Hartwick, E.R. Peressini and M.T. Weiss: Phys. Rev. Lett. **6**, 176–177 (1961)
7.10 V.V. Zautkin, V.S. L'vov, B.I. Orel and S.S. Starobinets: Sov. Phys. - JETP **45**, 143–149 (1977)
7.11 V.V. Zautkin and B.I. Orel: Fiz. Tverd. Tela **20**, No 2, 593–595 (1978)
7.12 A.M. Frishman: Fiz. Nizkikh Temp. **8**, No 5, 554–556 (1982)
7.13 V.I. Ozhogin, S.M. Syleymanov and A.Yu. Yakubovsky: Zh. Eksp. Teor. Fiz., Lett. **34**, No 11, 606–608 (1981)
7.14 V.V. Zautkin and B.I. Orel: Zh. Eksp. Teor. Fiz. **79**, No 1(7), 281–287 (1980)
7.15 V.E. Zakharov, V.S. L'vov, S.L. Musher: Sov. Phys.–Solid State **14**, 710 (1972)
7.16 V.S. Zhitnyuk and G.A. Melkov: Zh. Eksp. Teor. Fiz.,Lett **32**, 149–152 (1980)
7.17 H.Le Gall and J.P. Jamet: "Spin Wave Investigation by Means of Faraday Rotation" in *Proc. of Int. Conf. on Magnetism, Moscow 1973*, Vol.1 (Nauka, Moscow 1974) pp. 20–35
7.18 V.N. Venitsky, V.V. Eremenko and E.V. Matyushkin: Zh. Eksp. Teor. Fiz.,Lett. **27**, No 4, 239–241 (1978)
7.19 E.V. Podivilov and V.B. Cherepanov: Zh. Eksp. Teor. Fiz. **90**, 767–780 (1986)
7.20 E. Janke, F. Emde and F. Losch: *Tafeln Höherer Funktionen*, (Teubner, Stuttgart 1960)
7.21 V.V. Zautkin, V.S. L'vov, B.I. Orel, E.V. Podivilov and V.B. Cherepanov: Sov. Phys.–JETP **66** 717–724 (1987)

Chapter 8

8.1 Ya.A. Monosov: *Non–Linear Ferromagnetic Resonance*, (Nauka Publ., Moscow 1971), [in Russian]
8.2 W.E. Courtney, T. Claricoats: Elestron and Control **16**, No 1, 1 (1964)
8.3 V.L. Grankin, V.S. L'vov, V.I. Motorin and S.L. Musher: Zh. Eksp. Teor. Fiz. **81**, No 2(8) 757 (1981)
8.4 S. Wang, S. Thomas and Ta-Lin Hsu: J.Appl.Phys. **39**, 2719 (1968)
8.5 S. Wang and Ta-Lin Hsu: Appl. Phys. Lett. **16**, 537 (1970)
8.6 J.J. Green and E. Schlömann: Appl. Phys. Lett., **33**, 1358 (1968)
8.7 V.S. L'vov, S.L. Musher and S.S. Starobinets: Sov. Phys. - JETP **37**, 546 (1973)
8.8 V.V. Zautkin and S.S. Starobinets: Fiz. Tverd. Tela **16**, No 3, 678–686 (1974)
8.9 A.J. Lichtenberg and M.A. Lieberman: *Regular and Stochastic Motion*, (Springer-Verlag, New York 1983)
8.10 L.D. Landau and E.M. Lifshitz: *Course of Theoretical Physics, Vol.5, Fluid Mechanics* Sect. 30–32 (Pergamon, Oxford 1986)

8.11 H. Yamazaki: *Oscillations and Period–Doubling of Magnon Amplitude Under Parallel Pumping in Antiferromagnetic $CuCl_2$*, J.Phys. Soc. Japan, **53**, 1155 (1984)

8.12 G. Gibson and C. Jeffries: *Observation of Period-Doubling and Chaos in Spin-Wave Instabilities in Yttrium Iron Garnet*, Phys. Rev.A. **29**, No 2, 811–818 (1984)

8.13 F. Waldner, D.R. Barberis and H. Yamazaki: *Route to Chaos by Irregular Period: Simulation of Parallel Pumping in Ferromagnets*, Phys. Rev. A, **31**, No 1, 420 (1985)

8.14 S.M. Rezende, O.F.de Alcantara Bonfim and F.M.de Aguiar: *Model for Chaotic Dynamics of the Perpendicular Pumping Spin-Wave Instability*, Phys. Rev. B. **33**, No 7, 5153–5157 (1986)

8.15 H. Yamazaki, M. Mino, H. Nagashima and M. Warden: *Strange Attractor of Chaotic Magnons Observed in Ferromagnetic $(CH_3NH_3)CuCl_4$*, J. Phys. Soc. Japan **56**, No 2, 742–750 (1987)

8.16 A.I. Smirnov: *Study of Chaotic Regime of Parametric–Magnon Density*, Zh. Eksp. Teor. Fiz.**90**, 385–397 (1986); A.I. Smirnov: *Study of Geometry of Spin-Wave Turbulence Attractors in Antiferromagnetic $CsMnF_3$*, Zh. Eksp. Teor. Fiz. **94**, No 5,185—193 (1988)

8.17 P. Bryant, C. Jeffiers and K. Nakamura: "Spin–wave turbulence" in *International Conference on the Physics of Chaos and Systems far from Equilibrium* ed. by M. Duong-Van (North Holland, Amsterdam 1987)

8.18 X. Zhang and H. Suhl: Phys. Rev. B **38**, No 7, 4893 (1987); H. Suhl and X.Y.Zhang: J.Appl.Phys. **63**, No 8, 4147 (1988)

8.19 S.M. Resende, F.M. de Augur, *Spin-wave instabilities, auto-oscillations and chaos in yttrium-iron-garnet*, Proc. of the IEEE, **78** No. 6 893–908 (1990); S.M. Resende, F.M. de Aguiar and O.F. de Alcantara Bonfim: J. Mag. Materials, **54–57** 1127 (1986); S.M. Resende, Phys. Rev. Lett. **56** 1070 (1986)

8.20 E.N. Lorens: *Deterministic Nonperiodic Flow*, J.Atmos. Sci. **20**, 130 (1963)

8.21 M.J. Feigenbaum: J. Stat. Phys. **19**, 25 (1978); **21**, 669 (1979)

8.22 M.J. Feigenbaum: - Los Alamos Science, **1**, 4 (1980)

8.23 E.N. Lorenz: "Nonlinear Dynamics", *Ann. N.Y. Acad. Sci., Vol. 357* (N.Y. Ac. Sci., N.Y. 1980) p. 282

8.24 P. Collet and J.-P. Echman: "Iterated Maps on the Interval as Dynamical Systems",in *Progress in Physics, Vol. 1* (Birkhäuser Verlag, Basel 1983)

8.25 R.H.G. Helleman: in *Nonequelibrium Problems in Statistical Mechanics, Vol. 2*, ed. by W. Horton, L. Reichl and V. Szebehely (Wiley, N.Y. 1981)

8.26 S. Grossmann and S. Thomae: Z. Naturforsch., **32A**, No 1 353 (1977)

8.27 P. Manville and Y. Pomeau: *Intermittency and the Lorenz Model*, Phys.Lett. **75A**, No 1, 1–2 (1979)

8.28 A.I. Smirnov: *Spin-Wave Instabilities in Excited Spin System of Antiferromagnets*, D. Sc. Thesis, (Inst. for Physical Problems, Moscow 1987)

8.29 A. Wolf, J. Swift: Phys.Lett. A **83**, 184 (1981)

8.30 V.I. Arnold: *Additional Chapters to the Theory of Ordinary Differential Equations* (Nauka, Moscow 1978)

8.31 N. Packard, J. Gruitchfield, J. Farmer and R. Show: Phys. Rev. Lett.**45**, No 9, 712–715 (1980).

8.32 F. Takens: in *Lecture Notes in Mathematics* , Vol. 898, ed. by D. Rard and L. Young (Springer, Berlin 1981)

8.33 S.N. Lukaschuk, A.A. Predteshensky , G.E. Falkovich and A.I. Chernykh: *Determination of Attractor Dimensionality Using Experimental Data*. Preprint No 280 (Institute of Automation and Electrometry, Novosibirsk 1985)

8.34 S.N. Lukaschuk, V.S. L'vov, A.A. Predtechensuky and A.I. Chernykh: "First Bifurcation in Circular Cuvette Flow: Laboratory and Numerical Experiments", in *Laminar-Turbulent Transition IUTAM Symposium Novosibirsk 1984* , ed by V.V. Kozlov (Springer, Berlin, Heidelberg 1985) pp. 653–658
8.35 V.S. L'vov: Sov. Phys.–Solid State **13**, 2949 (1972)
8.36 V.S. L'vov and A.M. Rubenchik: Preprint No 1–72 (Institute for Nuclear Physics, Novosibirsk 1972)
8.37 S.A. Akhmanov, A.P. Sukhorukov, R.V. Khokhlow: Usp. Fiz. Nauk **93**, 19 (1967)
8.38 B.B. Kadomtsev and V.I. Karpman: Usp. Fiz. Nauk **103**, 193 (1970)
8.39 V.I. Talanov: Izv. Vuzov, Radiofizika **7**, 564 (1964)
8.40 V.E. Zakharov: Zh. Eksp. Teor. Fiz. **53**, 1710 (1967)
8.41 V.E. Zakharov, V.V. Sobolev, V.S. Sunakh: Sov. Phys.–JETP, Lett. **14**, 390 (1971)
8.42 V.N. Vlasov, V.A. Petrishchev and V.I. Talanov: Izv. Vuzov, Radiofizika **14**, 1353–1364 (1971)
8.43 V.S. L'vov, A.M. Rubenchik, V.V. Sobolev and V.S. Synakh: Sov. Phys.–Solid State **15**, 550 (1973)
8.44 V.E. Zakharov, V.V. Sobolev and V.S. Sunakh: Zh. Eksp. Teor. Fiz. **60**, 136 (1971)

Chapter 9

9.1 F.R.Morgenthaler: J. Appl. Phys. **31S**, 95S–97S (1960)
9.2 E. Schlömann, J.H.Saunder, and M.H.Sirvetz: *L-Band Ferromagnetic Resonance at High Peak Power Level*, Trans. IRE, MTT **8**, No 8, 96–100 (1960)
9.3 E. Schlömann, J.Green and V.Milano: *Recent Developments in Ferromagnetic Resonance at High Power Levels*, J. Appl. Phys. **31**, Suppl., 386S–395S (1960)
9.4 B.Ya.Kotyuzhansky and L.A. Prozorova: Zh. Eksp. Teor. Fiz. **62**, 2199 (1972)
9.5 W.J.Jantz, B.Anglaer and J.Schneider: Solid State Commun. **10**, 937 (1972)
9.6 B.Ya.Kotuyzhansky,L.A. Prozorova: Zh. Eksp. Teor. Fiz. Lett. **32**, 254 (1980)
9.7 E.M.Turner: Phys. Rev. Lett. **5**, 100 (1960)
9.8 M.H.Seavey: Phys. Rev. Lett. **23**, 132 (1969)
9.9 V.V.Kveder, B.Ya.Kotyuzhansky and L.A. Prozorova: Zh. Eksp. Teor. Fiz. **63**, 2205 (1972)
9.10 B.Ya. Kotuyzhansky, L.A. Prozorova: Zh. Eksp. Teor. Fiz. **62**, No 6, 1199 (1972)
9.11 B.Ya.Kotuyzhansky and L.A. Prozorova: Zh. Eksp. Teor. Fiz. **65**, No (6) 12, 2470–2478 (1973)
9.12 A.N. Anisimov and A.G.Gurevitch: Zh. Eksp. Teor. Fiz. **68**, 677 (1975)
9.13 A.N.Anisimov and A.G.Gurevitch: Fiz. Tverd. Tela **18**, 38 (1976)
9.14 I.V. Kolokolov, V.S. L'vov, V.B. Cherepanov: Sov. Phys.–JETP **57**, 605 (1983)
9.15 I.V. Kolokolov, V.S. L'vov, V.B. Cherepanov: Sov. Phys.–JETP **59**, 1131 (1984)
9.16 T.Kasuya and R.Le Craw: Phys. Rev. Lett. **6**, 223 (1961)
9.17 V.L. Grankin, G.A. Melkov and S.M.Ruabchenko: Fiz. Tverd. Tela **17**, 358 (1975)
9.18 B.Ya.Kotuyzhansky and L.A. Prozorova: Zh. Eksp. Teor. Fiz. **81**, No 5, 1931 (1981)
9.19 V.S. Lutovinov: Fiz. Tverd. Tela **20**, 1807 (1978)
9.20 R.B.Woolsey and P.M.White: Phys. Rev. **188**, 813 (1969)
9.21 A.V.Andrienko and L.A. Prozorova: Zh. Eksp. Teor. Fiz. **88**, 213 (1985)
9.22 V.V. Zautkin, V.E. Zakharov, V.S. L'vov, S.L. Musher and S.S. Starobinets: Sov. Phys.–JETP **35**, 926 (1972)
9.23 B. Lax and K.J. Button: *Microwave Ferrites and Ferrimagnets*, (Mc-Graw-Hill, New York, London 1962)
9.24 G.A. Melkov and I.V. Krutsenko: Sov. Phys.–JETP **45**, 295 (1977)
9.25 T.S. Hartwick, E.R. Peressini and M.T. Weiss: Phys. Rev. Lett. **6**, 176–177 (1961)
9.26 E. E. Schlömann: *Longitudinal Susceptibility of Ferromagnets in Strong RF Field*, J.Appl.Phys., **33**, No 2, 527–534 (1962)

9.27 P. Gottlib, H. Suhl: *Saturation of Ferrimagnetic Resonance with Parallel Pumping*, J.Appl. Phys, **33**, No 4, 1508 (1962)
9.28 Ya.A. Monosov: *Non-Linear Ferromagnetic Resonance*, (Nauka Publ., Moscow 1971), [in Russian]
9.29 J.J. Green, B.J.Healy: Apple. Phys. **34**, No 4, 1285–1286 (1963)
9.30 V.S. L'vov and M.I.Shirokov: *Nonlinear Theory of Parametric Excitation of Spin Waves in Antiferromagnets* Sov. Phys.–JETP **67**, 960–967 (1974)
9.31 L.A. Prozorova and A.I. Smirnov: Zh. Eksp. Teor. Fiz. **74**, 1554–1561 (1978)
9.32 V.S. L'vov and S.S. Starobinets: Sov. Phys.– Solid State **13**, 418–425 (1971)
9.33 A.G. Gurevich and S.S. Starobinets: Fiz. Tverd. Tela **3**, No 7, 1995–1988 (1961)
9.34 H. Suhl: Phys. Chem. Sol. **1**, 209–227 (1957)
9.35 V.V. Zautkin, B.I. Orel, V.B. Cherepanov: Zh. Eksp. Teor. Fiz. **85**, 708 (1983)
9.36 L.A. Prozorova and A.I. Smirnov: Zh. Eksp. Teor. Fiz. **67**, 1952 (1974)
9.37 L.A. Prozorova and A.I. Smirnov: Zh. Eksp. Teor. Fiz. **69**, No 2(8), 758–763 (1975)
9.38 V.N. Venitsky, V.V.Eremenko and E.V.Matyushkin: Zh. Eksp. Teor. Fiz.,Lett. **27**, No 4, 239–241 (1978)
9.39 V.G.Zhotikov and N.M.Kreines: Zh. Eksp. Teor. Fiz. **77**, 2486 (1979)
9.40 B.Ya.Kotuyzhansky, L.A. Prozorova and L.E. Svistov: Zh. Eksp. Teor. Fiz., Lett. **37**, No 12, 585–588 (1983)
9.41 B.Ya.Kotuyzhansky and L.A. Prozorova: Zh. Eksp. Teor. Fiz. **84**, 658–664 (1984)
9.42 A.S. Mikhailov and A.V.Chubikov: Zh. Eksp. Teor. Fiz. **86** 1401 (1984)
9.43 V.S. L'vov: Sov. Phys.–JETP **42**, 1057 (1976)
9.44 I.V. Krutsenko, V.S. L'vov and G.A. Melkov: Sov. Phys.–JETP **48**, 561 (1978)
9.45 V.V. Zautkin: D. Sc. hesis, (Institute for Physics and Techics, Ac. Sci SSSR, Leningrad 1988)
9.46 V.V. Zautkin, V.S. L'vov and S.S. Starobinets: *Resonance Phenomena in Parametric Spin Wave System*, Sov. Phys.–JETP **36**, No 1, 96–99 (1973)
9.47 A.I.Akhiezer, V.G.Bar'yakhtar and S.V. Peletmisky: *Spin Waves*, (North-Holland, Amsterdam 1968)
9.48 B.I. Orel and S.S. Starobinets: Sov. Phys.–JETP **41**, 154 (1975)
9.49 V.V. Zautkin, V.S. L'vov, S.L. Musher, S.S. Starobinets: *Proof of Stage-by-Stage Excitation of Parametric Spin Waves*, Sov. Phys.–JETP, Lett. **14**, 206 (1971)
9.50 G.A.Petrakovsky and V.N.Berzhansky: Zh. Eksp. Teor. Fiz., Lett. **12**, 429 (1970)
9.51 V.V. Zautkin, V.S. L'vov and E.V. Podivilov, Sov. Phys.–JETF **69**, No 1, 177–185 (1989)
9.52 V.V. Zautkin and S.S. Starobinets: Fiz. Tverd. Tela **16**, No 3, 678–686 (1974)
9.53 H.Le Gall, B.Lemair and D.Sere: Solid State Com. **5**, No 12, 919 (1967)
9.54 V.S. L'vov: *On the Interaction of Parametrically Excited Spin Waves with Thermal Spin Waves* Prep. No 69–72 (Inst. for Nuclear Physics, Novosibirsk 1972)
9.55 V.V. Zautkin and B.I. Orel: Fiz. Tverd. Tela **20**, No 2, 593–595 (1978)
9.56 V.V. Zautkin, V.S. L'vov, B.I. Orel and S.S. Starobinets: Sov. Phys.–JETP **45**, 143–149 (1977)
9.57 V.V. Zautkin and B.I. Orel: Zh. Eksp. Teor. Fiz. **79**, No 1(7), 281–287 (1980)
9.58 V.V. Zautkin, B.I. Orel, V.B. Cherepanov: Zh. Eksp. Teor. Fiz. **85**, 708 (1983)

Chapter 10

10.1 V.S. L'vov: *Nonlinear Theory of Parametric Excitation of Waves* Sov. Phys.–JETP **42**, 1057 (1976)
10.2 H.W. Wyld: Ann. Phys. **14** , 143, (1961)
10.3 L.D. Landau and E.M. Lifshitz: *Course of Theoretical Physics. Vol. 10* E.M. Lifshitz and L.P. Pitaevsky: *Physical Kinetics* (Pergamon, Oxford 1981)
10.4 V.S. L'vov: *Nonlinear Spin Waves* (Nauka, Moscow 1987) [in Russian]

- 10.5 V.S. L'vov: "Solitons and Nonlinear Phenomena in Parametricaly Excited Spin Waves", in *Solitons*, ed. by S.E. Trullinger, V.E. Zakharov and V.L. Pokrovsky, (Elsevier Science Publishers B.L. 1986) Ch.5 pp. 241–300
- 10.6 V.S. L'vov, L.A. Prozorova: "Spin Waves Above the Threshhold of Paramagnetic Exitation", in *Spin Waves and Magnetic Exitation 1*, ed. by A.S. Borovik-Romanov and S.K. Sinha, (Elsevier, Amsterdam 1988) Ch.4 pp. 233–285
- 10.7 V.E. Zakharov and V.S. L'vov: Izv. Vuzov, Radiofizika **18** 1470–1487 (1965)
- 10.8 M. Sparks: *Ferromagnetic Relaxation Theory*, (New York 1964)
- 10.9 E. Schlömann: Phys. Rev. **182**, 632 (1969)
- 10.10 A.G. Gurevich: *Magnetic Resonance in Ferro– and Antiferromagnets* (Nauka, Moscow 1973) [in Russian]
- 10.11 E. Schlömann: Phys. Rev. **16**, No 4, 828–837 (1959)
- 10.12 H. Suhl: *Note on the Ferromagnets*, J. Appl. Phys. **30**, No 12, 1961–1964 (1959)
- 10 13 G.A. Melkov and V.L. Grankin: Sov. Phys.– JETP **42**, 721 (1975)
- 10.14 S.V. Petrov and A.I. Smirnov: Zh. Eksp. Teor. Fiz. **80**, No 4, 1628–1638 (1981)
- 10.15 V.E. Zakharov and V.S. L'vov: Sov. Phys.– Solid State **14**, 2513 (1973)
- 10.16 V.S. L'vov and M.I. Shirokov: *Nonlinear Theory of Parametric Excitation of Spin Waves in Antiferromagnets*, Sov. Phys. - JETP **40**, 960 (1975)
- 10.17 V.S. L'vov and V.B. Cherepanov: Sov. Phys.–JETP, **48** No. 5 822–828 (1978)
- 10.18 I.V. Krutsenko, V.S. L'vov and G.A. Melkov: Sov. Phys.– JETP **48**, 561–569 (1978)
- 10.19 V.S. L'vov and V.B. Cherepanov: *Highly Anisotropic Distributions of Parametrically Excited Waves in Near-Isotropic Media* Sov. Phys.–JETP **49**, 1145–1151 (1979)

Subject Index

Amplitude of interaction of two-waves 304
— — of three-waves 8, 12, 20–27
— — of three-magnons 60–62, 68
— — of four-waves 8–15
— — of pairs of waves (S) 105–106, 110–117, 123–126, 167, 179–181
— — of four-magnons 15, 60–62, 63–65, 69
— — of waves with pumping 87–94
— of waves 4–34
Anisotropy energy 46
— of the "easy-axis type" 44
— of the "easy-plane type" 44
Antiferromagnets 37, 52, 53
— with anisotropy of the "easy-axis type" 65
— — of the "easy-plane type" 66, 68–70, 92–94, 249–250 254, 262
— with cubic anisotropy 67
Asymmetrical S-theory 161–172
Attractors of secondary parametric turbulence in CsMnF$_3$ 226–234
Auto-oscillations 214–218, 226–234, 281–289

Basic equations of S-theory 102–107
— — — under frequency sweeping 204–205
— — of asymmetrical S-theory 165–167
— — of noise S-theory 173–175
— — of spatially-inhomogeneous S-theory 155–157
— — of temperature S-theory 148–149
— — of S, g^2-theory 308–310
— — of S, T^2-theory 311

Canonical variables for spin waves (magnons) 72–73
— — for waves 5–11
— motion equaton for waves *see* Motion equations for waves
Chaos in dynamic systems 218–234
— of parametric spin waves (magnons) 215–217, 226–231
Collapse 28, 31–34
Collective oscillations 179–189, 266–276
Correlation auto-oscillations 171
— instability of parametric waves 169–171

Damping decrement of waves 17–20
— — of spin waves 245–250
Decay instability 22–24, 78–79
— — increment 22–24
— processes 22 –24

Subject Index

Dimensionality of inclusion 229–230
Double parametric resonance 193–196, 293–294
Dynamic equation of motion for waves *see* Motion equations for waves

Exchange interaction 38–44
Explosive instability 25–26
— processes 25–26
External stability 108, 169

Feigenbaum scenario 232–234
Ferromagnets 37, 45–49, 52, 57
Ferrimagnets 37, 49–52, 245–250, 252–254, 258–260, 263-266
First-order Suhl processes (instability of the homogeneous precession of magnetization) 87–90, 143–146
Frequency of collective oscillations *see* Spectrum of collective oscillations
— of spin waves (magnons) *see* Spectrum of spin waves
— of waves *see* Spectrum of waves

Green's functions (normal and abnormal) 148–149, 302–307
Ground state (of parametric wave system) 108–119, 167–172

Hamiltonian 3–15, 31
— of crystallographic anisotropy
— of non-interacting waves 5–8, 67, 165
— of interaction 8–17
— — of two-wave scattering by inhomogeneities 306
— — of three-waves 8, 20–27
— — of three-magnons 61–62, 68
— — of four-waves 8, 10–13, 27–31, 166
— — of four-magnons 15, 60, 63–65, 69
— — of waves with pumping 87–94, 99
— of exchange interaction 14
— of magnetic dipole–dipole interaction 43, 48–49
— of basic S-theory 107
— of asymmetrical S-theory 165
— equations of motion for waves *see* Motion equations for waves
— — of basic S-theory 106

Internal stability 108
Invariant phase of the pairs 110

Landau–Hopf scenario 221
Linear wave damping *see* Damping of wave 17

Magnetodielectrics 35–53
Magnons *see* Spin-waves
Mean-field approximation 103–106
Modulation instability 28–30
— — increment 28–30, 81
— — of spin waves 80–82
Motion equations for waves 16, 19, 22, 26, 27, 100, 141, 194, 304

Noise pumping 173, 294–299

— S-theory 171–175
Nonlinear damping of waves 18–19, 101, 132–139
— ferromagnetic resonance 139–146, 257–258
— frequency shift 28, 104, 106, 156–157, 167, 255–257
— susceptibility of collective oscillations 186–187, 272–273
— — of parametric waves 118–119, 131, 133–134, 144–145, 250–255, 258–260
Non-simultaneous correlator (of wave amplitudes) 147–151, 152
Nuclear magnons (spin waves) 69–70, 94
Numerical simulation of auto-oscillation 215–217

"Oblique" Pumping of Spin Waves 91
One-frequency turbulence 313–316

Parallel pumping 90–91, 92–93
Parametric excitation of collective oscillations see Double parametric resonance
— — of waves see "Oblique" Pumping, Parallel pumping, Noise pumping, Transverse pumping,
Phase relations 100–101
— of pairs of waves 260–261
— mechanism of amplitude limitation 102, 113–114, 176–177
— — — under sweeping of their frequency 200–211, 291–293

Resonance excitation of collective oscillations 184–187, 191–192
— surface 109
— processes 8–9, 20–21
Route to chaos in parametric waves turbulence see Chaos of parametric waves

Second group of pairs 114–116, 125–131, 278–281
Second-order Suhl processes (instability of the homogeneous precession of magnetization) 91–92, 146–147
Second threshold (threshold of generation of second group of pairs) 114–116, 125–128
Secondary parametric wave turbulence 213–242
Self-consistent pumping 104, 106, 110–117, 140, 156, 167
Self-focusing 32–34,
— mechanism for amplitude limitation 238–242
— of magnetoelastic waves 82–86
Spatially inhomogeneous S-theory 155–164
Spectrum (dispersion law) of collective oscillations 179–184, 189, 269–271
— of waves of different nature 12—15
— of magnons in ferromagnetics 56–60
— — in antiferromagnetics 64–66, 244–245
— of nuclear magnons 68–69
Spectrum (energy distribution) of parametric waves 153, 313–316
Spin-waves 15, 54–77
Soliton of the envelopes 32–34, 86, 237–238,
— mechanism of amplitude limitation 241–242
Stepwise excitation of parametric waves 114–116, 125–132, 276–281
Strange attractors 222–223, 228–233
Structure of spinels 50
— of garnets 51–52
— of magnets with hexagonal symmetry 52
— — with rhombohedral symmetry 53
Susceptibility see Nonlinear susceptibility

Temperature S-theory 148–155
Tensor of demagnetizing factors 48, 57
Thermal noise 18
— waves (thermally excited waves) 17–18, 151, 308
Transverse pumping of spin waves 87–90

Printing: Saladruck, Berlin
Binding: Buchbinderei Lüderitz & Bauer, Berlin

D. Park
Classical Dynamics and Its Quantum Analogues

2nd enl. and updated ed. 1990. IX, 333 pp. 101 figs. Hardcover ISBN 3-540-51398-1

The primary purpose of this textbook is to introduce students to the principles of classical dynamics of particles, rigid bodies, and continuous systems while showing their relevance to subjects of contemporary interest. Two of these subjects are quantum mechanics and general relativity. The book shows in many examples the relations between quantum and classical mechanics and uses classical methods to derive most of the observational tests of general relativity. A third area of current interest is in nonlinear systems, and there are discussions of instability and of the geometrical methods used to study chaotic behaviour. In the belief that it is most important at this stage of a student's education to develop clear conceptual understanding, the mathematics is for the most part kept rather simple and traditional.

A. Hasegawa
Optical Solitons in Fibers

2nd enl. ed. 1990. XII, 79 pp. 25 figs. Softcover ISBN 3-540-51747-2

The most recent developments associated with two topical and very important theoretical and practical subjects are combined: **Solitons** as analytical solutions of nonlinear partial differential equations and as lossless signals in dielectric **fibers.** Starting from an elementary level readily accessible to undergraduates, this pioneer in the field provides a clear and up-to-date exposition of the prominent aspects of the theoretical background and most recent experimental results in this new and rapidly evolving branch of science.

S. Nettel
Wave Physics
Oscillations – Solitons – Chaos

1992. Approx. 260 pp. 56 figs. Softcover ISBN 3-540-53295-1

This is a lecture course for 2nd and 3rd year students intending to graduate in physics, mathematics, or engineering. After elucidating the mathematical foundations, the author proceeds to discuss applications to physical systems: classical oscillators, waves on strings, electromagnetic waves, Schrödinger's equation and the postulates of quantum mechanics. Modern concepts like solitons and chaos are also dealt with. The step-by-step development leads the student from the traditional Fourier analysis to advanced concepts such as the method of inverse scattering in soliton theory. The final part contains an essay by Prof. M. C. Gutzwiller on chaos and associated phenomena.
Wave Physics provides an excellent preparation for an understanding of high-frequency waves, quantum mechanics, and even nonlinear phenomena such as trans-Pacific telecommunications via optical solitons or the Great Red Spot of Jupiter.

A. G. Sitenko
Scattering Theory
1991. XI, 294 pp. 32 figs. (Springer Series in Nuclear and Particle Physics)
Hardcover ISBN 3-540-51953-X

This book is an introduction to nonrelativistic scattering theory. The presentation is mathematically rigorous, but is accessible to upper level undergraduates in physics. The relationship between the scattering matrix and physical observables, i. e. transition probabilities, is discussed in detail. Among the emphasized topics are the stationary formulation of the scattering problem, the inverse scattering problem, dispersion relations, three-particle bound states and their scattering, collisions of particles with spin and polarization phenomena. The analytical properties of the scattering matrix are discussed. Problems round off this volume.

B. N. Zakhariev, A. A. Suzko
Direct and Inverse Problems
Potentials in Quantum Scattering
1990. XIII, 223 pp. 42 figs.
Softcover ISBN 3-540-52484-3

This textbook can almost be viewed as a "how-to" manual for solving quantum inverse problems, that is, for deriving the potential from spectra or scattering data and also, as somewhat of a quantum "picture book" which should enhance the reader's quantum intuition. The formal exposition of inverse methods is paralleled by a discussion of the direct problem. Differential and finite-difference equations are presented side by side. The common features and (dis)advantages of a variety of solution methods are analyzed. To foster a better understanding, the physical meaning of the mathematical quantities are discussed explicitly. Wave confinement in continuum bound states, resonance and collective tunneling, energy shifts and the spectral and phase equivalence of various interactions are some of the physical problems covered.

R. M. Dreizler, E. K. U. Gross
Density Functional Theory
An Approach to the Quantum Many-Body Problem
1990. XI, 302 pp. 18 figs.
Hardcover ISBN 3-540-51993-9

Density Functional Theory is a rapidly developing branch of many-particle physics that has found applications in atomic, molecular, solid state and nuclear physics. This book describes the conceptual framework of density functional theory and discusses in detail the derivation of explicit functionals from first principles as well as their application to Coulomb systems. Both non-relativistic and relativistic systems are treated. The connection of density functional theory with other many-body methods is highlighted. The presentation is self-contained; the book is thus well suited for a graduate course on density functional theory.

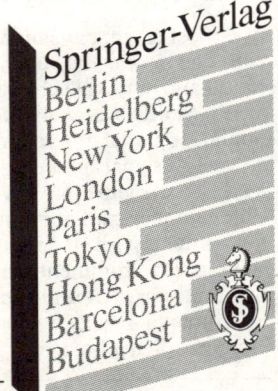

Springer-Verlag
Berlin
Heidelberg
New York
London
Paris
Tokyo
Hong Kong
Barcelona
Budapest